日本産オサムシ図説

The *Carabus* of Japan

井村有希・水沢清行
Yûki IMURA & Kiyoyuki MIZUSAWA

September, 2013 by Roppon-Ashi Entomological Books (Tokyo, JAPAN)　昆虫文献 六本脚

序　PREFACE

　1990年代後半から2000年代初頭にかけて，分子系統の分野で空前の盛り上がりを見せたオサムシ．その一方で，日本産の全種，全亜種を網羅した専門の図鑑と呼べるものが久しく存在しなかったのは，考えてみれば不思議な話である．

　1962年に出版された中根猛彦の日本昆虫分類圖説（第2集第3部鞘翅目・オサムシ科[I]）－以下，中根圖説と略す－は，当時知られていた邦産オサムシの大半を網羅し，詳細な解説に加え，豊富な付図とカラープレートを駆使した，当時としては画期的な専門書であり，長年に亘って我が国のオサムシ学の基礎的資料として活用された．とは言え，出版からすでに半世紀以上が経過し，その間新たに発見，記載された種や亜種も膨大な数に上る．中根圖説では，今回上梓される運びとなった本書の分類基準に照らせば，計63亜種（単型種は便宜上，1亜種と数えた）ほどが解説され，うち57亜種が図示されたが，同圖説以降に発見，記載されたタクソンは120以上にも及び，本書では実に計39種193亜種（全て図示）を扱うことになった．単純計算で，亜種の数にして3倍強に増えたことになるが，このデータ一つを見ても，過去半世紀の間に邦産オサムシに関する状況が如何に大きな変化を遂げたか－よく言えば知見の増大とそれに伴う研究の進展，悪く言えば過度の細分化－を窺い知ることができよう．

　中根圖説のカラープレートを転用した北隆館の原色昆虫大圖鑑II（甲虫篇）（1963年初版発行）や，1955年の初版以来，ロングセラーとなった保育社の原色日本昆虫図鑑（上）・甲虫編も，手軽に入手できる利便性の高い図鑑ではあったが，原色図鑑シリーズの中の一セクションという性格上，内容的には中根圖説に比ぶべくもないものであった．

　中根圖説から20余年を経た1985年，保育社から原色日本甲虫図鑑が出版された．各分類群はそれぞれの専門家による分担執筆制（第2巻のオサムシ亜科の担当は石川良輔）で，同図鑑に掲載された他の分類群に比し，オサムシには格段に詳細な解説が与えられており，他に適当な図鑑がなかったこともあって，オサムシ屋にとっても必携の書となったが，如何せん，これも原色図鑑の一部に過ぎないという構成上の縛りが大きく響き，その内容には当然，限界があった．図示された総亜種数は100と，数から言えば中根圖説の2倍近くに増えはしたものの，この段階においてもまだ現段階の半数強に過ぎない．すなわち，翌1986年以降，新たに発見，記載されたタクソンが計2種90亜種もあるうえ，分子系統解析の研究も進んだため，日本産オサムシの全貌を知るためのツールとしてはすでに陳腐化してしまった感は否めない．1996年の拙著「世界のオサムシ大図鑑」（むし社）において，その時点で知られていた邦産オサムシの全種全亜種に触れはしたが，とりあえず成虫の写真を示したというに過ぎず，1986年以降に発見，記載されたタクソンの詳細について知るには結局，個々の原記載論文を当たるしかないという状況が続いていた．ネットや幾つかの成書上に，個人的に編纂された邦産オサムシ目録の類も散見されるが，所詮は名前の羅列に過ぎず，実用的ではない．日本のオサムシは今，いったいどうなっているのだ，という声をしばしば耳にするにつけ，これまでの知見をまとめた総括的なオサムシ図鑑の必要性を平素より強く感じていた．

　本書の執筆にあたり，筆者が目指したのは，現時点で知られている全種，全亜種を網羅した，新しいスタンダートとなりうる邦産オサムシの総合的な図説である．外部形態のみならず，交尾器までを含んだ形態全般に亘る詳細な描写，原記載論文の出典，研究史からタイプ産地，タイプ標本の所在，分子系統，分布，生息環境，生態，地理的変異，保全，そして種，亜種名の由来に至るまで，広く浅くではあるが，必要事項をなるべく多く盛り込むよう心掛けた．保育社の原色日本甲虫図鑑が出版されてから既に30年近い歳月が経過し，前述の如く，その間に発見，記載されたタクソンは90ほどにも及ぶ．大半は新亜種だが，21世紀に入ってから発見された利尻島のマックレイセアカオサムシや，本書でベールを脱ぐ北四国のホウオウオサムシなど，種レベルで新たに我が国のファウナに加わったものも二つある（亜種から昇格されたものは除く）．これらすべてを図示し，和文のみならず，国際的な使用に配慮して英文でも解説を加えた．こうした点が本書の大きな特徴である．

　♂交尾器内嚢を完全に反転膨隆させた状態で各方向から描写し，記載した点も，これまでの書物になかった特徴と言えよう．個々の論文では，分類形質として重視される陰茎先端や指状片の形態が図示，記載されることが多いが，同一の基準で示された完全反転下の内嚢を含む交尾器の全貌を，全種に亘り，一冊の書物の中で比較できることの意義は大きい．種や亜種の特徴が交尾器，特に指状片の形態に反映される代表的なグループとしてしばしば取り上げられるオホモプテルス亜属各種にしても，これまでは指状片のみが単独で図示される場合が殆どで，反転膨隆させた内嚢にどのような状態で付着しているのかまでを全種に亘り図示，記載したものは見当たらない．ちなみに，交尾器を初めとするこうした微小で立体的な器官を克明に図示できたのは，ひとえにデジタルカメラ技術の進歩と深度合成ソフトの存在に依るところが大きい．一昔前のように，双眼実体顕微鏡にとりつけた描画装置を用いて一つ一つ手描きのペン画で仕上げていたら，全種分の付図完成までにおそらく数倍の労力と時間を要したであろう．もっとも，深度合成を行う場合にも，多いもので100枚近い連続写真を撮らねばならず，加えて事後の繊細なレタッチ作業が必要となるため，やはりそれなりの手間はかかっているのだが．

　こうして出来上がった本書を眺めてみると，図鑑としてはそれなりに詳しいが，いわゆるモノグラフには遠く及ばず，言うなれば帯に短し襷に長し，いかにも中途半端な内容で，著者としては不満に感じる点も多い．本邦最大，とりわけ離島に産するオサムシ標本の充実ぶりでは他の追随を許さない水沢コレクションを自由に活用できる立場にありながら，紹介できたのは結局，ほんの一握りの標本に過ぎなかった．全種に亘る詳細なsynonymy（変遷名リスト）を挙げることも叶わなかったし，幼虫の形態に関しても割愛せざるを得なかった．生態や生息環境についてはもっと詳しいことがわかっている種も多いし，分子系統の項では文章だけでなく実

させたため，古い標本を軟化して使わざるを得なかったものも多い．交通機関の発達した現在，飛行機と車を使えば大抵の場所へは週末の土日で行って来られるのだから，どうせならもう一息気合を入れて，すべて新鮮な標本で統一することもできたはずである．

とは言え，プライベートに使える時間やページ数との兼ね合いを考えると，このあたりが当面の限界と諦めるしかないのかもしれない．そもそも，こうした図鑑作りというものは，クリエイティブな面よりも，内容的には先人の業績や過去の知見のまとめに徹した，いわば後ろ向きの面を多く抱えた作業の連続である．面倒でえらく手間がかかる割に，諸文献の引用は元より，地名の正確な読み方からその行政区画上の所属に至るまで，誤りは許されず，非常に神経を使う，全くもって「割に合わない」仕事なのである．そのため，これまで何冊か手掛けた昆虫関係の本の中で，今回はモチベーションの維持に最も苦労した．水沢清行氏の強い勧めがなければ，自分一人でこうした日本産オサムシの図鑑を手掛けようとは考えなかったかもしれない．

筆者がオサムシを求めて日本各地を盛んに飛び回っていたのは10代の後半から20代の頃，今から30~40年近くも昔のことである．その後，興味の対象は世界のオサムシへと移り，1980年代には韓国，1990年代以降は中国が主なフィールドとなった．タイプ標本調査のため，欧州を中心とする海外の博物館や研究所巡りもできる限り行った．国立科学博物館の上野俊一博士のご指導を賜りながら，新種や新亜種の原記載を含む本格的な学術論文の執筆を始めたのは1989年である．つい昨日のことのように思われるが，そこから早くも四半世紀近い歳月が流れたことになる．21世紀に入ってからの数年間は，中国産ルリクワガタ類の調査研究に熱を上げていたため，オサムシ，とりわけ邦産種に関してはすっかりご無沙汰の状態が続いていた．ここ数年，本書の執筆に合わせて再度，国内での採集や調査にも出るようにはなったが，長年のブランクが祟り，オサムシを扱う上での独特の勘のようなものが鈍ってしまったことは否めないし，地方同好会誌から本格的な学術論文に至るまで，最新の文献をこまめに渉猟するといった作業も十分にはできなかった．邦産オサムシ図鑑の主著者としては本来，「旬を過ぎた」立場であり，役不足もいいところであろうと自認している．

このように，色々と満足のいかない点や，やり残した点も多いが，なにはともあれ，この本を世に送り出せたことを，先ずは素直に喜びたい．オサムシは付き合うほどに味の出る，奥深く楽しい甲虫である．なにも研究だ調査だと堅苦しいことばかり言わなくてもよい．コレクションが目的でも構わないし，オサ掘りやトラップ採集を楽しむため，というスタンスだって「アリ」である．虫と触れ合う方法も，その目的も，人によって千差万別．法に触れず，人に迷惑をかけない範囲であれば，自分が一番楽しいと思う方法で付き合えばよい．まずは本書を，良い意味で「オサムシを遊び倒す」ためのツールだと思って大いにご活用いただきたい．そのうえで，オサムシに興味を持つ内外の方々に基礎的資料の一つとして多少なりともお役に立つことがあるならば，本書はその役割の大半を果たしたことになるだろう．

2013年盛夏
主著者 井村有希（いむらゆうき）

謝辞　ACKNOWLEDGEMENTS

本書の作成にあたり，以下の方々からは様々な形でご協力をいただいた．ここに記して厚くお礼を申し上げる（敬称略）．

青木俊明, 秋山美文, 荒井充朗, 荒木哲, 池崎善博, 石田幸生, 石谷正宇, 伊藤雅男, 伊東善之, 井ノ上健, 井村正行, 岩橋正, 岩本常夫, 岩本道彦, 梅木要, 遠藤正浩, 大久保憲秀, 大谷規夫, 大坪修一, 大野勝示, 大野正男, 岡義人, 岡島秀治並びに東京農業大学昆虫学研究室関連の諸氏, 岡田裕之, 岡本巌, 奥村尚, 尾崎俊寛, 小鹿亨, 遅沢恭二, 越智恒夫, 小野田晃治, 笠原須磨生, 葛信彦, 門脇久志, 神吉正雄, 茅橋輝昭, 鳥山邦夫, 河原宏幸, 河原正和, 河原安孝, 菅晃, 木野田毅, 木村欣二, 草刈広一, 草野憲二, 国兼信之, 国兼正明, 久米加寿徳, 黒澤良彦, 毛塚尚利, 小岩屋敏, 小阪敏和, 児玉雅一, 小宮次郎, 斎藤修司, 坂本繁夫, 坂本充, 櫻井俊一, 櫻谷鎮雄, 佐々木邦彦, 佐藤誠吾, 佐藤福男, 佐藤雅彦, 佐野信雄, 潮崎正浩, 塩見和之, 四方圭一郎, 七野芳彦, 篠永哲, 渋谷誠, 清水勝幸, 清水清市, 白石正人, 蕭嘉廣, 鈴木邦雄, 鈴木俊夫, 鈴木裕, 須田亨, 瀬谷幸次郎, 惣名実, 高井泰, 多賀敏正, 高野敏明, 高橋進, 武田卓明, 田添京二, 舘野鴻, 田中昭太郎, 田中勇, 田中馨, 谷口明, 田村昭夫, 堤内雄二, 鶴巻洋志, 出嶋利明, 手束喜洋, 土岐和多瑠, 鳥羽明彦, 富取満, 冨永修, 中峰空, 永井虚, 長尾康, 中込正男, 中田博臣, 中村裕之, 永幡嘉之, 生川展行, 新居悟（長尾悟）, 西田光康, 西村正賢, 野崎敦士, 野田正美, 野中勝, 野村禎男（山脇好之）, 春沢圭太郎, 伴一利, 平野幸彦, 廣川典範, 広美栄, 深田晋一, 藤川匠, 藤本博文, 堀繁久, 前川和則, 正木清, 松尾照男, 松永正光, 松原至, 松本堅一, 三木将義, 三木武司, 水沢孝, 水沢洋, 水沢美環, 水沼哲郎, 溝上誠司, 南雄二, 宮川続, 三宅誠治, 宮下公範, 椋木博昭, 森正人, 森田誠司, 保田信紀, 矢田直之, 山岡幸雄, 山口就平, 山崎一夫, 山迫淳介, 山地治, 山添寛治, 山元一裕, 横山裕正, 吉越肇, 淀江賢一郎, 吉松定昭, 和田潤, 渡辺崇, 渡辺康之, I. BELOUSOV, M. BRENDELL, B. BRUGGE, Th. DEUVE, I. KABAK, B. KATAEV, N. NIKITSKIJ, H. SCHÜTZE, H. SCHÖNMANN.

目次　CONTENTS

凡例 ··· 9
用語解説 1 成虫の外部形態 ·· 11
用語解説 2 ♂交尾器形態(1)陰茎と完全反転下の内嚢 ·· 12
用語解説 3 ♂交尾器形態(2)カラブス亜属群の内嚢 ·· 13
図解検索 1 亜属群・亜属・種の検索(1) ·· 14
図解検索 2 亜属群・亜属・種の検索(2) ·· 15
図解検索 3 種の検索(1) ·· 16
図解検索 4 種の検索(2) ·· 17
図解検索 5 種の検索(3) ·· 18
図解検索 6 種の検索(4) ·· 19
日本の地方区分と県 ·· 20
日本の島 ·· 21

本編

Genus *Carabus* Linnaeus, 1758　カラブス属

I. *Limnocarabus* subgenus-group　リムノカラブス亜属群

  I-1. Subgenus *Limnocarabus* Géhin, 1876　リムノカラブス亜属

  1. *Carabus* (*Limnocarabus*) *maacki* Morawitz, 1862　マークオサムシ ················· 24
   1) *C.* (*L.*) *m. aquatilis* Bates, 1883　マークオサムシ本州亜種 ·· 27

  I-2. Subgenus *Euleptocarabus* Nakane, 1955　エウレプトカラブス亜属

  2. *Carabus* (*Euleptocarabus*) *porrecticollis* Bates, 1883　アキタクロナガオサムシ ······ 30
   1) *C.* (*E.*) *p. porrecticollis* Bates, 1883　アキタクロナガオサムシ基亜種 ······················· 33
   2) *C.* (*E.*) *p. pacificus* (Imura et Matsunaga, 2011)　アキタクロナガオサムシ富士亜種 ············ 34
   3) *C.* (*E.*) *p. kansaiensis* Nakane, 1961　アキタクロナガオサムシ岩湧亜種 ························· 35

II. *Hemicarabus* subgenus-group　ヘミカラブス亜属群

  II-1. Subgenus *Hemicarabus* Géhin, 1885　ヘミカラブス亜属

  3. *Carabus* (*Hemicarabus*) *macleayi* Dejean, 1826　マックレイセアカオサムシ ······ 38
   1) *C.* (*He.*) *m. amanoi* (Imura, 2004)　マックレイセアカオサムシ利尻島亜種 ···················· 41

  4. *Carabus* (*Hemicarabus*) *tuberculosus* Dejean et Boisduval, 1829　セアカオサムシ ····· 44

  II-2. Subgenus *Homoeocarabus* Reitter, 1896　ホモエオカラブス亜属

  5. *Carabus* (*Homoeocarabus*) *maeander* Fischer, 1822　セスジアカガネオサムシ ········ 50
   1) *C.* (*Ho.*) *m. paludis* Géhin, 1885　セスジアカガネオサムシカムチャツカ北海道亜種 ··········· 53
   2) *C.* (*Ho.*) *m. nobukii* (Imura, 2003)　セスジアカガネオサムシ大雪山パルサ湿原亜種 ············ 54

III. Subgenus *Pentacarabus* Ishikawa, 1972　ペンタカラブス亜属

  6. *Carabus* (*Pentacarabus*) *harmandi* Lapouge, 1909　ホソヒメクロオサムシ ············ 58
   1) *C.* (*P.*) *h. yudanus* Nakane, 1977　ホソヒメクロオサムシ奥羽山脈亜種 ························· 62
   2) *C.* (*P.*) *h. adatarasanus* Ishikawa, 1966　ホソヒメクロオサムシ東北地方南西部亜種 ········ 63
   3) *C.* (*P.*) *h. syui* (Imura et Mizusawa, 2002)　ホソヒメクロオサムシ阿武隈高地亜種 ········· 63
   4) *C.* (*P.*) *h. kojii* (Imura et Mizusawa, 2002)　ホソヒメクロオサムシ八溝山亜種 ················ 64
   5) *C.* (*P.*) *h. harmandi* Lapouge, 1909　ホソヒメクロオサムシ基亜種 ······························· 64
   6) *C.* (*P.*) *h. quinquecatellatus* (Ishikawa, 1986)　ホソヒメクロオサムシ飛騨山脈北部亜種 ··· 64
   7) *C.* (*P.*) *h. mizunumai* Ishikawa, 1972　ホソヒメクロオサムシ白山飛騨御嶽木曽亜種 ········ 65
   8) *C.* (*P.*) *h. karasawai* Ishikawa, 1966　ホソヒメクロオサムシ八ヶ岳亜種 ······················· 65
   9) *C.* (*P.*) *h. okutamaensis* Ishikawa, 1966　ホソヒメクロオサムシ奥秩父亜種 ················· 66
   10) *C.* (*P.*) *h. tanzawaensis* (Ishikawa, 1986)　ホソヒメクロオサムシ丹沢亜種 ··················· 66
   11) *C.* (*P.*) *h. fujisan* Ishikawa, 1972　ホソヒメクロオサムシ富士山亜種 ··························· 66
   12) *C.* (*P.*) *h. akaishiensis* (Ishikawa, 1986)　ホソヒメクロオサムシ赤石山脈亜種 ············· 67
   13) *C.* (*P.*) *h. moritai* (Imura et Mizusawa, 2002)　ホソヒメクロオサムシ安倍峠亜種 ········· 67

IV. *Leptocarabus* subgenus-group　レプトカラブス亜属群

  IV-1. Subgenus *Aulonocarabus* Reitter, 1896　アウロノカラブス亜属

7. *Carabus* (*Aulonocarabus*) *kurilensis* Lapouge, 1913 チシマオサムシ ---------- 70
   1) *C.* (*A.*) *k. sugai* Ishikawa, 1966 チシマオサムシ礼文島亜種 ---------- 73
   2) *C.* (*A.*) *k. rishiriensis* Nakane, 1957 チシマオサムシ利尻島亜種 ---------- 73
   3) *C.* (*A.*) *k. kurilensis* Lapouge, 1913 チシマオサムシ基亜種 ---------- 74
   4) *C.* (*A.*) *k. rausuanus* Ishikawa, 1966 チシマオサムシ道央道東亜種 ---------- 74
   5) *C.* (*A.*) *k. daisetsuzanus* Kôno, 1936 チシマオサムシ大雪山亜種 ---------- 75
   6) *C.* (*A.*) *k. yoteizanus* (Ishikawa et Miyashita, 2000) チシマオサムシ羊蹄山亜種 ---------- 75

IV-2. Subgenus *Leptocarabus* Géhin, 1885 レプトカラブス亜属
  8. *Carabus* (*Leptocarabus*) *kyushuensis* Nakane, 1961 キュウシュウクロナガオサムシ ---------- 78
   1) *C.* (*L.*) *k. nakatomii* Ishikawa, 1966 キュウシュウクロナガオサムシ中国地方亜種 ---------- 81
   2) *C.* (*L.*) *k. cerberus* Ishikawa, 1962 キュウシュウクロナガオサムシ広島県東部亜種 ---------- 81
   3) *C.* (*L.*) *k. kawaharai* (Imura, 2011) キュウシュウクロナガオサムシ山口県亜種 ---------- 82
   4) *C.* (*L.*) *k. kyushuensis* Nakane, 1961 キュウシュウクロナガオサムシ基亜種 ---------- 82
  9. *Carabus* (*Leptocarabus*) *hiurai* Kamiyoshi et Mizoguchi, 1960 シコククロナガオサムシ ---------- 84
  10. *Carabus* (*Leptocarabus*) *kumagaii* (Kimura et Komiya, 1974) オオクロナガオサムシ ---------- 90
   1) *C.* (*L.*) *k. kumagaii* (Kimura et Komiya, 1974) オオクロナガオサムシ基亜種 ---------- 93
   2) *C.* (*L.*) *k. hida* (Kubota et Ishikawa, 2004) オオクロナガオサムシ飛騨高山亜種 ---------- 93
   3) *C.* (*L.*) *k. nishi* (Kubota et Ishikawa, 2004) オオクロナガオサムシ近畿・中部地方亜種 ---------- 93
  11. *Carabus* (*Leptocarabus*) *procerulus* Chaudoir, 1862 クロナガオサムシ ---------- 96
   1) *C.* (*L.*) *p. procerulus* Chaudoir, 1862 クロナガオサムシ基亜種 ---------- 99
   2) *C.* (*L.*) *p. miyakei* Nakane, 1961 クロナガオサムシ九州亜種 ---------- 99
  12. *Carabus* (*Leptocarabus*) *arboreus* Lewis, 1882 コクロナガオサムシ ---------- 102
   1) *C.* (*L.*) *a. pararboreus* (Ishikawa, 1992) コクロナガオサムシ道央道東道北亜種 ---------- 107
   2) *C.* (*L.*) *a. ishikarinus* (Ishikawa, 1992) コクロナガオサムシ石狩亜種 ---------- 108
   3) *C.* (*L.*) *a. karibanus* (Ishikawa, 1992) コクロナガオサムシ渡島半島北部亜種 ---------- 108
   4) *C.* (*L.*) *a. arboreus* Lewis, 1882 コクロナガオサムシ基亜種 ---------- 109
   5) *C.* (*L.*) *a. nepta* (Ishikawa, 1984) コクロナガオサムシ青森県亜種 ---------- 109
   6) *C.* (*L.*) *a. pronepta* (Ishikawa, 1992) コクロナガオサムシ岩手県北部亜種 ---------- 110
   7) *C.* (*L.*) *a. shimoheiensis* (Ishikawa, 1992) コクロナガオサムシ下閉伊亜種 ---------- 110
   8) *C.* (*L.*) *a. kitakamisanus* (Ishikawa, 1992) コクロナガオサムシ北上山地亜種 ---------- 111
   9) *C.* (*L.*) *a. akitanus* (Ishikawa, 1992) コクロナガオサムシ秋田県亜種 ---------- 111
   10) *C.* (*L.*) *a. parexilis* Nakane, 1961 コクロナガオサムシ東北地方南部亜種 ---------- 111
   11) *C.* (*L.*) *a. tenuiformis* Bates, 1883 コクロナガオサムシ北関東上越亜種 ---------- 112
   12) *C.* (*L.*) *a. exilis* Bates, 1883 コクロナガオサムシ佐渡島亜種 ---------- 113
   13) *C.* (*L.*) *a. babai* Ishikawa, 1968 コクロナガオサムシ妙高連峰亜種 ---------- 114
   14) *C.* (*L.*) *a. shinanensis* Born, 1922 コクロナガオサムシ八ヶ岳亜種 ---------- 114
   15) *C.* (*L.*) *a. ogurai* Ishikawa, 1969 コクロナガオサムシ奥秩父亜種 ---------- 115
   16) *C.* (*L.*) *a. fujisanus* Bates, 1883 コクロナガオサムシ富士箱根丹沢亜種 ---------- 115
   17) *C.* (*L.*) *a. horioi* Nakane, 1961 コクロナガオサムシ赤石山脈亜種 ---------- 116
   18) *C.* (*L.*) *a. gracillimus* Bates, 1883 コクロナガオサムシ飛騨御嶽木曽亜種 ---------- 116
   19) *C.* (*L.*) *a. hakusanus* Nakane, 1961 コクロナガオサムシ白山亜種 ---------- 117
   20) *C.* (*L.*) *a. ohminensis* Harusawa, 1978 コクロナガオサムシ大峰山脈亜種 ---------- 117

V. *Tomocarabus* subgenus-group トモカラブス亜属群
 V-1. Subgenus *Asthenocarabus* Lapouge, 1932 アステノカラブス亜属
  13. *Carabus* (*Asthenocarabus*) *opaculus* Putzeys, 1875 ヒメクロオサムシ ---------- 120
   1) *C.* (*A.*) *o. kurosawai* Breuning, 1957 ヒメクロオサムシ道央道東道北亜種 ---------- 124
   2) *C.* (*A.*) *o. opaculus* Putzeys, 1875 ヒメクロオサムシ基亜種 ---------- 125
   3) *C.* (*A.*) *o. apoi* (Imura, 2011) ヒメクロオサムシ日高山脈南端部亜種 ---------- 125
   4) *C.* (*A.*) *o. shirahatai* Nakane, 1961 ヒメクロオサムシ東北地方亜種 ---------- 126

VI. *Carabus* subgenus-group カラブス亜属群
 VI-1. Subgenus *Carabus* Linnaeus, 1758 カラブス亜属
  14. *Carabus* (*Carabus*) *arvensis* Herbst, 1784 コブスジアカガネオサムシ ---------- 128
   1) *C.* (*C.*) *a. hokkaidensis* Lapouge, 1924 コブスジアカガネオサムシ北海道亜種 ---------- 131

15. *Carabus (Carabus) granulatus* Linnaeus, 1758 アカガネオサムシ -- 134
   1) *C. (C.) g. yezoensis* Bates, 1883 アカガネオサムシ北海道亜種 --- 137
   2) *C. (C.) g. telluris* Lewis, 1882 アカガネオサムシ本州亜種 --- 138
16. *Carabus (Carabus) vanvolxemi* Putzeys, 1875 ホソアカガネオサムシ --- 140
   1) *C. (C.) v. vanvolxemi* Putzeys, 1875 ホソアカガネオサムシ基亜種 -- 143
   2) *C. (C.) v. noesskei* Van Emden, 1932 ホソアカガネオサムシ佐渡島亜種 -------------------------------- 143

VI-2. Subgenus *Ohomopterus* Reitter, 1896 オホモプテルス亜属

17. *Carabus (Ohomopterus) yamato* (Nakane, 1953) ヤマトオサムシ --- 146
   1) *C. (O.) y. shikatai* (Imura et Mizusawa, 2002) ヤマトオサムシ下伊那亜種 --------------------------------- 150
   2) *C. (O.) y. ojikai* (Imura et Mizusawa, 2002) ヤマトオサムシ三河亜種 ------------------------------------ 150
   3) *C. (O.) y. takanonis* (Imura et Mizusawa, 2002) ヤマトオサムシ北陸地方亜種 ------------------------- 151
   4) *C. (O.) y. yamato* (Nakane, 1953) ヤマトオサムシ基亜種 --- 151
   5) *C. (O.) y. kinkimontanus* (Imura et Mizusawa, 2002) ヤマトオサムシ近畿地方中東部亜種 ----------- 152
   6) *C. (O.) y. kitai* (Imura, 2004) ヤマトオサムシ熊野亜種 -- 152
18. *Carabus (Ohomopterus) kimurai* (Ishikawa, 1966) スルガオサムシ -- 154
19. *Carabus (Ohomopterus) lewisianus* Breuning, 1932 ルイスオサムシ -- 158
   1) *C. (O.) l. lewisianus* Breuning, 1932 ルイスオサムシ基亜種 -- 161
   2) *C. (O.) l. awakazusanus* Ishikawa, 1981 ルイスオサムシ房総半島南部亜種 ---------------------------- 161
20. *Carabus (Ohomopterus) albrechti* Morawitz, 1862 クロオサムシ -- 164
   1) *C. (O.) a. itoi* Ishikawa et Takami, 1996 クロオサムシ道東亜種 -- 168
   2) *C. (O.) a. hidakanus* Ishikawa et Takami, 1996 クロオサムシ日高亜種 ------------------------------------ 169
   3) *C. (O.) a. albrechti* Morawitz, 1862 クロオサムシ基亜種 -- 169
   4) *C. (O.) a. tohokuensis* Ishikawa, 1984 クロオサムシ東北地方東部亜種 ---------------------------------- 170
   5) *C. (O.) a. yamauchii* Takami et Ishikawa, 1997 クロオサムシ白神山地亜種 ---------------------------- 170
   6) *C. (O.) a. hagai* Takami et Ishikawa, 1997 クロオサムシ東北地方中部亜種 ------------------------------ 171
   7) *C. (O.) a. echigo* Ishikawa et Takami, 1996 クロオサムシ越後亜種 ------------------------------------- 172
   8) *C. (O.) a. awashimae* Ishikawa et Takami, 1996 クロオサムシ粟島亜種 -------------------------------- 172
   9) *C. (O.) a. freyi* Van Emden, 1932 クロオサムシ佐渡島亜種 --- 173
   10) *C. (O.) a. esakianus* (Nakane, 1960) クロオサムシ関東地方北西部亜種 --------------------------------- 173
   11) *C. (O.) a. tsukubanus* Takami et Ishikawa, 1997 クロオサムシ関東地方北東部亜種 ------------------- 174
   12) *C. (O.) a. okumurai* (Ishikawa, 1966) クロオサムシ山梨長野亜種 --- 175
21. *Carabus (Ohomopterus) daisen* (Nakane, 1953) ダイセンオサムシ --- 178
   1) *C. (O.) d. daisen* (Nakane, 1953) ダイセンオサムシ基亜種 --- 181
   2) *C. (O.) d. okianus* (Nakane, 1961) ダイセンオサムシ隠岐亜種 --------------------------------------- 181
22. *Carabus (Ohomopterus) chugokuensis* (Nakane, 1961) アキオサムシ ----------------------------------- 184
   1) *C. (O.) c. chugokuensis* (Nakane, 1961) アキオサムシ基亜種 --- 187
   2) *C. (O.) c. seizaburoi* Imura, Dejima et Mizusawa, 1993 アキオサムシ小豆島亜種 ---------------------- 187
   3) *C. (O.) c. mikianus* (Imura, 2003) アキオサムシ讃岐山脈東部亜種 ----------------------------------- 187
   4) *C. (O.) c. mochizukii* Imura et Mizusawa, 1994 アキオサムシ芸予諸島大島亜種 ------------------------ 188
   5) *C. (O.) c. umekii* Imura et Mizusawa, 1999 アキオサムシ周防亜種 -------------------------------------- 189
23. *Carabus (Ohomopterus) sue* Imura, 2012 ホウオウオサムシ --- 192
24. *Carabus (Ohomopterus) japonicus* Motschulsky, 1857 ヒメオサムシ -------------------------------------- 198
   1) *C. (O.) j. awajiensis* Imura, Dejima et Mizusawa, 1993 ヒメオサムシ淡路島四国亜種 --------------------- 203
   2) *C. (O.) j. yoshiyukii* Imura et Mizusawa, 1999 ヒメオサムシ鷲尾山亜種 ---------------------------------- 203
   3) *C. (O.) j. okinoshimanus* Imura, Dejima et Mizusawa, 1993 ヒメオサムシ沖の島亜種 --------------------- 204
   4) *C. (O.) j. japonicus* Motschulsky, 1857 ヒメオサムシ基亜種 --- 204
   5) *C. (O.) j. tsushimae* Breuning, 1932 ヒメオサムシ対馬亜種 -- 205
   6) *C. (O.) j. ikiensis* (Nakane, 1968) ヒメオサムシ壱岐亜種 -- 205
   7) *C. (O.) j. hiradonis* Imura, Dejima et Mizusawa, 1993 ヒメオサムシ平戸島亜種 -------------------------- 205
   8) *C. (O.) j. chotaroi* Imura, Dejima et Mizusawa, 1993 ヒメオサムシ生月島亜種 ---------------------------- 206
   9) *C. (O.) j. nozakicola* Imura et Mizusawa, 1994 ヒメオサムシ野崎島亜種 -------------------------------- 206
   10) *C. (O.) j. wakamatsuensis* Imura et Mizusawa, 1994 ヒメオサムシ若松島亜種 ----------------------------- 207
   11) *C. (O.) j. onodai* Imura, Dejima et Mizusawa, 1993 ヒメオサムシ下甑島亜種 ----------------------------- 207

25. *Carabus (Ohomopterus) kawanoi* (Kamiyoshi et Mizoguchi, 1960) アワオサムシ --- 210
   1) *C. (O.) k. kawanoi* (Kamiyoshi et Mizoguchi, 1960) アワオサムシ基亜種 --- 213
   2) *C. (O.) k. botchan* Imura et Mizusawa, 1995 アワオサムシ高縄半島亜種 --- 213
26. *Carabus (Ohomopterus) tosanus* (Nakane, Iga et Uéno, 1953) トサオサムシ --- 216
   1) *C. (O.) t. tosanus* (Nakane, Iga et Uéno, 1953) トサオサムシ基亜種 --- 219
   2) *C. (O.) t. ishizuchianus* (Nakane, 1953) トサオサムシ石鎚山脈亜種 --- 220
27. *Carabus (Ohomopterus) dehaanii* Chaudoir, 1848 オオオサムシ --- 222
   1) *C. (O.) d. punctatostriatus* Bates, 1873 オオオサムシ本州中部亜種 --- 227
   2) *C. (O.) d. dehaanii* Chaudoir, 1848 オオオサムシ基亜種 --- 227
   3) *C. (O.) d. katsumai* Imura et Mizusawa, 1995 オオオサムシ四国東部亜種 --- 228
   4) *C. (O.) d. strenuus* Breuning, 1932 オオオサムシ熊本県南西部亜種 --- 228
   5) *C. (O.) d. kumaso* Imura et Mizusawa, 1995 オオオサムシ九州山地南部亜種 --- 229
   6) *C. (O.) d. ishidai* Imura et Mizusawa, 1995 オオオサムシ御所浦島亜種 --- 229
   7) *C. (O.) d. koshikicola* Imura et Mizusawa, 1995 オオオサムシ上甑島亜種 --- 229
   8) *C. (O.) d. nakagomei* Kubota, 2010 オオオサムシ大隅半島亜種 --- 230
28. *Carabus (Ohomopterus) iwawakianus* (Nakane, 1953) イワワキオサムシ --- 232
   1) *C. (O.) i. narukawai* Ishikawa et Kubota, 1995 イワワキオサムシ布引山地亜種 --- 235
   2) *C. (O.) i. shima* Ishikawa et Kubota, 1995 イワワキオサムシ志摩半島亜種 --- 235
   3) *C. (O.) i. iwawakianus* (Nakane, 1953) イワワキオサムシ基亜種 --- 236
   4) *C. (O.) i. muro* Ishikawa et Kubota, 1995 キイオサムシ室生山地亜種 --- 236
   5) *C. (O.) i. kiiensis* (Nakane, 1953) イワワキオサムシ紀伊半島亜種 --- 237
29. *Carabus (Ohomopterus) yaconinus* Bates, 1873 ヤコンオサムシ --- 240
   1) *C. (O.) y. blairi* (Breuning, 1934) ヤコンオサムシ北陸地方亜種 --- 244
   2) *C. (O.) y. sotai* Ishikawa et Kubota, 1994 ヤコンオサムシ近畿地方北部亜種 --- 244
   3) *C. (O.) y. cupidicornis* Ishikawa et Kubota, 1994 ヤコンオサムシ近畿地方中部亜種 --- 245
   4) *C. (O.) y. yaconinus* Bates, 1873 ヤコンオサムシ基亜種 --- 245
   5) *C. (O.) y. oki* Ishikawa, 1994 ヤコンオサムシ隠岐亜種 --- 245
   6) *C. (O.) y. maetai* Ishikawa, 1994 ヤコンオサムシ山陰地方亜種 --- 246
   7) *C. (O.) y. yamaokai* Ishikawa, 1994 ヤコンオサムシ四国和歌山亜種 --- 247
   8) *C. (O.) y. seto* Ishikawa, 1994 ヤコンオサムシ忽那諸島西部亜種 --- 247
30. *Carabus (Ohomopterus) komiyai* (Ishikawa, 1966) カケガワオサムシ --- 250
   1) *C. (O.) k. komiyai* (Ishikawa, 1966) カケガワオサムシ基亜種 --- 253
   2) *C. (O.) k. matsunagai* (Imura, 2008) カケガワオサムシ磐田原亜種 --- 253
   3) *C. (O.) k. yamazumiensis* (Imura et Matsunaga, 2010) カケガワオサムシ山住峠亜種 --- 253
31. *Carabus (Ohomopterus) esakii* Csiki, 1927 シズオカオサムシ --- 256
   1) *C. (O.) e. esakii* Csiki, 1927 シズオカオサムシ基亜種 --- 259
   2) *C. (O.) e. suruganus* Ujiie, Ishikawa et Kubota, 2010 シズオカオサムシ富士川流域以西亜種 --- 259
32. *Carabus (Ohomopterus) insulicola* Chaudoir, 1869 アオオサムシ --- 262
   1) *C. (O.) i. kita* Ishikawa et Ujiie, 2000 アオオサムシ東北地方亜種 --- 266
   2) *C. (O.) i. awashimaensis* Ishikawa et Ujiie, 2000 アオオサムシ粟島亜種 --- 267
   3) *C. (O.) i. kantoensis* Ishikawa et Ujiie, 2000 アオオサムシ関東平野多摩川以北亜種 --- 267
   4) *C. (O.) i. nishikawai* (Ishikawa, 1966) アオオサムシ房総半島南部亜種 --- 268
   5) *C. (O.) i. insulicola* Chaudoir, 1869 アオオサムシ基亜種 --- 268
   6) *C. (O.) i. okutonensis* Ishikawa et Ujiie, 2000 アオオサムシ奥利根亜種 --- 268
   7) *C. (O.) i. sado* Ishikawa et Ujiie, 2000 アオオサムシ佐渡島亜種 --- 269
   8) *C. (O.) i. shinano* Ishikawa et Ujiie, 2000 アオオサムシ信濃亜種 --- 269
   9) *C. (O.) i. kiso* Ishikawa et Ujiie, 2000 アオオサムシ木曽亜種 --- 270
   10) *C. (O.) i. komaganensis* Ishikawa et Ujiie, 2000 アオオサムシ駒ヶ根亜種 --- 270
33. *Carabus (Ohomopterus) uenoi* (Ishikawa, 1960) ドウキョウオサムシ --- 274
34. *Carabus (Ohomopterus) arrowianus* (Breuning, 1934) ミカワオサムシ --- 278
   1) *C. (O.) a. nakamurai* (Ishikawa, 1966) ミカワオサムシ天竜川中流域亜種 --- 281
   2) *C. (O.) a. shichinoi* (Imura et Matsunaga, 2010) ミカワオサムシ佐久間亜種 --- 282
   3) *C. (O.) a. hidaosa* Ishikawa et Kubota, 1994 ミカワオサムシ岐阜県中北部亜種 --- 283
   4) *C. (O.) a. arrowianus* (Breuning, 1934) ミカワオサムシ基亜種 --- 283
   5) *C. (O.) a. minoensis* Ishikawa et Kubota, 1994 ミカワオサムシ岐阜県南部亜種 --- 284
   6) *C. (O.) a. murakii* Ishikawa et Kubota, 1984 ミカワオサムシ志摩半島北部亜種 --- 284
   7) *C. (O.) a. kirimurai* Kubota et Yahiro, 2003 ミカワオサムシ御浜町亜種 --- 284

35. *Carabus* (*Ohomopterus*) *maiyasanus* Bates, 1873 マヤサンオサムシ ---- 288
   1) *C. (O.) m. hokurikuensis* Ishikawa et Kubota, 1994 マヤサンオサムシ北陸地方亜種 ---- 291
   2) *C. (O.) m. maiyasanus* Bates, 1873 マヤサンオサムシ基亜種 ---- 291
   3) *C. (O.) m. yoroensis* Ishikawa et Kubota, 1994 マヤサンオサムシ養老山地亜種 ---- 292
   4) *C. (O.) m. suzukanus* Ishikawa et Kubota, 1994 マヤサンオサムシ鈴鹿山脈南東部亜種 ---- 292
   5) *C. (O.) m. shigaraki* (Hiura et Katsura, 1971) マヤサンオサムシ信楽亜種 ---- 293
   6) *C. (O.) m. takiharensis* (Katsura et Tominaga, 1978) マヤサンオサムシ滝原亜種 ---- 293
   7) *C. (O.) m. ohkawai* (Nakane, 1968) マヤサンオサムシ鵜方亜種 ---- 294

VII. *Procrustes* subgenus-group プロクルステス亜属群
 VII-1. Subgenus *Megodontus* Solier, 1848 メゴドントゥス亜属
  36. *Carabus* (*Megodontus*) *kolbei* Roeschke, 1897 アイヌキンオサムシ ---- 298
   1) *C. (M.) k. hanatanii* Imura, 1991 アイヌキンオサムシ利尻島亜種 ---- 304
   2) *C. (M.) k. futabae* (Ishikawa, 1966) アイヌキンオサムシ道北亜種 ---- 305
   3) *C. (M.) k. chishimanus* (Nakane, 1961) アイヌキンオサムシ択捉島亜種 ---- 305
   4) *C. (M.) k. aino* Rost, 1908 アイヌキンオサムシ道央道東亜種 ---- 306
   5) *C. (M.) k. mitsumasai* Imura, 1994 アイヌキンオサムシ増毛山地南西部亜種 ---- 306
   6) *C. (M.) k. taikinus* (Ishikawa, 1993) アイヌキンオサムシ大樹町亜種 ---- 307
   7) *C. (M.) k. nitidipunctatus* (Ishikawa, 1966) アイヌキンオサムシトマム亜種 ---- 307
   8) *C. (M.) k. yubariensis* (Ishikawa, 1966) アイヌキンオサムシ夕張山地亜種 ---- 308
   9) *C. (M.) k. hidakamontanus* (Ishikawa, 1966) アイヌキンオサムシ日高山脈亜種 ---- 308
   10) *C. (M.) k. kolbei* Roeschke, 1897 アイヌキンオサムシ基亜種 ---- 308
   11) *C. (M.) k. kosugei* (Nakane, 1955) アイヌキンオサムシ積丹半島亜種 ---- 309
   12) *C. (M.) k. kuniakii* (Ishikawa, 1971) アイヌキンオサムシニセコ亜種 ---- 309
   13) *C. (M.) k. yasudai* (Ishikawa, 1971) アイヌキンオサムシ狩場山亜種 ---- 310
   14) *C. (M.) k. munakataorum* (Ishikawa, 1969) アイヌキンオサムシ大千軒岳亜種 ---- 310

 VII-2. Subgenus *Acoptolabrus* Morawitz, 1886 アコプトラブルス亜属
  37. *Carabus* (*Acoptolabrus*) *gehinii* Fairmaire, 1876 オオルリオサムシ ---- 314
   1) *C. (A.) g. aereicollis* (Hauser, 1921) オオルリオサムシ道北亜種 ---- 321
   2) *C. (A.) g. konsenensis* (Ishikawa, 1968) オオルリオサムシ根釧台地亜種 ---- 322
   3) *C. (A.) g. manoianus* (Imura, 1989) オオルリオサムシ南富良野亜種 ---- 322
   4) *C. (A.) g. radiatocostatus* (Ishikawa, 1968) オオルリオサムシ日高山脈南東部亜種 ---- 323
   5) *C. (A.) g. sapporensis* (Uchida et Tamanuki, 1927) オオルリオサムシ日高山脈南西部亜種 ---- 323
   6) *C. (A.) g. gehinii* Fairmaire, 1876 オオルリオサムシ基亜種 ---- 324
   7) *C. (A.) g. shimizui* Imura, 1994 オオルリオサムシ積丹半島亜種 ---- 326
   8) *C. (A.) g. nishijimai* Imura, 1991 オオルリオサムシ大平山亜種 ---- 326
   9) *C. (A.) g. munakatai* (Ishikawa, 1968) オオルリオサムシ渡島半島亜種 ---- 327

 VII-3. Subgenus *Damaster* Kollar, 1836 ダマステル亜属
  38. *Carabus* (*Damaster*) *blaptoides* (Kollar, 1836) マイマイカブリ ---- 330
   1) *C. (D.) b. rugipennis* (Motschulsky, 1861) マイマイカブリ北海道亜種 ---- 336
   2) *C. (D.) b. viridipennis* (Lewis, 1880) マイマイカブリ東北地方北部亜種 ---- 336
   3) *C. (D.) b. babaianus* (Ishikawa, 1985) マイマイカブリ東北地方南部亜種 ---- 337
   4) *C. (D.) b. fortunei* (Adams, 1861) マイマイカブリ粟島亜種 ---- 338
   5) *C. (D.) b. capito* (Lewis, 1881) マイマイカブリ佐渡島亜種 ---- 339
   6) *C. (D.) b. oxuroides* (Schaum, 1862) マイマイカブリ関東・中部地方亜種 ---- 340
   7) *C. (D.) b. brevicaudus* Imura et Mizusawa, 1995 マイマイカブリ隠岐亜種 ---- 340
   8) *C. (D.) b. braptoides* (Kollar, 1836) マイマイカブリ基亜種 ---- 341

 VII-4. Subgenus *Coptolabrus* Solier, 1848 コプトラブルス亜属
  39. *Carabus* (*Coptolabrus*) *fruhstorferi* (Roeschke, 1900) ツシマカブリモドキ ---- 344

文献 ---- 350
索引 ---- 359
学名 ---- 366
著者 ---- 367

凡例　EXPLANATORY NOTES

1. 本書の構成

　1) 本書は，日本から知られているオサムシの全種全亜種(2013年春現在)について，外部形態と交尾器形態をカラー画像で示し，解説を加えたものである．

　2) ここでいうオサムシとは，subtribe Carabinaオサムシ亜族に属する甲虫類の総称である．本書では同亜族を1属 *Carabus* (genus *Carabus* (s. lat.) 広義のカラブス属)とみなし，その下に亜属を設ける扱いとした．これは，世界で現在最も一般的に採用されている方式，いわゆるグローバルスタンダードであり，筆者らの前著「世界のオサムシ大図鑑」(むし社，1996)との整合性を図る意味もある．オサムシの場合，属と亜属の間に更にいくつかの任意の階級(節・群等)が設けられ，各階級には固有の名称が与えられて，研究者ごとに異なる様々な分類体系が提唱されているが，同一の著者による体系ですら数年を経ずして目まぐるしく変遷しているのが実状で，大方のコンセンサスを得られている安定した体系は今の所存在しない．本書ではこうした国際動物命名規約の適用を受けない階級の使用は控え，形態並びに分子系統双方の所見を考慮に入れたうえで，系統学的に類縁が近いと思われる複数の亜属からなるグループを便宜上，「亜属群」としてまとめた．亜属群名には，記載年において(最)先行する亜属の名称を採用した(例：リムノカラブス亜属(1876年設立)とエウレプトカラブス亜属(1956年設立)からなるグループはリムノカラブス亜属群とする)．世界レベルでの分類なので，日本に産しない亜属が亜属群名として採用されている場合もある(トモカラブス亜属群とプロクルステス亜属群)．

　3) 本書では，日本産のオサムシを1属14亜属39種193亜種(単型種は便宜上，1亜種と数えた)に分類した．亜属，種の分類と配列は，現時点において筆者らが最も妥当と考える類縁関係を反映させたものとし，亜種の配列は原則として分布域北東方に産するものから南西方に産するものの順とした(一部例外あり)．

　4) 巻頭の総論部分には，用語解説(交尾器各部の用語は基本的にDeuve(1994, 2004)に従う)と亜属・種までの図解検索，並びに国際的な使用に配慮して日本国内の地方区分名と県名，代表的島嶼名をローマ字で表記した地図を，巻末には文献一覧と索引を示した．各論部分は種ごとの図版(1~4ページのカラープレート)とそれに続く解説，分布図及び生態写真(ページの関係で省略した種もある)によって構成される．

　5) 各分類階級の和名は以下のように表記した．
　　(1) 属階級群名：属，亜属の学名を片仮名に転記し，語尾に属，亜属を付ける(例：カラブス属，リムノカラブス亜属)．
　　(2) 種名：最も普遍的に用いられている固有の種和名により表記する(例：マークオサムシ，アキタクロナガオサムシ)．
　　(3) 亜種名：「種和名＋地域(その亜種の主たる分布域)名＋亜種」という形で表記する(例：マークオサムシ本州亜種)．慣用的に用いられている固有の亜種和名(通称)は，各亜種の解説末尾の「亜種小名の由来と通称」の項に記した．

2. 図版(カラープレート)

　1) 図版の1ページ目には，上段に外部形態(♂♀成虫(背面観)，頭部と前胸背板の拡大画像(同)，♂右前跗節(腹面観)，♂右前脛節(斜背面観)，♂右前腿節(左側面観；一部の種のみ)，左鞘翅中央部の拡大画像(背面観))を，下段に交尾器形態(内嚢完全反転下の陰茎(右側面観)，陰茎先端(2方向ないし3方向)，完全反転下の内嚢(基本的に尾側面観・頭側面観・背面観・左斜側面観の4方向)，内嚢先端の頂板(一部の種のみ)，♀交尾器膣底部節片内板(一部の種では省略))を示した．複数の亜種を擁する種では，極力その種のタイプ産地から得られた基亜種の標本を用いたが，タイプ産地の不明な種や，基亜種が国外に分布する種についてはこの限りではない．また，下段の交尾器標本は必ずしも上段に示した個体のものとは限らないが，少なくとも同一産地から得られたものを使用した．

　2) 本書では，公共機関に保管されているタイプ標本(将来納入される予定のものを含む)を除き，すべて井村有希・水沢清行両著者のコレクション中にある標本を使用した．

　3) 各画像に付した番号は，カラープレートの後に続く分布図，生態写真を含め，種ごとの通し番号とした．

　4) カラープレート最下段には，英文と和文による脚注を付した．使用した標本がタイプ標本である場合には，その由を略号(HT = holotypeホロタイプ，PT = paratypeパラタイプ)で示した．

　5) 図版の2ページ目以降(種によって最大3ページ，ただし単型種の場合，省略したものもある)には，原則としてその種に含まれる邦産の全亜種の♂♀成虫(ただし，ヤコンオサムシ近畿地方北部亜種と同近畿地方中部亜種のみは♀を省略)と交尾器の画像を示した．ただし，交尾器に亜種の特徴が殆ど現れない一部の種では，交尾器画像を示さなかったものもある．成虫の拡大・縮小率は図版ごとに異なるが，ページ内における全個体の相対的な大きさは実際の比率を反映させたものとなっている．実際の体長については解説文中に示した値を参照されたい．

3. 解説

　1) 各種の解説は英文と和文の二言語により示した．両者は内容的にほぼ対訳関係にあるが，細部では必ずしも一致しておらず，和文部分の内容のほうがやや詳しい場合もある．

　2) 種の解説は，原記載(英文のみ)，研究史，タイプ産地，タイプ標本の所在，形態，分子系統，分布，生息環境，生態，地理的変異，種小名の由来の11項目からなり，必要に応じて保全に関する項目を加えた．

3）各タクソンの原記載は英文の冒頭に学名，著者名，記載年及び雑誌名と掲載ページ，付図番号を示し，synonymy（変遷名リスト）は原則として省略した．ただし，同名関係や異名関係等，現行の学名が用いられるに至った経緯を特に示す必要のあるものに関しては，最小限の変遷名リストを付した．

4）研究史以下の和文部分では，外国人名を各種の初出箇所で現綴りの後に括弧に入れた片仮名で示した．外国人名を片仮名で正確に表記することは困難な場合も多いが，本書では必ずしもこれまでの慣習にとらわれず，母国語の発音になるべく近い表記を心掛けた．難読と思われる地名，邦人名ないし固有名詞の一部については，種ごとの解説の初出箇所において，漢字の後に括弧に入れた平仮名で読み方を併記した．ただし，英文部分にローマ字表記のあるものについてはこれを省いた．

※外国人名の中で特記しておきたいものがある．それはフランスの現役オサムシ研究家 Deuve の発音である．我が国では「ダーヴ」と片仮名表記されることがあり（おそらく石川（1991）以来，筆者（井村）も以前，迂闊にもそのように記してしまったことがあるが，これには Deuve 本人からクレームが寄せられている．彼を含む複数のフランス人，並びに翻訳を生業とするフランス語に堪能な日本人に確認した結果，最も適切な片仮名表記は「ドゥーヴ」である．彼は現在，世界で最もアグレッシブなオサムシ研究家の一人であり，和文で書かれた文章にも今後，登場する機会が多いと思われるので，一言触れて関係諸氏の注意を喚起しておきたい．他の人名については本文中のそれぞれの箇所を参照されたい．

5）タイプ産地名は，原記載にある表現をそのまま転写した．そのため，本文中の他の箇所で用いられているローマ字表記と異なるもの，長音記号のないもの，また標高を表す数字の千と百の位の間にコンマのないもの等がある．記載当時の行政区画が現在のそれと異なる場合，必要に応じて随時，現在の所属を括弧に入れて示した．市町村名は 2013 年現在で最新のものを用いるよう心掛けたが，いわゆる平成の大合併後も市町村の合併・分離は行われているので，あくまでも本書の原稿を執筆した時点での行政区画名と理解されたい．

6）各タクソンの体長は，原則として大顎先端から鞘翅端までの長さを小数点以下を四捨五入した値により示した．

7）分布域については，原則としてまず日本の四大島嶼（北海道，本州，四国，九州）における分布を記した後，スラッシュで区切ってそれら以外の小島嶼における分布を記した．以前は独立した二つの島であったが現在は地続きになっているもの（広島県の江田島と能美島等）は二つの島名を並列し「＋」で結んだ．また，以前は独立した島であったが現在は埋め立てられて本土（ここでは北海道，本州，四国，九州の四大島嶼を指す）と陸続きになっているもの（千葉県館山市の沖ノ島や高ノ島等），ごく狭い海峡により隔てられてはいるものの，実質的に本土ないし本体となる島の一部に等しいもの（香川県小豆島南西端の前島等），或いはかつて狭い海峡で本土と隔てられていたが，開発により現在はほぼ陸続きになっており，現行の法定区分では本土の一部とされているもの（香川県の屋島等）は，島名を括弧で括って示した．海外の分布もこれにほぼ準じ，大陸における分布の後に，スラッシュで区切って島嶼における分布を記した．

8）亜種の解説は，原則として原記載（英文のみ），タイプ産地，タイプ標本の所在，亜種の特徴，分布，亜種小名の由来と通称の 6 項目からなるが，必要に応じて種の解説に用いた項目を加えた．

4. 分布図

1）各種の分布図では，原則としてページ左上の日本全図（ただし，オサムシを産しない南西諸島は省いた）に日本国内におけるその種の分布域を，右下の拡大図にはより詳細な亜種の分布域を示した．ただし，日本のほぼ全域に分布する種（セアカオサムシとマイマイカブリ），或いは逆に種としての国内における分布域が極めて狭いもの（マックレイセアカオサムシ）に関しては，単一の日本全図のみを用いた．小島嶼の分布は，色彩だけでは認識しづらいものも多いので，必要に応じ赤色の矢印で指し示すようにしたが，縮小率の関係で，近接する複数の島嶼を単一の矢印で表した場合もある．全ての分布図は過去の文献と筆者らの所蔵標本を元に作成した．離島の分布に関しては，本書で初めて発表される新知見が多数含まれているが，その一方で，既記録地点を点で表示した分布図ではないので，とりわけ四大島嶼における分布域については細部で正確さを欠く部分がある．あくまでも，それぞれの種・亜種のおおよその分布域を把握するためのものとみなしていただきたい．

5. 生態写真

各種の末尾には，原則として 3 葉（ダイセンオサムシのみ 2 葉）の生態写真を示し，簡単な解説を加えた．ただし，ページ数の都合により生態写真を省いた種も多い．用いた写真は，セスジアカガネオサムシの図 31（保田信紀氏撮影）を除き，すべて主著者の井村有希によって撮影されたものである．また，利尻礼文サロベツ国立公園の特別保護地区内に分布し，環境省第 4 次レッドリスト（2012）で絶滅危惧 II 類に指定されているマックレイセアカオサムシ利尻島亜種に関しては，環境省の認可（環北地国許第 110322001 号，平成 23 年 3 月 22 日交付）を受けて行った調査の際に撮影したもので，同個体は撮影後，捕殺することなく現地にリリースした．

6. 文献

和文で書かれた文献には，その末尾にローマ字による著者名と出版年を鉤括弧に入れて付記した．

7. 索引

索引は，原則として本書で解説を与えた亜属・種・亜種の学名と和名（通称を含む）に限定して作成したが，変遷名リストや研究史の項で触れたタクソンについても，重要と思われるものは取り上げた．

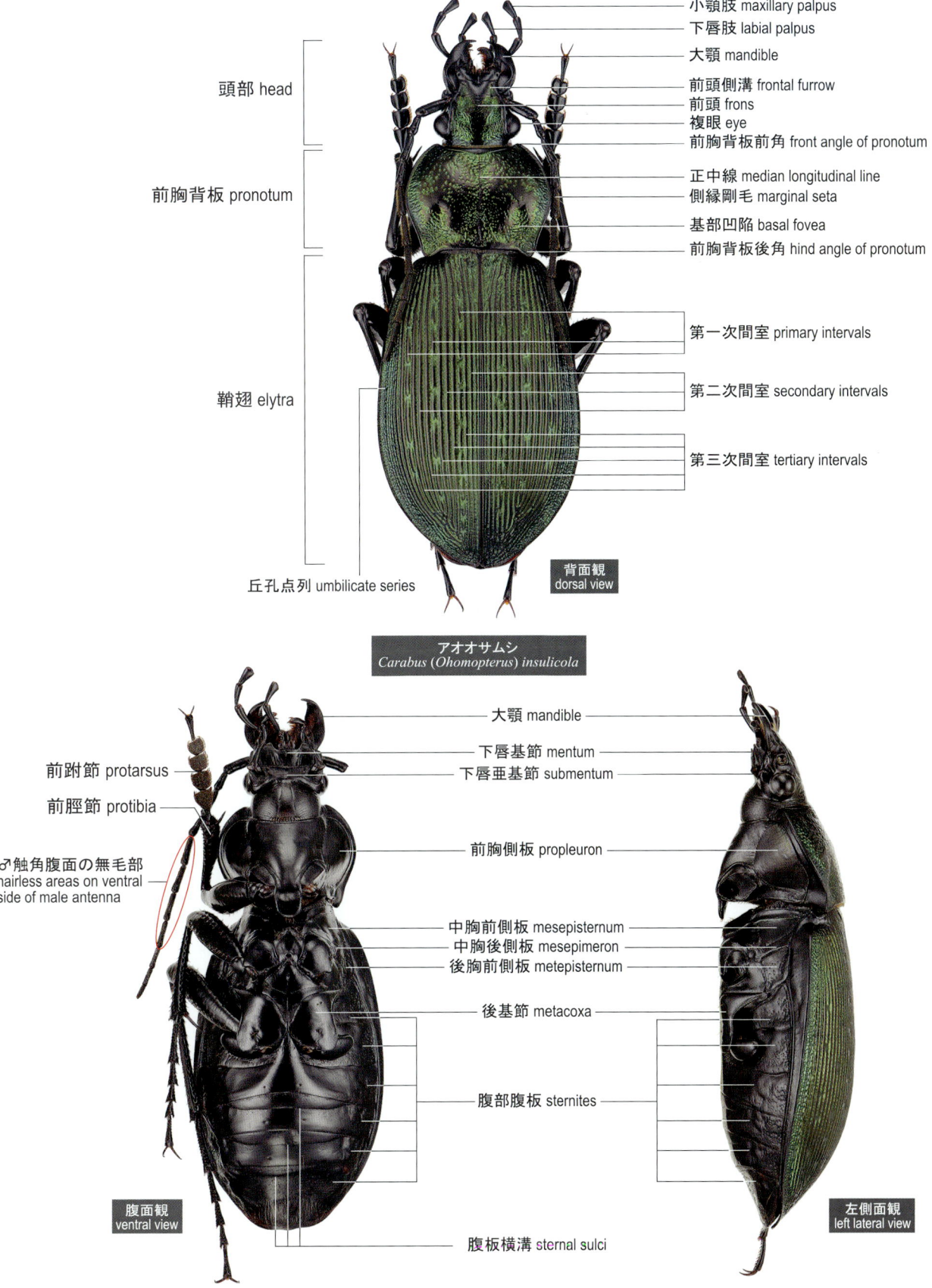

用語解説2 ♂交尾器形態 (1) 陰茎と完全反転下の内嚢　Male genital morphology (1) Penis and fully everted internal sac

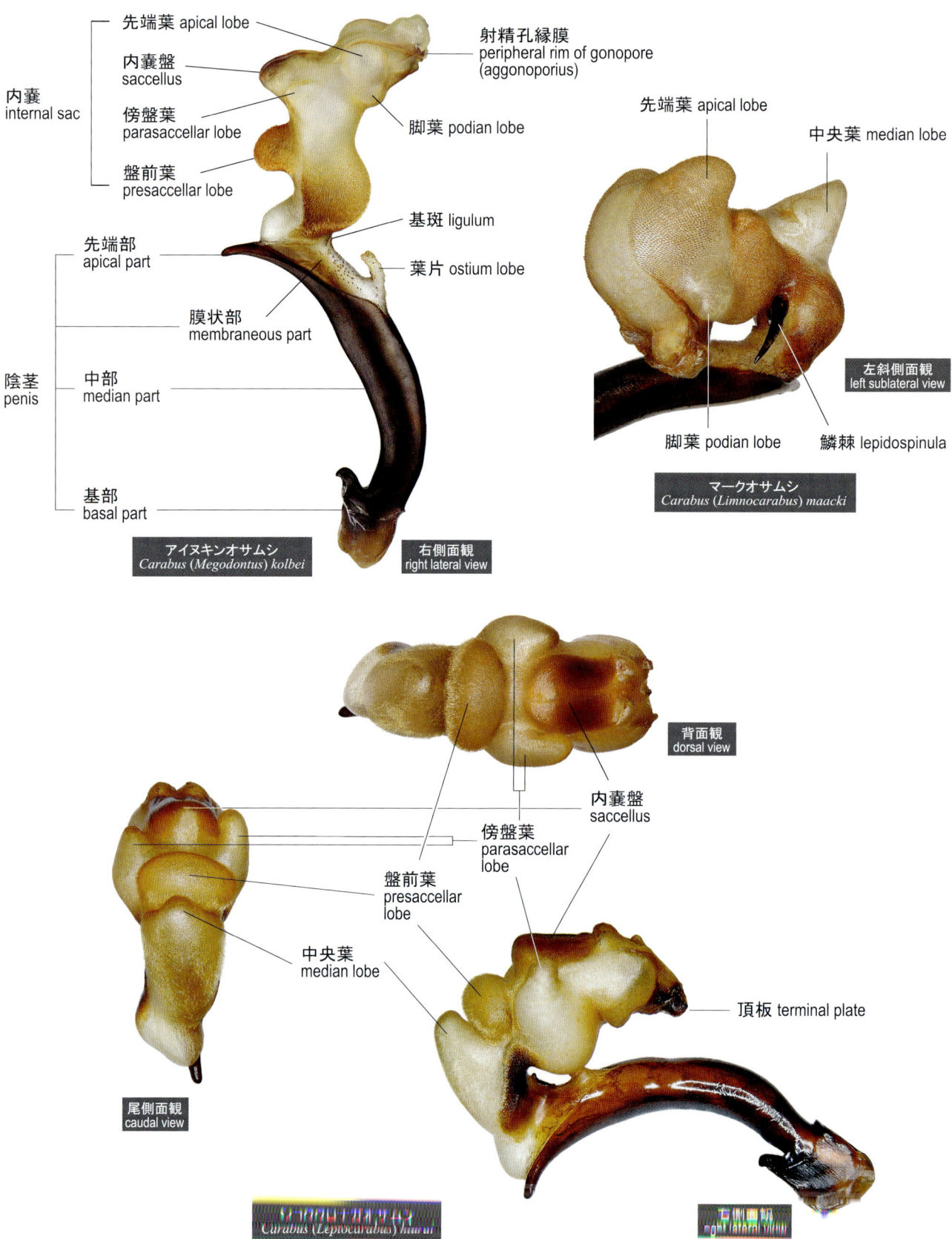

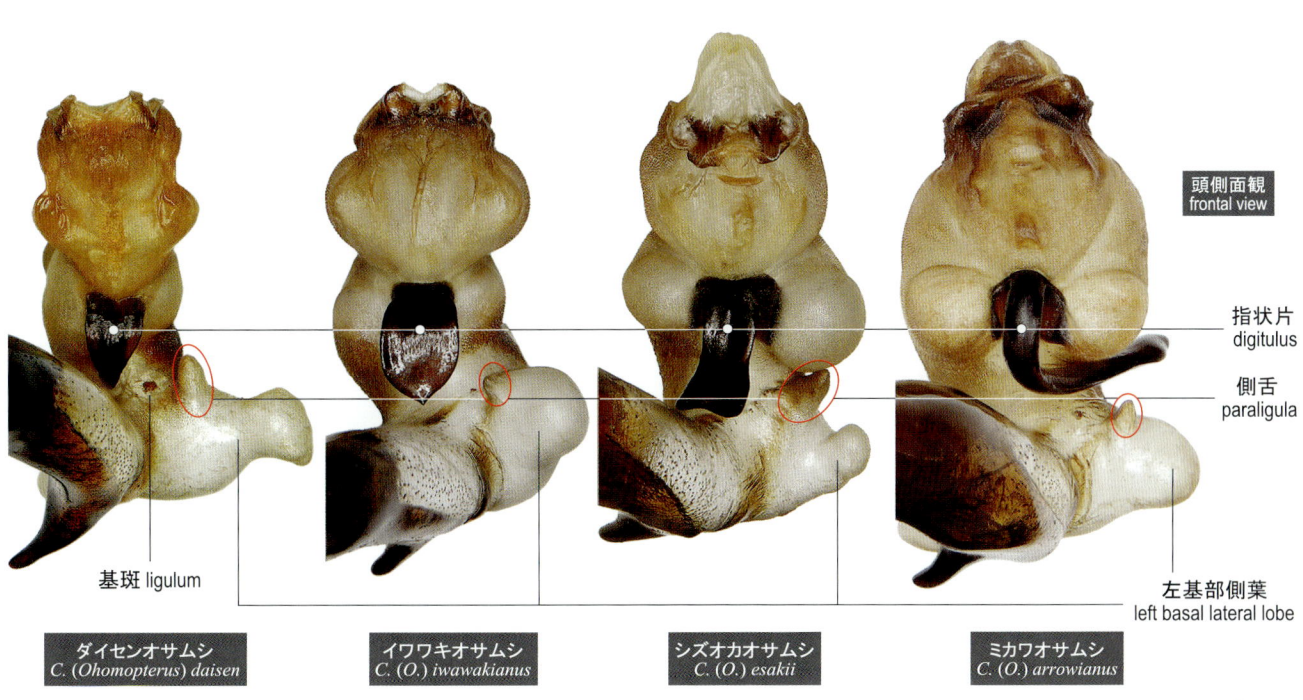

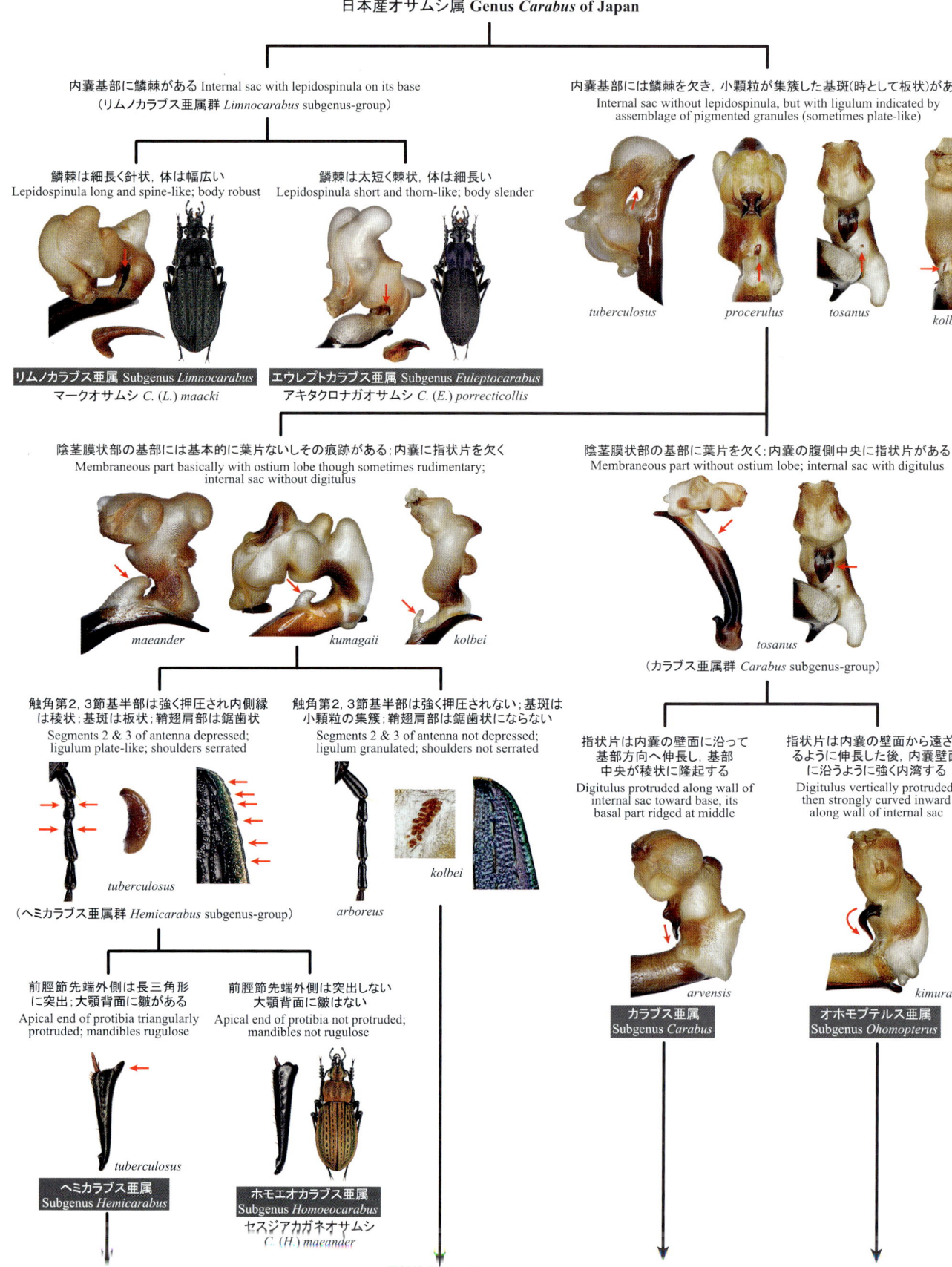

図解検索2　亜属群・亜属・種の検索 (2)　Key to the subgenus-group, subgenus and species (2)

図解検索1の最下行中央より続く

小顎外葉の先端節は匙状に窪まない Apical segment of galea not distinctly concave / *kumagaii*

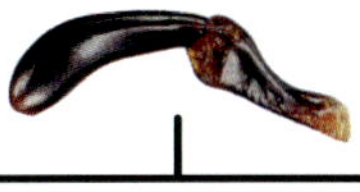

小顎外葉の先端節は長く，匙状に窪む Apical segment of galea long and distinctly concave / *blaptoides*

下唇肢亜端節の剛毛は2本；背面は基本的に黒色で光沢に乏しい
Penultimate segment of labial palpus bisetose; body above dark and mat / *arboreus*

下唇肢亜端節は多毛（剛毛は3本以上）；背面は多彩で，強い金属光沢がある
Penultimate segment of labial palpus multisetose; body above colorful and metallic

メゴドントゥス亜属 Subgenus *Megodontus*
アイヌキンオサムシ *C. (M.) kolbei*

右大顎抱歯は前方突起＞後方突起；鞘翅端は尾状突起を形成しない；第三次間室が第一次間室に融合する；腹部腹板先端3節に横溝を欠く；後基節は内側剛毛を欠く
Retinaculum of right mandible bidentate, with anterior tooth larger than posterior; elytral apices not forming mucros; tertiaries fused with primaries; sternal sulci absent; metacoxa lacking inner setae

右大顎抱歯は前方突起≦後方突起；鞘翅端は突出し，尾状突起を形成する；第三次間室は第一次間室に融合しない；腹部腹板先端3節に横溝がある；後基節は前方剛毛を欠く
Retinaculum of right mandible bidentate, with anterior tooth as large as, or smaller than posterior; elytral apices forming mucros; tertiaries not fused with primaries; sternal sulci present; metacoxa lacking anterior setae

fruhstorferi

アコプトラブルス亜属 Subgenus *Acoptolabrus*
オオルリオサムシ *C. (A.) gehinii*

blaptoides

鞘翅彫刻は基本的に五元的；♀鞘翅端のえぐれが比較的顕著
Elytral sculpture basically pentaploid; female elytra emarginated before apices

鞘翅彫刻は基本的に三元的；鞘翅端は殆どえぐれない
Elytral sculpture basically triploid; female elytra hardly emarginated before apices

ペンタカラブス亜属 Subgenus *Pentacarabus*
ホソヒメクロオサムシ *C. (P.) harmandi*

♂触角腹面に無毛部はない；丘孔点列は光沢帯を欠く；後基節は内側剛毛を欠くことはない
Male antennae without hairless areas on ventral side; umbilicate series without glossy stripe; metacoxa lacking inner setae

♂触角腹面に無毛部がある；丘孔点列に光沢帯がある；後基節は内側剛毛を欠くことはない
Male antennae with hairless areas on ventral side; umbilicate series with glossy stripe; metacoxa not lacking inner setae

procerulus

（レプトカラブス亜属群 *Leptocarabus* subgenus-group）

前胸前側板と腹板側面は滑らか；頸部は長く，複眼長径の1.7倍以上；前胸背板は細長く，少なくとも長幅等しく，側縁剛毛を欠く；
Propleura & sides of sternites smooth; neck long, at least 1.7 times as long as eye; pronotum slender, at least as broad as long, lacking marginal setae

前胸前側板と腹板側面は強く点刻される；頸部は短く，複眼長径とほぼ同長；前胸背板は通常幅広く，側縁には通常，剛毛がある
Propleura & sides of sternites strongly punctate; neck short, almost as long as eye; pronotum wider, usually with marginal setae

アステノカラブス亜属 Subgenus *Asthenocarabus*
ヒメクロオサムシ *C. (A.) opaculus*

ダマステル亜属 Subgenus *Damaster*
マイマイカブリ *C. (D.) blaptoides*

コプトラブルス亜属 Subgenus *Coptolabrus*
ツシマカブリモドキ *C. (C.) fruhstorferi*

鞘翅第一次間室は高い隆条，第二，三次間室は低く弱い隆条ないし顆粒列
Primaries indicated by highest carinae; secondaries & tertiaries are low ridge or rows of granules

鞘翅第一次間室は高い隆条にならず，第二，三次間室とほぼ同高
Primaries much reduced, indicated by low costae or rows of granules as in secondaries & tertiaries

arboreus

レプトカラブス亜属 Subgenus *Leptocarabus*

アウロノカラブス亜属 Subgenus *Aulonocarabus*
チシマオサムシ *C. (A.) kurilensis*

図解検索3 (p.16) へ

図解検索 3   種の検索 (1) Key to the species (1)

ヘミカラブス亜属 Subgenus *Hemicarabus*

鞘翅は金赤緑色；第一次間室は断続する隆条；第二次間室は減退
Elytra reddish green and metallic; primaries indicated by segmented costae; secondaries reduced

マックレイセアカオサムシ *C. (H.) macleayi*

鞘翅は黒色で僅かに赤〜緑がかる；第一次間室は紡錘形の瘤状隆起列；第二次間室は隆条ないし顆粒列
Elytra black with weak reddish or greenish tinge; primaries indicated by rows of large tubercles; secondaries are costae or rows of granules

セアカオサムシ *C. (H.) tuberculosus*

レプトカラブス亜属 Subgenus *Leptocarabus*

♂前跗節基部4節はほぼ同幅；内嚢先端の縁膜は殆ど硬化せず，突出しない
Basal four segments of male protarsus almost equally dilated; aggonoporius hardly sclerotized and not protruded apicad

コクロナガオサムシ *C. (L.) arboreus*

♂前跗節基部4節は先端部に向けて細くなる　内嚢先端の縁膜は硬化し，突出する
Basal four segments of male protarsus gradually narrowed toward apex; aggonoporius sclerotized and protruded apicad

hiurai　*kumagaii*

カラブス亜属 Subgenus *Carabus*

陰茎先端は直線的で腹側に屈曲しない；指状片は基部が太く，中央の隆起は高く先端は丸まり右へ曲がる
Apical part of penis straight, not hooked ventrad; digitulus wide and strongly ridged at base, with apex bent rightward and rounded at tip

陰茎先端は腹側に屈曲する；指状片は基部が細く，中央の隆起は低く先端はより尖り左右に曲がらない
Apical part of penis hooked ventrad; digitulus narrow and gently convex at base, with apex straight and more sharply pointed

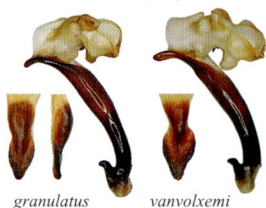
granulatus　*vanvolxemi*

コブスジアカガネオサムシ *C. (C.) arvensis*

前胸背板は方形で鞘翅は肩が張り出す；第三次間室は退化的で，せいぜい不規則な顆粒列；指状片先端は背側へ屈曲しない
Pronotum quadrate; shoulders distinct; tertiaries vestigial, at most forming irregular rows of granules; digitulus not flexed dorsad

アカガネオサムシ *C. (C.) granulatus*

前胸背板は心臓形で鞘翅は肩の張り出しが弱い；第三次間室は明らかな隆条；指状片先端は背側へ屈曲する
Pronotum cordiform; shoulders effaced; tertiaries recognized as costae; digitulus flexed dorsad

ホソアカガネオサムシ *C. (C.) vanvolxemi*

♂前腿節が相対的に太い；♂交尾器内嚢の中央葉は小さく退化的
Male profemur relatively thick; median lobe of internal sac small and vestigial

キュウシュウクロナガオサムシ *C. (L.) kyushuensis*

♂前腿節が相対的に細い；♂交尾器内嚢の中央葉は大きく発達する
Male profemur relatively narrow; median lobe of internal sac large

hiurai

陰茎先端部は円筒形；中・後胸前側板の点刻は明瞭
Apical part of penis cylindrical; meso- & metepisternum punctate

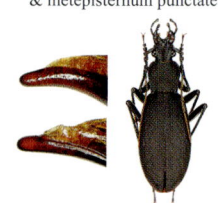

シコククロナガオサムシ *C. (L.) hiurai*

陰茎先端は背腹方向に押圧される　中・後胸前側板（特に前者）は滑らか
Apical part of penis depressed; meso- & metepisternum, above all the former rather smooth

procerulus

丘孔点列の顆粒は鞘翅端近くに達する；内嚢頂板はより短く，外角が鋭角をなす
Granules of umbilicate series reach elytral apices; terminal plate of internal sac shorter, its outer corner acute-angled

クロナガオサムシ *C. (L.) procerulus*

丘孔点列の顆粒は通常，鞘翅前半部のみ；内嚢頂板はより長く，先端は丸い
Granules of umbilicate series recognized apical half of elytra; terminal plate of internal sac longer, with apex obtusely rounded

オオクロナガオサムシ *C. (L.) kumagaii*

図解検索 4

種の検索 (2) Key to the species (2)

オホモプテルス亜属 Subgenus *Ohomopterus*

鞘翅には各4条の第一次間室が鎖線として認められ，丘孔点列は第4第一次間室の外側にある；触角は長く，先端が♀でも鞘翅中央を越える

Each elytron with 4 rows of primary intervals as chain striae; umbilicate series is recognized at outer side of 4th primaries; antennae long, extending middle of elytra even in female

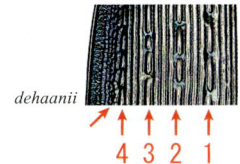 

（オオオサムシ種群 *Dehaanii* species-group）

鞘翅には各3条の第一次間室が鎖線として認められ，丘孔点列は第3第一次間室の外側にある；触角は相対的に短く，♀では先端が鞘翅中央を越えることはない

Each elytron with 3 rows of primary intervals as chain striae; umbilicate series is recognized at outer side of 3rd primaries; antennae shorter, not extending middle of elytra in female

陰茎先端部の背面には溝状の窪みがある；♂交尾器内嚢の左基部側葉はほぼ水平に突出し，腹側へ湾曲しない

Apical part of penis with longitudinal gutter on dorsal surface; left basal lateral lobe of internal sac horizontally protruded, not curved ventrad

アワオサムシ *C. (O.) kawanoi*

陰茎先端部の背面に溝状の窪みはない；♂交尾器内嚢の左基部側葉は腹側へ湾曲する

Apical part of penis without longitudinal gutter on dorsal surface; left basal lateral lobe of internal sac curved ventrad

---

指状片は小さく，左右ほぼ対称

Digitulus small and nearly symmetric

指状片はより大きく，左右非対称で歪矩形，鉤形，細長い棒状ないしヘラ形

Digitulus larger, asymmetric, with the shape deformed rectangular or elongated hook-like

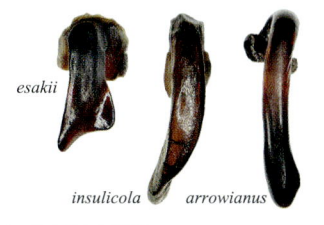

（アオオサムシ種群 *Insulicola* species-group）

図解検索6 (p.19) へ

背面は緑，緑青，銅，緑銅色を帯び，紺色や紫菫色を帯びることはない；左基部側葉はより太く内湾が弱い；指状片は左右やや非対称で先端は丸みを帯びる

Body above greenish, green-bluish, coppery or greenish coppery, never bearing dark bluish or purplish tinge; left basal lateral lobe robuster and less strongly bent inward; digitulus asymmetric, with apex roundish

トサオサムシ *C. (O.) tosanus*

背面は紺〜紫菫色を帯び，緑色を帯びることは通常ない；左基部側葉はより細く内湾が強い；指状片は左右ほぼ対称で，先端はより鋭く尖る

Body above dark bluish or purplish, usually not bearing greenish tinge; left basal lateral lobe slenderer and more strongly bent inward; digitulus nearly symmetric, with apex more sharply pointed

オオオサムシ *C. (O.) dehaanii*

指状片は長三角形

Digitulus long triangular

（ヒメオサムシ種群 *Japonicus* species-group）

図解検索5 (p.18) へ

指状片は桃の実形もしくは五角形

Digitulus peach-shaped or pentagonal

 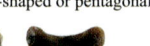

（ヤコンオサムシ種群 *Yaconinus* species-group）

より小型で触角が長く，陰茎先端はより太短く，指状片は桃の実形

Size smaller; antennae longer; apical part of penis robuster; digitulus peach-shaped

イワワキオサムシ *C. (O.) iwawakianus*

より大型で触角が短く，陰茎先端はより細長く，指状片は五角形

Size larger; antennae shorter; apical part of penis slenderer; digitulus pentagonal

ヤコンオサムシ *C. (O.) yaconinus*

図解検索 5

種の検索 (3) Key to the species (3)

(ヒメオオサムシ種群 *Japonicus* species-group)

♂触角腹面に無毛凹陥部がある
Male antennae with hairless ventral depressions

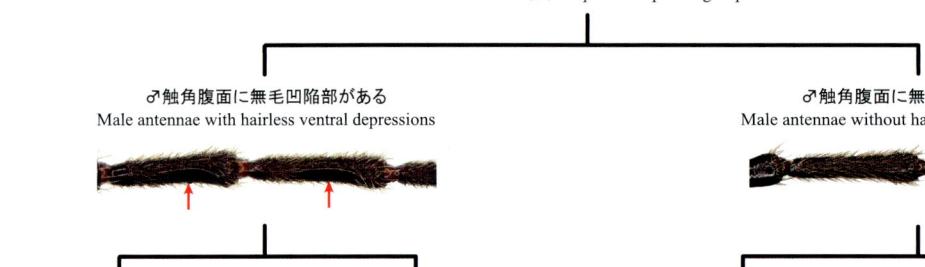

♂触角腹面に無毛凹陥部がない
Male antennae without hairless ventral depressions

陰茎先端は側面から見て顕著に膨らみ、内嚢左基部側葉は鉤状に大きく腹側へ屈曲する
Apical part of penis remarkably dilated in lateral view; left basal lateral lobe of internal sac strikingly hooked ventrad

ヤマトオサムシ *C. (O.) yamato*

内嚢先端の射精孔縁膜は硬化しない
Peripheral rim of gonopore not sclerotized

内嚢先端の射精孔縁膜は硬化し左右一対の頂板を形成する
Peripheral rim of gonopore sclerotized to form a pair of terminal plates

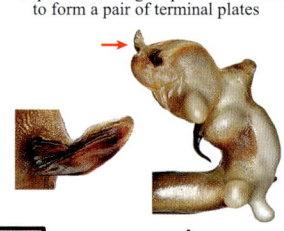

陰茎先端は非常に細長く、内嚢左基部側葉は短く、先端は僅かに腹側へ突出する
Apical part of penis very long and slender; left basal lateral lobe of internal sac short, robust, and faintly protruded ventrad at tip

クロオサムシ *C. (O.) albrechti*

陰茎先端はより太く、内嚢左基部側葉は水平に伸長し、先端手前で顕著にくびれる
Apical part of penis robuster; left basal lateral lobe of internal sac protruded horizontally, and remarkably constricted before tip

ダイセンオサムシ *C. (O.) daisen*

陰茎先端は側面から見て太短く、先端手前で背側に膨らみ、内嚢左基部側葉には腹側突起があり、指状片先端は鋭く尖る
Apical part of penis short and robust in lateral view; left basal lateral lobe of internal sac with a ventral process; digitulus sharply pointed at tip

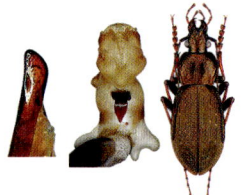

スルガオサムシ *C. (O.) kimurai*

陰茎先端は側面から見て細長く、内嚢左基部側葉は単葉で、指状片先端は鈍く丸い
Apical part of penis slender in lateral view; left basal lateral lobe of internal sac unilobed; digitulus rounded at tip

ルイスオサムシ *C. (O.) lewisianus*

内嚢の頂板は小さく短く、左基部側葉は鶏の手羽先形で、腹側突起がない
Terminal plate of internal sac short and small; left basal lateral lobe of internal sac chicken wing tip-like, without ventral process

アキオサムシ *C. (O.) chugokuensis*

内嚢の頂板は小さく短く、左基部側葉には細長く、内湾する腹側突起がある
Terminal plate of internal sac short and small; left basal lateral lobe of internal sac with a ventral process which is long and bent inward

ホウオウオサムシ *C. (O.) sue*

内嚢の頂板は大きく、細長く突出し、基部側葉の腹側突起は短く、内湾しない
Terminal plate of internal sac elongate and longer than wide; left basal lateral lobe of internal sac with a ventral process which is short and not bent inward

ヒメオサムシ *C. (O.) japonicus*

図解検索 6 種の検索（4）Key to the species (4)

（アオオサムシ種群 Insulicola species-group）

指状片は歪方形で先端右角は突出し，左角は広く丸まる
Digitulus deformed rectangular, its apical righ corner protruded, apical left corner broadly rounded

シズオカオサムシ *C. (O.) esakii*

指状片は鉤形で細長く伸長する
Digitulus hook-like and narrowly elongated

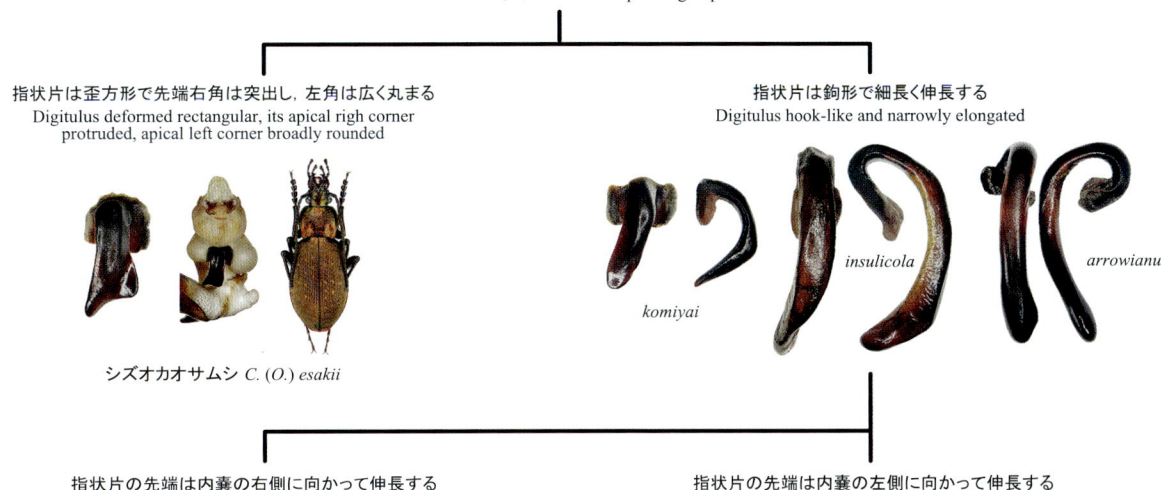

指状片の先端は内嚢の右側に向かって伸長する
Apex of digitulus turns to right side of internal sac

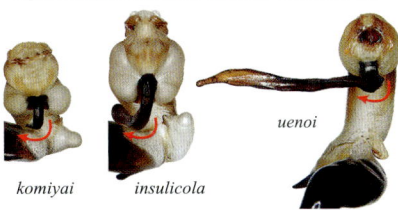
komiyai　insulicola　uenoi

指状片の先端は内嚢の左側に向かって伸長する
Apex of digitulus turns to left side of internal sac

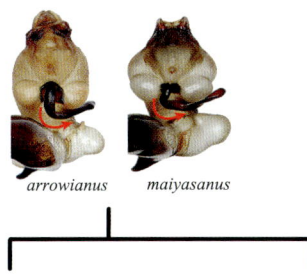
arrowianus　maiyasanus

指状片は短く，先端は背腹方向に押圧され，指状片両側の葉は左右非対称（右＜左）
Digitulus short, with apical part depressed dorso-ventrad; lobes on both sides of digitulus asymmetric (right＜left)

カケガワオサムシ *C. (O.) komiyai*

陰茎は極めて大きく，鞘翅長の3/5に及び，指状片は陰茎長の7/8に達する 内嚢は筒状で変形が著しい
Penis monstrously large, about three-fifths as long as elytra; digitulus also large, reaching seven-eights as long as penis; internal sac tubular and deformed

ドウキョウオサムシ *C. (O.) uenoi*

指状片は細長く，先端部は側圧されず，背腹方向に押圧されるか棒状
Digitulus long and slender, with apical part not compressed laterad, but either depressed dorso-ventrad or cylindrical

ミカワオサムシ *C. (O.) arrowianus*

指状片は長く，先端部は側圧されて垂直板を形成し，その腹面は硬化が弱く淡色
Digitulus long, compressed bilaterad in apical portion to form vertical plate, its ventral side weakly sclerotized and light-colored

マヤサンオサムシ *C. (O.) maiyasanus*

指状片は大きく，先端は側圧され，垂直板を形成する 指状片両側の葉は左右対称
Digitulus large, with apical part compressed bilaterad to form vertical plate; lobes on both sides of digitulus symmetric

アオオサムシ *C. (O.) insulicola*

日本の地方区分と県

海外の読者に対する利便性を考慮し，ここでは47都道府県名（ただし，最南端の沖縄県にはオサムシが分布しないので，本書では除外した）と地方区分名をローマ字表記により地図上に示した．

我が国では一般に，北海道・東北・関東・中部・近畿・中国・四国・九州の八地方に分ける地方区分が用いられ，北海道はさらに道北・道東・道央・道南の四つに，東北地方は北東北と南東北に，関東地方は北関東と南関東に，中部地方は北陸・東山・東海の三つに，中国地方は山陰と山陽に区分される．

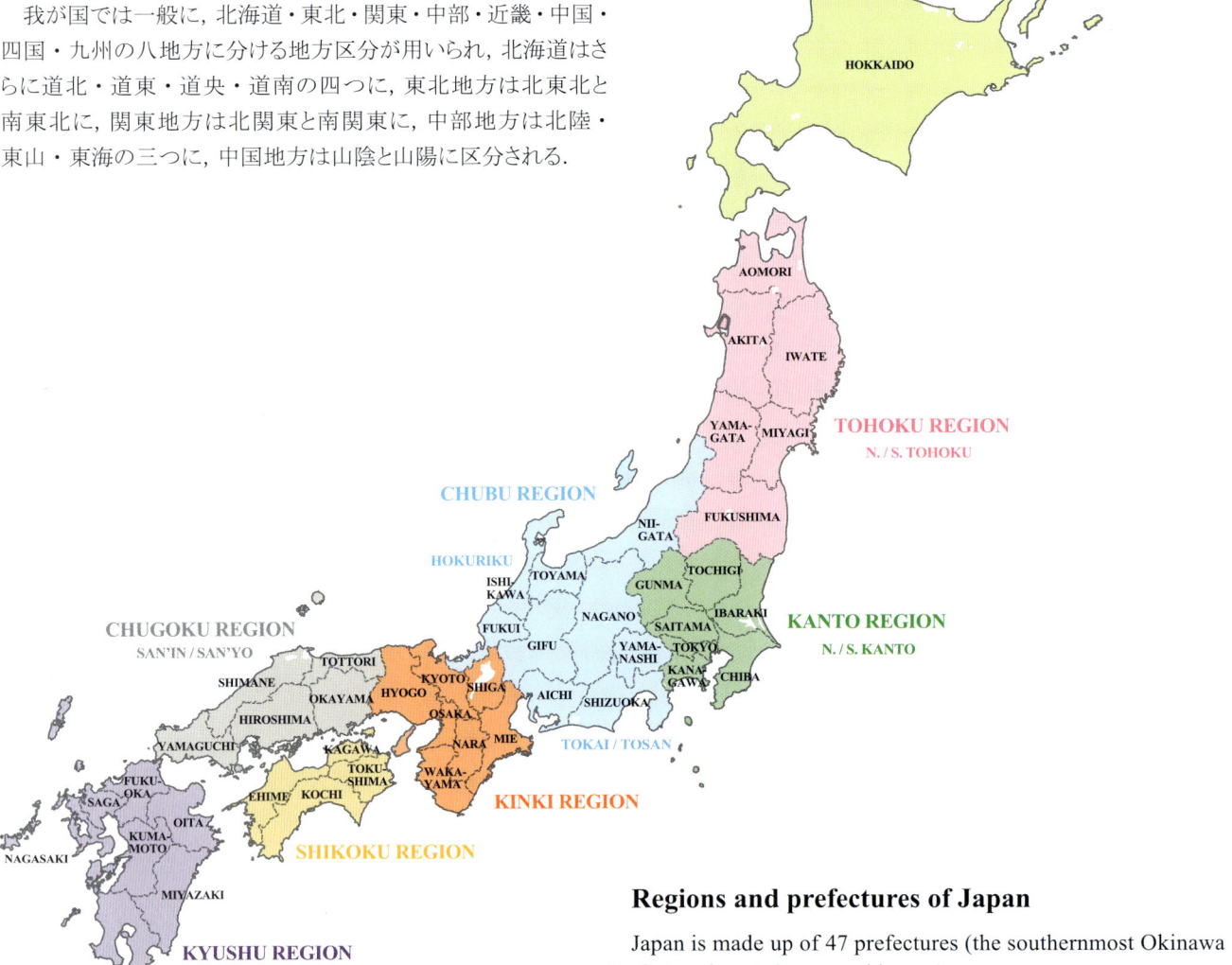

Regions and prefectures of Japan

Japan is made up of 47 prefectures (the southernmost Okinawa Prefecture is not shown on this map).

But the prefecture is sometimes too small as a unit to use as a basis for dividing the nation into regions. We must therefore use units consisting of several prefectures grouped together.

Each grouping of prefectures will be called a Region, and each Region is shown in different color on this map.

The most commonly used eight Regions are: Hokkaido, Tohoku, Kanto, Chubu, Kinki, Chugoku, Shikoku and Kyushu.

The Hokkaido Region is further divided into four subregions, Dohoku, Doto, Do'o and Donan, which means northern, eastern, central and southern Hokkaido, respetively.

The Tohoku and Kanto Regions are subdivided into northern part and southern part, respectively.

The Chubu Region is usually divided into three subregions, Hokuriku, Tosan and Tokai.

The Chugoku Region is divided into San'in (north) and San'yo (south)

The prefectural names are shown in black capital letter.

日本の島

島国日本は，北海道・本州・四国・九州の4島と，それらに付随する6,850余の島嶼からなる．ここでは，主に2種以上のオサムシが分布する，あるいは固有の亜種が記載されている代表的な島嶼の名をローマ字表記により地図上に示した．

Islands of Japan

Japan is an island nation consisting of four large islands - Hokkaido, Honshu, Shikoku and Kyushu, as well as numerous, more than 6,850 small islands. On this map are shown the name of representative islands on which two or more (with some exceptions) *Carabus* species are distributed or from which an endemic subspecies is described.

Holotype (♂) of *Carabus* (*Damaster*) *blaptoides* (Kollar, 1836)
(preserved in the collection of Naturhistorisches Museum Wien).

マイマイカブリ基亜種ホロタイプ（♂, ウィーン自然史博物館所蔵）.

図版PLATE／解説TEXT

1. *Carabus* (*Limnocarabus*) *maacki* Morawitz, 1862
マークオサムシ

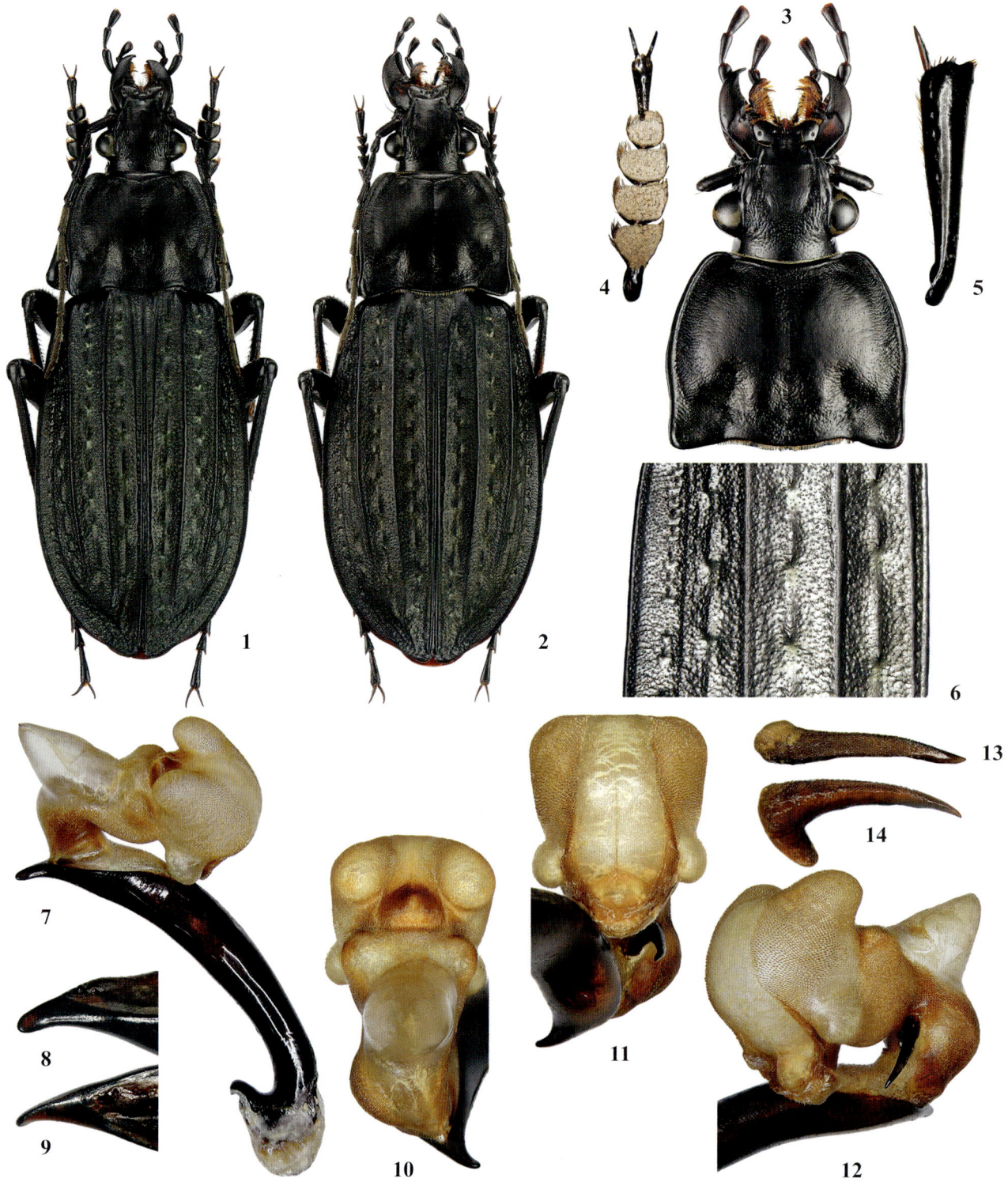

Figs. 1-14. *Carabus* (*Limnocarabus*) *maacki aquatilis* Bates, 1883 (Riverside marsh of Riv. Iwaki-gawa, Nakadomari-machi, Aomori Pref.). — 1, ♂; 2, ♀; 3, male head & pronotum; 4, male right protarsus in ventral view; 5, male right protibia in subdorsal view; 6, male left elytron (median part); 7, penis & fully everted internal sac in right lateral view; 8, apical part of penis in right lateral view; 9, ditto in dorsal view; 10, internal sac in caudal view; 11, ditto in frontal view; 12, ditto in left sublateral view; 13, lepidospinula in dorsal view; 14, ditto in lateral view.

1. *Carabus* (*Limnocarabus*) *maacki*

Figs. 15-26. ***Carabus*** (***Limnocarabus***) ***maacki aquatilis***. — 15, 16, 18, 19, 22, 24, ♂, 17, 20, 21, 23, 25, 26, ♀; 15, Hotoké-numa Moor, Lake Ogawara-ko, Misawa-shi, Aomori Pref.; 16, Yagisawa, Miyako-shi, Iwate Pref.; 17, Matsuyama, Miyako-shi, Iwate Pref.; 18, Kamiôdo, Daisen-shi, Akita Pref.; 19-20, Lake Kejo-numa, Ôsaki-shi, Miyagi Pref.; 21, Akaishi, Tomiya-machi, Miyagi Pref.; 22-23, Lake Hakuryû-ko, Nan'yô-shi, Yamagata Pref.; 24-25, Akai-yachi Moor, Aizu-wakamatsu-shi, Fukushima Pref.; 26, Yorii, Nasu-machi, Tochigi Pref.

図 15-26. **マークオサムシ本州亜種**. — 15, 16, 18, 19, 22, 24, ♂, 17, 20, 21, 23, 25, 26, ♀; 15, 青森県三沢市小川原湖仏沼; 16, 岩手県宮古市八木沢, 17, 岩手県宮古市松山; 18, 秋田県大仙市上大戸; 19-20, 宮城県大崎市化女沼; 21, 宮城県富谷町明石; 22-23, 山形県南陽市白竜湖; 24-25, 福島県会津若松市赤井谷地; 26, 栃木県那須町寄居.

1. *Carabus* (*Limnocarabus*) *maacki* MORAWITZ, 1862 マークオサムシ

Original description *Carabus Maacki* MORAWITZ, 1862, Mélang. Biol. Bull. Acad. Sci. St. Pétersbourg, 4, p. 191.

History of research This taxon was described by MORAWITZ (1862a) based on the specimen collected by Richard MAACK from the Lake Khanka of the Russian Maritime Provinces. Although originally described as an independent species, it has often been regarded as a subspecies of European *Carabus clathratus* (type species of *Limnocarabus* GÉHIN, 1876). Adopted subgenus for *C. clathratus* and *C. maacki* has been different according to authors, i.e., *Limnocarabus* by REITTER (1896), LAPOUGE (1931) and NAKANE (1962), *Carabus* in the narrow sense by BREUNING (1932), etc. ISHIKAWA (1973) redefined the genera and subgenera of the subtribe Carabina based on the male genitalic morphology, and placed *C. maacki* (+ *C. clathratus*), together with *C. porrecticollis*, in the genus *Apotomopterus*. Based on the mitochondrial ND5 gene analysis of the Carabina on a world-wide basis, IMURA et al. (1998) separated *C. maacki* (+ *C. clathratus*) and *C. porrecticollis* from *Apotomopterus*, and placed them in the genus *Limnocarabus* (*Euleptocarabus* as its subgenus for *L. porrecticollis*). According to the analysis of nuclear genes, however, different results have been obtained, suggesting an affinity between *Limnocarabus* and *Apotomopterus* (SOTA & ISHIKAWA, 2004; DEUVE et al., 2012, etc.). IMURA et al. (1998) proposed a new name "lepidospinula" for basal sclerite of the internal sac of *Limnocarabus* to distinguish it from "spinula" of *Apotomopterus*. BATES (1883) described *C. aquatilis* from Japan, which is generally regarded as a subspecies of *C. maacki*, though morphological difference from the nominotypical *C. maacki* is small.

Type locality Kenka-See (= Lake Khanka).

Type depository Zoological Institute of Academy of Sciences (St. Petersburg) [Holotype, ♂].

Morphology 20~38 mm. Body above black with dark bronze or greenish tinge on depressed parts of pronotum and elytra. External features: Figs. 1~6. Description on details: penultimate segment of labial palpus bisetose; median tooth of mentum much shorter than lateral lobes, with tip sharply pointed; submentum bisetose; male antennae without hairless areas on ventral side; pronotum basically with two pairs of marginal setae, one in median portion and one near hind angle; elytra with weak preapical emargination in both sexes; hind wings are developed in some individuals; elytral sculpture heterodyname; primaries indicated by rows of short costae segmented by large and shallow primary foveoles; secondaries higher than primaries, indicated by continuous narrow costae; tertiaries vestigial and areas between intervals coarsely scattered with small granules; propleura sporadically scattered with small granules and vaguely punctate; sides of sternites irregularly rugoso-punctate; metacoxa bisetose, lacking anterior seta; sternal sulci clearly carved. Male genitalic features: Figs. 7~14. Penis a little shorter than half the elytral length; membraneous area without ostium lobe; internal sac with a spine-like lepidospinula on its base to the left, its surface fully covered with scale-like microstructure; median lobe very large and conically shaped; apical part of internal sac strongly inflated, both apical and podian lobes well-developed.

Molecular phylogeny On a molecular phylogenetic tree of the subtribe Carabina constructed from the mitochondrial ND5 gene sequences, *Carabus* (*Limnocarabus*) *maacki aquatilis* is most closely related to *C.* (*Euleptocarabus*) *porrecticollis*, and European *C.* (*L.*) *clathratus* was recognized as their outgroup. All the analyzed species belonging to *Apotomopterus* constitute another cluster, and there is no indication of close phylogenetic relationship between *Apotomopterus* and *Limnocarabus* + *Euleptocarabus* (for the results using nuclear genes, see history of research of this species). *Carabus* (*L.*) *maacki* of eastern Eurasia would have differentiated from *C.* (*L.*) *clathratus* of eastern Eurasia, then the former immigrated into a restricted area of ancient Japan with its separation from the continent, differentiated again into an endemic subspecies, *aquatilis*. Following this, *C.* (*E.*) *porrecticollis* would have diverged from *C.* (*L.*) *m. aquatilis* in the ancient Japan area (IMURA et al., 1998).

Distribution NE. Honshu. Outside Japan: SE. Siberia, Russian Maritime Provinces, NE. China & N. Korea.

Habitat As in the close relative, *C.* (*L.*) *clathratus* of western Eurasia, this species is stenotopic, strongly hygrophilous, preferring to inhabit moors, peaty lands or marshy environments near a river or lake. In Japan, it is sometimes found from fallow rice fields which once existed as wetlands.

Bionomical notes Little has been known on the biology of the nominotypical *maacki*. As to subsp. *aquatilis*, see explanation of that subspecies.

Geographical variation Classified into two subspecies, the nominotypical *maacki* of the Eurasian Continent and subsp. *aquatilis* of Honshu, Japan, though morphological differences between them are not so large.

Etymology This species is named after Richard MAACK (1825~1886), the Russian geographer and naturalist who visited the Ussuri Region in the late 1880s as a member of the research team organized by the Russian Geographical Society, and collected the first specimen of this species from the Lake Khanka.

研究史　本タクソンは，1800年代後半にロシア沿海地方のハンカ湖でRichard MAACK（リチャード（リヒャルト）・マーク）により採集された標本に基づいて，MORAWITZ（モラウィッツ）（1862a）により記載された．独立種として記載されたが，基本的形態の酷似したヨーロッパマークオサムシの亜種として扱われることも多い．広義のオサムシ属の中のリムノカラブス亜属（基準種：ヨーロッパマークオサムシ）に置かれることが多かったが，著者によりその扱いは一定していなかった．♂交尾器内嚢の基本形態に基づいてオサムシ亜族内の属・亜属の再定着を行ったISHIKAWA（1973）は，本種の内嚢基部にある硬化片（舌状片）を，中国に分化の中心を持つアポトモプテルス類のそれと相同の形質とみなし，アオタクロナガオサムシらと共に本種（＋ヨーロッパマークオサムシ）を同一属アポトモ

1. *Carabus* (*Limnocarabus*) *maacki*

と移した. 世界のオサムシのミトコンドリアDNAを分析した生命誌研究館のチームは, 分子系統樹の上でマークオサムシ（＋ヨーロッパマークオサムシ）, アキタクロナガオサムシの両者（本書で言うリモノカラブス亜属群）とアポトモプテルス類との間に直接の類縁関係を証明することは困難であるとして, これらを独立属リムノカラブスに置き, ♂交尾器内嚢基部の硬化片に対してはアポトモプテルス類のそれに対する狭義の名称（基棘spinula）との混同を避けるため, 鱗棘lepidospinulaという名称を与えた（IMURA *et al.*, 1998）. 核遺伝子の分析による追試（SOTA & ISHIKAWA, 2004; DEUVE（ドゥーヴ）*et al.*, 2012等）では, リムノカラブス亜属群とアポトモプテルス類との近縁性を示唆する結果も得られている. BATES（ベイツ）（1883）により長野県の諏訪湖から独立種として記載された*C. aquatilis*は本種の一亜種とみなされることが多いが, 形態学的に見た両者の差異はそれほど大きくない.

タイプ産地　ハンカ湖（ロシア沿海地方）.

タイプ標本の所在　科学アカデミー動物学研究所（サンクトペテルブルク）［ホロタイプ, ♂］.

形態　20~38 mm. 背面は黒色に近く, 光沢は弱く, 前胸背板や鞘翅の凹陥部には弱い暗緑~銅色の金属光沢がある. 外部形態：図1~6. 細部の記載：下唇肢亜端節の剛毛は2本. 下唇基節の中央歯は側葉よりはるかに短く, 先端は鋭く尖る. 下唇亜基節は有毛. ♂触角腹面に無毛部を欠く. 前胸背板の側縁剛毛は2対（中央と後角付近に各1対）. 鞘翅端部は♂♀共に僅かにえぐれる. 時に後翅の発達した個体が見られる（尾形, 1980, 1982; 平山, 1985等）. 鞘翅彫刻は異規的で, 第一次間室は浅く広い第一次凹陥によってやや不規則に分断され鎖線となり, 第二次間室は最も強く隆起した連続する隆条で, 第三次間室は退化的. 間室間の基面には小顆粒を密に装う. 前胸側板には小顆粒と小点刻を疎に装い, 腹部腹板の側面には不規則な皺と小点刻を装う. 後基節の剛毛は2本で前方剛毛を欠く. 腹部腹板の横溝は細いが明瞭に刻まれる. ♂交尾器形態：図7~14. 陰茎は鞘翅長の1/2よりやや短く, 膜状部基部に葉片を欠き, 内嚢基部左側に細長い針状の鱗棘がある. 鱗棘の表面は鱗状の微細構造で被われる. 内嚢基部外側には三角錐形の巨大な中央葉があり, 内嚢先端部は大きく膨らんで先端葉と脚葉が発達する.

分子系統　ミトコンドリアND5遺伝子を用いて作成された世界のオサムシ亜族の分子系統樹上で, 本種（分析されたのは本州亜種のみで, 大陸の基亜種は未分析）はアキタクロナガオサムシと共にまとまったクラスターを形成し, ヨーロッパマークオサムシはその外群に来る. 中国のアポトモプテルス類は, これとは異なる独立性の高いクラスターを形成し, 両者の間に直接的な類縁関係を証明することは困難であった（ただし, 前述の如く, 核遺伝子の分析からは両者の類縁の近さを示唆する結果も得られている）. また, しばしば同種として扱われるヨーロッパマークオサムシは, 本種と祖先を共有する関係にはあるものの, 両者の分岐は深く, 分化の歴史は比較的長いことが判明している. 本種はまず, 西ヨーロッパのヨーロッパマークオサムシから分化したのち, 古日本列島が大陸から分離する時期と前後して東ユーラシアの基亜種が古日本列島に隔離されて亜種分化を起し, その後, 日本列島内で同亜種からアキタクロナガオサムシが分化したと考えられる（IMURA *et al.*, 1998）.

分布　本州（北東部）. 国外：極東ロシア~朝鮮半島北部.

生息環境　湿地やそれに準じた環境に強く依存する.

生態　本州亜種の項で詳述. 基亜種の生態に関する知見は極めて乏しいが, 基本的には湿地を好むものと思われる.

地理的変異　基亜種と本州亜種の2亜種に分類され, 本邦には後者を産するが, 両者の形態学的差異は軽微である.

種小名の由来　種小名は, 地理学者にして博物学者でもあったRichard MAACK（1825~1886）に因む. 現在のエストニアにあるクレサーレ（当時のロシア帝国アレンスブルク）出身のロシア人で, ファーストネームRichardの発音はエストニア語では英語のそれに近いリチャードもしくはリチャルドだが, ドイツ系の人間であったというからリヒャルトに近い発音が正しいかもしれない. 本職は教員で, サンクトペテルブルク, イルクーツク, 東シベリア等各地で教職に就く傍ら, 1800年代後半にロシア地理学会により組織されたシベリアから沿海地方の学術調査に複数回参加し, 1859年から1860年にかけてウスリーを訪れた際にハンカ湖畔で本種を採集. その標本がMORAWITZの手に渡って記載された.

1-1. *C.* (*L.*) *m. aquatilis* BATES, 1883　マークオサムシ本州亜種

Original description　*Carabus aquatilis* BATES, 1883, Trans. ent. Soc. Lond., 1883, p. 224.

Type locality　Shimonosuwa Lake (= presumably corresponding to the Lake Suwa-ko of Nagano Pref. in central Honshu).

Type depository　Museum Nationale d'Histoire Naturelle, Paris [Lectotype, ♂].

Diagnosis　25~34 mm. Closely allied to *C.* (*L.*) *maacki maacki*, and barely distinguished from that race in the following points: 1) depressed parts of elytra less strongly greenish; 2) pronotum with lateral margins a little more narrowly rimmed and less strongly turned up, above all in posterior halves; 3) propleura a little more coarsely scattered with punctures and small granules; 4) elytra less robuster, more gently narrowed toward apices; 5) elytral sculpture a little more vaguely impressed, with elevated parts of secondaries narrower and more weakly raised, secondary foveoles shallower; 6) apical part of penis not so different in shape from that of *C.* (*L.*) *m. maacki*.

Distribution　NE. Honshu (Tohoku Region & a part of Chubu Region; Aomori Pref., Iwate Pref., Akita Pref., Yamagata Pref., Miyagi Pref., Fukushima Pref., Niigata Pref. & Nagano Pref., though the Nagano population is presumed to be extinct). TSUCHIYA (1982) recorded the left elytron of this species from Is. Sado-ga-shima, though it is doubtful whether this record is true or not.

Habitat　Moors, peaty lands or marshy environments.

Bionomical notes　Spring breeder. Adult has its main activity peak early in summer, from the end of May or the beginning of June to July in the Tohoku Region of northern Honshu. Newly emerged adult is found from the beginning of September, though some

1. *Carabus* (*Limnocarabus*) *maacki*

individuals are already hibernating in the soil as early as mid-September. In captivity, adults prefer to feed on small frogs, snails in the water and earthworms (OGATA, 1982), but larvae prefer to eat larvae of paper wasp (OSANAI, 1986).

Conservation   This subspecies is endemic to the northeastern part of Honshu, and is strongly dependent on moors, peaty lands or marshy environments. It was fairly abundant until the 1960s, but has become much rarer during the past few decades. The cause of the drastic population decline is considered to have been the sharp decrease in the habitats as a result of high economic growth in Japan during the 1950s and 1960s. This subspecies is designated as the threatened species (VU; vulnerable species) in the 4th edition of Red List (Threatened Wildlife of Japan) of Ministry of the Environment (2012).

Etymology   The subspecific name means "aquatic".

タイプ産地　シモノスワ湖（＝おそらく長野県の諏訪湖を指すものと思われる）．

タイプ標本の所在　パリ自然史博物館［レクトタイプ，♂］．英国人研究者BATESにより記載された本タクソンのレクトタイプがなぜパリにあるのかについては，アキタクロナガオサムシの「タイプ標本の所在」の項を参照．

亜種の特徴　25~34 mm．基亜種のホロタイプ（♂）と日本産の複数の標本を比較したところ，以下のような差異が見られた．日本産は，1) 鞘翅凹陥部の緑色味が弱い；2) 前胸背板側縁の縁取りが特に後方部分でやや幅狭く，上方への反りがやや弱い傾向にある．また表面の点刻はより密に刻まれるものが多いが個体変異も大きい；3) 前胸側板にはより密に点刻と小顆粒を装うものが多いが，これも個体変異があり，例外も多い；4) 鞘翅全体の輪郭が，やや太短くずんぐりした箱型の基亜種に比し，やや細長く，後方に向けての狭まりがより緩やかなものが多い；5) 上翅の彫刻は基亜種のそれよりもやや凹凸に乏しく，とりわけ第二次間室の鎖線隆起部がやや低く，幅もやや狭く，第二次凹陥がより浅いため，遠目で見ると鞘翅全体がより滑らかに見える；6) 陰茎先端の形態には特に明らかな差異を見出しえない．ただし，こうした差異はいずれも軽微なものでり，個体変異の範疇に没してしまうものも多いので，形態学的には本州産の集団を敢えて異なる亜種に分ける必要もないのかもしれない．

分布　本州北東部（青森，岩手，秋田，山形，宮城，福島，新潟の各県）．湿地帯に強く依存するため，こうした環境が激減した今日において，本亜種の分布は著しく不連続になっている．本亜種のタイプ産地は長野県の諏訪湖と考えられるが，ルイス以後，同地からの採集記録はなく，すでに絶滅したものと思われる．また，佐渡（相川町（あいかわまち）濁川（にごりかわ）＝現佐渡市相川濁川町）から左鞘翅のみの記録（1982年3月27日に「水田の付近の崖から，サドマイマイカブリとともに掘り出した」という）（土屋，1982）があるが，同島からはこれ以外に確実な記録がない．記録自体の信憑性に対する検証が必要と思われ，少なくとも現時点において本種が佐渡に生息している可能性は極めて低いと見なさざるを得ない．秋田・山形・新潟の各県においても激減しており，近年における確実な採集記録は殆ど見られなくなっている．

生息環境　主として泥炭地や湿地，湿田，休耕田，河川敷などの湿地的環境に生息するが，開発とともにその生息範囲が急激に狭められつつある．

生態　春繁殖・成虫越冬型のオサムシ．尾形（1982）によれば，岩手県宮古市付近では，越冬から覚めた成虫の出現は5月下旬から6月上旬で，他のオサムシより半月以上遅く，外気温がおよそ10°C前後になると活動を開始するらしい．その後，交尾・産卵が行われ，新成虫は9月上旬に出現し，9月中旬にはすでに土中で越冬態勢に入っている個体も見られることから，越冬に入る時期は他のオサムシに比べて早い傾向にあるという．飼育下で，成虫は食餌としてカエルを最も好み，タニシ，ミミズ，オカモノアラガイなども食するという．また，水際における活動性をアオオサムシのそれと比較した場合，本種は水泳・潜水能力とも明らかに勝っており，水域への適応を示していることが観察されている．また，幼虫に関しても，通常のトラップに落ちにくいことや，蛹化場所（周囲を水に囲まれた島状の盛り土部分に複数の蛹室が観察された例がある）などから，水中での活動機会が多いのではないかと推察されている．幼虫の食性について，長内（1986）はミミズとアシナガバチの幼虫を与えたところ，後者を食したというから，基本的には他の昆虫の幼虫食であろう．

保全　本種は近年，生息環境の悪化により，各地で個体数の減少が著しく，環境省第4次レッドリスト（絶滅のおそれのある野生生物の種のリスト）（2012年公表）では絶滅危惧II類（カテゴリーVU）に指定されている．以前は，東北地方各地の湿地や水田に比較的多く生息していたが，高度経済成長に伴う都市化や農村の基盤整備に伴い，その後，急速に生息地が失われた．本種が衰退した最大の要因は，宅地化や護岸工事などによって湿地そのものが消滅したことであろう．各種の記録によれば，1950~1960年代初頭までは水田でも多数の個体が見られたようなので，湿地帯の水田への転化自体は，必ずしも本種の生息に悪影響を及ぼしたとは限らず，むしろ水田の基盤整備（乾田化および水路整備）と農薬の使用が本種衰退の主要因になったと考えられる．こうした変化の起きた時期は，わが国の高度経済成長期によく一致している．こうして一旦整備されてしまった環境を元の湿地に戻すことは殆ど不可能に近いため，現存する本種の生息環境をこれ以上破壊しない努力を続けるとともに，残された生息地に大量のトラップを仕掛けて本種を根絶やしにするような乱獲を慎むことなどが，本種の保全に対するせめてもの具体策と言えよう．

亜種小名の由来と通称　亜種小名は「水生の」を意味する．タイプ標本が英国のLEWIS（ルイス）によって「シモノスワ湖」の湖岸の藻を排して採集されたことに因むものであろう．一般にマークオサムシと呼ばれるが，この名は種和名であると同時に本州亜種，大陸産の基亜種双方に対しても用いられるため，種，亜種いずれを指すものか，和名だけで判断することは困難である．台湾産の昆虫であろうとみなされて，タカサゴカタビロオサムシという名が与えられていた時期もある．

1. *Carabus* (*Limnocarabus*) *maacki*

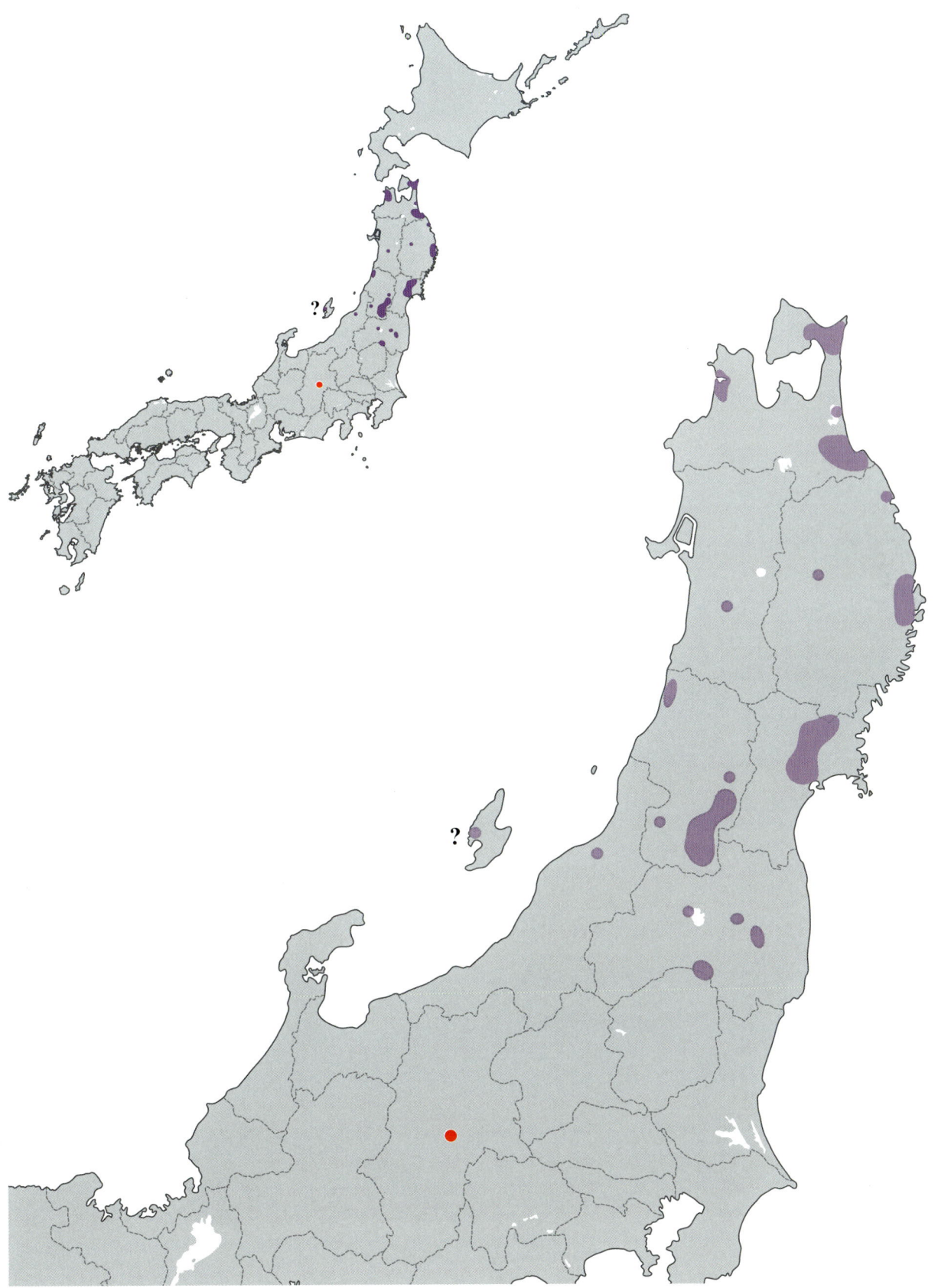

Fig. 27. Map showing the distributional range of ***Carabus*** (***Limnocarabus***) ***maacki aquatilis*** BATES, 1883. ●: Type locality of the subspecies (Lake Suwa ko). ?: Record of left elytron from Is. Sado-ga-shima, though it is not certain whether this record is true or not.
図 27. **マークオサムシ本州亜種**分布図. ●: 亜種のタイプ産地（諏訪湖）. ?: 佐渡島からの左鞘翅のみの記録（真偽不詳）.

2. *Carabus* (*Euleptocarabus*) *porrecticollis* Bates, 1883
アキタクロナガオサムシ

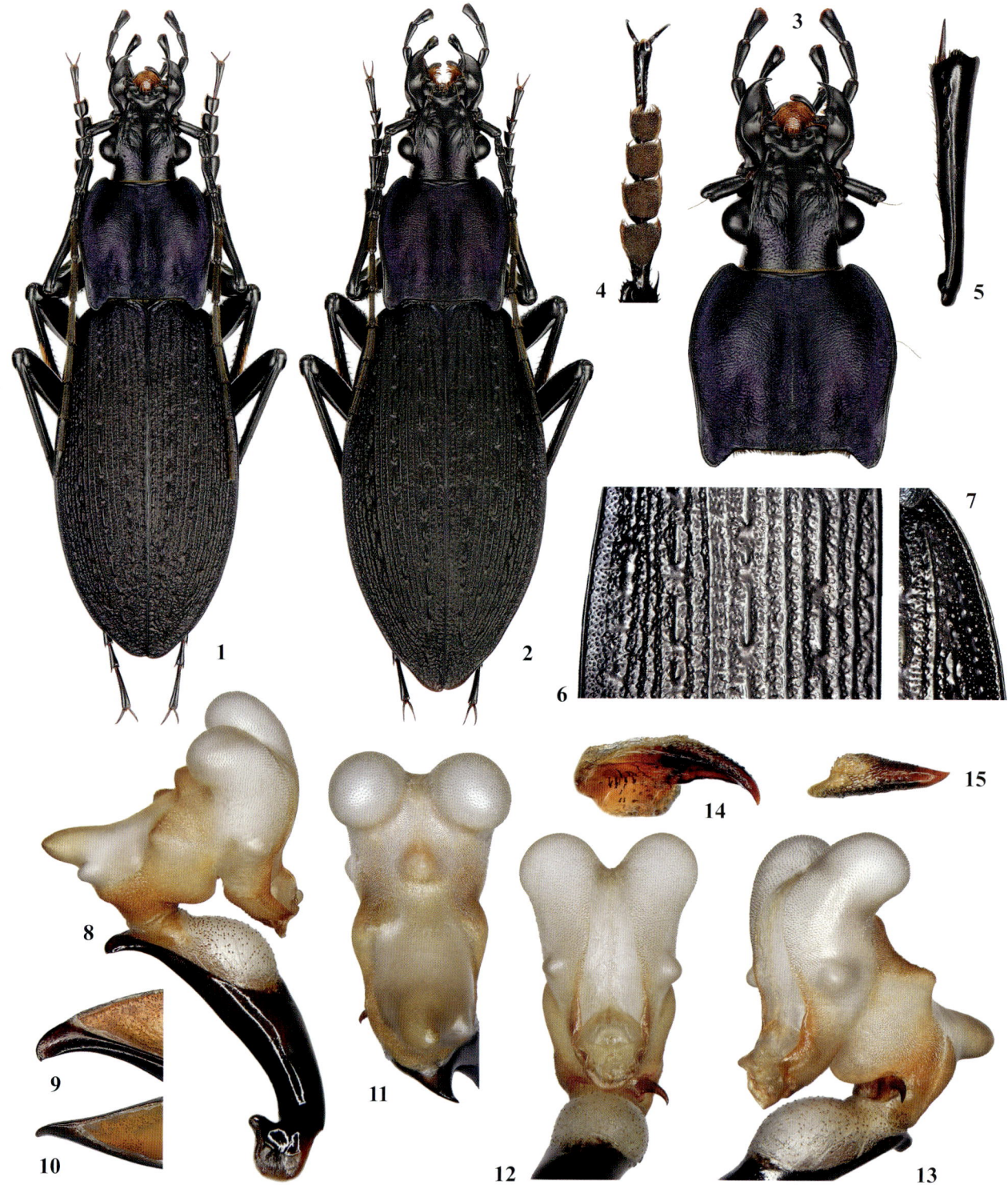

Figs. 1-15. *Carabus* (*Euleptocarabus*) *porrecticollis porrecticollis* Bates, 1883 (Urasa, Minami-uonuma-shi, Niigata Pref.). — 1, ♂; 2, ♀; 3, male head & pronotum; 4, male right protarsus in ventral view; 5, male right protibia in subdorsal view; 6, male left elytron (median part); 7, humeral serration of male right elytron; 8, penis & fully everted internal sac in right lateral view; 9, apical part of penis in right lateral view; 10, ditto in dorsal view; 11, internal sac in caudal view; 12, ditto in frontal view; 13, ditto in left sublateral view; 14, lepidospinula in lateral view; 15, ditto in dorsal view.

2. *Carabus* (*Euleptocarabus*) *porrecticollis*

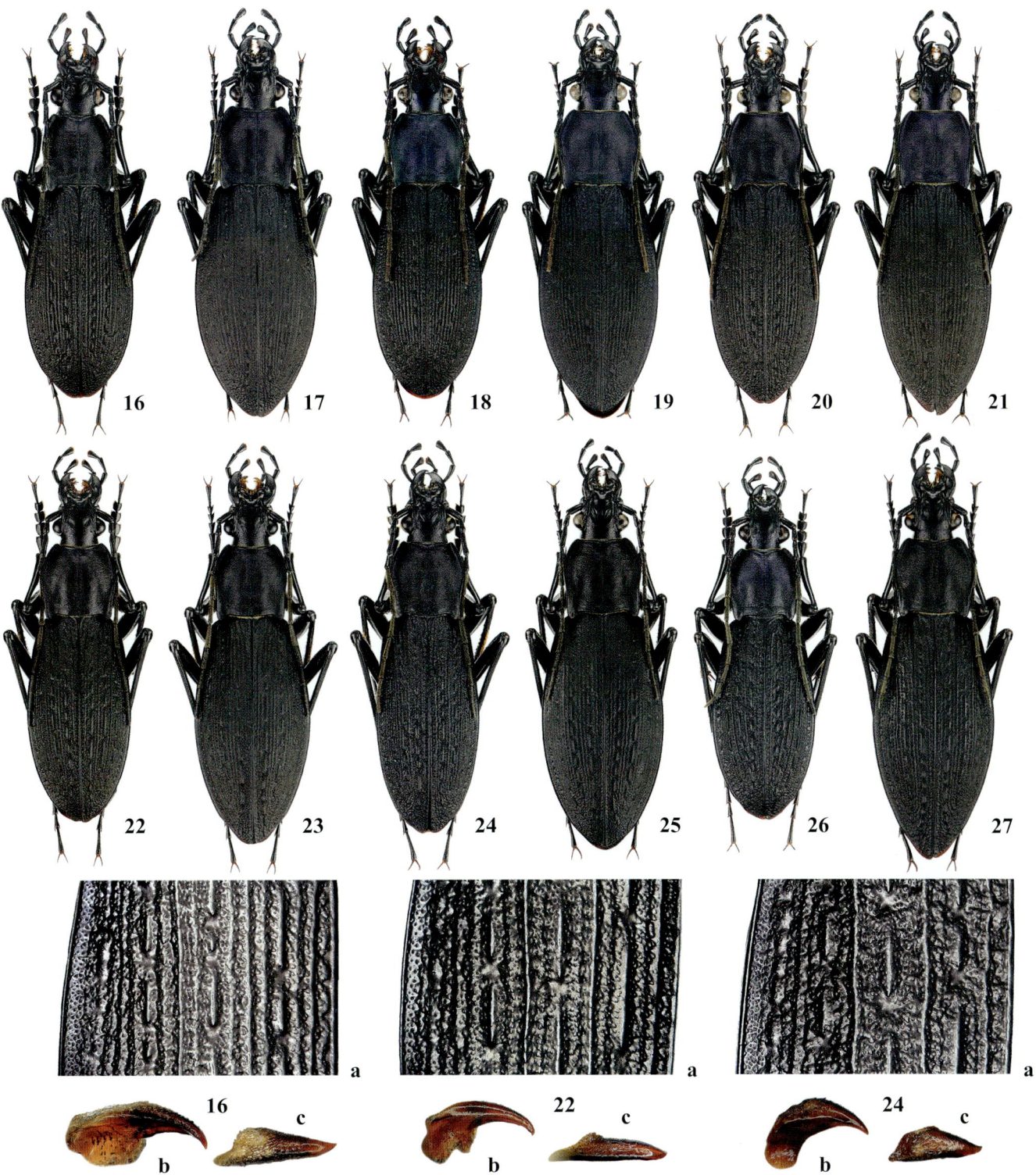

Figs. 16-27. *Carabus* (*Euleptocarabus*) *porrecticollis* subspp. — 16-21, *C.* (*E.*) *p. porrecticollis* (16, 18, 20, ♂, 17, 19, 21, ♀; 16, Kyôwa-funasawa, Daisen-shi, Akita Pref.; 17, Ôbata, Yokoté-shi, Akita Pref.; 18, Ômori-machi, Tochigi-shi, Tochigi Pref.; 19, Kami-nanma-machi, Kanuma-shi, Tochigi Pref.; 20-21, Azô-chô, Toyota-shi, Aichi Pref.); 22-23, *C.* (*E.*) *p. pacificus* (22, ♂ (HT), 23, ♀ (PT), Ôbuchi, Fuji-shi, Shizuoka Pref.;); 24-25, *C.* (*E.*) *p. kansaiensis* (24, ♂, 25, ♀, Mt. Iwawaki-san, Osaka Pref.); 26-27, *C.* (*E.*) *p. porrecticollis* ? (26, ♂, 27, ♀, Tarumi, Niimi-shi, Okayama Pref.). — a, Male left elytron (median part); b, lepidospinula in lateral view; c, ditto in dorsal view.

図 16-27. アキタクロナガオサムシ各亜種. — 16-21, **基亜種**(16, 18, 20, ♂, 17, 19, 21, ♀; 16, 秋田県大仙市協和船沢; 17, 秋田県横手市大畑; 18, 栃木県栃木市大森町; 19, 栃木県鹿沼市上南摩町; 20-21, 愛知県豊田市阿蔵町); 22-23, **富士亜種**(22, ♂ (HT), 23, ♀ (PT), 静岡県富士市大渕); 24-25, **岩湧亜種**(24, ♂, 25, ♀, 大阪府岩湧山); 26-27, **基亜種**?(26, ♂, 27, ♀, 岡山県新見市足見). — a, ♂左鞘翅中央部; b, 鱗棘(側面観); c, 同(背面観).

2. *Carabus* (*Euleptocarabus*) *porrecticollis* BATES, 1883 アキタクロナガオサムシ

Original description　*Carabus porrecticollis* BATES, 1883, Trans. ent. Soc. Lond., 1883, p. 228.

History of research　This species was described by the British coleopterologist H. W. BATES (1883) based on the specimens collected by G. LEWIS. BREUNING (1932) placed it in the section *Eucarabus* of the subgenus *Carabus* of the grand genus *Carabus*. NAKANE (1955) proposed for this species a new subgenus *Euleptocarabus* (adopted genus is *Carabus*), and later described a new subspecies *C*. (*E*.) *porrecticollis kansaiensis* from Iwawaki of Osaka Prefecture (NAKANE, 1961). Recently, an isolated population of this species was unexpectedly found from the southern foot of Mt. Fuji-san in Shizuoka Prefecture, and was described as a new subspecies under the name of *Limnocarabus* (*E*.) *p*. *pacificus* (IMURA & MATSUNAGA, 2011). As to the systematic position of *Euleptocarabus*, refer to the same section of *C*. (*Limnocarabus*) *maacki*.

Type locality　Urasa, and on the north-west coast at Akita and Sakata. For the location of each locality, see the distribution map of this species.

Type depository　Natural History Museum, London [Syntypes].

Morphology　23~33 mm. Body above black, more or less with blue-purplish tinge on head, pronotum and elytral base and margins. External features: Figs. 1~7. Description on details: penultimate segment of labial palpus bisetose; median tooth of mentum shorter than lateral lobes, sharply pointed at apex in ventral view, protruded ventrad and divergent toward apex in lateral view; submentum bisetose; male antennae without hairless areas on ventral side; pronotal margins basically bisetose; elytral shoulders often weakly serrated; preapical emargination of elytra faintly recognized, though often unclear in male; elytral sculpture triploid heterodyname as shown in Figs. 6, 16-a, 22-a and 24-a; propleura vaguely and sporadically punctate, sides of sternites irregularly rugoso-punctate; metacoxa trisetose; sternal sulci clearly carved. Male genitalic features: Figs. 8~15. Penis small, about one-third as long as elytra; membraneous part markedly inflated in fully everted condition, ostium lobe absent; lepidospinula much smaller than that of *C*. (*L*.) *maacki*, thorn-shaped, markedly emarginated, and covered with scale-like microstructure except for apical part and ventral surface; internal sac narrowly constricted at base, strikingly inflated in median portion, then acutely tapered and turned up toward apex; right basal lateral lobe either weakly developed or vestigial, median lobe large and conically shaped, with a small accessory lobe lying at its right side (= right paramedian lobe); saccellus rudimentary; apical lobe extraordinarily developed to form a pair of egg-shaped protrusions; podian lobe small, mush shifted to dorsal side; peripheral rim of gonopore weakly sclerotized, semi-circularly or triangularly protruded.

Molecular phylogeny　Viewed from the molecular phylogenetic tree of the mitochondrial ND5 gene, this species is most closely related to *C*. (*L*.) *maacki aquatilis*, from which the former would have branched off in the Japanese Archipelago. On the same tree, this species is divided into three major lineages whose distributional areas are the Tohoku-Kanto-Chubu Regions, the Kinki Region and the Chugoku Region, respectively. The first lineage is further divided into three sublineages, each distributed in the Kanto-Tohoku, northern Chubu and southern Chubu. The second lineage is also divided into three sublineages distributed in Hokuriku-western Kinki, eastern Kinki and south-central Kinki, respectively (KIM *et al*., 1999; OSAWA *et al*., 2002, 2004).

Distribution　Honshu (rather sporadically distributed mainly on the Japan Sea side, from N. Akita Pref. to NW. Hiroshima Pref.) / Is. Oki-no-shima [Lake Biwa-ko] (Shiga Pref.).

Habitat　Forest species, usually distributed on hills or low mountains, and prefers moist places.

Bionomical notes　Spring breeder. Larvae prefer to feed on other insects' larvae.

Geographical variation　Classified into three subspecies, though the distributional borders between the nominotypical subspecies and subsp. *kansaiensis* has not been clearly defined as yet.

Etymology　The specific name means "elongated thorax".

研究史　英国の甲虫研究者，H. W. BATES（ヘンリー・ウォルター・ベイツ）により記載された，古くから知られている本邦特産種．BREUNING（ブロイニング）（1932）はこれをカラブス属カラブス亜属の一員とみなし，その中のエウカラブス節に位置付けたが，NAKANE（1955）は本種に対しエウレプトカラブス亜属を設け，さらに大阪の岩湧からアキタクロナガオサムシ岩湧亜種を記載した（NAKANE, 1961）．本種は各集団間ではっきりとした形態学的差異を見出しにくいため，岩湧亜種以外に亜種分類はなされていなかったが，ごく近年になって富士亜種が発見，記載された（IMURA & MATSUNAGA, 2011）．エウレプトカラブス亜属の系統上の位置についてはマークオサムシの研究史の項を参照．

タイプ産地　「浦佐，および秋田と酒田の北西海岸」— 原記載に記された産地を直訳すればこのようになる．このうち，浦佐（新潟県南魚沼市）については特定することが可能だが，後二者はどこを指すのかがはっきりしない．おそらく，採集者ルイスからの伝聞に基づいて記載者のベイツがこのように記述したものであろう．

タイプ標本の所在　ロンドン自然史博物館［シンタイプ］．本種の記載者であるBATESのコレクションは，フランスの富豪，René OBERTHÜR（ルネ・オーベルチュール）によって買い取られ，彼の死後，パリ自然史博物館に収められた．従って，我が国のオサムシ研究の黎明期にBATESによって記載された種のタイプシリーズの大半は現在，パリにあり，その中からレクトタイプが指定されている

2. *Carabus* (*Euleptocarabus*) *porrecticollis*

にも残されており，またBatesによって記載された日本産オサムシの標本の元々の所属先であるLewis（ルイス）コレクションは現在，ロンドン自然史博物館に保管されているので，レクトタイプ指定された標本がパリにないものもある．本種はこうした例外の一つと思われ，ロンドンに保管されている標本がシンタイプに相当するものと考えてよいであろう．タイプ産地の特定を含め，レクトタイプの指定作業が必要であるが，井村が検した限りでは，現存する標本のラベルに産地の詳細は記されていない．

形態 23～33 mm．背面は黒色で，頭部，前胸背板，鞘翅基部および側縁部には青紫色の光沢を帯びる場合が多い．外部形態：図1～7．細部の記載：下唇肢亜端節の剛毛は2本．下唇基節の中央歯は側葉より短く，先端は腹面から見て鋭く尖り，側面から見て腹側に突出し，先端に向けて広がる．下唇亜基節には2本の剛毛を有する．♂触角腹面に無毛部を欠く．前胸背板の側縁剛毛は通常2対で，中央と後角付近に各1本．鞘翅肩部にはしばしば弱い鋸歯状の切れ込みがあるが，時に消失する．翅端部は僅かにえぐれるが，♂ではしばしば不明瞭．鞘翅彫刻は図6, 16-a, 22-a, 24-aに示すように三元異規的で，基本的にはマークオサムシのそれと同じく，第一次間室は強く隆起した細い鎖線，第二次間室は連続する隆条となる．第三次間室はマークオサムシのそれより発達し，隆条または顆粒列として認められるが，第二次間室よりは弱く，時に不明瞭．間室間の基面にはやや粗大な顆粒を密に装う．前胸側板には浅い小点刻が散在し，腹部腹板の側面には不規則な皺と点刻を装う．後基節の剛毛は3本で．腹部腹板の横溝は明瞭．♂交尾器形態：図8～15．陰茎は小さく，鞘翅長の1/3程度に過ぎない．膜状開口部の基部は完全膨隆下では強く膨らみ，葉片を欠く．鱗棘はマークオサムシのそれよりはるかに小さく，棘状で腹側は強くえぐれ，先端部と腹面を除き鱗状の微細構造で覆われる．内嚢は基部で強く狭まり，中部で著しく膨らみ，先端に向けて細くなり，背側へ反る．右基部側葉は，これを全く欠くものから弱く発達するものまで変異がある．中央葉は著しく発達し，三角形の巨大な突起となり，その右基部にしばしば小さな傍中央葉がある．内嚢盤は未発達．先端葉は巨大な卵形の突起となって著しく発達する．脚葉は内嚢先端部の変形に伴い背側へとシフトし，射精口縁膜は腹側中央が弱く硬化し，半円形ないし三角形に弱く突出する．

分子系統 ミトコンドリアND5遺伝子の解析結果から，本種はマークオサムシに最も類縁が近く，日本列島内で同種から分化したものであろうと推測されている．分子系統樹上では，分布域にリンクした3系統に大きく分かれ，各系統はそれぞれ主に1）中部～東北地方，2）近畿地方，3）中国地方に分布する．1）はさらに三つの亜系統（それぞれ主に東北～関東地方，中部地方南部，中部地方北部に分布）に，2）もさらに三つの亜系統（それぞれ主に北陸～近畿地方西部，近畿地方東部，近畿地方中南部に分布）に分かれる（Kim *et al*., 1999; 大澤ら, 2002, Osawa *et al*., 2004）．

分布 本州（秋田県北部～広島県北西部の，主として日本海側）／沖島［琵琶湖］（滋賀県）．本州の固有種で，秋田県北部（北限は田代岳）から広島県北西部に至る日本海側斜面の，平野に面した丘陵地から低山地にかけて断続的に分布する．脊梁山脈を越えて太平洋側斜面にも進出している場所は，宮城平野中部，関東平野北縁，富士山南麓，濃尾平野北東縁，紀伊半島北部などの地域に限られる．

生息環境 分布域の北東部では低地～低山地に産し，日照の少ないやや湿潤な環境を好む（山谷, 1989）．近畿地方における生息地は主として山地に限られ，一般に第四紀層には見られないが，滋賀県北西部の饗庭野（あえばの）など近畿北部では段丘にも進出する傾向があり，疎林に生息する（近畿オサムシ研究グループ, 1979）．

生態 春繁殖・成虫越冬型のオサムシで，幼虫は基本的に節足動物食（昆虫類の幼虫など）と考えられるが，本種の生態に関する報告は極めて少ない．土中・朽木中いずれにおいても越冬し，同一の越冬窩内に複数の個体が集団で越冬しているケースもしばしば観察される．分布は断続的だが，産地における個体数は一般に多い．

地理的変異 基亜種の他に岩湧亜種と富士亜種が記載されており，形態学的には3亜種に分類される．しかし，細かく見れば他にも特徴のある地域集団が存在し，例えば分布域の西限に近い岡山県においては，北東部に産する集団は前胸背板の青色味が弱く，鞘翅の彫刻も弱いのに対し，西部の新見市（にいみし），高梁市（たかはしし）などに産する個体は前胸背板の青味が強く，鞘翅の彫刻も強く粗いことなどが指摘されている（山地・脇本, 1979）．いずれにせよ，本種の地理的変異や亜種分類に関する研究は不十分で，♂交尾器形態や分子系統解析の結果などを考慮に入れたうえでの更なる検討が必要であろう．

種小名の由来 種小名は「長く伸びた胸の」を意味する．種和名として広く使われている「アキタクロナガオサムシ」は，戸沢信義（1933）により与えられたものとされているが，これは*porrecticollis*という種小名によって代表される分類群に対してではなく，コクロナガオサムシ東北地方南部亜種 *Carabus* (*Leptocarabus*) *arboreus parexilis* に対して与えられた名称である可能性の高いことが黒沢（1988）により指摘されている．

2-1. *C.* (*E.*) *p. porrecticollis* Bates, 1883 アキタクロナガオサムシ基亜種

Original description, type locality & type depository　See explanation of the species.

Diagnosis　23～31 mm.　Dorsal color variable according to localities or individuals, though usually bearing rather remarkable blue-purplish tinge on head, pronotum and elytral margins. 1) Serration of shoulders variable, either remarkably recognized or almost vestigial; 2) tertiary intervals of elytra indicated by rows of granules to form narrow ridges and usually more prominently recognized than in two other subspecies; 3) lepidospinula larger on an average than in two other subspecies.

Distribution　Honshu (from N. Akita Pref. in the northeast to NW. Hiroshima Pref. in the southwest, excepting the distributional range of the following two subspecies).

原記載・タイプ産地・タイプ標本の所在　種の項を参照.

亜種の特徴　23~31 mm. 背面の色彩には地域変異, 個体変異が見られ, 頭胸部と鞘翅側縁は青紫色の光沢を帯びる場合が多いが, 暗色の個体が多くを占める集団もある. 1) 鞘翅肩部が鋸歯状となる頻度は産地や個体により大きく変化し, 顕著に刻まれるものから退化的なものまで幅が大きい; 2) 第三次間室は密な顆粒列で基本的に連続する隆条として認められる; 3) ♂交尾器内嚢基部の鱗棘が比較的大きい.

分布　本州 (下記2亜種の分布域を除く秋田県北部~広島県北西部の主として日本海側).

2-2. *C. (E.) p. pacificus* (IMURA et MATSUNAGA, 2011) アキタクロナガオサムシ富士亜種

Original description　*Limnocarabus* (*Euleptocarabus*) *porrecticollis pacificus* IMURA & MATSUNAGA, 2011, Elytra, Tokyo, N. S., 1, p. 2, figs. 1~3, 8.

Type locality　Ôbuchi in Fuji-shi, 360 m in altitude, on the southern foot of Mt. Fuji, in Shizuoka Prefecture, central Honshu, Central Japan.

Type depository　Hokkaido University Museum (Sapporo) [Holotype, ♂: specimen number 30633].

Diagnosis　24~33 mm. Distinguishable from two other subspecies in combination of the following characters: 1) dorsal coloration stably darker, above all in elytra which are mat black without purplish tinge except for lateral margins; 2) sides of pronotum less remarkably cordate in anterior halves, less remarkably sinuate posteriorly, with lateral portions less strongly reflex above; 3) propleura coarsely and evidently punctate; 4) epipleura a little less coarsely granulate than in nominotypical subspecies; 5) humeral serration of elytra barely recognizable in most specimens examined; 6) primary foveoles of elytra averagingly shallower; 7) secondary intervals of elytra indicated by narrow and independent costae, while they are often adhesive to adjacent granules to form reticular- or radiate pattern in subsp. *kansaiensis*; 8) tertiary interval of elytra not costate as in nominotypical subspecies but indicated by irregularly set rows of granules as in subsp. *kansaiensis*, though areas between intervals much less coarsely scattered with granules which are smaller than those of subsp. *kansaiensis*; 9) internal sac of male genitalia with right paramedian lobe and right basal lateral lobe more prominently protruded; 10) lepidospinula unique in size and shape, apparently smaller than that of nominotypical subspecies and slenderer than in subsp. *kansaiensis*, sometimes indistinctly tuberculate a little behind tip on ventral or inflexed side.

Distribution　Narrowly restricted to part of Fuji-shi & Fujinomiya-shi of Shizuoka Pref., at the southern foot of Mt. Fuji-san, 230~560 m in altitude.

Habitat　Forest of deciduous broadleaved tree and/or planted cedar- or cypress tree.

Bionomical notes　Sympatric with *Carabus* (*Ohomopterus*) *lewisianus*, *C*. (*O*.) *esakii* and *C*. (*Damaster*) *blaptoides*. Two *Leptocarabus* species, namely, *C*. (*L*.) *procerulus* and *C*. (*L*.) *kumagaii*, parapatrically inhabit the surrounding area, and they are essentially parapatric with *C*. (*E*.) *porrecticollis pacificus*.

Conservation　This subspecies is designated as the threatened species (VU; vulnerable species) in the 4th edition of Red List (Threatened Wildlife of Japan) of Ministry of the Environment (2012).

Etymology　Named after the location of its habitat which faces the Pacific Ocean in a short distance.

タイプ産地　中部日本, 本州中部, 静岡県, 富士山南麓, 富士市大渕, 標高360 m.

タイプ標本の所在　北海道大学総合博物館 (札幌) [ホロタイプ, ♂: 標本番号30633].

亜種の特徴　24~33 mm. 1) 背面は他亜種に比しより暗色で, 特に鞘翅は側縁部を除き青紫色光沢を欠くものが多い; 2) 前胸背板側縁は前方がより直線的で, 後方の波曲は弱い; 3) 前胸側板の点刻は密で強い; 4) 前胸背板側片の顆粒は基亜種に比しやや疎; 5) 鞘翅肩部に鋸歯状の切れ込みを持たない個体が大半を占める; 6) 鞘翅第一次回陥は平均してより浅い; 7) 第二次間室は連続する隆条となるが, 岩湧亜種のように間室間の顆粒と融合して放射状の突起を形成することは殆どない; 8) 第三次間室は岩湧亜種同様, 発達が弱いが, 間室間の顆粒は明らかにより小さく, 疎; 9) ♂交尾器内嚢の右傍中央葉と右基部側葉は他亜種に比しより強く突出する; 10) 鱗棘は基亜種のそれよりも明らかに小さく, 岩湧亜種のそれよりも細く, 時に先端近くの屈側が弱く, 同部において結節状に膨らむ.

分布　静岡県東部 (富士山南麓, 富士市と富士宮市の一部), 標高230~560 m. 富士市から富士宮市にかけての, ごく狭い範囲のみから知られている. 周辺には他にアキタクロナガオサムシの生息地は知られておらず, 不自然にかけ離れた孤立分布を示すことから, 人為的に移入されたものが定着している可能性も考慮しなければならない. ただ, 仮にそうだとしても, 現在の生息範囲が少なくとも数キロ四方に及び, かつ生息密度も高いことから, ごく最近運び込まれた少数の雌雄の子孫が一時的に繁殖しているとは考えにくい. 開墾などに伴い, 一定の年数に亘って連続的に親虫が供給され続けるといったような状況がなければ無理であろうが, いずれにせよ, そのことを証明する術はないに等しい.

生息環境　ナラ, クリ等からなる落葉広葉樹林やスギ・ヒノキの植林の林床に生息する.

生態　ルイスオサムシ, シズオカオサムシ, マイマイカブリの3種と混生しているが, クロナガオサムシ, オオクロナガオサムシの2種は本亜種の分布域の周辺部に局所的に見られ, 基本的には棲み分けている.

(2012年公表)においてカテゴリーVU(絶滅危惧II類)に指定されている．現地では宅地開発と森林の伐採が著しい速度で進行しており，また，訪れることの容易な場所であるため，生息環境の維持と共に，過度の採集圧をかけないようにすることが，本集団存続のためには必要であろう．

亜種小名の由来と通称　亜種小名は生息地が太平洋に面していることに因む．原記載論文中で提唱された通称はフジアキタクロナガオサムシ．

2-3. *C. (E.) p. kansaiensis* Nakane, 1961 アキタクロナガオサムシ岩湧亜種

Original description　*Carabus (Euleptocarabus) porrecticollis kansaiensis* Nakane, 1961, Fragm, coleopterol., Kyoto, (1), p. 2.
Type locality　Iwawaki, Osaka Pref., Honshu.
Type depository　Hokkaido University Museum, Sapporo [Holotype, ♂].
Diagnosis　25~32 mm. 1) Body slender; 2) tertiary intervals of elytra weaker, not forming costae but indicated by irregularly set rows of granules; 3) granules between intervals relatively large and coarsely set; 4) lepidospinula short and robust.
Distribution　SC. Kinki Region (distributional range of this subspecies is not yet defined properly).
Etymology　Named after the Kansai Region, which roughly corresponds to the Kinki Region.

　タイプ産地　本州，大阪府，岩湧(大阪府南部の岩湧山ないしその周辺を指すものと思われる)．
　タイプ標本の所在　北海道大学総合博物館(札幌)[ホロタイプ，♂]．
　亜種の特徴　25~32mm．1)体形は細い；2)第三次間室は連続する隆条とならずに不規則な顆粒列へと減退する；3)間室間の顆粒は粗大かつ比較的密；4)♂交尾器鱗棘は太短い．
　分布　近畿地方中南部．タイプ産地の岩湧山を含む近畿地方中南部一帯に産するものを便宜上，本亜種に充てておくが，どの範囲のものまでを本亜種に含めるべきかについては定見がない．
　亜種小名の由来と通称　亜種小名は「関西地方の」を意味する．ホソアオクロナガオサムシという通称で呼ばれることが多いが，ホソアキタクロナガオサムシという名が提唱されたこともある．

2. *Carabus* (*Euleptocarabus*) *porrecticollis*

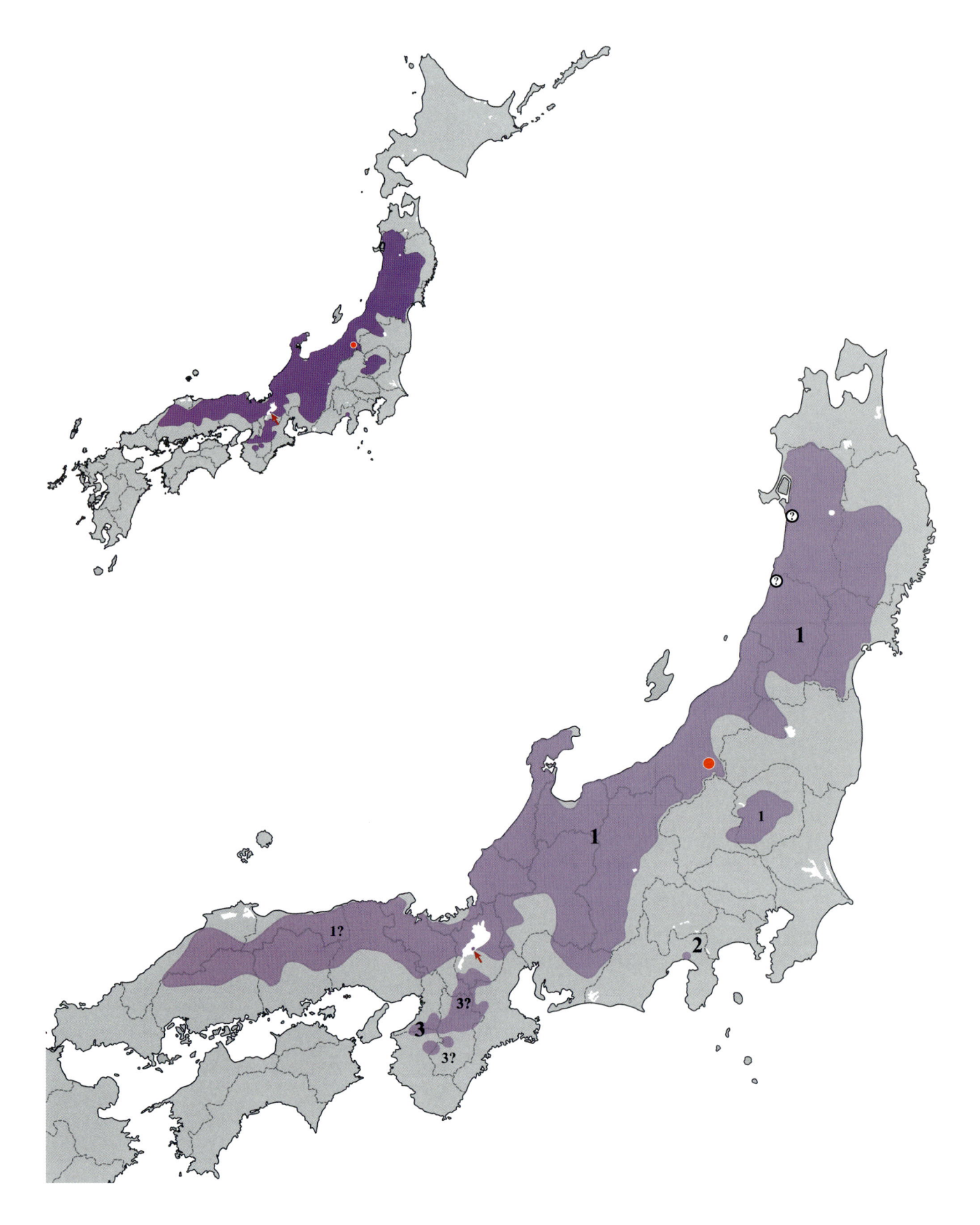

Fig. 28. Map showing the distributional range of ***Carabus*** (***Euleptocarabus***) ***porrecticollis*** BATES, 1883. — 1, Subsp. ***porrecticollis***; 2, subsp. ***pacificus***; 3, subsp. *iwayuushi*. ●& ⊙: Type localities of the onesies designated in the original description (●, Urasa; ⊙, "northwestern coast of Akita and Sakata").
図 28. アキタクロナガオサムシ分布図. — 1, 基亜種; 2, 富士亜種; 3, 岩湧亜種. ●と⊙: 原記載に示されたタイプ産地 (●, 浦佐; ⊙, 「秋田と酒田の北東海岸」).

36

2. *Carabus* (*Euleptocarabus*) *porrecticollis*

Fig. 29. Urasa of Minami-uonuma-shi, Niigata Pref., one of the type localities of *Carabus* (*Euleptocarabus*) *porrecticollis*.

図29. アキタクロナガオサムシのタイプ産地の一つ, 新潟県南魚沼市浦佐.

Fig. 30. Two adults of *Carabus* (*Euleptocarabus*) *porrecticollis porrecticollis* hibernating in rotten wood (Urasa, Minami-uonuma-shi, Niigata Pref.).

図30. 朽木内で越冬中のアキタクロナガオサムシ基亜種(新潟県南魚沼市浦佐).

Fig. 31. An adult of *Carabus* (*Euleptocarabus*) *porrecticollis pacificus* passing the winter in rotten cedar tree (Ôbuchi, Fuji-shi, Shizuoka Pref.).

図31. 杉倒木の腐朽部分で越冬中のアキタクロナガオサムシ富士亜種(静岡県富士市大渕).

3. *Carabus* (*Hemicarabus*) *macleayi* Dejean, 1826
マックレイセアカオサムシ

Figs. 1-15. *Carabus* (*Hemicarabus*) *macleayi amanoi* (Imura, 2004) (Mt. Rishiri-zan, Is. Rishiri-tô, Hokkaido). — 1, ♂; 2, ♀; 3, male head & pronotum; 4, male right protarsus in ventral view; 5, male right protibia in subdorsal view; 6, male left elytron (median part); 7, humeral serration of female right elytron; 8, penis & fully everted internal sac in right lateral view; 9, apical part of penis in right lateral view; 10, ditto in dorsal view; 11, internal sac in caudal view; 12, ditto in frontal view; 13, ditto in dorsal view; 14, ditto in left sublateral view; 15, ligulum in left sublateral view.

図 1-15. マックレイセアカオサムシ利尻島亜種（北海道利尻島利尻山）. — 1, ♂; 2, ♀; 3, ♂の頭部と前胸背板; 4, ♂右前跗節（腹面観）; 5, ♂右前脛節（斜背面観）; 6, ♂左鞘翅中央部; 7, ♀右鞘翅肩部の鋸歯状辺縁; 8, 陰茎と完全反転下の内嚢（右側面観）; 9, 陰茎先端（右側面観）; 10, 同（背面観）; 11, 内嚢（尾側面観）; 12, 同（頭側面観）; 13, 同（背面観）; 14, 同（左斜側面観）; 15, 基斑（左斜側面観）.

3. *Carabus* (*Hemicarabus*) *macleayi*

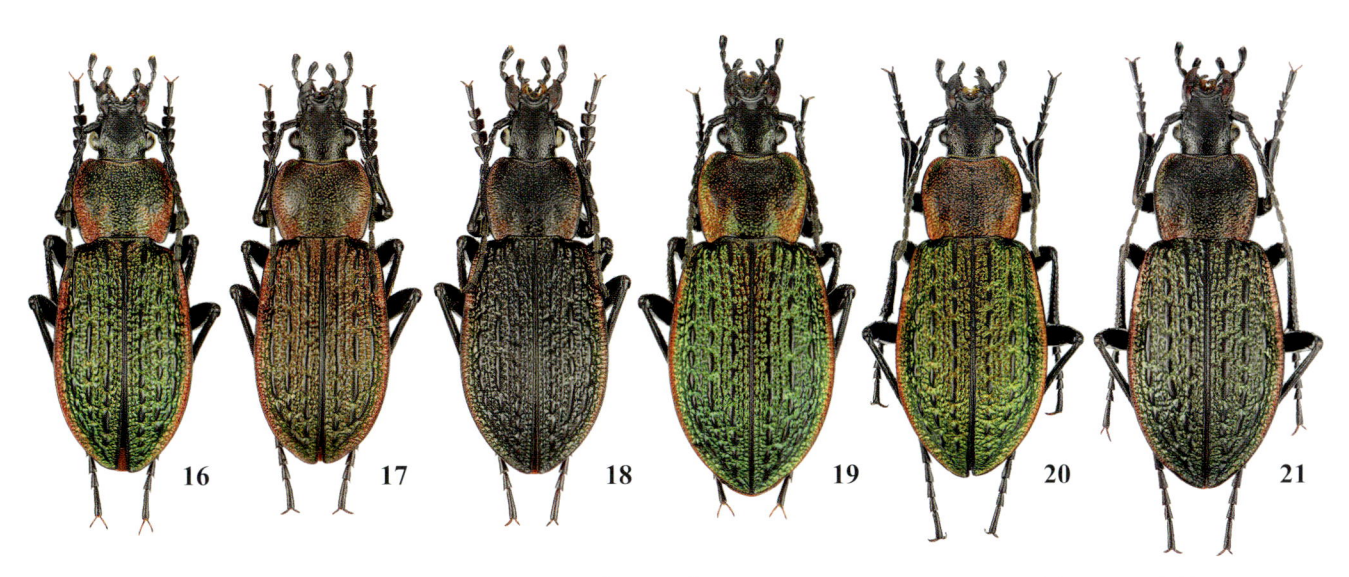

Figs.16-21. *Carabus* (*Hemicarabus*) *macleayi amanoi*. — 16-18, ♂, 19-21, ♀ (20, PT; 21, HT), Mt. Rishiri-zan, Is. Rishiri-tô, Hokkaido.
図 16-21. マックレイセアカオサムシ利尻島亜種. — 16-18, ♂, 19-21, ♀ (20, PT; 21, HT), 北海道利尻島利尻山.

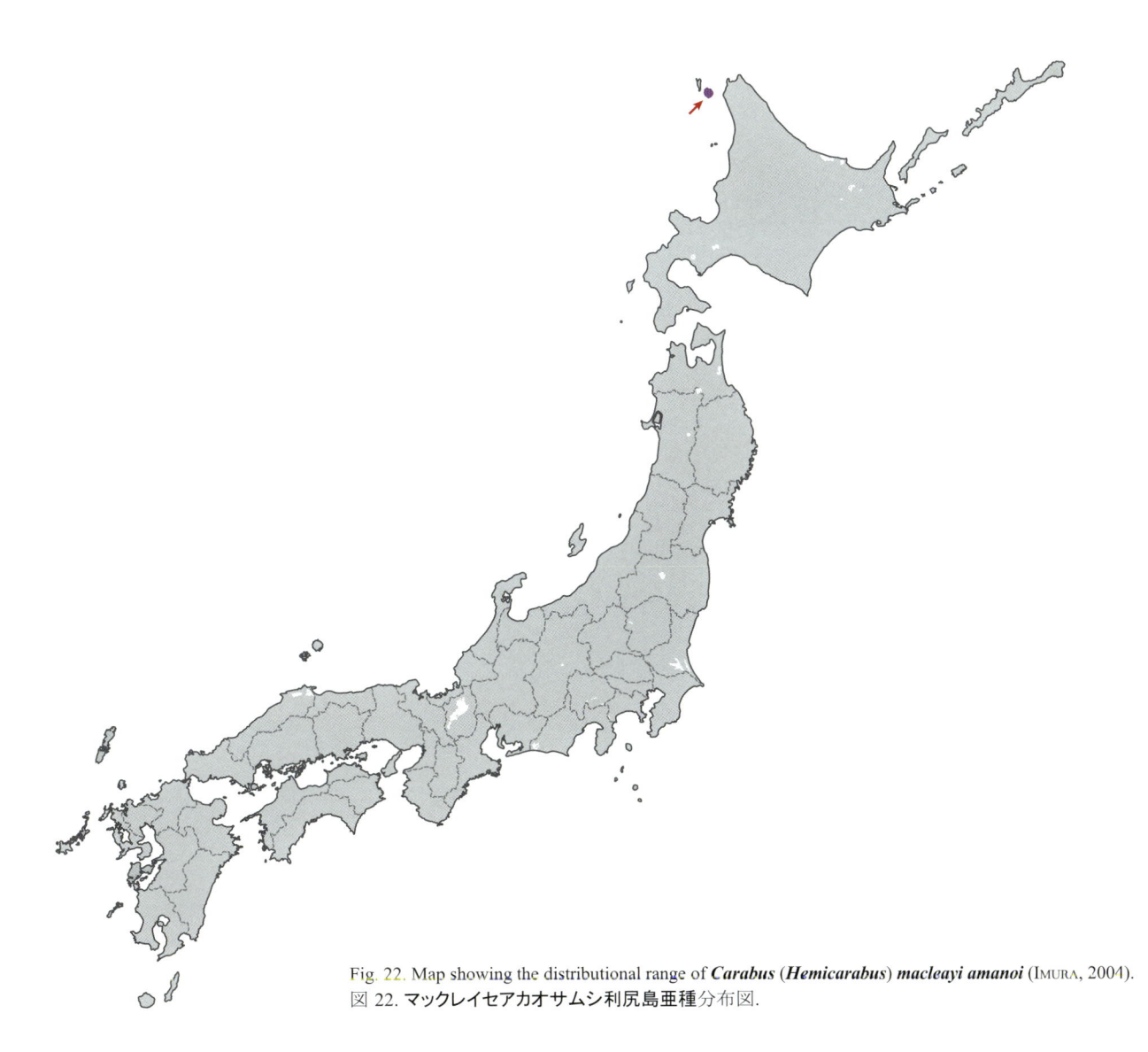

Fig. 22. Map showing the distributional range of *Carabus* (*Hemicarabus*) *macleayi amanoi* (IMURA, 2004).
図 22. マックレイセアカオサムシ利尻島亜種分布図.

3. *Carabus* (*Hemicarabus*) *macleayi* DEJEAN, 1826 マックレイセアカオサムシ

Original description   CARABUS MAC LEAYI DEJEAN, 1826, Species général des Coléoptères de la collection de M. le comte DEJEAN, Paris-Crenot, 2, p. 485.

History of reseach   This species was originally described by DEJEAN (1826) from Daourie (= Dauria) of East Siberia, and is distributed from southeastern Siberia to the Kamchatka Peninsula. Geographical variation is not so remarkable, and it has usually been divided into two subspecies, *Carabus* (*Hemicarabus*) *macleayi macleayi* occupying the greater part of its distributional range and *C.* (*H.*) *m. coreensis* BREUNING, 1933 of North Korea. In 2001, an isolated population of this species was unexpectedly found from the alpine zone of Is. Rishiri-tô in northernmost Japan, and described as a new subspecies under the name of *Hemicarabus macleayi amanoi* IMURA, 2004.

Type locality   Daourie, dans la Sibérie orientale.

Type depository   Muséum Nationale d'Histoire Naturelle, Paris [Holotype, ♀].

Morphology   15~20 mm. Body above dark bluish purple excepting pronotum and elytral margins which are iridescent and strongly metallic above all in the latter, or body above greenish golden excepting pronotum and elytral margins which are reddish coppery; appendages black. External features: Figs. 1~7. Description on details: mandibles with dorsal surface punctate and roughly rugulose; penultimate segment of labial palpus with two setae near base; median tooth of mentum shorter than lateral lobes, with tip sharply pointed and only slightly protruded ventrad; submentum bi- to quadrisetose; antennae with segment two and basal half of segment three depressed so that their inner margins are somewhat carinate; male antennae with flat hairless ventral areas from segment five to ten, widely so in segment five to seven, more narrowly so occupying two-third to half of each segment in segment eight to ten; pronotum with three to five, usually four pairs of marginal setae; elytral shoulders with or without weak serration; preapical emargination of elytra faintly recognized only in female; elytral sculpture triploid heterodyname; primaries the strongest, indicated by rows of short costae or oval-shaped tubercles segmented by small and shallow primary foveoles; secondaries weaker than primaries, indicated by irregularly and frequently segmented short costae or rows of large granules; tertiaries the weakest, indicated by irregularly set rows of granules which are often fused with adjacent intervals to form reticular pattern above all around sutural part; propleura smooth, at most weakly rugulose; sides of sternites irregularly rugoso-punctate; metacoxa bisetose, lacking inner seta; sternal sulci clearly carved. Male genitalic features: Figs. 8~15. Penis about 0.4 times as long as elytra; ostium lobe large and bilobed at tip; ligulum well developed, indicated by strongly pigmented oblong-shaped sclerite, with distal tip hooked and sharply pointed; internal sac with parasaccellar lobes small and symmetrical, saccellus strongly protruded dorsad, asymmetrically bilobed at tip as shown in Fig. 11~12, peripheral rim of gonopore weakly sclerotized and pigmented, with median part triangularly protruded. Outer plate of ligular apophysis of female genitalia with the shape of nock of an arrow, and weakly sclerotized, inner plate vestigial.

Molecular phylogeny   The subgenera *Hemicarabus* and *Homoeocarabus* are morphologically similar to each other, and they have often been combined into a single category (e.g. Crenolimbi REITTER, 1896). On the molecular phylogenetic trees of the mitochondrial ND5 gene, *C.* (*Ho.*) *maeander* is sharply separated from all the four species of the subgenus *Hemicarabus*, and diversification of these two subgenera seems to be considerably old. Within the subgenus *Hemicarabus*, *C.* (*He.*) *tuberculosus*, *C.* (*He.*) *macleayi* and *C.* (*He.*) *nitens* are very close in their ND5 gene sequences, and presumed to be derived from a common ancestor. The remaining one, *C.* (*He.*) *serratus* of North America separated fairly long ago from other Eurasian *Hemicarabus* species. Presumably, the common ancestor of all the *Hemicarabus* species was distributed throughout the Eurasian Continent and North America when they were still connected by a land bridge. Upon the formation of the Bering Straits, the Eurasian population and the North American one evolved in different directions, resulting in differentiation of *C.* (*He.*) *serratus* in North America (SU, IMURA *et al.*, 2000).

Distribution   Is. Rishiri-tô (Hokkaido). Outside Japan: NE. Eurasia (SE. Siberia, Amur Region, Khabarovsk Region, Russian Maritime Provinces, North Korea, Yakut, Magadan & Kamchatka Penins.) / Is. Sakhalin.

Habitat   Our knowledge is still very poor on the habitat of this species inhabiting the Eurasian Continent. For that of subsp. *amanoi*, see the following section concerning that subspecies.

Bionomical notes   Our knowledge is also very poor on the ecology of this species inhabiting the Eurasian Continent. For that of subsp. *amanoi*, also see the section of that subspecies.

Geographical variation   This species is currently classified into three subspecies, namely, subspp. *macleayi* (NE. Eurasia & Sakhalin), *coreensis* (N. Korea) and *amanoi* (Is. Rishiri-tô of N. Japan).

Etymology   This species is named after Alexander MACLEAY (1767~1848), the colonial administrator and entomologist who was a controversial and important figure of Australian colonial society. His entomological collection formed the basis of the Macleay Museum at the University of Sydney.

　研究史　本種はDEJEAN（ドゥジョン）（1826）によりシベリア東部のダウリヤから記載された美麗種で，シベリア南東部から極東ロシア，北朝鮮，ヤクート，マガダン，カムチャツカ半島にかけて比較的広く分布することが知られている．いくつかの地域変異集団に対して亜種名が与えられているが，北鮮亜種 *C.* (*H.*) *m. coreensis* BREUNING, 1933 を除くその他の軽微な地域変異集団は一般に

3. *Carabus* (*Hemicarabus*) *macleayi*

タイプ産地　シベリア東部, ダウリヤ.
タイプ標本の所在　パリ自然史博物館［ホロタイプ, ♀］.
形態　15~20 mm. 一般に背面が暗青紫色で, 前胸背板と鞘翅の側縁部は外側から内側に向けて赤銅色~黄色~緑色~緑青色へと虹色に変化する強い金属光沢を持つものが多いが, 鞘翅背面が金緑色で前胸背板と鞘翅の側縁が赤銅色を呈する個体や集団も知られている. 外部形態：図1~7. 細部の記載：大顎背面には点刻と粗い皺がある. 下唇肢亜端節には基部よりに2本の剛毛を有し, 下唇基節の中央歯は側葉より短く, 三角形で先端は鋭く尖り, 腹側へは殆ど突出しない. 下唇亜基節には2~4本の長毛を持つ. 触角第2節と第3節の基部1/2は背腹方向に押圧され, 内側縁は稜状. ♂触角腹面には, 第5節から第7節には全長に亘り, 第8節から第10節には各節長の2/3~1/2に亘り, 平坦な無毛部がある. 前胸背板の側縁剛毛は3~5対（通常4対）. 鞘翅肩部の鋸歯状突起はごく弱く, これを欠く個体もある. 鞘翅端部のえぐれは♀でごく僅かに認められる程度. 鞘翅彫刻は三元異規的. 第一次間室は最も強く発達し, 小さく浅い第一次凹陥によりやや不規則に分断された隆条片列あるいは長楕円形の結節列を形成する. 第二次間室は第一次間室よりも弱く, 頻繁かつ不規則に分断された短い隆条列ないし粗大な顆粒列となる. 第三次間室は最も弱く, 不規則な顆粒列で, しばしば隣接する間室隆起部と融合し, 不規則な網眼状構造を呈する. この網眼状構造は鞘翅会合線付近で特に顕著である場合が多い. 前胸側板は滑らかで, せいぜい不規則な弱い皺を認める程度. 腹部腹板の側面には不規則な皺と小点刻を装う. 後基節の剛毛は2本で, 内側剛毛を欠く. 腹部腹板には横溝がある. 交尾器形態：図8~15. 陰茎は鞘翅長の0.4倍前後. 葉片は大きく, 先端は顕著に分葉する. 基斑は色素沈着を伴う細長い硬化片で, その遠位端は明らかに内嚢膜面から遊離し, 先端は鉤爪状に鋭く尖る. 内嚢は基部側葉と中央葉を欠き, 盤前葉は左右対称で小さく, 内嚢盤は先端が分葉し, 左右非対称で右葉は左葉より大きい. 先端葉の膨隆は弱く, 脚葉は不明瞭. 射精孔縁膜は経度に硬化, 色素沈着し, 先端中央は三角形に突出する. ♀交尾器膣底部節片外板は矢筈形で硬化は弱く, 内板は細く痕跡的.
分子系統　オサムシ亜族（広義のオサムシ属）の分子系統樹上において, 本種を含むヘミカラブス亜属4種は単一のクラスターを形成する. このうち, ユーラシア産の3種（ヨーロッパセアカオサムシ, セアカオサムシ, マックレイセアカオサムシ）は互いにごく近縁で, 北米のホクベイセアカオサムシのみ, やや類縁が遠い. 本種はユーラシア産の他の2種と共通の祖先種から比較的最近になって分化したものと思われる（Su, Imura *et al.*, 2000）. 利尻島の集団は, 恐らく氷期の海退により大陸~サハリン~利尻島が陸続きになっていた時期に南進し, その後, 同島の高所に取り残されたものと考えられる.
分布　利尻島（北海道）. 国外：ユーラシア大陸北東部（シベリア南東部, アムール地方, ハバロフスク地方, ロシア沿海地方, 北朝鮮北部, ヤクート地方, マガダン地方, カムチャツカ半島）／サハリン.
生息環境　ユーラシア北東部の亜寒帯~周極地域に分布する種. 近隣諸外国における知見は少ないが, 同属他種と同様, 草地や荒地, 裸地に近い環境を好むものと思われる. ロシア沿海地方では林縁や貯木場から得られたという個体を検しており, カムチャツカ半島南部のバシュカヘッツ山麓では雪融け直後の湖岸に近い湿った環境から見つかっている. またサハリンにおいて本種を採集された川田光政氏（札幌）からの私信によれば, 日中ハナゴケの上を歩行していたそうで, 一見ある種のゴミムシと見紛うような外見であったという. 利尻島亜種に関しては同亜種の項を参照.
生態　近隣諸国における本種の生態に関する報告は殆ど見られない. 利尻島亜種の生態については同亜種の項を参照.
地理的変異　基亜種, 北鮮亜種, 利尻島亜種の3亜種に分類される. このうち, 本邦には利尻島亜種のみを産する.
種小名の由来　種小名は, イギリス植民地時代（1700年代末~1800年代）のオーストラリアで植民地行政官として大きな足跡を残したAlexander Mackleay（アレクサンダー・マックレイ）（1767~1847）に因む. 彼は同時に昆虫学者でもあり, そのコレクションはシドニー大学にあるマックレイ博物館の基礎資料となっている.

3-1. *C.* (*H.*) *m. amanoi* (Imura, 2004) マックレイセアカオサムシ利尻島亜種

Original description　*Hemicarabus macleayi amanoi* Imura, 2004, Elytra, Tokyo, 32, p. 236, figs. 1~2.
History of research　This subspecies was discovered by Masaharu Amano (Seto-shi in Aichi Prefecture) in 2002 from the alpine zone of the Island of Rishiri-tô in northernmost Japan. It was described by Imura in 2004 as a new subspecies under the name of *amanoi*, and sensationally debuted into the Japanese carabine fauna as s remarkable newcomer. For the story related to its discovery, see Imura (2004c, 2005c) and Imura & Nagahata (2006).
Type locality　Alpine zone of Mt. Rishiri-zan, on Is. Rishiri-tô, off the western coast of the northern tip of Hokkaido, Northeast Japan.
Type depository　National Museum of Nature and Science, Tokyo [Holotype, ♀].
Diagnosis　16~19 mm. Differs from other subspecies of *C.* (*H.*) *macleayi* in the following points: 1) dorsal coloration more strongly reddish and no purple-colored individual has been found; 2) elytra slenderer and less acutely convergent toward apices, pronotum and legs also slenderer and longer; 3) primary costae of elytra more narrowly but prominently convex, showing a striking contrast to much reduced tertiary intervals which are usually recognized as irregularly set rows of granules; 4) apical lobe of penis a little shorter and robuster.
Distribution　Alpine zone of Mt. Rishiri-zan on Is. Rishiri-tô, off the northwestern coast of Hokkaido.
Habitat　The main habitat of *C.* (*H.*) *m. amanoi* is a narrow gravely slope near the summit of Mt. Rishiri-zan. The slope is very steep

3. *Carabus* (*Hemicarabus*) *macleayi*

and frequently accompanying small-scale landslides here and there. In such a harsh condition, this race selectively inhabits the environment with the ground surface covered by low grasses, alpine mosses and lichens, so that the top soil is rather stable.

Bionomical notes   Spring breeder. This race has its main activity peak early in summer, usually between the mid-June to July in the alpine zone of Mt. Rishiri-zan. Larvae prefer to feed on fresh insects, above all on small scarabaeid beetles such as *Sericania sachalinensis*, one of the dominant beetles in the alpine zone of Mt. Rishiri-zan. Also they prefer to eat *Popillia japonica* or maggot (larva of *Phaenicia cuprina*) as a substitute food *in vitro*.

Conservation   The present subspecies is endemic to Is. Rishiri-tô, and is narrowly restricted to the alpine zone of Mt. Rishiri-zan. Though its habitat is located in the special protection zone of the national nature conservation area called the Rishiri-Rebun-Sarobetsu National Park, a special attention should be paid for its protection mainly against disturbance caused by poachers aimed at this beetle, since it is a beautifully colored newcomer added to the Japanese carabid fauna in recent times. This subspecies is designated as the threatened species (VU; vulnerable species) in the 4th edition of Red List (Threatened Wildlife of Japan) of Ministry of the Environment (2012).

Etymology   Named after Masaharu AMANO (Seto-shi, Aichi Pref.) who first found the dead female specimen of this subspecies on Mt. Rishiri-zan.

研究史　本亜種は愛知県瀬戸市の天野正晴により2001年の夏，利尻山登山道上で偶然1♀の死骸が拾われたことが発端となり，3年後の2004年，井村有希，戸田尚希，永幡嘉之の調査によって利尻島における土着が確認された．本種の発見と調査にまつわる一連の経緯は，「リシリノマックレイセアカオサムシ－さいはての島の小さな奇跡－」（井村・永幡, 2006）に詳しくまとめられている．世界的に見てもオサムシの調査精度が極めて高い日本国内から数十年ぶりに発見された本邦未記録種として，大きな話題となった．

タイプ産地　北東日本, 北海道北端の西方海上, 利尻島, 利尻山高所（北海道利尻町／利尻富士町）.

タイプ標本の所在　国立科学博物館［ホロタイプ, ♀］.

亜種の特徴　16~19 mm. 他亜種に比し，背面の赤味がより強く（基亜種においてしばしば見られる，鞘翅が紫色を呈する型は知られていない），鞘翅はより細長く，翅端部に向けての狭まりがより緩やかで，前胸背板と脚もより細長く，鞘翅の第一次原線は細いが強く隆起し，第二次，第三次間室の隆起部と連結して網眼状構造を呈することがなく，不規則な顆粒列に減退した第三次間室と顕著な対照をなす．陰茎先端は他亜種のそれに比し，やや太短いものが多い．

分布　利尻島（利尻山高所）．

生息環境　利尻山高所に局在する，ごく小規模な，地衣類に覆われた砂礫地や草地等に生息する．高山雪潤草原や高茎草原（いわゆるお花畑），ハイマツや矮小化したミヤマハンノキ・ダケカンバ群落等の林床には見られない．

生態　春繁殖・成虫越冬型のオサムシ．成虫は比較的雑食性で，飼育下において各種の昆虫や蜘蛛類，カタツムリ，果実などを食するが，幼虫は基本的に節足動物（昆虫）食で，コガネムシ科昆虫の成虫や双翅目の幼虫などを食する．本種の繁殖活動期にカラフトチャイロコガネが発生のピークを迎え，極めて多数の個体が見られるので，野外では，これらの成虫・幼虫等が食餌資源として活用されている可能性が高い．

保全　本亜種の生息地は利尻礼文サロベツ国立公園の特別保護地区内にあり，同所において許可なく動植物を採取することは固く禁じられているが，ごく最近になって日本産のファウナに加わったオサムシであり，なおかつ極めて限られた分布範囲を持つ美麗種であることから，密猟に対する対応が最も重要である．また，特有の地形に伴い，現地では山肌の自然崩落が年々著しい速度で進行しており，近い将来，生息地の少なくとも一部が自然消滅してしまう可能性も高い．環境省第4次レッドリスト（2012）では，どういうわけか以前よりも評価が下がってダウンリスト種となり，絶滅危惧II類（カテゴリー VU：絶滅の危険が増大している種）に指定されているが，本亜種の発見後も環境省の許可を得て定期的に現地調査を行っている井村の印象では，ここ数年でその個体数が増加したわけでも，新たに確実な繁殖地が発見されたわけでもなく，利尻山高所の慢性的な崩落に伴い，その稀少性はむしろ高まっている．同カテゴリーに挙げられているアカガネオサムシ！（本書で言うアカガネオサムシ本州亜種，以下同じ），マークオサムシ（マークオサムシ本州亜種），タカネセスジアカガネオサムシ（セスジアカガネオサムシ大雪山パルサ湿原亜種），ウガタオサムシ（マヤサンオサムシ鵜方亜種），フジアキタクロナガオサムシ（アキタクロナガオサムシ富士亜種），ドウキョウオサムシ！（末尾に感嘆符を付したものは，地域によっては未だ絶滅には程遠い潤沢な個体数が維持されているか，あるいは最近になって新たな生息地が発見されているタクソン）等と同列にランクされるようなものとは思えない．

亜種小名の由来と通称　亜種小名は，本亜種の第一発見者である天野正晴（愛知県瀬戸市在住の自然愛好家・貝類収集家）に因む．通称はリシリノマックレイセアカオサムシ．

3. *Carabus* (*Hemicarabus*) *macleayi*

Fig. 23. Habitat of *Carabus* (*Hemicarabus*) *macleayi amanoi* (alpine zone of Mt. Rishiri-zan, Is. Rishiri-tô, Hokkaido).

図23. マックレイセアカオサムシ利尻島亜種の生息地（北海道利尻島利尻山高所）.

Fig. 24. *Carabus* (*Hemicarabus*) *macleayi amanoi* (alpine zone of Mt. Rishiri-zan, Is. Rishiri-tô, Hokkaido).

図24. マックレイセアカオサムシ利尻島亜種（北海道利尻島利尻山高所）.

Fig. 25. *Carabus* (*Hemicarabus*) *macleayi amanoi* (alpine zone of Mt. Rishiri-zan, Is. Rishiri-tô, Hokkaido).

図25. マックレイセアカオリムシ利尻島亜種（北海道利尻島利尻山高所）.

4. *Carabus* (*Hemicarabus*) *tuberculosus* Dejean et Boisduval, 1829
セアカオサムシ

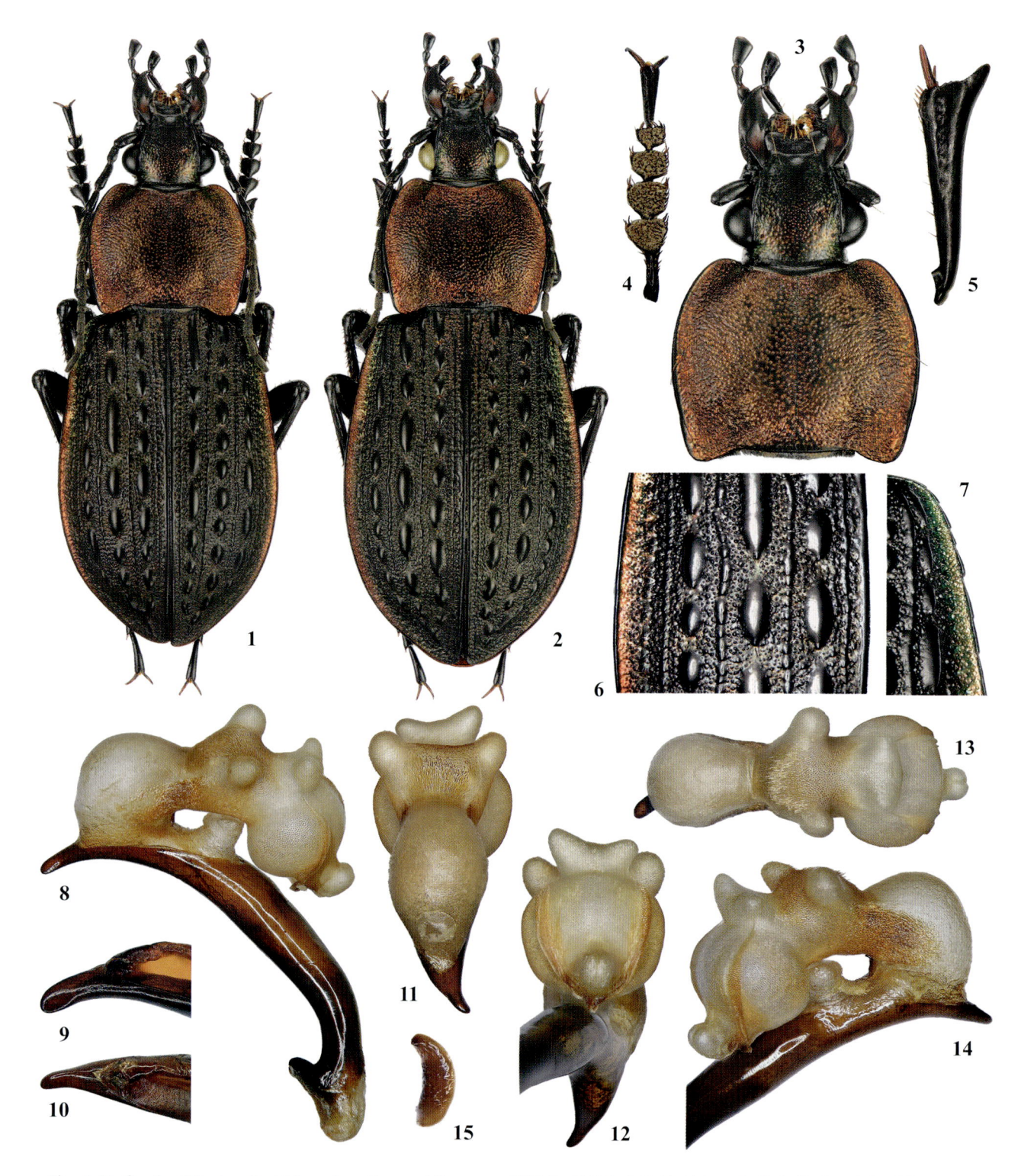

Figs. 1-15. *Carabus* (*Hemicarabus*) *tuberculosus* Dejean et Boisduval, 1829 (Is. Rishiri-tô, Hokkaido). — 1, ♂; 2, ♀; 3, male head & pronotum; 4, male right protarsus in ventral view; 5, male right protibia in subdorsal view; 6, male left elytron (median part); 7, humeral serration of male right elytron; 8, penis & fully everted internal sac in right lateral view; 9, apical part of penis in right lateral view; 10, ditto in dorsal view; 11, internal sac in caudal view; 12, ditto in frontal view; 13, ditto in dorsal view; 14, ditto in left sublateral view; 15, ligulum in left sublateral view.

図 1-15. セアカオサムシ (北海道利尻島). — 1, ♂; 2, ♀; 3, ♂の頭部と前胸背板; 4, ♂右前跗節 (腹面観); 5, ♂右前脛節 (斜背面観); 6, ♂左

4. *Carabus* (*Hemicarabus*) *tuberculosus*

Figs.16-33. ***Carabus*** (***Hemicarabus***) ***tuberculosus***. — 17-19, 21, 23, 25-27, 29, 30, 32, 33, ♂, 16, 20, 22, 24, 28, 31, ♀, 16, Mt. Kita-no-miné, Furano-shi, Hokkaido; 17, Benten, Is. Teuri-tô, Hokkaido; 18, Mt. Kamui-yama, Is. Okushiri-tô, Hokkaido; 19, Is. Oshima-ô-shima, Matsumae-chô, Hokkaido; 20, Is. Oshima-ko-jima, Matsumae-chô, Hokkaido; 21-22, Tazawako-kôgen, Senboku-shi, Akita Pref.; 23, Uchiura, Is. Awa-shima, Niigata Pref.; 24, Momura, Nasu- shiobara-shi, Tochigi Pref.; 25, Koga-shi, Ibaraki Pref. ; 26, Riv. Tama-gawa, Tokyo; 27, Hachi-no-shiri, Is. Izu-ô-shima, Tokyo; 28, Wanouchi-chô, Gifu Pref. ; 29, Riv. Kizu-gawa, Yawata-shi, Kyoto Pref.; 30, Mt. Dai-sen, Tottori Pref.; 31, Mt. Shiozuka-miné, Miyoshi-shi, Tokushima Pref.; 32, Hijûdai, Kusu-machi, Oita Pref.; 33, Te-arai, Kirishima-shi, Kagoshima Pref.

図 16-33. **セアカオサムシ**. — 17-19, 21, 23, 25-27, 29, 30, 32, 33, ♂, 16, 20, 22, 24, 28, 31, ♀; 16, 北海道富良野市北の峰; 17, 北海道天売島弁天; 18, 北海道奥尻島神威山; 19, 北海道松前町渡島大島; 20, 北海道松前町渡島小島; 21-22, 秋田県仙北市田沢湖高原; 23, 新潟県粟島内浦; 24, 栃木県那須塩原市百村; 25, 茨城県古河市; 26, 東京都多摩川; 27, 東京都伊豆大島蜂の尻; 28, 岐阜県輪之内町; 29, 京都府八幡市木津川; 30, 鳥取県大山; 31, 徳島県三好市塩塚峰; 32, 大分県玖珠町日出生台; 33, 鹿児島県霧島市手洗.

4. *Carabus* (*Hemicarabus*) *tuberculosus*

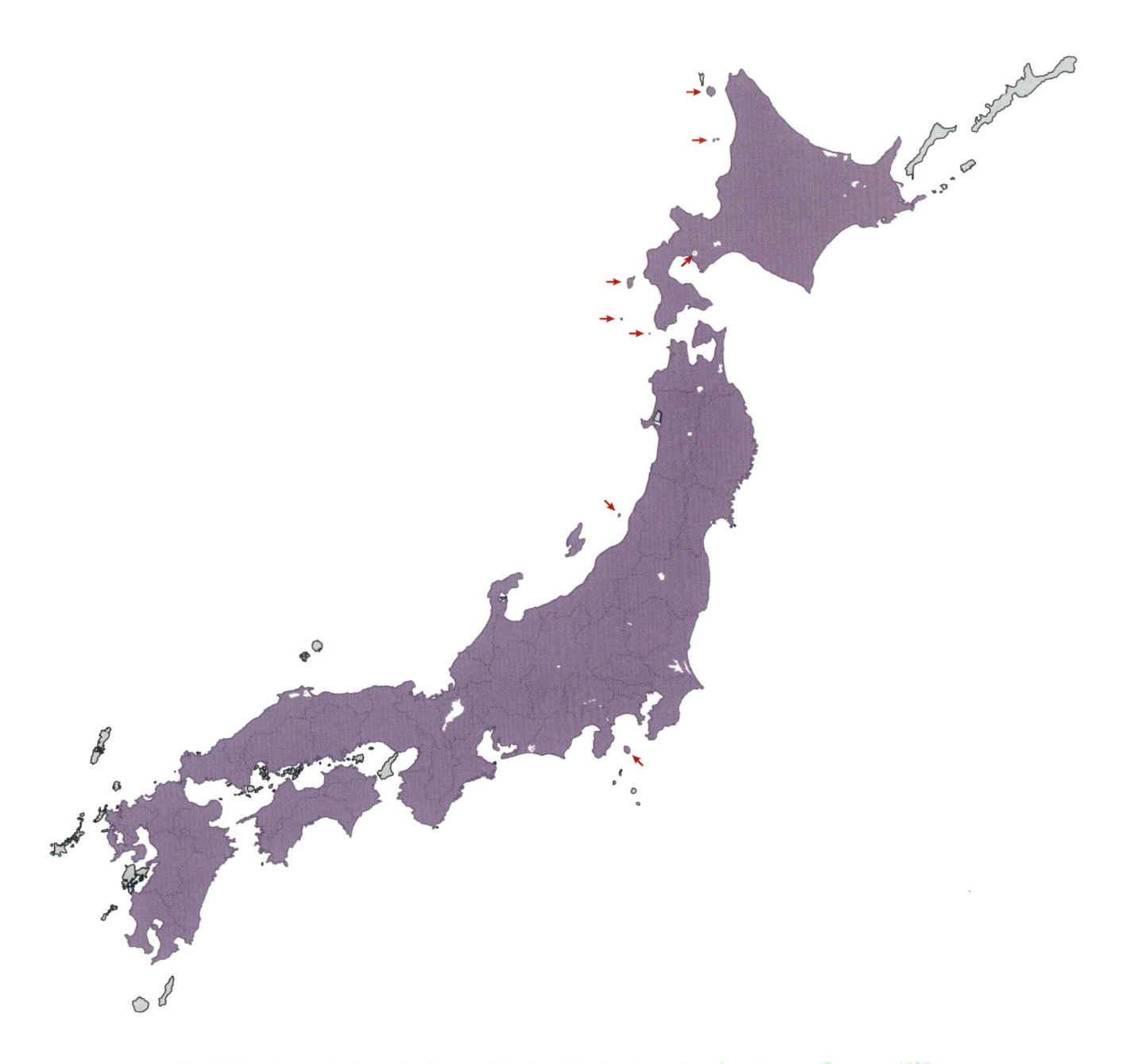

Fig. 34. Map showing the distributional range of *Carabus* (*Hemicarabus*) *tuberculosus* Dejean et Boisduval, 1829
図 34. セアカオサムシの分布図

48

4. *Carabus* (*Hemicarabus*) *tuberculosus*

Fig. 35. Habitat of *Carabus* (*Hemicarabus*) *tuberculosus* (Is. Rishiri-tô, Hokkaido).
図35. セアカオサムシの生息地（北海道利尻島）.

Fig. 36. *Carabus* (*Hemicarabus*) *tuberculosus* (Is. Rishiri-tô, Hokkaido).
図36. セアカオサムシ（北海道利尻島）.

Fig. 37. *Carabus* (*Hemicarabus*) *tuberculosus* (Is. Rishiri-tô, Hokkaido)
図37. セアカオサムシ（北海道利尻島）.

5. *Carabus* (*Homoeocarabus*) *maeander* Fischer, 1822
セスジアカガネオサムシ

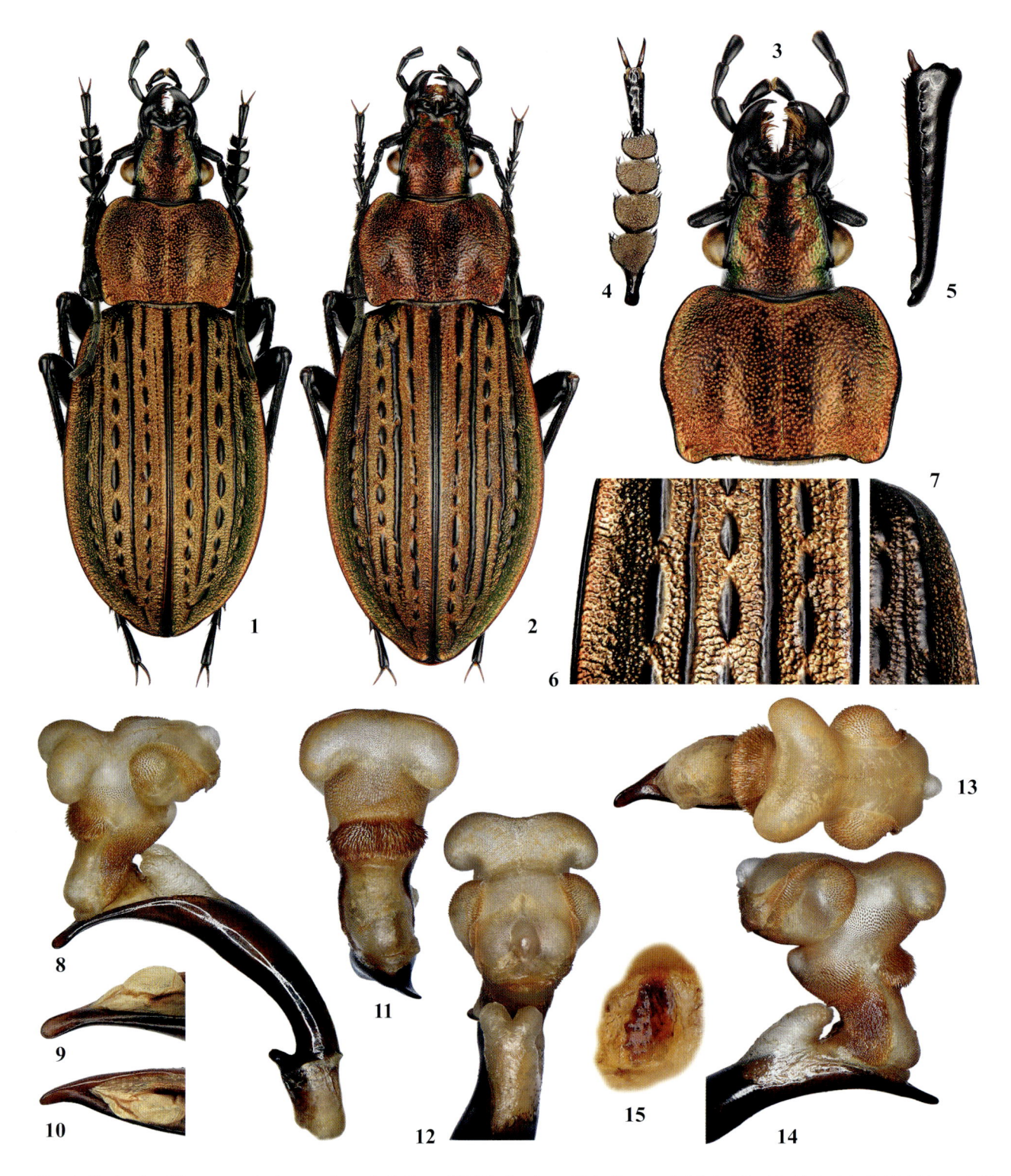

Figs. 1-15. *Carabus* (*Homoeocarabus*) *maeander paludis* Géhin, 1885 (1, Sarobetsu Plain, Toyotomi-chô; 2, Akkeshi-chô; both in Hokkaido). — 1, ♂; 2, ♀; 3, male head & pronotum; 4, male right protarsus in ventral view; 5, male right protibia in subdorsal view; 6, male left elytron (median part); 7, humeral serration of male right elytron; 8, penis & fully everted internal sac in right lateral view; 9, apical part of penis in right lateral view; 10, ditto in dorsal view; 11, internal sac in caudal view; 12, ditto in frontal view; 13, ditto in dorsal view; 14, ditto in left sublateral view; 15, ligulum in frontal view.

図 1-15. セスジアカガネオサムシカムチャツカ北海道亜種 (1, 北海道豊富町サロベツ原野; 2, 同厚岸町). — 1, ♂; 2, ♀; 3, ♂の頭部と前胸背板; 4, ♂右前跗節(腹面観); 5, ♂右前脛節(斜背面観); 6, ♂左鞘翅中央部; 7, ♂右鞘翅肩部の鋸歯状辺縁; 8, 陰茎と完全反転下の内嚢(右側面観); 9, 陰茎先端(右側面観); 10, 同(背面観); 11, 内嚢(尾側面観); 12, 同(頭側面観); 13, 同(背面観); 14, 同(左側側面観); 15, 舌雉(舌側面観).

5. *Carabus* (*Homoeocarabus*) *maeander*

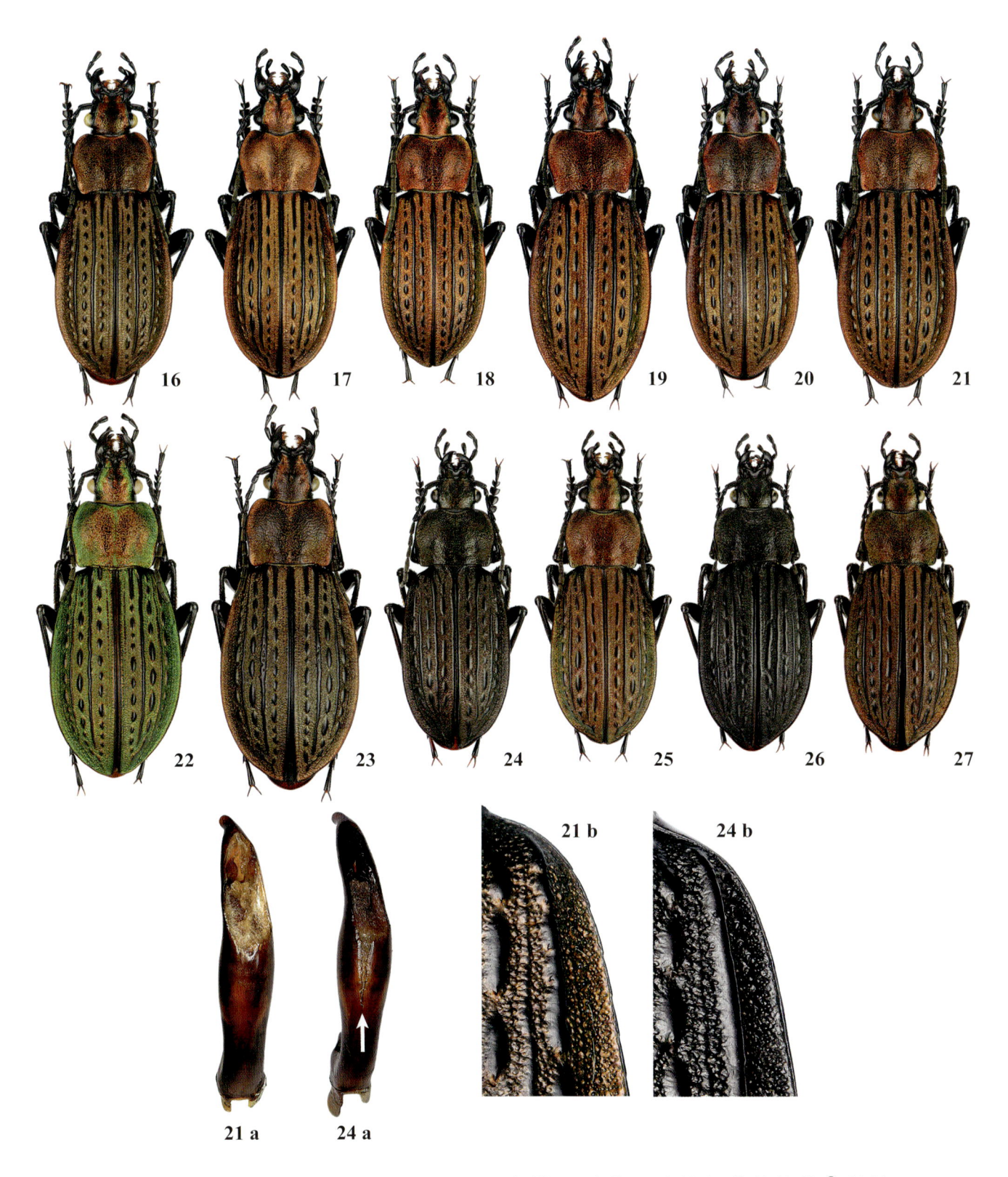

Figs. 16-27. *Carabus* (*Homoeocarabus*) *maeander* subspp. — 16-23, *C*. (*H*.) *m. paludis* (16-18, 20, 21, ♂, 19, 22, 23, ♀; 16, Meguma-numa Moor, Wakkanai-shi; 17, Sarobetsu Plain, Toyotomi-chô; 18, Katamusari, Akkeshi-chô; 19, Kushiro Moor, Kushiro-shi; 20, Ôtsu wetlands, Toyokoro-chô; 21-22, lakeside marsh of Utonai-ko, Tomakomai-shi; 23, ♀, Mourai coastal area, Atsuta-ku, Ishikari-shi); 24-27, *C*. (*H*.) *m. nobukii* (24, 25, ♂, 26, 27, ♀ (24, HT, 25-27, PT); palsa bog, S. of Mt. Hira-ga-daké, Daisetsu-zan Mts.). — a, Penis in dorsal view (arrow indicates a membraneous furrow on the dorsal wall of the penis in subsp. *nobukii*); b, shoulder of male right elytron. All from Hokkaido.

図 16-27. セスジアカガネオサムシ各亜種. — 16-23, カムチャツカ北海道亜種 (16-18, 20, 21, ♂, 19, 22, 23, ♀; 16, 稚内市メグマ沼湿原; 17, 豊富町サロベツ原野; 18, 厚岸町片無去; 19, 釧路市釧路湿原; 20, 豊頃町大津湿地; 21-22, 苫小牧市ウトナイ湖; 23, 石狩市厚田区望来海岸); 24-27, 大雪山パルサ湿原亜種 (24, 25, ♂, 26, 27, ♀ (24, HT, 25-27, PT); 大雪山平ヶ岳南方パルサ湿原). — a, 陰茎(背面観)(矢印は大雪山パルサ湿原亜種に特有の陰茎背面の膜状溝を示す); b, ♂右鞘翅肩部. すべて北海道産.

5. *Carabus* (*Homoeocarabus*) *maeander*

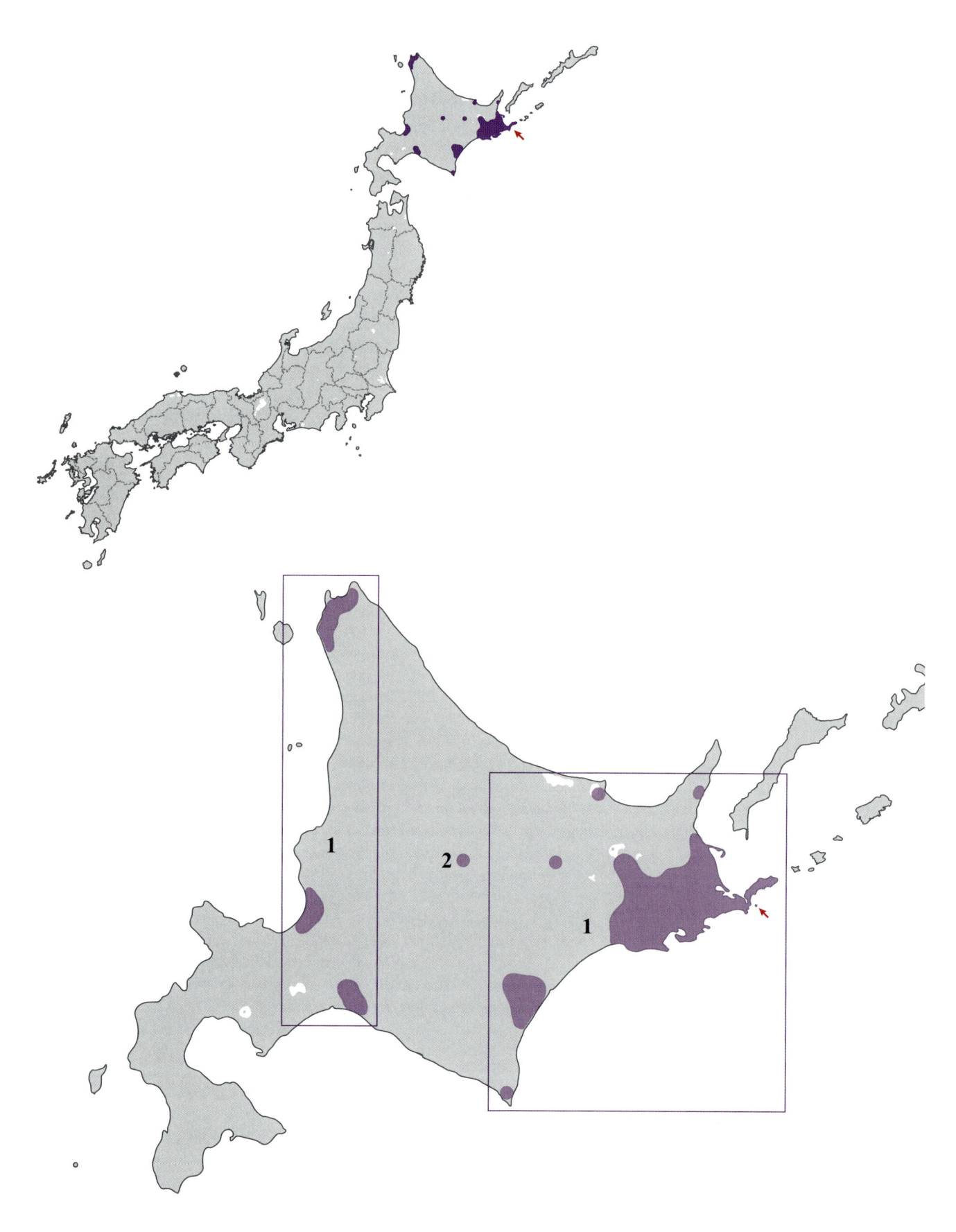

Fig. 28. Map showing the distributional range of ***Carabus*** (***Homoeocarabus***) ***maeander*** Fischer von Waldheim, 1822. — 1, Subsp. ***paludis***; 2, subsp. ***nobukii***.

5. Carabus (Homoeocarabus) maeander

Fig. 29. Habitat of *Carabus* (*Homoeocarabus*) *maeander paludis* (Benten-numa Moor, Tomakomai-shi, Hokkaido).

図29. セスジアカガネオサムシカムチャツカ北海道亜種の生息地(北海道苫小牧市弁天沼湿原).

Fig. 30. *Carabus* (*Homoeocarabus*) *maeander paludis* (Benten-numa Moor, Tomakomai-shi, Hokkaido).

図30. セスジアカガネオサムシカムチャツカ北海道亜種(北海道苫小牧市弁天沼湿原).

Fig. 31. Habitat of *Carabus* (*Homoeocarabus*) *maeander nobukii* (Palsa bog, S. of Mt. Hira-ga-daké, Daisetsu-zan Mts., Hokkaido; taken by Nobuki YASUDA).

図31. セスジアカガネオサムシ大雪山パルサ湿原亜種の生息地(北海道大雪山平ヶ岳南方パルサ湿原; 保田信紀撮影).

6. *Carabus* (*Pentacarabus*) *harmandi* Lapouge, 1909
ホソヒメクロオサムシ

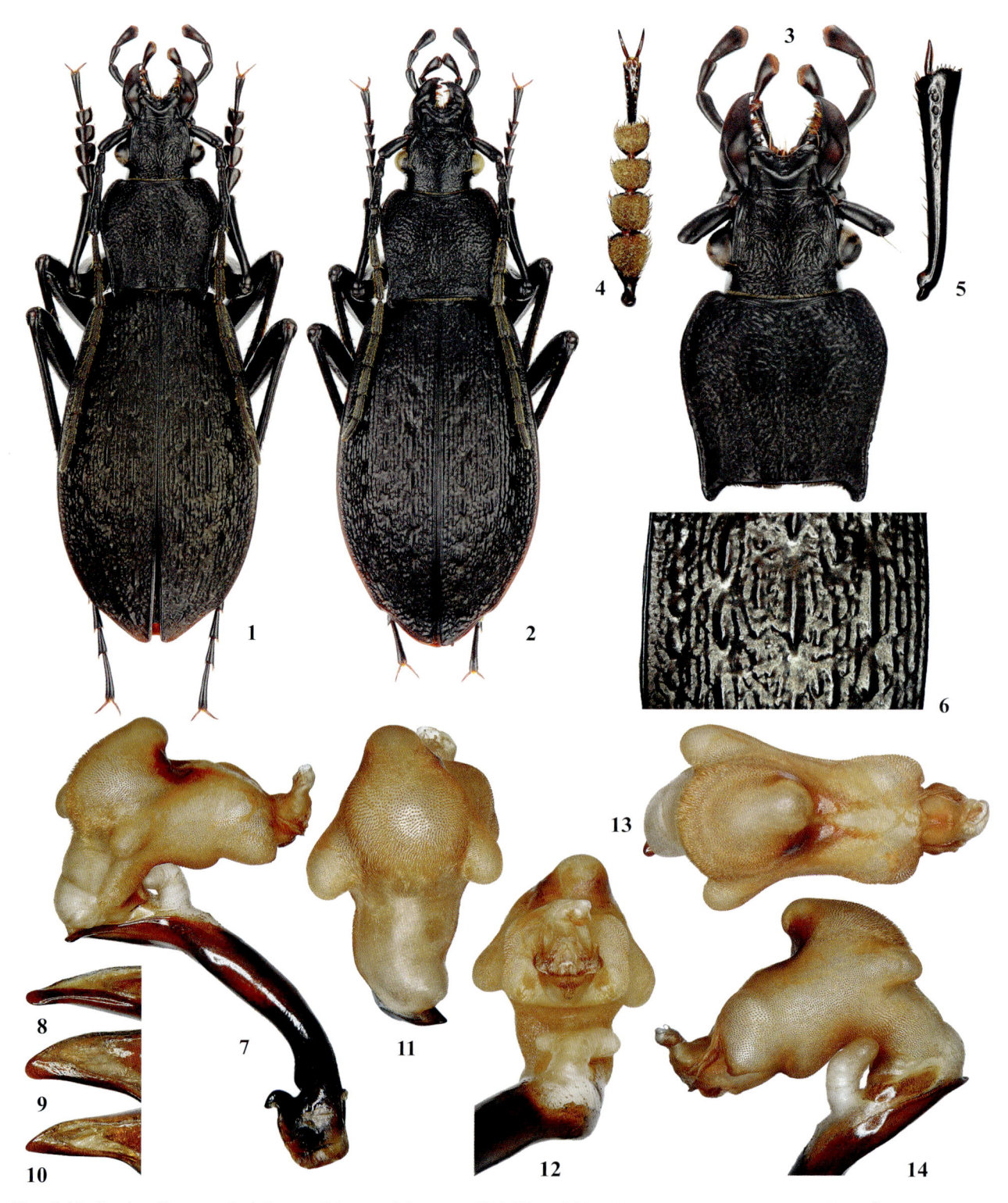

Figs. 1-14. *Carabus* (*Pentacarabus*) *harmandi harmandi* Lapouge, 1909 (Oku-nikkô, Nikkô-shi, Tochigi Pref.). — 1, ♂; 2, ♀; 3, male head & pronotum; 4, male right protarsus in ventral view; 5, male right protibia in subdorsal view; 6, male left elytron (median part); 7, penis & fully everted internal sac in right lateral view; 8, apical part of penis in right lateral view; 9, ditto in right subdorsal view; 10, ditto in dorsal view; 11, internal sac in caudal view; 12, ditto in frontal view; 13, ditto in dorsal view; 14, ditto in left sublateral view.

6. *Carabus* (*Pentacarabus*) *harmandi*

Figs. 15-20. ***Carabus*** (***Pentacarabus***) ***harmandi*** subspp. — 15, *C.* (*P.*) *h. yudanus* (Washiai-mori, Nishi-waga-machi, Iwate Pref.); 16, *C.* (*P.*) *h. adatarasanus* (Mt. Adatara-yama, Nihonmatsu-shi, Fukushima Pref.); 17, *C.* (*P.*) *h. syui* (a, HT, b, PT, Mt. Takatsuka-yama, Kawauchi-mura, Fukushima Pref.); 18, *C.* (*P.*) *h. kojii* (a, HT, b, PT, Mt. Yamizo-san, Tanagura-machi, Fukushima Pref.); 19, *C.* (*P.*) *h. quinquecatellatus* (Rengé-onsen, Itoigawa-shi, Niigata Pref.); 20, *C.* (*P.*) *h. mizunumai* (Mt. Haku-san, Hakusan-shi, Ishikawa Pref.). — a, ♂; b, ♀; c, penis in right subdorsal view; d, ditto in right lateral view; e, apical part of penis in right subdorsal view.

図 15-20. **ホソヒメクロオサムシ**各亜種. — 15, 奥羽山脈亜種（岩手県西和賀町鷲合森）; 16, 東北地方南西部亜種（福島県二本松市安達太良山）; 17, 阿武隈高地亜種 (a, HT, b, PT, 福島県川内村高塚山); 18, 八溝山脈亜種 (a, HT, b, PT, 福島県棚倉町八溝山); 19, 飛騨山脈北部亜種（新潟県糸魚川市蓮華温泉）; 20, 白山飛騨御嶽木曽亜種（石川県白山市白山）. — a, ♂; b, ♀; c, 陰茎（右斜背面観）; d, 同（右側面観）; e, 陰茎先端部（右斜背面観）.

6. *Carabus* (*Pentacarabus*) *harmandi*

Figs. 21-26. *Carabus* (*Pentacarabus*) *harmandi* subspp. — 21, *C.* (*P.*) *h. karasawai* (Shibu-no-yu-onsen, Chino-shi, Nagano Pref.); 22, *C.* (*P.*) *h. okutamaensis* (Mt. Daibosatsu, Kôshû-shi, Yamanashi Pref.); 23, *C.* (*P.*) *h. tanzawaensis* (Mt. Hinokibora-maru, Yamakita-machi, Kanagawa Pref.); 24, *C.* (*P.*) *h. fujisan* (Mt. Fuji-san, Fujiyoshida-shi, Yamanashi Pref.); 25, *C.* (*P.*) *h. akaishiensis* (Niken-goya, Shizuoka-shi, Shizuoka Pref.); 26, *C.* (*P.*) *h. moritai* (Pass Abé-tôgé, Shizuoka-shi, Shizuoka Pref.). — a, ♂; b, ♀; c, penis in right subdorsal view; d, ditto in right lateral view; e, apical part of penis in right subdorsal view.

図 21-26. ホソヒメクロオサムシ各亜種. — 21, 八ヶ岳亜種(長野県茅野市渋の湯温泉); 22, 奥秩父亜種(山梨県甲州市大菩薩); 23, 丹沢亜種(神奈川県山北町檜洞丸); 24, 富士山亜種(山梨県富士吉田市富士山); 25, 赤石山脈亜種(静岡県静岡市二軒小屋); 26, 安倍峠亜種(静岡県静岡市安倍峠). — a, ♂; b, ♀; c, 陰茎(右斜め背面観), d, 同(右側面観), e, 陰茎先端部(右斜め背面観).

6. *Carabus* (*Pentacarabus*) *harmandi* Lapouge, 1909 ホソヒメクロオサムシ

Original description *Carabus Procerulus* v. *Harmandi* Lapouge, 1909, L'Echange, 25, p. 190.

History of research This taxon was described by Georges Vacher de Lapouge (1909) as a variety of *Carabus Procerulus* [sic]. It was combined with *Carabus* (s. str.) by Breuning (1932) or with *Carabus* (*Asthenocarabus*) by Nakane (1962). Ishikawa (1972) proposed a new subgenus *Pentacarabus* for this species, and combined it with the genus *Leptocarabus*. Based on the morphology of the male genital organ and mitochondrial ND5 gene analysis, Imura (2002b) upgraded *Pentacarabus* to a distinct genus in the subtribe Carabina. This species is polytypical, and totally 13 subspecies, including the nominotypical one, have been described (Ishikawa, 1966b, 1968b, 1972, 1986b; Imura & Mizusawa, 2002a).

Type locality Alpes de Nikko.

Type depository Muséum Nationale d'Histoire Naturelle, Paris [Lectotype, ♂].

Morphology 16~24 mm. Body above black and mat, sometimes with faint dark coppery tinge. External features: Figs. 1~6. Description on details: penultimate segment of labial palpus basically bisetose; median tooth of mentum shorter than lateral lobes, triangular in shape, with apex sharply pointed and not remarkably protruded ventrad; submentum bisetose; male antennae with hairless ventral depressions from segment six to eight; pronotum basically with two pairs of marginal setae, one before middle and another near hind angle; preapical emargination of elytra weakly recognized in both sexes, though often unclear in male; elytral sculpture tri- or pentaploid heterodyname; primaries the strongest, indicated by rows of narrow costae rather irregularly segmented by shallow primary foveoles; areas between primaries with three to five rows of contiguous weak costae or rows of granules; propleura smooth; sides of sternites weakly rugulose; metacoxa bisetose, inner seta absent; sternal sulci usually unclear excepting those of populations distributed in southern area, whose sternal sulci are more clearly carved. Male genitalic features: Figs. 7~14. Penis a little shorter than half the elytral length; ostium lobe large and markedly bilobed at tip; internal sac very unique in shape as shown in Figs. 7, 11~14, much different from that of *Asthenocarabus* or *Leptocarabus*, bearing hemispherical median lobe, asymmetrical parasaccellar lobes (left＞right), and extraordinarily developed, asymmetrically shaped saccellus; apical lobes not so large, podian lobes shifted dorsad; apical part of internal sac with a hemispherically shaped inflation on inflexed side; aggonoporius weakly pigmented though not strongly sclerotized.

Molecular phylogeny In the molecular phylogenetic tree of the world Carabina, this species belongs to an highly independent lineage and no close affinity is recognized to the species belonging to the subgenera *Asthenocarabus* and *Leptocarabus*. The tree suggests that there are two major lineages within the cluster of *C*. (*P*.) *harmandi*, the distributional range of which are separated by the Itoigawa-Shizuoka tectonic line (Osawa et al., 2002, 2004).

Distribution C.~ NE. Honshu.

Habitat The main habitat of this species is the deciduous broadleaved forest in the upper temperate zone, mainly composed of *Fagus* and *Quercus*, or sometimes extending to subalpine mixed forest, from less than 500 m to more than 2,000 m in altitude. It is often sympatric with *C*. (*L*.) *arboreus* in the greater part of its distributional range.

Bionomical notes Spring breeder. Adult has its main activity peak twice in a year, spring (May) and autumn (September to October). Larvae prefer to feed on larvae of other insects.

Geographical variation Classified into 13 subspecies including the nominotypical one.

Etymology Named after F. J. Harmand, a minister of pleniopotentiary of France, who stayed in Japan from 1894 to 1905 and collected this species from Nikkô for the first time.

研究史　本種は1909年にGeorges Vacher de Lapouge (ジョルジュ・ヴァッシェー・ドゥ・ラプージュ)(ラプージュと略されることが多いが，ヴァッシェー・ドゥ・ラプージュが正式な姓)によって著されたクロナガオサムシ(原著論文中での綴りは*Carabus Procerulus*)に関する短い報文の中で，*arboreus, tenuiformis, gracillimus, fujisanus, exilis*といった，今日ではコクロナガオサムシの亜種として扱われている各タクソンや*porrecticollis*(アキタクロナガオサムシ)等と共に，クロナガオサムシの一変種(var.)として検索表の中で記載された．我が国では狭義のカラブス属あるいはレプトカラブス属の一員として扱われることが多かったが，♂交尾器内嚢形態を含む形態形質ならびに分子系統解析の結果などから，オサムシ亜族の中でもかなり独立性の高い固有の系統に属する種と見なすべきであろう．従って，本書ではIshikawa (1972)によりレプトカラブス属の亜属として設立されたペンタカラブスを広義のカラブス属内における独立した亜属とみなし，1亜属1種を構成するものとして扱う．多型種で，Ishikawa (1966b, 1968b, 1972, 1986b)，井村・水沢(2002a)らにより，基亜種の他に計12亜種が記載されている．

タイプ産地　日光山地．パリ自然史博物館に保管されている本種のレクトタイプ標本のラベルには "Japon, Chûzenji, Edme Gallois" と記されている．

タイプ標本の所在　パリ自然史博物館［レクトタイプ，♂］．本種を記載したLapougeは元々，人類学の研究家として知られ，弁護士や司書等の職にも就いていたようだが，甲虫学の分野にも多くの足跡を残している．彼の所蔵標本は，オサムシのコレクションとしては当時，世界有数のものであり，多数のタイプ標本を含んでいたが，第二次世界大戦前後の混乱期を経た後，パリ自然史博物館に寄贈を申し出たものの受け入れられず(このあたりの経緯については諸説あるようで，真相は不明)，標本は長年，高湿度の劣悪な環境下で昆虫学には縁のない彼の子孫の家に放置されていたという．1980年代に入ってようやく，パリ自然史博

6-4. *C. (P.) h. kojii* (IMURA et MIZUSAWA, 2002) ホソヒメクロオサムシ八溝山亜種

Original description　*Pentacarabus harmandi kojii* IMURA & MIZUSAWA, 2002, Gekkan-Mushi, Tokyo, (380), p. 5, figs. 2 (p. 3) & 5 (p. 4).
Type locality　Mt. Yamizo-san (ca. 1,000 m in alt.), Tanagura Town, Higashi-shirakawa Co., Fukushima Pref., eastern Honshu, central Japan.
Type depository　National Museum of Nature and Science, Tokyo [Holotype, ♂].
Diagnosis　20~23 mm. Most closely allied to subsp. *syui*, but differs from it in the following points: 1) pronotum slenderer and differently shaped; 2) elytra with primary intervals narrower and less frequently interrupted, secondaries and tertiaries hardly interrupted, preapical emargination in female deeper; 3) penis with median portion hardly inflated, apex narrower in dorsal view.
Distribution　Mt. Yamizo-san (Fukushima Pref. / Ibaraki Pref.), EC. Honshu.
Habitat　Narrowly restricted to the beech forest near the summit of Mt. Yamizo-san.
Etymology　Named after Koji ONODA (Oyama-shi) who collected the type specimens.

　　タイプ産地　中部日本，本州東部，福島県，東白川郡，棚倉町，八溝山，標高約1,000 m.
　　タイプ標本の所在　国立科学博物館［ホロタイプ，♂］.
　　亜種の特徴　20~23 mm. 阿武隈高地亜種に近いが，1) 前胸背板はより細長く，幅が長さの1.1倍以下 (阿武隈高地亜種では1.1倍以上) で，側縁は最広部の膨らみが弱く，後角前方で側方により強く拡がり，後角はより強く外側後方へ突出し，その先端は鋭く尖る；2) 鞘翅第一次原線はやや細く，阿武隈高地亜種の場合ほど頻繁に分断されない．第二次，第三次原線は共に連続性の高い隆条で，殆ど分断されず，第三次原線が更に2列の隆条に分かれて5元的な彫刻を呈する部分は殆ど見られず，間室間には部分的に弱い顆粒列が認められるにすぎない．♀鞘翅端は東北地方南西部亜種，阿武隈高地亜種の場合よりもさらに強くえぐれる；3) 陰茎中央部は円筒状で細長く，中央腹面は殆ど膨隆せず，先端部は背面から見て幅が狭い．
　　分布　八溝山地 (八溝山々頂付近). 久保田 (1993) により初めて記録されたもので，井村・水沢 (2002a) により新亜種として記載された．八溝山の山頂付近に残存するブナ林のみから記録されており，個体数は極めて少ない．
　　亜種小名の由来と通称　亜種小名はタイプ標本を採集した小野田晃治 (小山市) に因む．原記載論文中で提唱された通称はヤミゾホソヒメクロオサムシ．

6-5. *C. (P.) h. harmandi* LAPOUGE, 1909 ホソヒメクロオサムシ基亜種

Original description, type locality & type depository　See explanation of the species.
Diagnosis　17~22 mm. Distinguished from other subspecies by coarser punctuation of pronotum, deeper preapical emargination of elytra, characteristically featured penis, etc.
Distribution　Echigo Mts., Taishaku Mts., Mikuni Mts. & mountains in N. Kanto Region, C. Honshu.

　　原記載・タイプ産地・タイプ標本の所在　種の項を参照．
　　亜種の特徴　17~22 mm. 頭部背面の皺は強い．前胸背板は後方に向け強く狭まり，側縁は多少とも波曲して心臓形をなし，後角は細長く，後方に強く突出し，表面は密に点刻される．鞘翅第二次隆条は強く隆起し，結節状になる．第一次間室間には5列の不規則な顆粒列を認め，これらは互いに連なりあい，不規則に断続する細い隆条を形成するが，時に3列の鎖線に加え，しばしば痕跡的な顆粒列からなる第四次間室を認める．鞘翅端のえぐれは雌雄とも深く，その前縁はしばしば角張る．陰茎は比較的細長く，右側縁中央は弱いながら顕著に突出し，同所から先端に向けて強く内湾する．先端部はやや細長い．比較的広い範囲に分布し，鞘翅彫刻の変化が大きいが，鞘翅端のえぐれが深いこと，前胸背板表面の点刻が密なこと，および特徴的な陰茎の形態などにより他亜種から識別される．
　　分布　阿賀野川以南の越後山脈，帝釈山地，三国山地および関東地方北部山地の高所．

6-6. *C. (P.) h. quinquecatellatus* (ISHIKAWA, 1986) ホソヒメクロオサムシ飛騨山脈北部亜種

Original description　*Leptocarabus* (*Pentacarabus*) *harmandi quinquecatellatus* ISHIKAWA, 1986, Trans. Shikoku Ent. Soc., 17 (4), p. 224, figs. 3, 4 (p. 223).
Type locality　Route to Renge-onsen, 1300 m, on the northern slope of Mt. Shiroumadake, Niigata Pref.
Type depository　Systematic Zoology Laboratory, Department of Biological Sciences, Tokyo Metropolitan University (Hachiôji) [Holotype, ♂].
Diagnosis　19~23 mm. Allied to the nominotypical *harmandi* or subsp. *adatarasanus*, but is unique in having five, nearly equally catenulated ridges between primaries almost constantly. Apical part of penis short and abruptly narrowed near tip.
Distribution　N. Hida Mts. (southern edge of distribution almost coincides with the left bank of Riv. Kago-gawa of Ômachi-shi), Myôkô Mts. & Higashi-kubiki Hills, C. Honshu.

Etymology Named after characteristically carved elytral sculpture, which is indicated by five rows of chain-like striae between primary intervals.

タイプ産地　新潟県, 白馬岳の北麓, 蓮華温泉への道, 1300 m.
タイプ標本の所在　首都大学東京動物系統分類学研究室(八王子)［ホロタイプ, ♂］.
亜種の特徴　19~23 mm. 前頭部から頚部にかけての皺は強く密に刻まれる. 前胸背板側縁の波曲は弱く, 後角の後方への突出は顕著ながら必ずしも強くはない. 前胸背板表面は比較的規則的に点刻される. 鞘翅第一次間室は細く, 隆起は弱い. 第一次原線間にはほぼ常に5条の均等な鎖状の隆条を具える. 陰茎は基亜種や東北地方南西部亜種のそれに似るが, やや太短く, 先端に向けて急激に細くなるため, 先端部左側縁は通常, 顕著に張り出して見える. 基亜種ならびに東北地方南西部亜種に近いが, 鞘翅第一次間室間にはほぼ常に5条の均等な隆条を具える点, および陰茎先端部が短く, 先端に向けて急激に狭まる点により識別される.
分布　飛騨山脈北部（南限は大町市の篭川（かごがわ）左岸), 妙高連峰, 東頸城丘陵.
亜種小名の由来と通称　亜種小名は「5本の鎖線を持つ」を意味し, 本亜種が鞘翅第一次原線間に5条の隆条を具えることに因む. ゴスジホソヒメクロオサムシ, ゴスジアルマンオサムシ等の通称がある.

6-7. *C. (P.) h. mizunumai* Ishikawa, 1972　ホソヒメクロオサムシ白山飛騨御嶽木曽亜種

Original description　*Carabus* (*Pentacarabus*) *harmandi mizunumai* Ishikawa, 1972, Bull. natn. Sci. Mus., Tokyo, 15 (1), p. 25.
Type locality　Mt. Hakusan, at 2100 m. alt.
Type depository　National Museum of Nature and Science, Tokyo [Holotype, ♀].
Diagnosis　16~22 mm. Distinguished from other subspecies by characteristically shaped penis which is slender with very short apex.
Distribution　Ryôhaku Mts., Hida Highland, S. Hida Mts., Mt. Ontaké-san & Kiso Mts., C. Honshu.
Etymology　Named after Tetsuo Mizunuma (Osaka).

タイプ産地　白山, 標高 2100 m.
タイプ標本の所在　国立科学博物館［ホロタイプ, ♀］.
亜種の特徴　16~22 mm. 頭部背面の皺は弱いながらも密. 前胸背板は比較的幅広く, 側縁はあまり波曲せず, ほぼ直線的に後方に向けて狭まり, 後角の突出は弱い. 第一次原線の隆起は強く, やや深い第一次凹陥によって断続し, 亜結節状を呈する. 第一次間室間には通常3列の断続する不規則な稜ないし顆粒列が認められ, 第二次間室はしばしば不規則. 第四次間室は通常痕跡的な顆粒列. 鞘翅端のえぐれは浅い. 陰茎は細く, 右側腹面の突出は弱く, 先端は極めて短い. 陰茎が細く, 先端が短いことにより他亜種から識別される. 両白山地に産するものは陰茎先端の丸みが強く, より短いものが多いが, 飛騨高地や御嶽山など分布域の東方に産するものでは先端がやや長く, 三角形に尖るものが多い. 岐阜県揖斐川町（いびがわちょう）の冠山峠（かんむりやまとうげ), 飛騨市の楢峠（ならとうげ), 白木峰（しらきみね）などのものはより高所に産するものに比べ, やや大型である.
分布　両白山地, 飛騨高地, 飛騨山脈南部（大町市の篭川（かごがわ）右岸以南), 御嶽山, 木曽山地. 山塊ごとに軽微な差異が認められ, 形態的に均質な集団であるとは必ずしも言えない.
亜種小名の由来と通称　亜種小名は大阪の昆虫標本業者, 水沼哲郎に因む. ハクサンホソヒメクロオサムシ, ニシホソヒメクロオサムシ, ニシアルマンオサムシ等の通称がある.

6-8. *C. (P.) h. karasawai* Ishikawa, 1966　ホソヒメクロオサムシ八ヶ岳亜種

Original description　*Carabus* (*Asthenocarabus*) *harmandi karasawai* Ishikawa, 1966, Bull. natn. Sci. Mus., Tokyo, 9 (4), p. 454, figs. 8 (p. 454) & 5, 6 (pl. 1).
Type locality　Shibu Spa (more precisely, Shibu-no-yu Onsen (= Spa)), on Mt. Yatsugadake, Nagano. Pref.
Type depository　National Museum of Nature and Science, Tokyo [Holotype, ♂].
Diagnosis　17~20 mm. Allied to subsp. *okutamaensis*, but differs from it in having differently sculptured elytra: primaries strongly elevated, broken into short segments and usually subtuberculate; interprimary areas coarse and uneven, usually with three rows of irregular granules.
Distribution　Mt. Yatsu-ga-také, C. Honshu.
Etymology　Named after Yasumi Karasawa of Tokyo.

タイプ産地　長野県, 八ヶ岳, 渋（の湯）温泉.
タイプ標本の所在　国立科学博物館［ホロタイプ, ♂］.
亜種の特徴　17~20 mm. 奥秩父亜種に近いが, 主として鞘翅彫刻の違いにより識別される. 頭部背面の皺は強く密. 前胸背板は四角形に近く, やや横長. 側縁は後方に向けて強く狭まり, 後角の幅は比較的狭い. 鞘翅第一次原線は強く隆起し, 短く破

断され，結節状となる．第一次間室間の鞘翅基面は粗く，通常3列の不規則な顆粒列を装う．鞘翅端のえぐれは浅いながら両性とも顕著に認められる．陰茎は奥秩父亜種のそれに近い．

分布 八ヶ岳．タイプ産地の渋の湯温泉のほか，蓼科山（たてしなやま），大河原峠（おおがわらとうげ），地獄谷などから記録がある．

亜種小名の由来と通称 亜種小名は東京のアマチュア甲虫研究家，唐沢安美に因む．ヤツホソヒメクロオサムシ，ヤツアルマンオサムシ等の通称がある．

6-9. *C. (P.) h. okutamaensis* ISHIKAWA, 1966 ホソヒメクロオサムシ奥秩父亜種

Original description *Carabus* (*Asthenocarabus*) *harmandi okutamaensis* ISHIKAWA, 1966, Bull. natn. Sci. Mus., Tokyo, 9 (4), p. 453, figs. 7 (p. 454) & 3, 4 (pl. 1).
Type locality Mt. Daibosatsu, Yamanashi Pref.
Type depository National Museum of Nature and Science, Tokyo [Holotype, ♂].
Diagnosis 18~23 mm. Distinguished from other subspecies mainly by having less strongly uneven interprimary areas of elytra.
Distribution Chichibu Mts., Kantô Mts. & Misaka Mts., C. Honshu.
Etymology Named after Oku-tama, a regional name in which this subspecies is distributed.

タイプ産地 山梨県，大菩薩．
タイプ標本の所在 国立科学博物館［ホロタイプ，♂］．
亜種の特徴 18~23 mm. 鞘翅彫刻の特徴により，他亜種から比較的容易に識別される．頭部の皺は強く密．前胸背板は後方に向けてやや狭まり，側縁の波曲は弱く，後角は細く，強く後方に突出する．鞘翅第一次原線は細いが強く隆起し，やや結節状．第一次間室間の鞘翅基面は凹凸が比較的少なく，5列の原線は主として小顆粒列により構成される．陰茎は富士山亜種のそれに似るが，右側縁中央部の隆起はあまり強くなく，先端部はより細長い．
分布 秩父山地〜関東山地〜御坂山地．タイプ産地は大菩薩で，関東山地から秩父山地，御坂山地にかけて分布する．南限は三つ峠山（みつとうげやま）．
亜種小名の由来と通称 亜種小名は奥多摩に因む．オクタマホソヒメクロオサムシ，オクタマアルマンオサムシ等の通称で呼ばれることもある．

6-10. *C. (P.) h. tanzawaensis* (ISHIKAWA, 1986) ホソヒメクロオサムシ丹沢亜種

Original description *Leptocarabus* (*Pentacarabus*) *harmandi tanzawaensis* ISHIKAWA, 1986, Trans. Shikoku Ent. Soc., 17 (4), p. 234, fig. 21 (p. 234).
Type locality Mt. Hinokiboramaru, 1300 m.
Type depository Systematic Zoology Laboratory, Department of Biological Sciences, Tokyo Metropolitan University (Hachiôji) [Holotype, ♂].
Diagnosis 20~23 mm. Allied to subsp. *fujisan* or subsp. *okutamaensis*, but distinguished from them by a little longer elytra, very irregularly sculptured areas between third primaries and umbilicate series, shallow but recognizable preapical emargination of elytra in both sexes, longer and slenderer penis, etc.
Distribution Tanzawa Mts., SC. Honshu.
Etymology Named after the Tanzawa Mountains, distributional range of this subspecies.

タイプ産地 檜洞丸，1300 m．
タイプ標本の所在 首都大学東京動物系統分類学研究室（八王子）［ホロタイプ，♂］．
亜種の特徴 20~23 mm. 頭部背面の皺は強く密．前胸背板はやや縦長で，後方への狭まりはごく僅かで，側縁の波曲もごく弱い．後角の突出は顕著．鞘翅は他亜種に比しより長く，第一次原線は強く隆起し，稀ならず結節状．第一次原線間には通常3列の顆粒列があり，第三次間室と丘孔点列の間にある顆粒は極めて不規則．鞘翅端のえぐれはごく浅いが，両性共に認められる．陰茎は奥秩父亜種のそれに似るが，全体により細長く，膜状開口部の左側縁は強く膨らみ，先端部は幅広く，長い．
分布 丹沢山地．檜洞丸，大室山（おおむろやま／おおむろざん），丹沢山堂平（どうだいら）など，標高1,000 m以上の地域に産する．
亜種小名の由来と通称 亜種小名は丹沢に因む．タンザワホソヒメクロオサムシ，タンザワアルマンオサムシ等の通称がある．

6-11. *C. (P.) h. fujisan* ISHIKAWA, 1972 ホソヒメクロオサムシ富士山亜種

Original description *Carabus* (*Pentacarabus*) *harmandi fujisan* ISHIKAWA, 1972, Bull. natn. Sci. Mus., Tokyo, 15 (1), p. 26.
Type locality Mt. Fujisan 1500 m. on the southern slope.
Type depository National Museum of Nature and Science, Tokyo [Holotype, ♂].
Diagnosis 18~21 mm. This subspecies is peculiar in having strongly cordate pronotum, very irregularly contiguous granules in

interprimary areas, almost unrecognizable preapical emargination of elytra, strongly carved sternal sulci and characteristically shaped penis.

Distribution　Mt. Fuji-san, C. Honshu.

Etymology　Named after its type locality, Mt. Fuji-san.

タイプ産地　富士山南麓, 1500 m.

タイプ標本の所在　国立科学博物館［ホロタイプ, ♀］.

亜種の特徴　18~21 mm. 頭部背面の皺は密で強い. 前胸背板はやや幅が狭く, 後方へ向けて顕著に狭まり, 側縁は顕著に波曲し, 後方で顕著にえぐれるため, 後角はやや外側方に突出して見える. 鞘翅第一次原線は顕著な稜線をなし, 第一次凹陥は大きく深い. 第一間室間には極めて不規則に連続する顆粒列が並び, 間室間の基面は凹凸が著しい. 鞘翅端のえぐれは♀においても殆ど認められないか, あるいは欠く. 腹部腹板の横溝は末端の3節において顕著に刻まれる. 陰茎は太短く, 基部腹面にある結節状の隆起は顕著で, 右側縁中央部は広範囲に亘り強く膨隆し, そこから先端に向けて急激に深く陥入する. 膜状開口部の左側縁は弧を描いて強く膨隆する. 先端部は幅広く, 丸まって終わる. 特徴的な前胸背板の形態と鞘翅第一次間室間の彫刻, 鞘翅端のえぐれが殆ど認められないこと, 腹部横溝が明瞭に刻まれること, および特徴的な陰茎の形態等により, 他亜種からの識別は容易である.

分布　富士山. 富士山中腹の標高およそ1,300~2,000 mの範囲に分布する. タイプ産地は同山南麓の1,500 m地点.

亜種小名の由来と通称　亜種小名は富士山に因む. フジホソヒメクロオサムシ, フジアルマンオサムシ等の通称がある.

6-12. *C. (P.) h. akaishiensis* (ISHIKAWA, 1986) ホソヒメクロオサムシ赤石山脈亜種

Original description　*Leptocarabus* (*Pentacarabus*) *harmandi akaishiensis* ISHIKAWA, 1986, Trans. Shikoku Ent. Soc., 17 (4), p. 235, fig. 19 (p. 233).

Type locality　Nikengoya, ca 1400 m, Shizuoka Pref.

Type depository　Systematic Zoology Laboratory, Department of Biological Sciences, Tokyo Metropolitan University (Hachiôji) [Holotype, ♂].

Diagnosis　17~21 mm. Allied to subsp. *fujisan*, but differs from it as follows: pronotum more plainly narrowed posteriad, with lateral sides less remarkably sinuous, hind angles shorter and not directed postero-laterad, discal punctures finer; elytra with primary foveae shallower, areas between primaries not so uneven with granulated intervals not so irregularly contiguous, granules of umbilicate series more numerous, preapical emargination shallowly but distinctly concave; penis with right lateral margin more gently convex, that in left side back of the membraneous area more remarkably swollen, apex longer and narrower.

Distribution　Akaishi Mts. excepting the southeastern part (Pass Abé-tôgé and its nearby regions), C. Honshu.

Etymology　Named after the Akaishi Mountains, the distributional range of this subspecies.

タイプ産地　静岡県, 二軒小屋, 約1400 m.

タイプ標本の所在　首都大学東京動物系統分類学研究室(八王子)［ホロタイプ, ♂］.

亜種の特徴　17~21 mm. 頭部背面の皺は強く, 前胸背板側縁は波曲しないか, してもごく軽度. 前胸背板後角の後方への突出は弱い. 鞘翅第一次原線は細く, 第一間室間の顆粒は部分的に不規則な列をなすが, 富士山亜種の場合ほど連続性は高くない. 鞘翅端のえぐれは浅いが, 両性とも明瞭に認められる. 陰茎は比較的太短いが, 右側縁中央部はそれほど強く膨隆しない. 膜状開口部の左側縁は強く膨らむ. 陰茎先端は非常に細長い.

分布　赤石山脈(南東部の安倍峠付近を除く). タイプ産地は同山脈中央部の二軒小屋で, 北限は山梨県南アルプス市西部にある野呂川(のろがわ)上流の広河原(ひろがわら)付近と思われ, ほかに櫛形山(くしがたやま), 伝付峠(でんつくとうげ), 荒川小屋などから記録がある. これまでに確認されている南限は静岡県静岡市の山伏(やんぶし)(山伏岳), 井川峠, 蕎麦粒山(そばつぶやま)から水窪川(みさくぼがわ)源流にかけての高所(井村・平井, 2003)だが, これらの地域に産するものは二軒小屋のものともやや異なっており, 井村と松永により現在, 詳しい調査と形態学的検討が行われている. 山脈の南東端に位置する安倍峠のものは形態学的に本亜種と異なっているため, 別亜種に分類されている(次亜種の項参照).

亜種小名の由来と通称　亜種小名は赤石山脈に因む. アカイシホソヒメクロオサムシ, アカイシアルマンオサムシ等の通称がある.

6-13. *C. (P.) h. moritai* (IMURA et MIZUSAWA, 2002) ホソヒメクロオサムシ安倍峠亜種

Original description　*Pentacarabus harmandi moritai* IMURA & MIZUSAWA, 2002, Gekkan-Mushi, Tokyo, (380), p. 6, figs. 3 (p. 3) & 6 (p. 6).

Type locality　Pass Abe-tôgé (ca. 1,300 m in alt., on the border between Minobu Town in Yamanashi Pref. and Shizuoka City in Shizuoka Pref.) at the southern part of the Akaishi Mountains, south-central Honshu, central Japan.

Type depository　National Museum of Nature and Science, Tokyo [Holotype, ♂].

6. *Carabus* (*Pentacarabus*) *harmandi*

Diagnosis　20~24 mm. Most closely allied to subsp. *akaishiensis*, but differs from it in having slenderer pronotum and elytra, sharper and more strongly protruded pronotal hind angles, weaker elytral sculptures and differently shaped penis.
Distribution　Pass Abé-tôgé, SE. Akaishi Mts., C. Honshu.
Etymology　Name after Seiji Morita (Tokyo), who collected the type series of this subspecies.

　タイプ産地　中部日本, 本州中南部, 赤石山脈南部, 山梨県身延町と静岡県静岡市の境界上, 安倍峠, 標高約1,300 m.
　タイプ標本の所在　国立科学博物館［ホロタイプ, ♂］.
　亜種の特徴　20~24 mm. 赤石山脈亜種に近いが, 前胸背板はより細長く, 前角はより鋭角で, 後角はより鋭く尖り, 細長く後方に突出する. 鞘翅は非常に細長く, 側縁は直線的で膨らみが弱く, 各原線の膨隆が弱いため, 表面の彫刻は全体として滑らかに見える. 陰茎は中央腹面の隆起がやや弱く, 先端部はより細長く伸長し, より鋭く尖って終わり, 背面から見て左側面がより明瞭に角張る.
　分布　南アルプス南東部, 安倍峠付近.
亜種小名の由来と通称　亜種小名は, タイプシリーズの採集者である, 歯科医でゴミムシ研究家の森田誠司（東京）に因む. 原記載論文中で与えられた通称はミナミホソヒメクロオサムシ.

6. *Carabus* (*Pentacarabus*) *harmandi*

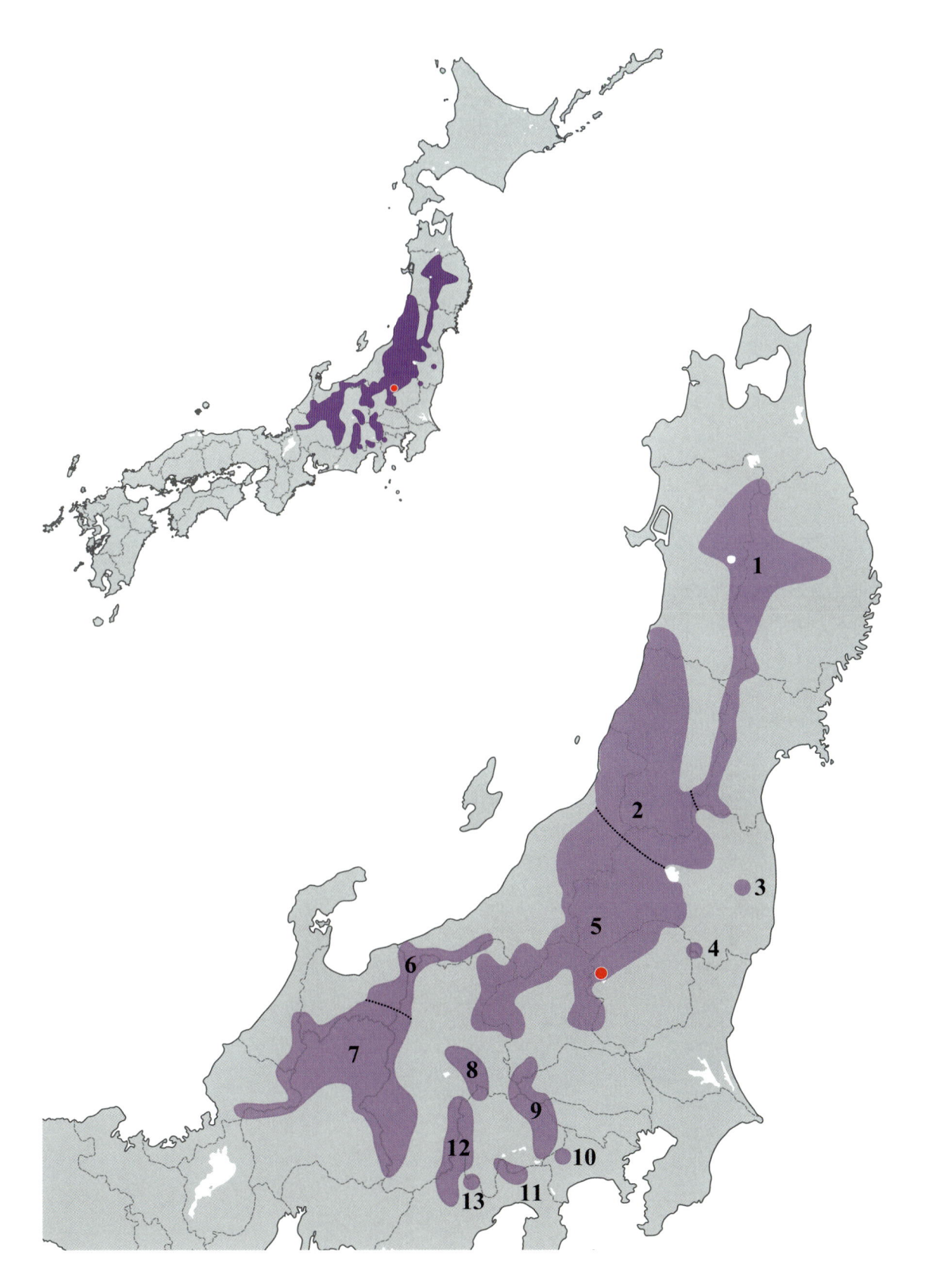

Fig. 27. Map showing the distributional range of ***Carabus*** (***Pentacarabus***) ***harmandi*** Lapouge, 1909. — 1, Subsp. ***yudanus***; 2, subsp. ***adatarasanus***; 3, subsp. ***syui***; 4, subsp. ***kojii***; 5, subsp. ***harmandi***; 6, subsp. ***quinquecatellatus***; 7, subsp. ***mizunumai***; 8, subsp. ***karasawai***; 9, subsp. ***okutamaensis***; 10, subsp. ***tanzawaensis***; 11, subsp. ***fujisan***; 12, subsp. ***akaishiensis***; 13, subsp. ***moritai***. ●: Type locality of the species (Nikko).

図 27. ホソヒメクロオサムシ分布図. — 1, 奥羽山脈亜種; 2, 東北地方南西部亜種; 3, 阿武隈高地亜種; 4, 八溝山亜種; 5, 基亜種; 6, 飛騨山脈北部亜種; 7, 白山飛騨御嶽木曽亜種; 8, 八ヶ岳亜種; 9, 奥秩父亜種; 10, 丹沢亜種; 11, 富士山亜種; 12, 赤石山脈亜種; 13, 女怡峠亜種. ●: 種のタイプ産地（日光）.

7. *Carabus* (*Aulonocarabus*) *kurilensis* Lapouge, 1913
チシマオサムシ

Figs. 1-14. *Carabus* (*Aulonocarabus*) *kurilensis rishiriensis* Nakane, 1957 (Is. Rishiri-tô, Hokkaido). — 1, ♂; 2, ♀; 3, male head & pronotum; 4, male right protarsus in ventral view; 5, male right protibia in subdorsal view; 6, male left elytron (median part); 7, penis & fully everted internal sac in right lateral view; 8, apical part of penis in right lateral view; 9, ditto in right subdorsal view; 10, ditto in dorsal view; 11, internal sac in caudal view; 12, ditto in frontal view; 13, ditto in dorsal view; 14, ditto in left sublateral view.

図 1-14. チシマオサムシ利尻島亜種(北海道利尻島). — 1, ♂; 2, ♀; 3, ♂の頭部と前胸背板; 4, ♂右前附節(腹面観); 5, ♂右前脛節(斜背面観); 6, ♂左鞘翅中央部; 7, 陰茎と完全反転下の内囊(右側面観); 8, 陰茎先端(右側面観); 9, 同(右斜背面観); 10, 同(背面観); 11, 内囊(尾側面観); 12, 同(頭側面観); 13, 同(背面観); 14, 同(左斜側面観).

7. *Carabus* (*Aulonocarabus*) *kurilensis*

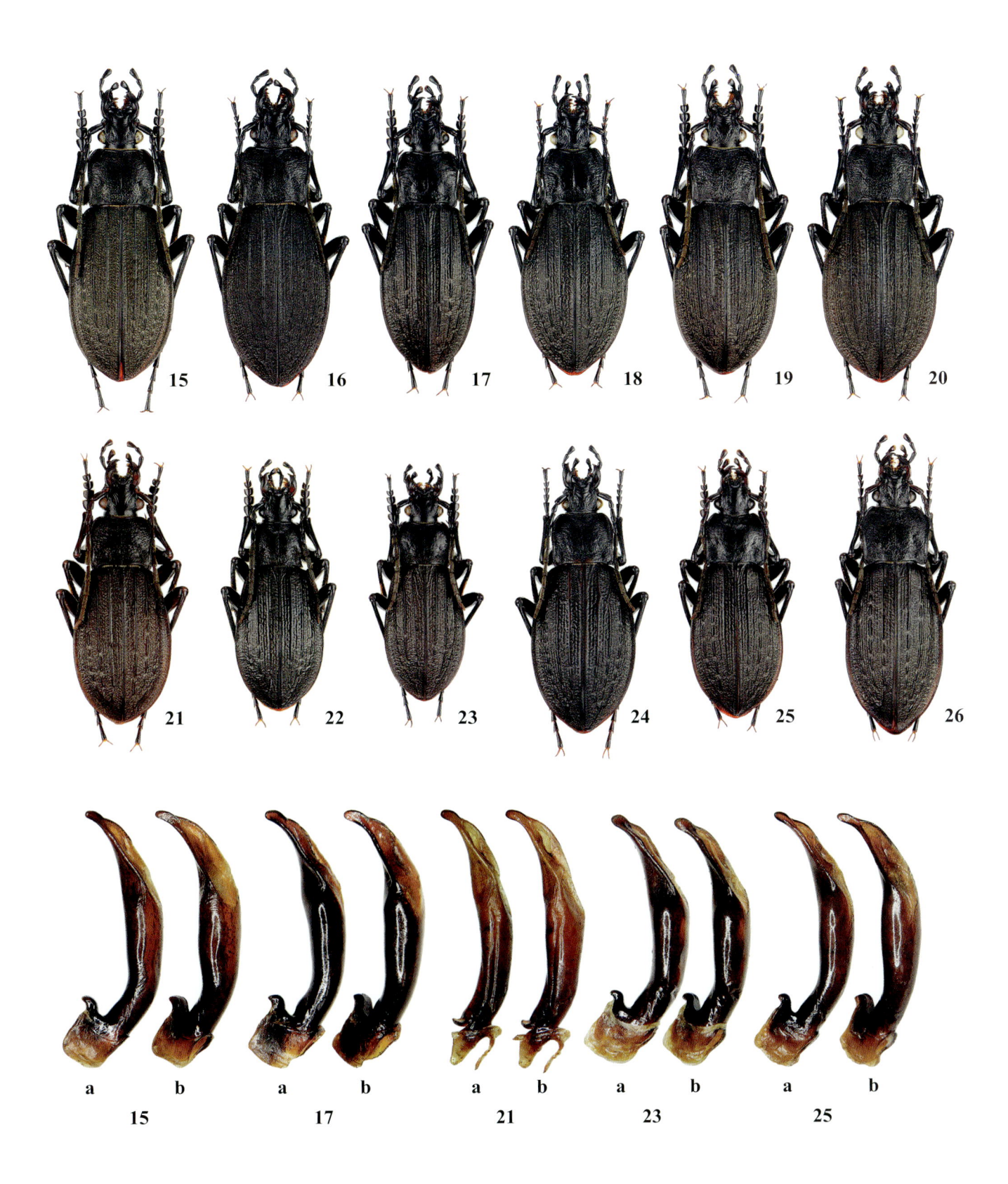

Figs. 15-26. *Carabus* (*Aulonocarabus*) *kurilensis* subspp. — 15-16, *C.* (*A.*) *k. sugai* (15, ♂, 16, ♀, Is. Rebun-tô); 17-20, *C.* (*A.*) *k. rausuanus* (17, 19, ♂, 18, 20, ♀; 17-18, Mt. Rausu-daké, Rausu-chô; 19-20, Maruseppu-murii, Engaru-chô); 21, *C.* (*A.*) *k. kurilensis* (♂, Is. Etorofu-tô (= Iturup)); 22-24, *C.* (*A.*) *k. daisetsuzanus* (22, ♂, Mt. Chûbetsu-daké, Daisetsu-zan Mts.; 23, ♂, 24, ♀, Mt. Kuro-daké, Daisetsu-zan Mts.); 25-26, *C.* (*A.*) *k. yoteizanus* (25, ♂, 26, ♀, Mt. Yôtei-zan, Shiribeshi). — a, Penis in right lateral view; b, ditto in right subdorsal view. All from Hokkaido.

図 15-26. チシマオサムシ各亜種. — 15 16, 礼文島亜種 (15, ♂, 16, ♀, 礼文島); 17-20, 道央道東亜種 (17, 19, ♂, 18, 20, ♀; 17-18, 羅臼町羅臼岳; 19-20, 遠軽町丸瀬布武利); 21, 基亜種 (♂, 択捉島); 22-24, 大雪山亜種 (22, ♂, 大雪山忠別岳; 23, ♂, 24, ♀, 大雪山黒岳); 25-26, 羊蹄山亜種 (25, ♂, 26, ♀, 後志地方, 羊蹄山). — a, 陰茎(右側面観); b, 同(右斜背面観). すべて北海道産.

7. *Carabus* (*Aulonocarabus*) *kurilensis* LAPOUGE, 1913 チシマオサムシ

Original description *Carabus canaliculatus Kurilensis* LAPOUGE, 1913, Misc. Ent., 21 (1), p.15.

History of research This taxon was originally described by VACHER DE LAPOUGE (1913) as a subspecies of *Carabus canaliculatus*. BREUNING (1932) raised its rank to a full species, and it is currently regarded as an independent species *C. kurilensis* by most authors. KÔNO (1936) described *C. k. daisetsuzanus* from the Daisetsu-zan Mountains of central Hokkaido, and placed it in the subgenus *Eucarabus* following the BREUNING's system. NAKANE (1957) described *C.* (*E.*) *kurilensis rishiriensis* from Is. Rishiri-tô. ISHIKAWA (1966b) placed this species in the subgenus *Aulonocarabus*, and described two new subspecies *sugai* from Is. Rebun-tô of northernmost Hokkaido and *rausuanus* from Mt. Rausu-daké of eastern Hokkaido. ISHIKAWA & MIYASHITA (2000) revised this species (*Leptocarabus* (*Aulonocarabus*) in their sense), and classified it into seven subspecies including two Sakhalin races, *diamesus* (described by SEMENOV & ZNOJKO, 1932) and *pseudodiamesus* (IVANOVS, 1994, *ibid*.). A new subspecies *yoteizanus* was described from Mt. Yôtei-zan of southwestern Hokkaido, and subsp. *sugai* was synonymized with subsp. *rishiriensis* in the same paper.

Type locality Iles Urup et Iturup, montagnes de l'intérieur.

Type depository Zoölogisch Museum Amsterdam [Lectotype, ♂].

Morphology 17~28 mm. Body above black, often with weak bronze tinge and not so shiny. External features: Figs. 1~6. Description on details: penultimate segment of labial palpus bisetose; median tooth of mentum as long as lateral lobes, sharply pointed at tip and protruded ventrad; submentum bisetose; male antennae with hairless ventral depressions from segment five to eight, sometimes to nine; pronotum with three to four marginal setae, two to three in median portion and one near hind angle; preapical emargination of elytra hardly recognized in both sexes; elytral sculpture triploid heterodyname; primaries the strongest, indicated by contiguous costae though often segmented by small primary foveoles in posterior halves of inner two costae and segmented throughout in outermost costa; secondaries and tertiaries much weaker, indicated by rows of granules; areas between intervals coarsely scattered with small granules; propleura smooth; sides of sternites irregularly and weakly rugulose; metacoxa trisetose; sternal sulci not recognized or at most rudimentary. Male genitalic features: Figs. 7~14. Penis a little shorter than half the elytral length; ostium lobe small and unilobed; internal sac simple, no marked lobes excepting parasaccellar lobes are developed.

Molecular phylogeny On a molecular phylogenetic tree of the subtribe Carabina of the world constructed from the mitochondrial ND5 gene sequences, this taxon belongs to the same lineage as that constructed by so-called *Leptocarabus* subgenus-group distributed in Japan, Korea, northeast China and Russian Maritime Provinces (*Leptocarabus*, *Adelocarabus*, *Aulonocarabus*, *Weolseocarabus*, *Sinoleptocarabus* etc.). In the same tree, this taxon is intermingled with several races of *C.* (*Aulonocarabus*) *canaliculatus* from eastern Eurasia, and cannot be discriminated as an independent lineage (KIM, ZHOU *et al*., 2000).

Distribution Hokkaido / Iss. Rebun-tô, Rishiri-tô, Kunashiri-tô & Etorofu-tô (Hokkaido). Outside Japan: Iss. Urup (C.~S. Kurils) & Sakhalin.

Habitat Distribution of this species in Hokkaido is very restricted and sporadical, and the habitat is different according to the subspecies or population. Habitats of subspp. *daisetsuzanus* and *yoteizanus* are narrowly restricted to the alpine region. The former inhabits exclusively the barren summit area about 2,000 m alt. of the Daisetsu-zan Mountains, and the latter inhabits the alpine crater at the summit of Mt. Yôtei-zan above 1,700 m alt. Although subsp. *rausuanus* has wider ranges than the others, it is found only at the small permafrost areas that remain sporadically in the mountain forests with a few exceptional records from the alpine region. Thus, all the three subspecies of Hokkaido proper are confined to the alpine region or the permafrost areas, showing that they are relics of the past cold time. On the other hand, both subsp. *sugai* and subsp. *rishiriensis* widely inhabit from the forest of lower altitude up to the alpine region, and are predominant in greater part of their distributional ranges.

Bionomical notes Autumn breeder. Larvae most probably feed on larvae of other insects.

Geographical variation Classified into six subspecies as to the population distributed in Hokkaido, Northern Japan.

Etymology Named after the Kuril Islands to which both the islands designated as type localities belong.

研究史　本タクソンは，フランスのVACHER DE LAPOUGE（ヴァッシェー・ドゥ・ラプージュ）（1913）によってCarabus canaliculatus（セスジクロナガオサムシ）の亜種として記載されたもので，タイプ産地は千島列島南部のウルップ，エトロフ両島である．BREUNING（ブロイニング）（1932）により種に昇格されて以来，一般に独立種として扱われることが多い．河野（1936）は大雪山からC. (Eucarabus) kurilensis daisetsuzanusを，NAKANE（1957）は利尻島からC. (E.) k. rishiriensisを記載した（それぞれ，本書におけるチシマオサムシ大雪山亜種と同利尻島亜種）．ISHIKAWA（1966b）は本種をアウロノカラブス亜属に移し，C. (Aulonocarabus) k. sugai（礼文島）とC. (A.) k. rausuanus（羅臼岳）の2亜種を記載した．ISHIKAWA & MIYASHITA（2000）は本種（彼らの扱いはLeptocarabus (Aulonocarabus) kurilensis）の再検討を行い，羊蹄山から亜種yoteizanusを記載し，礼文島の亜種sugaiを利尻島亜種のシノニムとみなし，サハリンの2亜種（L. (A.) k. diamesus (SEMENOV et ZNOJKO, 1932)とL. (A.) k. pseudodiamesus (IVANOVS, 1994)）を加えて本種を7亜種に分類した．

タイプ産地　ウルップ島とエトロフ島，内陸部の山地．

タイプ標本の所在　アムステルダム動物学博物館［レクトタイプ，♂］．

形態　17~28 mm．背面は一様に黒色で，しばしば銅褐色を帯び，光沢は鈍い．外部形態：図1~6．細部の記載：下唇肢亜端

節の剛毛は2本．下唇基節の中央歯は側葉とほぼ同長で，前方に長く伸び，先端は鋭く尖り，腹側に突出する．下唇亜基節には2本の剛毛を有する．♂触角第5節から第8節，時に第9節の一部に顕著な無毛凹陥部がある．前胸背板の側縁剛毛は中央付近に2~3本，後角付近に1本．鞘翅端部のえぐれは♂♀共に殆ど認められない．鞘翅彫刻は三元異規的．第一次間室は強い隆条で，各鞘翅の内側の2条は前半部分が連続し，後半部分が断続するが，最外側の1条は全長を通して断続する．第一次凹陥は小さく浅い．第二次，第三次間室は共に不規則な顆粒列に減退し，間室間の基面には小顆粒を密に装う．前胸側板は滑らかで，腹部腹板の側面には不規則な弱い皺がある．後基節の剛毛は3本．腹部腹板の横溝はないか，あっても痕跡的．♂交尾器形態：図7~14．陰茎長は鞘翅長の1/2弱．葉片は小さく単葉．内嚢は単純で，傍盤葉のみが目立ち，他の突起は痕跡的．

分子系統　ミトコンドリア ND5 遺伝子による分子系統樹の上で，本種は日本や朝鮮半島，極東地域等に分布する広義のクロナガオサムシ類（*Leptocarabus* 亜属，*Adelocarabus* 亜属，*Aulonocarabus* 亜属，*Weolseocarabus* 亜属，*Sinoleptocarabus* 亜属等）によって構成される系統の一部に属するが，系統樹内では大陸の *C. (Aulonocarabus) canaliculatus*（セスジクロナガオサムシ）のクラスター中に埋没してしまい，独立した系統として区別することはできない（KIM, ZHOU et al., 2000）．交尾器を含む形態学的特徴においても，セスジクロナガオサムシと本種の間には共通点が多い．

分布　北海道／礼文島，利尻島，国後島，択捉島（以上，北海道）．国外：ウルップ島（千島列島中南部）（これ以外の島々にも生息している可能性はあるが，いわゆる北方領土を含む千島列島における分布状況についてはよくわかっていない），サハリン．

生息環境　北海道本島における分布は極めて局地的，散発的で，生息環境も地域により異なる．大雪山亜種と羊蹄山亜種は山頂付近の風衝地ないしそれに準じた環境に生息している．一方，道央道東亜種は中腹以下の森林帯にも生息しているが，その分布は永久凍土と密接に関係しており，分布は局地的である．これに対し，利尻・礼文両島では平地の林縁・森林内から高山帯まで幅広く生息し，個体数も多く，現地における優占種となっている．

生態　秋繁殖・幼虫越冬型のオサムシで，幼虫の基本食性は節足動物食（他種の昆虫やその幼虫など）であろうと思われる．道東亜種では，越冬成虫は6月下旬から7月下旬にかけて出現し，繁殖の後，8月から9月上旬に新成虫が出現するという．冬期には成虫以外に幼虫でも越冬し，翌年羽化した成虫は繁殖せずに越冬して次の年に繁殖活動に加わる，2年1化の生活環を持つのではないかと考えられている．

地理的変異　本邦（北方領土を含む）に産するものは6亜種に分類される．海外からは，サハリン中部の *C. (A.) k. diamesus* SEMENOV et ZNOJKO, 1932 と，同島南部の *C. (A.) k. pseudodiamesus* IVANOVS, 1994 の2亜種が知られている．

種小名の由来等　種小名は，タイプ産地の2島が属する千島列島の英名に因む．大雪山亜種と同じダイセツオサムシという種和名で呼ばれる場合も多い．

7-1. *C. (A.) k. sugai* ISHIKAWA, 1966 チシマオサムシ礼文島亜種

Original description　*Carabus (Aulonocarabus) kurilensis sugai* ISHIKAWA, 1966, Bull. natn. Sci. Mus., Tokyo, 9 (4), p. 458, figs. 9, 13 (p. 456) & 1, 2 (pl. 4).
Type locality　Mt. Rebundake, Is. Rebun.
Type depository　National Museum of Nature and Science, Tokyo [Holotype, ♂].
Diagnosis　21~27 mm. Closely allied to subsp. *rishiriensis*, and barely distinguishable from that race by showing a tendency of reduction of tertiary intervals. ISHIKAWA & MIYASHITA (2000) synonymized this race with subsp. *rishiriensis*.
Distribution　Is. Rebun-tô, off the northwestern coast of Hokkaido.
Etymology　Named after Kuniaki SUGA (Tokyo), a member of the research group on carabid beetles in Keihin Insect Lovers' Club, Tokyo.

　タイプ産地　礼文島，礼文岳．
　タイプ標本の所在　国立科学博物館［ホロタイプ，♂］．
　亜種の特徴　21~27 mm. 利尻島亜種に極めて近く，鞘翅第三次間室が減弱傾向を示す（かろうじて不規則な顆粒列として認められる場合が多い）ことにより識別されるが，利尻島亜種のシノニムとみなされる場合もある（ISHIKAWA & MIYASHITA, 2000）．
　分布　礼文島．
　亜種小名の由来と通称　亜種小名は京浜昆虫同好会オサムシグループの一員として活躍した東京の須賀邦輝に因む．通称はレブンオサムシ．

7-2. *C. (A.) k. rishiriensis* NAKANE, 1957 チシマオサムシ利尻島亜種

Original description　*Carabus (Eucarabus) kurilensis rishirienss* [sic] (error of *rishiriensis*) NAKANE, 1957, Sci. Rep. Saikyo Univ. (Nat. Sci. & Liv. Sci.), Kyoto, 2 (4), A Ser., p. 236, fig. 3 (p. 237).
Type locality　Mt. Rishiri, Rishiri Is. Hokkaido.
Type depository　Hokkaido University Museum (Sapporo) [Holotype, ♂].
Diagnosis　20~27 mm. Discriminated from other subspecies by densely punctate pronotal disc and narrower apical part of penis.

Distribution　Is. Rishiri-tô, off the northwestern coast of Hokkaido.
Etymology　Named after Is. Rishiri-tô, the type locality of this subspecies.

　タイプ産地　北海道, 利尻島, 利尻山.
　タイプ標本の所在　北海道大学総合博物館(札幌)[ホロタイプ, ♂].
　亜種の特徴　20〜27 mm. 前胸背板は中央部まで不規則かつ密に点刻され, それらは互いに連続して強い横皺となる. 第二次, 第三次間室は顆粒列となり, 前者は後者より強く明瞭だが隆条となることはない. 陰茎は細長く, 先端部の長さは幅の2.3〜2.8倍. 膜状開口部は短く, 長軸長は陰茎全長の1/2に遥かに及ばず, 同部右側面の窪みと皺は弱い. 陰茎背側縁は側面から見て基部近くで強く湾曲する. 前胸背板表面が密に点刻される点, 及び陰茎先端が非常に細くなる点により, 他亜種から識別される. 外見上はサハリン南部亜種 C. (A.) k. pseudodiamesus に近いが, 鞘翅第一次隆条の隆起が強く, 第二次, 第三次間室の顆粒列がより不規則になる.
　分布　利尻島.
　亜種小名の由来と通称　亜種小名はタイプ産地の利尻島に因む. 原記載論文中では "*rishirien**ss***" と綴られているが, *rishiriensis* の誤植であろう. 通称はリシリオサムシ.

7-3.　*C. (A.) k. kurilensis* Lapouge, 1913　チシマオサムシ基亜種

Original description, type locality & type depository　See explanation of the species.
Diagnosis　23〜28 mm. Allied to subsp. *rausuanus*, but differs from that race in the following points: 1) body above more strongly reddish; 2) pronotum more remarkably rugulose, bearing larger and stronger punctures; 3) lateral sides of pronotum more broadly rimmed before hind angles; 4) all intervals of elytra broader and more strongly raised; 5) penis with median portion less strongly inflated, apex slenderer, membraneous part shorter than half the length of penis, right subventral portion concave and rugulose.
Distribution　Iss. Kunashiri-tô & Etorofu-tô. Outside Japan: Is. Urup.

　原記載・タイプ産地・タイプ標本の所在　種の項を参照.
　亜種の特徴　23〜28 mm. 道央道東亜種に近いが, 以下の諸点で異なる：1)背面はより強く赤褐色味を強く帯びるものが多い；2)前胸背板中央の皺はより顕著に刻まれ, 同部から辺縁前方にかけての点刻は大きく強い；3)前胸背板側縁の縁取りは後角前方で幅広くなる；4)第一次間室の隆条はより幅広く, 第二次, 第三次間室の不規則に断続する隆条ないし顆粒列もやや幅広く, 隆起もやや強いものが多い；5)陰茎中央部は道央道東亜種ほど太くならず, 先端部がより細く, 膜状開口部の長軸長は陰茎全長の1/2以下. 同部の右側縁は窪み, 皺がある.
　分布　国後島, 択捉島. 国外：ウルップ島.

7-4.　*C. (A.) k. rausuanus* Ishikawa, 1966　チシマオサムシ道央道東亜種

Original description　*Carabus* (*Aulonocarabus*) *kurilensis rausuanus* Ishikawa, 1966, Bull. natn. Sci. Mus., Tokyo, 9 (4), p. 457, figs. 12, 16 (p. 456) & 5, 6 (pl. 4).
Type locality　Mt. Rausudake.
Type depository　National Museum of Nature and Science, Tokyo [Holotype, ♂].
Diagnosis　20〜26 mm. Pronotum subquadrate, though variable according to individuals; its lateral margins almost evenly rimmed and weakly curved dorsad; pronotal disk usually wrinkled transversely and not remarkably punctate; primary intervals of elytra strongly carinate; secondaries and tertiaries vary from carinae to rows of granules, though the former is usually stronger than the latter; penis with median portion rather strongly convex, concave and rugulose in right subventral portion, apical lobe 2.1 to 2.3 times as long as wide and broadest behind middle in lateral view, membraneous part shorter than half the total length of penis.
Distribution　Shiretoko Penins. including Mt. Shari-daké & Ishikari-chûô Mountain Range.
Etymology　Named after the type locality.

　タイプ産地　羅臼岳.
　タイプ標本の所在　国立科学博物館 [ホロタイプ, ♂].
　亜種の特徴　20〜26 mm. 前胸背板の輪郭は四角形に近いが, 後方に向けて僅かに狭まり, 側縁後方が波曲して心臓形に近いものから, 両側縁がほぼ平行で直線的になり, 後角が後側方に突出するものまで, 変異が大きい. 前胸背板側縁部はほぼ均等に縁取られ, 弱く反る. 前胸背板表面には通常強い横皺があり, 点刻は目立たない. 鞘翅第一次間室は強い隆条で, 第二次, 第三次間室は隆条から顆粒列まで変化するが, 第二次間室は一般に第三次間室よりも強い. 陰茎中央部は太く強壮. 先端部の長さは幅の2.1〜2.3倍で, 中央より先端よりで最も幅広い. 膜状開口部の長軸長は陰茎全長の1/2以下, 同部の右側縁は窪みがあり, 弱い皺がある.

分布　知床半島（斜里岳を含む）および石狩中央山地とその西・北東方．知床半島の羅臼岳から記載された亜種だが，道東〜道央にかけて散発的ながらも比較的広い範囲から知られており，垂直分布も標高500m程度の森林帯から高山帯に及ぶ．石狩山地北東方の低標高地帯では，自然林中に永久凍土層の残る場所にのみ分布しているようである．低標高地域のものは一般に大型で，高所のものは小型化し，後者は一見，大雪山亜種に似ているが，陰茎および体表面の彫刻の違いにより識別される．

亜種小名の由来と通称　亜種小名はタイプ産地の羅臼岳に因む．通称はラウスオサムシ．

7-5. *C. (A.) k. daisetsuzanus* KÔNO, 1936　チシマオサムシ大雪山亜種

Original description　*Carabus (Eucarabus) kurilensis daisetsuzanus* KÔNO, 1936, Biogeographica, 1, p. 78, fig. 5 (pl. 10).
Type locality　Berg Daisetsu.
Type depository　Hokkaido University Museum (Sapporo) [Syntype].
Diagnosis　17~23 mm. Smallest subspecies. Pronotum weakly punctate and rather smooth, though transversely rugulose near base; primary costae relatively weak, secondaries and tertiaries weaker than primaries, indicated by low crenulate costae or rows of large granules, sometimes connecting with large foveae between intervals to form reticular pattern; apical lobe of penis short and robust as in subsp. *rausuanus*.
Distribution　Uppermost part of the Daisetsu-zan Mts., C. Hokkaido.
Etymology　Named after the type locality.

タイプ産地　大雪山．原記載論文におけるタイプ産地の和文表記は「大雪山寒地帯」．

タイプ標本の所在　北海道大学総合博物館（札幌）[シンタイプ]．原記載は7♂1♀の標本に基づいてなされたが，ホロタイプは指定されていない．北大総合博物館には現在，少なくとも1♀の赤い色のラベルを付された本亜種の標本が保管されている．

亜種の特徴　17~23 mm．全亜種中，最も小型で，前胸背板表面の点刻は弱く，中央は滑らかだが，基部には強い横皺がある．第一次間室の隆条は他亜種に比しやや隆起が弱く，第二次，第三次間室はそれよりもさらに弱い断続する隆条ないし粗大顆粒列で，間室間線条の粗大点刻と連なり，所により網眼状を呈する．陰茎先端部は道央道東亜種のそれに似て太短い．

分布　大雪山連峰高所．

亜種小名の由来と通称　亜種小名はタイプ産地の大雪山に因む．通称はダイセツオサムシ．

7-6. *C. (A.) k. yoteizanus* (ISHIKAWA et MIYASHITA, 2000)　チシマオサムシ羊蹄山亜種

Original description　*Leptocarabus (Aulonocarabus) kurilensis yoteizanus* ISHIKAWA et MIYASHITA, 2000, Jpn. J. syst. Ent., 6 (1), p. 68, fig. 9 (p. 69).
Type locality　Mt. Yôteizan, at the alpine zone, in the crater, ca 1,850 m alt.
Type depository　Systematic Zoology Laboratory, Department of Biological Sciences, Tokyo Metropolitan University (Hachiôji) [Holotype, ♂].
Diagnosis　20~22 mm. Small subspecies. Allied to subsp. *daisetsuzanus*, but discriminated from it in the following points: 1) pronotum with lateral margins more remarkably rimmed and more strongly curved above, disk more roughly sculptured, basal foveae more clearly carved; 2) secondary and tertiary intervals of elytra weaker, indicated by rows of small tubercles or granules; 3) penis and its apical lobe slenderer.
Distribution　Summit of Mt. Yôtei-zan, SW. Hokkaido.
Etymology　Named after the type locality.

タイプ産地　羊蹄山，高山帯，噴火口内，標高約1,850m.

タイプ標本の所在　首都大学東京動物系統分類学研究室（八王子）[ホロタイプ, ♂]．

亜種の特徴　20~22 mm．小型の亜種で，大雪山亜種に似るが，以下の諸点で異なる：1) 前胸背板は側縁の縁取りが顕著で，全縁に亙り強く反り，表面の彫刻がより粗く，基部凹陥はより明瞭；2) 鞘翅第二次，第三次間室の発達が弱く，共に小結節列ないし小顆粒列となる；3) 陰茎及びその先端部が共により細長い．

分布　北海道南西部の羊蹄山山頂付近．

亜種小名の由来と通称　亜種小名はタイプ産地の羊蹄山に因む．ヨウテイオサムシという通称で呼ばれることもある．

8. *Carabus* (*Leptocarabus*) *kyushuensis* Nakane, 1961
キュウシュウクロナガオサムシ

Figs. 1-16. *Carabus* (*Leptocarabus*) *kyushuensis kyushuensis* Nakane, 1961 (Mt. Takachiho-no-miné, Takaharu-chô, Miyazaki Pref.). — 1, ♂; 2, ♀; 3, male head & pronotum; 4, male right protarsus in ventral view; 5, male right protibia in subdorsal view; 6, male right profemur in left lateral view; 7, male left elytron (median part); 8, penis & fully everted internal sac in right lateral view; 9, apical part of penis in right lateral view; 10, ditto in right subdorsal view; 11, ditto in dorsal view; 12, internal sac in caudal view; 13, ditto in frontal view; 14, ditto in dorsal view; 15, ditto in left sublateral view; 16, terminal plate of internal sac in frontal view.

図 1-16. キュウシュウクロナガオサムシ基亜種(宮崎県高原町高千穂峰). — 1, ♂; 2, ♀; 3, ♂の頭部と前胸背板; 4, ♂右前跗節(腹面観); 5, ♂右前脛節(斜背面観); 6, ♂右前腿節(左側面観); 7, ♂左鞘翅中央部; 8, 陰茎と完全反転した内嚢(右側面観); 9, 陰茎先端(右側面観); 10, 同(右斜背面観); 11, 同(背面観); 12, 内嚢(尾側面観); 13, 同(頭側面観); 14, 同(背面観); 15, 同(左斜側面観); 16, 項板(項側面観).

8. *Carabus* (*Leptocarabus*) *kyushuensis*

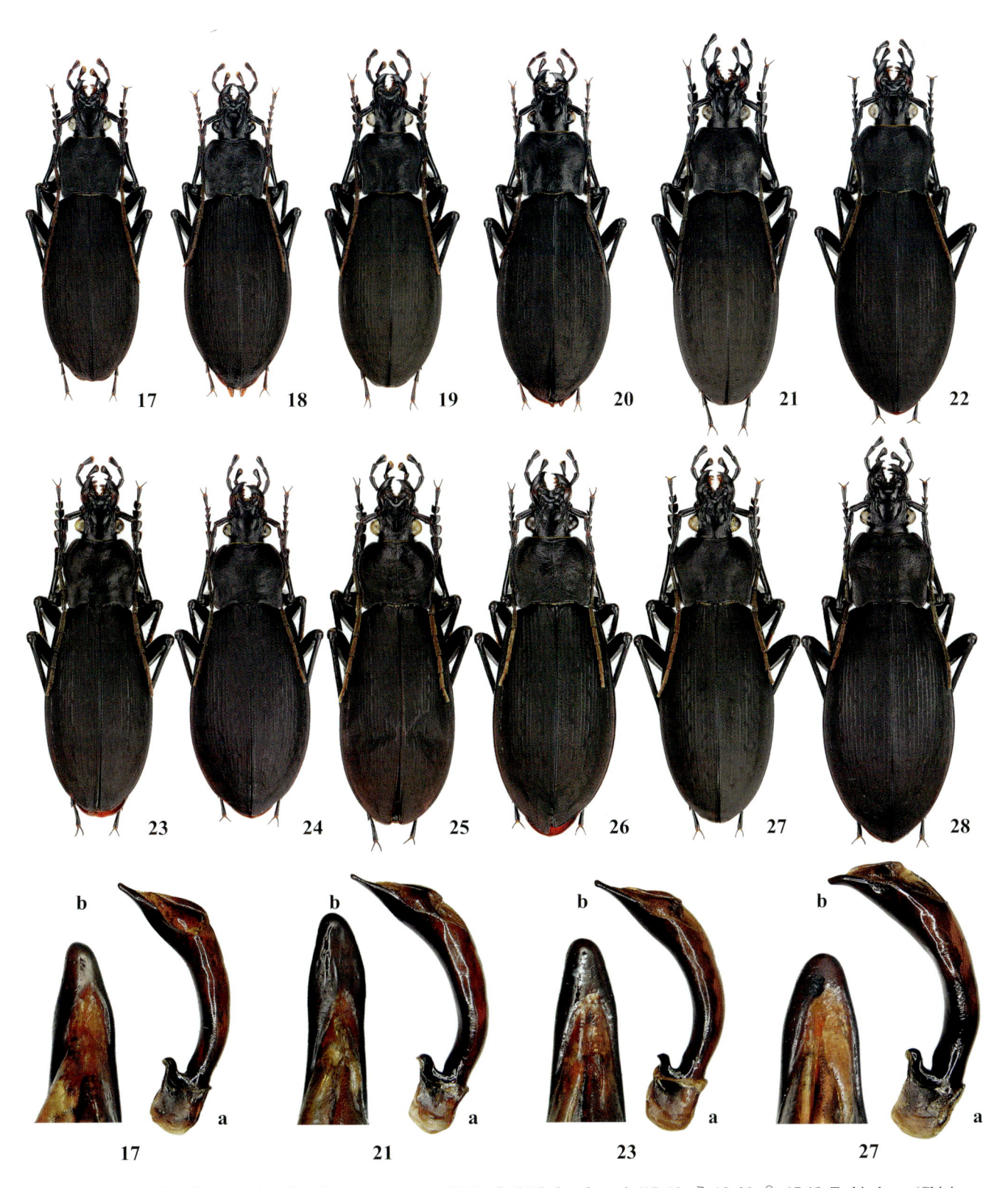

Figs. 17-28. *Carabus* (*Leptocarabus*) *kyushuensis* subspp. — 17-20, *C.* (*L.*) *k. kyushuensis* (17, 19, ♂, 18, 20, ♀; 17-18, Tashirobaru, (Chiziwa-chô), Unzen-shi, Nagasaki Pref.; 19-20, Mt. Futago-san, Kunisaki-shi, Oita Pref.); 21-22, *C.* (*L.*) *k. kawaharai* (21, ♂ (PT), 22, ♀ (PT); Isa-chô-kawara, Miné-shi, Yamaguchi Pref.); 23-26, *C.* (*L.*) *k. nakatomii* (23 (PT), 25, ♂, 24 (PT), 26, ♀; 23-24, Hamada, Shimane Pref.; 25-26, Mt. Ônadé-san, Sayô-chô, Hyogo Pref.); 27-28, *C.* (*L.*) *k. cerberus* (27, ♂, 28, ♀; Yano, Jôgé-chô, Fuchû-shi, Hiroshima Pref.). — a, Penis in right lateral view; b, apical part of penis in dorsal view.

図 17-28. キュウシュウクロナガオサムシ各亜種. — 17-20, 基亜種 (17, 19, ♂, 18, 20, ♀; 17-18, 長崎県雲仙市（千々石町）田代原; 19-20, 大分県国東市両子山); 21-22, 山口県亜種 (21, ♂ (PT), 22, ♀ (PT); 山口県美祢市伊佐町河原); 23-26, 中国地方亜種 (23 (PT), 25, ♂, 24 (PT), 26, ♀; 23-24, 島根県浜田; 25-26, 兵庫県佐用町大撫山)); 27-28, 広島県東部亜種 (27, ♂, 28, ♀; 広島県府中市上下町矢野). — a, 陰茎（右側面観); b, 同先端部（背面観).

8. *Carabus* (*Leptocarabus*) *kyushuensis* NAKANE, 1961　キュウシュウクロナガオサムシ

Original description　*Carabus* (*Leptocarabus*) *procerulus kyushuensis* NAKANE, 1961, Fragm, coleopterol., Kyoto, (1), p. 2.
History of research　This taxon was originally described by NAKANE (1961) as a subspecies of *Carabus* (*Leptocarabus*) *procerulus*. ISHIKAWA (1962) raised its rank to a full species mainly based on the difference in shape of the penis, profemur, etc., and described a new subspecies from Hiroshima Prefecture under the name of *cerberus*. Another subspecies was described also by ISHIKAWA (1966a) from Shimane Prefecture under the name of *nakatomii*. In 2011, a new subspecies was described by IMURA from Yamaguchi Prefecture under the name of *kawaharai*. Thus, this species is currently classified into four subspecies including the nominotypical race.
Type locality　Mt. Takachiho (Mt. Kirishima), Kyushu.
Type depository　Unknown. The holotype (♂) of this taxon is not found in the NAKANE collection now preserved in the Hokkaido University Museum (Sapporo).
Morphology　25~37 mm. Body above black and mat, sometimes a little brownish. External features: Figs. 1~7. Description on details: penultimate segment of labial palpus bisetose; median tooth of mentum almost as long as, or a little shorter than lateral lobes, narrowly rimmed by low ridge, with apex not so sharply pointed in ventral view, protruded ventrad in lateral view; submentum smooth and asetose; male antennae with hairless ventral depressions from segment five to eight; pronotal margins basically bisetose; preapical emargination of elytra sometimes faintly recognized in female; elytral sculpture with primaries indicated by rows of low costae rather irregularly segmented by shallow primary foveoles; both secondaries and tertiaries indicated by contiguous low costae or rows of granules; areas between intervals rather coarsely scattered with fine, small granules; intervals of granules in umbilicate series wider in posterior halves of elytra; propleura smooth; sides of sternites weakly rugoso-punctate; metacoxa basically bisetose, lacking anterior seta; sternal sulci clearly carved. Male genitalic features: Figs. 8~16. Penis a little shorter than half the elytral length; ostium lobe very small and weakly bilobed at tip; internal sac strongly pigmented in basal portion; ligulum indicated by oblong-shaped plate which is strongly sclerotized; pre- and parasaccellar lobes apparently smaller than other species of the same subgenus; apical part of internal sac rather short and strongly swollen; terminal plate short, spatulate and weakly protruded.
Molecular phylogeny　On the molecular genealogical tree of nuclear 28S ribosomal RNA, this species belongs to the same group as that constructed by four other *Leptocarabus* species of Japan (*hiurai*, *kumagaii*, *procerulus* and *arboreus*) and clearly separated from those of the Continental subgenera as *Adelocarabus*, *Aulonocarabus*, *Weolseocarabus* and *Sinoleptocarabus*, which revealed a monophyletic origin of all the Japanese *Leptocarabus* species (KIM, TOMINAGA et al., 2000; KIM, ZHOU et al., 2000). Within the Japanese species, however, there is serious discordance of mitochondrial gene genealogies with morphological species. Extensive sharing of haplotypes between geographically overlapping species has been observed, suggesting widespread introgressive hybridization. ZHANG & SOTA (2007) analyzed four nuclear genes, and the combined nuclear data revealed a species relationship of (((*kumagaii*, *procerulus*), *arboreus*), (*hiurai*, *kyushuensis*)). A notable trans-species polymorphisms in mitochondrial gene sequences observed in this group is considered to be caused by widespread introgressive hybridization between two parapatric species (SOTA, 2011).
Distribution　Honshu (Chugoku Region), Kyushu / Is. Ao-shima [Lake Koyama-iké] (Tottori Pref.).
Habitat　In the Chugoku Region of western Honshu, this species generally inhabits dry, thin forest around the basin, but it also inhabits matured forest of planted trees such as cedar or cypress. Our knowledge is still rather poor on the typical habitat of this species in Kyushu.
Bionomical notes　Autumn breeder. Larvae prefer to feed on larvae of other insects.
Geographical variation　Classified into four subspecies.
Etymology　Named after the Kyushu Region or Is. Kyushu in southwestern Japan.

研究史　本種はNAKANE(1961)によりクロナガオサムシの一亜種として記載されたが，ISHIKAWA(1962)は陰茎や♂前腿節等の形態上の相違に基づいて，これを独立種に昇格し，同時に広島県東部の矢野から亜種*cerberus*(本書で言うキュウシュウクロナガオサムシ広島県東部亜種)を記載した．ISHIKAWAは更に，島根県の浜田から亜種*nakatomii*を記載した(1966a)．本種は長年，これら3亜種(基亜種を含む)に分類されてきたが，2011年になってIMURAは，陰茎先端が著しく伸長するという特徴を持つ山口県産の集団を新亜種と認め，*kawaharai*という亜種小名を与えて記載した．

タイプ産地　九州，高千穂峰(霧島山)．

タイプ標本の所在　不明．北海道大学総合博物館(札幌)に保管されている中根コレクション中には，本タクソンのホロタイプに該当すると思われる標本は見当たらない．

形態　25~37 mm. 背面は光沢を欠いた黒色で，時に褐色味を帯びる．外部形態：図1~7. 細部の記載：下唇肢亜端節の剛毛は2本．下唇基節の中央歯は側葉とほぼ同長か，やや短く，辺縁は低い稜により顕著に縁取られ，先端はそれほど鋭く尖らず，腹側に突出する．下唇亜基節は無毛で表面は平滑．♂触角第5節から第8節の腹面に無毛部がある．前胸背板の側縁剛毛は2対(中央と後角付近に各1対)．鞘翅端は♀でごく僅かにえぐれる場合がある．鞘翅彫刻は三元異規的で，第一列間室は浅い陥凹によってやや不規則に分断された低い隆起，第二列，第三列間室は共に低い連続する隆条ないし顆粒列となり，間室間の

基面にはごく弱い小顆粒を比較的密に装う．丘孔点列の顆粒は鞘翅後半で間隔が大きくなる．前胸側板は滑らかで，腹部腹板の側面には弱い皺と点刻を装う．後基節の剛毛は2本（前方剛毛を欠く），時に3本．腹部腹板の横溝は極めて明瞭．♂交尾器形態：図8~16．陰茎は鞘翅長の1/2弱で，葉片は極めて小さく，先端は僅かに分葉し，内嚢基部は伸側を除き膜面の色素沈着と硬化が著しく，基斑は細長い板状で強く硬化・膨隆し，盤前葉と傍盤葉は近縁他種に比し小さく，内嚢先端部は太短く，強く膨隆し，頂板は短いヘラ状で突出は弱い．

分子系統　核28SリボゾームDNAによる分子系統樹の上で，本種は以下に述べる日本産の他の4種のレプトカラブス亜属（シコククロナガオサムシ（以下，各種名末尾のオサムシは略す），オオクロナガ，クロナガ，コクロナガ）と共に，大陸産のいわゆる広義のクロナガオサムシ類（*Adelocarabus* 亜属, *Aulonocarabus* 亜属, *Weolseocarabus* 亜属, *Sinoleptocarabus* 亜属等）から分かれた単系統群を形成する．一方，ミトコンドリア遺伝子の系統樹では，これら邦産レプトカラブス亜属において種間のハプロタイプ共有が広範に見られた（KIM, TOMINAGA *et al*., 2000; KIM, ZHOU *et al*., 2000）．その後，四つの核遺伝子の塩基配列を基に，同亜属の種間系統とミトコンドリアの種間共有についての検討が行われた（ZHANG & SOTA, 2007）結果，邦産の5種はやはり単系統で，日本国内でキュウシュウクロナガとシコククロナガの2種からなる群とオオクロナガ，クロナガ，コクロナガの3種からなる群に分化し，前者ではまず本州産のキュウシュウクロナガの亜種群が，次いで九州産のキュウシュウクロナガ基亜種とシコククロナガが分化，後者ではまずコクロナガが分化した後，オオクロナガとクロナガが分化したらしい．ミトコンドリアの種間共有は，側所的に分布する種同士の交雑によるものと推定されており，特に本州産のコクロナガ各亜種においてはクロナガからのミトコンドリアの移入・置換が著しい．ただし，現在のところ種間交雑帯が認められるのはクロナガとオオクロナガの間においてのみであり，多くの浸透性交雑は過去に起こったものか，ごく稀な交雑機会に由来するものであろうと考えられている（曽田, 2011）．

分布　本州（中国地方），九州／青島（鳥取県鳥取市湖山池）．

生息環境　中国地方では盆地周辺の乾燥した二次林を好む（近畿オサムシ研究グループ, 1979）が，成熟した杉・ヒノキなどの陰鬱な林床や林縁にもよく見られる．九州でもほぼ同様と思われるが，九州産の本種の生息地に関する報告は少ない．

生態　秋繁殖・幼虫越冬型のオサムシで，幼虫は基本的に昆虫幼虫食と考えられる．

地理的変異　九州に産する基亜種のほか，中国地方から3亜種が記載されており，形態学的には計4亜種に分類される．長崎県雲仙のものは，分子系統樹の上では大きく異なる系統に属することが判明しているが，形態学的には基亜種との識別が困難である．

種小名の由来　九州に因む．

8-1. *C. (L.) k. nakatomii* ISHIKAWA, 1966 キュウシュウクロナガオサムシ中国地方亜種

Original description　*Carabus (Leptocarabus) kyushuensis nakatomii* ISHIKAWA, 1966, Bull. natn. Sci. Mus., Tokyo, 9 (1), p. 22, figs. C (p. 21) & D~F (p. 23).

Type locality　Hamada, Shimane Pref.

Type depository　National Museum of Nature and Science, Tokyo [Holotype, ♂].

Diagnosis　28~37 mm. Black and mat. Closely allied to subsp. *cerberus*, but readily distinguished from that race by differently shaped apical lobe of penis which is much longer in both lateral and dorsal views, and hardly bent ventrad in lateral view.

Distribution　Chugoku Region (Hiroshima Pref., Shimane Pref., Tottori Pref., Okayama Pref. & SW. Hyogo Pref.), SW. Honshu / Is. Ao-shima [Lake Koyama-iké] (Tottori Pref.).

Etymology　Named after Kentaro NAKATOMI, a member of the research group on carabid beetles in Keihin Insect Lovers' Club, Tokyo, who first collected this subspecies.

タイプ産地　島根県，浜田．

タイプ標本の所在　国立科学博物館［ホロタイプ, ♂］．

亜種の特徴　28~37 mm．黒色で光沢は鈍い．広島県東部亜種に近いが，陰茎先端がはるかに長く，腹側に向かって殆ど湾曲しない点で容易に区別される．

分布　中国地方（広島，島根，鳥取，岡山，兵庫県南西部）／青島（鳥取県鳥取市湖山池）．次に述べる広島県東部亜種の分布域を取り囲むような形で広島，島根，鳥取，岡山各県に分布する．南東限は兵庫県南東部の佐用（さよう）地方．

亜種小名の由来と通称　亜種小名は，京浜昆虫同好会オサムシグループの一員で，本亜種を最初に採集した中臣 謙太郎に因む．チュウゴククロナガオサムシという通称がある．

8-2. *C. (L.) k. cerberus* ISHIKAWA, 1962 キュウシュウクロナガオサムシ広島県東部亜種

Original description　*Carabus (Leptocarabus) kyushuensis cerberus* ISHIKAWA, 1962, Kontyû, Tokyo, 30 (2), p. 114, figs. 4 (p. 111) & 6 (p. 112).

Type locality　Yano, Kônu co.(= present Fuchû-shi), Hiroshima pref.

Type depository　Systematic Zoology Laboratory, Department of Biological Sciences, Tokyo Metropolitan University (Hachiôji)

[Holotype, ♂].

Diagnosis 27~36 mm. Black and mat. Size large on an average; antennae relatively short, not reaching middle of elytra even in male; pronotum broad, with lateral furrows shallow but wide, their inner margins clearly bordered, hind angles broad and rounded at tips; apical lobe of penis short, apparently bent ventrad in lateral view.

Distribution E. Hiroshima Pref., SW. Honshu.

Etymology Cerberus is a three-headed dog guarding the underworld which appears in Greek Mythology, and is a nickname of Kentaro Nakatomi.

タイプ産地　広島県, 甲奴郡, 矢野 (甲奴郡は合併により消滅し, 矢野は現在, 府中市に属する).
タイプ標本の所在　首都大学東京動物系統分類学研究室 (八王子) [ホロタイプ, ♂].
亜種の特徴　27~36 mm. 黒色で光沢は鈍い. 平均して大型・強壮で, 触角は短く, ♂でも先端が鞘翅中央に達しない. 前胸背板は幅広く, 側溝は浅いが広く, その内縁は明瞭に縁どられ, 後角は幅広く丸い. 陰茎先端部は短く, 膜状部は先端近くまで達し, 側面から見て明らかに腹側に湾曲する.
分布　広島県東部. タイプ産地である広島県東部の矢野温泉付近のものが典型的で, 東・北・西の三方をキュウシュウクロナガオサムシ中国地方亜種に取り囲まれる形で分布するが, 両者の境界はまだ明確にされていない.
亜種小名の由来と通称　亜種小名は中臣謙太郎の仇名で, ギリシア神話に出てくるケルベロス (冥界の番犬) に因む. 通称はオニクロナガオサムシ.

8-3. *C. (L.) k. kawaharai* (Imura, 2011) キュウシュウクロナガオサムシ山口県亜種

Original description *Leptocarabus kyushuensis kawaharai* Imura, 2011, Coléoptères, Paris, 17 (3), p. 25, figs. 5, 6, 8-a, b.
Type locality Japan, western Honshu, Yamaguchi Pref., Miné-shi, Isachô-kawara, 120 m in altitude.
Type depository Muséum National d'Histoire Naturelle, Paris [Holotype, ♂].
Diagnosis 27~35 mm. Externally very similar to subsp. *nakatomii* and barely distinguished from that race by having a little slenderer body, but definitely different in shape of apical part of penis which is much slenderer in dorsal view, about 1.5 time as long as wide (1~1.1 times as long as wide in subsp. *nakatomii*), more remarkably narrowed toward apex, with dorsal wall weakly but apparently concave above.
Distribution Rather narrowly localized in the central part of Yamaguchi Pref. of westernmost Honshu, above all in the karst called Akiyoshidai and its surrounding area.
Etymology Named after Hiroyuki Kawahara (Hiroshima) who collected greater part of the type series of this subspecies.

タイプ産地　本州西部, 山口県, 美祢市, 伊佐町河原, 標高 120 m.
タイプ標本の所在　パリ自然史博物館 [ホロタイプ, ♂].
亜種の特徴　27~35 mm. キュウシュウクロナガオサムシ中国地方亜種に極めて近いが, 体形はより細く (♂においてより顕著), 陰茎先端部はより細長く, 長さが幅の約 1.5 倍で, 先端に向けてより強く細まり, 背面は弱く陥入することにより識別される.
分布　秋吉台を中心とする山口県中部.
亜種小名の由来　本亜種の分布域を積極的に調査し, タイプ標本の多くを採集した広島市の河原宏幸に因む.

8-4. *C. (L.) k. kyushuensis* Nakane, 1961 キュウシュウクロナガオサムシ基亜種

Original description, type locality & type depository See explanation of the species.
Diagnosis 25~32 mm. For diagnosis of the nominotypical subspecies, see those of the above three subspecies and explanation of the species.
Distribution Kyushu.

原記載・タイプ産地・タイプ標本の所在　種の項を参照.
亜種の特徴　25~32 mm. 上記 3 亜種および種の解説を参照.
分布　九州 (北限は福岡・佐賀県境の脊振 (せふり) 山地から大分県国東 (くにさき) 半島の両子 (ふたご) 火山群にかけての地域, 南限は宮崎・鹿児島県境の霧島火山群).

8. *Carabus* (*Leptocarabus*) *kyushuensis*

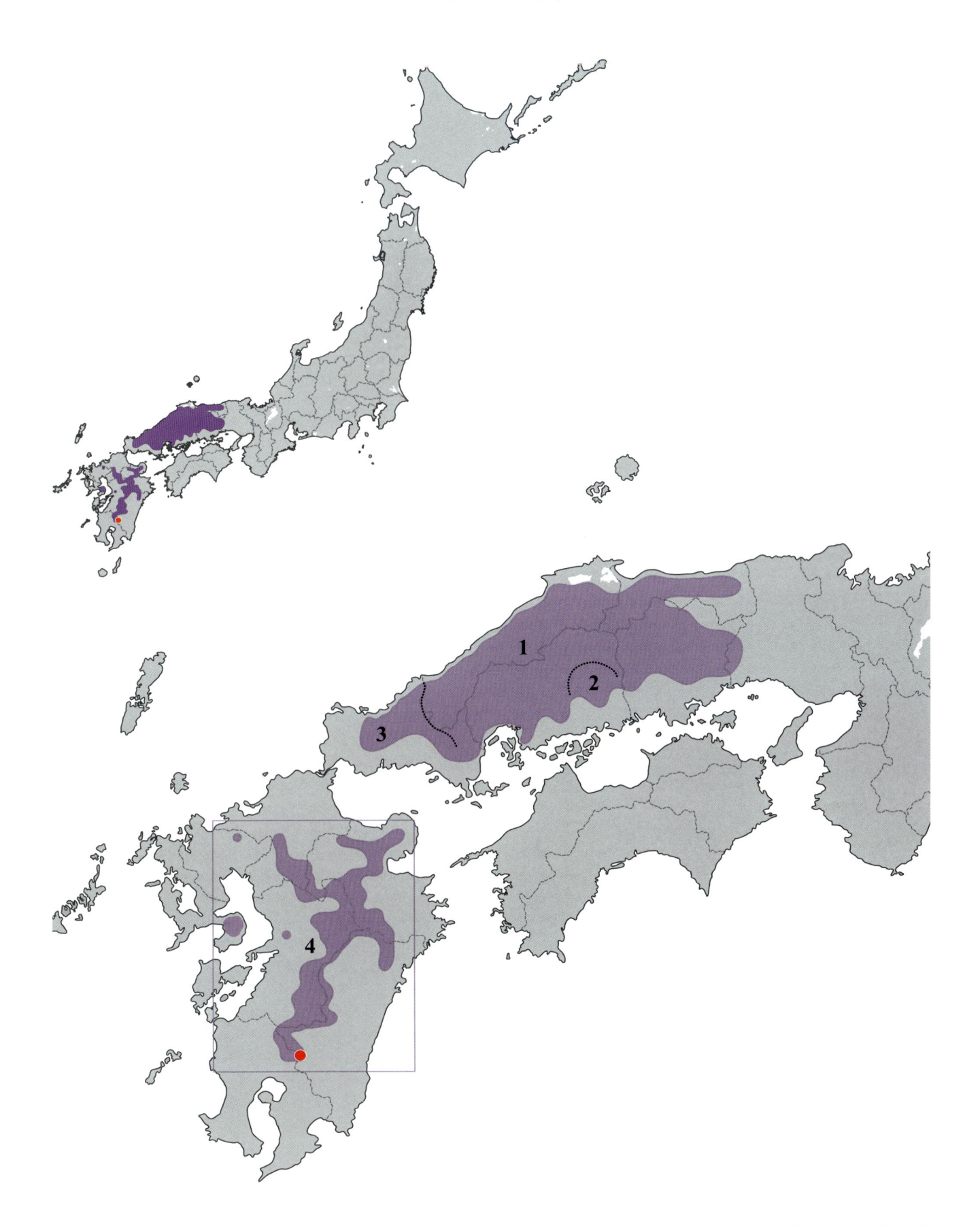

Fig. 29. Map showing the distributional range of ***Carabus*** (***Leptocarabus***) ***kyushuensis*** NAKANE, 1961. — 1, Subsp. ***nakatomii***; 2, subsp. ***cerberus***; 3, subsp. ***kawaharai***; 4, subsp. ***kyushuensis***. ●: Type locality of the species (Mt. Takachiho-no-miné).

図 29. キュウシュウクロナガオサムシ分布図. — 1, 中国地方亜種; 2, 広島県東部亜種; 3, 山口県亜種; 4, 基亜種. ●: 種のタイプ産地（高千穂峰）.

9. *Carabus* (*Leptocarabus*) *hiurai* Kamiyoshi et Mizoguchi, 1960
シコククロナガオサムシ

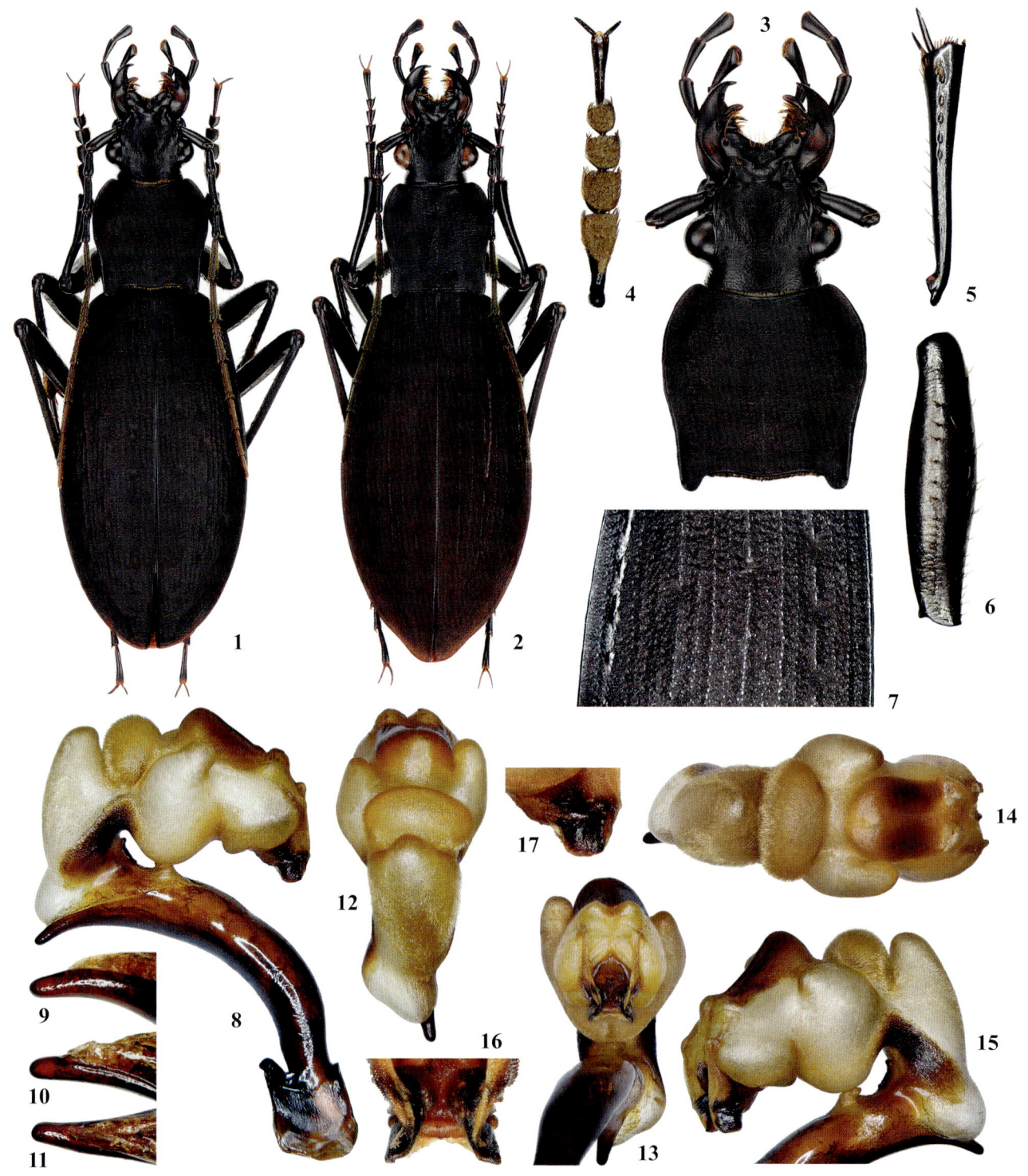

Figs. 1-17. *Carabus* (*Leptocarabus*) *hiurai* Kamiyoshi et Mizoguchi, 1960 (Mt. Tsurugi-san, Tokushima Pref.). — 1, ♂; 2, ♀; 3, male head & pronotum; 4, male right protarsus in ventral view; 5, male right protibia in subdorsal view; 6, male right profemur in left lateral view; 7, male left elytron (median part); 8, penis & fully everted internal sac in right lateral view; 9, apical part of penis in right lateral view; 10, ditto in right subdorsal view; 11, ditto in dorsal view; 12, internal sac in caudal view; 13, ditto in frontal view; 14, ditto in dorsal view; 15, ditto in left sublateral view; 16, terminal plate of internal sac in frontal view; 17, ditto in right lateral view.

図1-17．シコククロナガオサムシ（徳島県剣山）．— 1, ♂; 2, ♀; 3, ♂の頭部と前胸背板; 4, ♂右前跗節（腹面観）; 5, ♂右前脛節（斜背面観）; 6, ♂右前腿節（左側面観）; 7, ♂左鞘翅中央部; 8, 陰茎と完全反転下の内嚢（右側面観）; 9, 陰茎先端（右側面観）; 10, 同（右亜背面観）; 11, 同（背面観）; 12, 内嚢（尾側面観）; 13, 同（頭側面観）; 14, 同（背面観）; 15, 同（左亜側面観）; 16, 頂板（頭側面観）; 17, 同（右側面観）．

9. *Carabus* (*Leptocarabus*) *hiurai*

Figs. 18-25. *Carabus* (*Leptocarabus*) *hiurai*. — 18, ♂, 19, ♀, Mt. Kôtsu-san, Yoshinogawa-shi, Tokushima Pref.; 20, ♂, 21, ♀, Mt. Inamura-yama, Tosa-chô, Kôchi Pref.; 22, ♂, 23, ♀, Mt. Ishizuchi-san, Kumakôgen-chô, Ehime Pref.; 24, ♂, 25, ♀, Mt. Inoko-yama, Matsuyama-shi, Ehime Pref.

図 18-25. シコククロナガオサムシ. — 18, ♂, 19, ♀, 徳島県吉野川市高越山；20, ♂, 21, ♀, 高知県土佐町稲叢山；22, ♂, 23, ♀, 愛媛県久万高原町石鎚山；24, ♂, 25, ♀, 愛媛県松山市伊之子山.

9. *Carabus* (*Leptocarabus*) *hiurai* Kamiyoshi et Mizoguchi, 1960 シコククロナガオサムシ

Original description *Carabus hiurai* Kamiyoshi et O, Mizoguchi, 1960, Insect Science, p. 6, figs. G & H (p. 5).

History of research This taxon was described by Kamiyoshi & Mizoguchi (1960) as an independent species of the genus *Carabus*. *Carabus* (*Leptocarabus*) *procerulus shikokuensis* Nakane, 1961 is its junior synonym. Although once regarded as a mere local race of *C.* (*L.*) *procerulus*, it is currently regarded as a full, monotypical species endemic to the island of Shikoku.

Type locality Mt. Tsurugi, Tokushima Pref., Shikoku Is., Japan.

Type depository Depository of the holotype specimen (♂) designated in the original description is "Konchû Dantai Kenkyû-kai, Osamushi Group", but the holotype is now deposited in the Osaka Museum of Natural History.

Morphology 23~29 mm. Body above black or dark reddish black and mat. External features: Figs. 1~7. Description on details: penultimate segment of labial palpus bisetose; median tooth of mentum almost as long as lateral lobes, with tip sharply pointed and protruded ventrad; submentum asetose; male antennae with hairless ventral depressions from segment five to eight which are longer and more distinct than those of *C.* (*L.*) *kyushuensis*; pronotum basically with two pairs of marginal setae, one in median portion and another near hind angle; elytra with weak preapical emargination in female; elytral sculpture much more reduced than that of *C.* (*L.*) *kyushuensis*, secondaries and tertiaries recognized as rows of weak granules; propleura smooth; sides of sternites weakly rugoso-punctate; metacoxa bisetose, lacking anterior seta; sternal sulci clearly carved. Male genitalic features: Figs. 8~17. Penis a little shorter than half the elytral length; ostium lobe small and bilobed at tip; internal sac much more weakly pigmented and sclerotized than in *C.* (*L.*) *kyushuensis*, ligulum oblong, strongly sclerotized and convex above, pre- and parasaccellar lobes larger than those of *C.* (*L.*) *kyushuensis*, apical portion narrow, terminal plate a little longer than that of *C.* (*L.*) *kyushuensis*, square-shaped in lateral view and rhomboidal in frontal view.

Molecular phylogeny For the molecular phylogeny of this species, see the same section of *C.* (*L.*) *kyushuensis*.

Distribution Shikoku (Shikoku Mts. & Takanawa Penins.).

Habitat Forest species, mainly inhabiting a mature forest of the mountainous area within the range from 250 m to 1,900 m in the altitude, mainly around 1,000 m.

Bionomical notes Autumn breeder. Larvae most probably prefer to feed on larvae of other insects like.

Geographical variation Monotypical species.

Etymology Named after Isamu Hiura of Osaka Museum of Natural History.

研究史　本タクソンは関西のアマチュア昆虫研究家，神吉弘視(かみよしひろし)と溝口 修(みぞぐちおさむ)によって1960年，徳島県の剣山から独立種として記載された．原記載に記された著者名の正確な綴りは "Kamiyoshi et O, Mizoguchi" (O のあとのコンマは明らかにピリオドの誤植)だが，オサムシのタクソンを記載した同姓の著者は他に存在しないので，学名を表記する際にイニシャルは不要であろう．Nakane (1961)によりクロナガオサムシの亜種として記載された *Carabus* (*Leptocarabus*) *procerulus shikokuensis* は新参異名となるので使用できない．記載後しばらくはクロナガオサムシの一亜種に位置付けられることも多かったが，現在では一般に独立種とみなされている．

タイプ産地　日本，四国，徳島県，剣山．

タイプ標本の所在　大阪市立自然史博物館．原記載論文には，本種のホロタイプ(♂)の保管場所が「虫団研(昆虫団体研究会)オサムシグループ」と記されているが，著者らの意向によりその後，大阪市立自然史博物館に寄贈されたという．

形態　23~29 mm．背面は光沢を欠く黒色で，時に赤褐色味を帯びる．外部形態：図1~7．細部の記載：下唇肢亜端節の剛毛は2本．下唇基節の中央歯は側葉とほぼ同長で，先端は鋭く尖り，腹側に突出する．下唇亜基節は無毛で表面は平滑．♂触角第5節から第8節の腹面にはキュウシュウクロナガオサムシのそれよりも広汎で顕著な無毛部がある．前胸背板の側縁剛毛は2対（中央と後角付近に各1本）．♀の鞘翅端部はごく僅かにえぐれる．鞘翅彫刻は三つの間室がキュウシュウクロナガオサムシのそれよりもさらに減退し，第二次，第三次間室は通常，弱い顆粒列．丘孔点列は鞘翅端近くまで認められるが，後半部ではまばら．前胸側板は滑らかで，腹部腹板の側面には弱い皺と点刻を装う．後基節の剛毛は2本（前方剛毛を欠く）．腹部腹板の横溝は明瞭．♂交尾器形態：図8~17．陰茎は鞘翅長の1/2よりやや短い．葉片は小さく，先端は軽く分葉し，内嚢基部膜面の色素沈着と硬化はキュウシュウクロナガオサムシのそれよりも弱く，基斑は細長い板状で強く膨隆し，遠位端は膜面から遊離し，先端に向けてやや細くなる．盤前葉と傍盤葉はキュウシュウクロナガオサムシのものよりも大きく，内嚢先端部は細く，頂板はキュウシュウクロナガオサムシに比べてやや長く，側面から見ると長副ほぼ等しい四角形，正面から見ると菱形に見える．

分子系統　キュウシュウクロナガオサムシの分子系統の項を参照．4種の核遺伝子の解析結果から，本種は系統的に九州産のキュウシュウクロナガオサムシに最も近いことが指摘されている．

分布　四国(四国山地および高縄半島)．

生息環境　一般に山地性で，標高1,000 m 前後に多いが，標高500 m 以下の地点から2,000 m 近い地点まで記録がある．高縄半島では生息域の下限が低く，標高250 m から1,000 m 強まで分布するが，四国山地では1,000 m 弱から1,900 m の間に分布していることが多いようである．

9. *Carabus* (*Leptocarabus*) *hiurai*

れるが，高縄半島などでは朽木中に越冬している場合も多い．

地理的変異　これまでのところ亜種として記載されている集団はないが，地理的変異に関する研究は不十分である．高縄半島に産するものは一般にやや大型．

種小名の由来　種小名は，大阪市立自然史博物館の日浦 勇に因む．原記載論文中で神吉と溝口により提唱された最初の種和名はケンザンクロナガオサムシだが，この名はその後，殆ど使われていない．

9. *Carabus* (*Leptocarabus*) *hiurai*

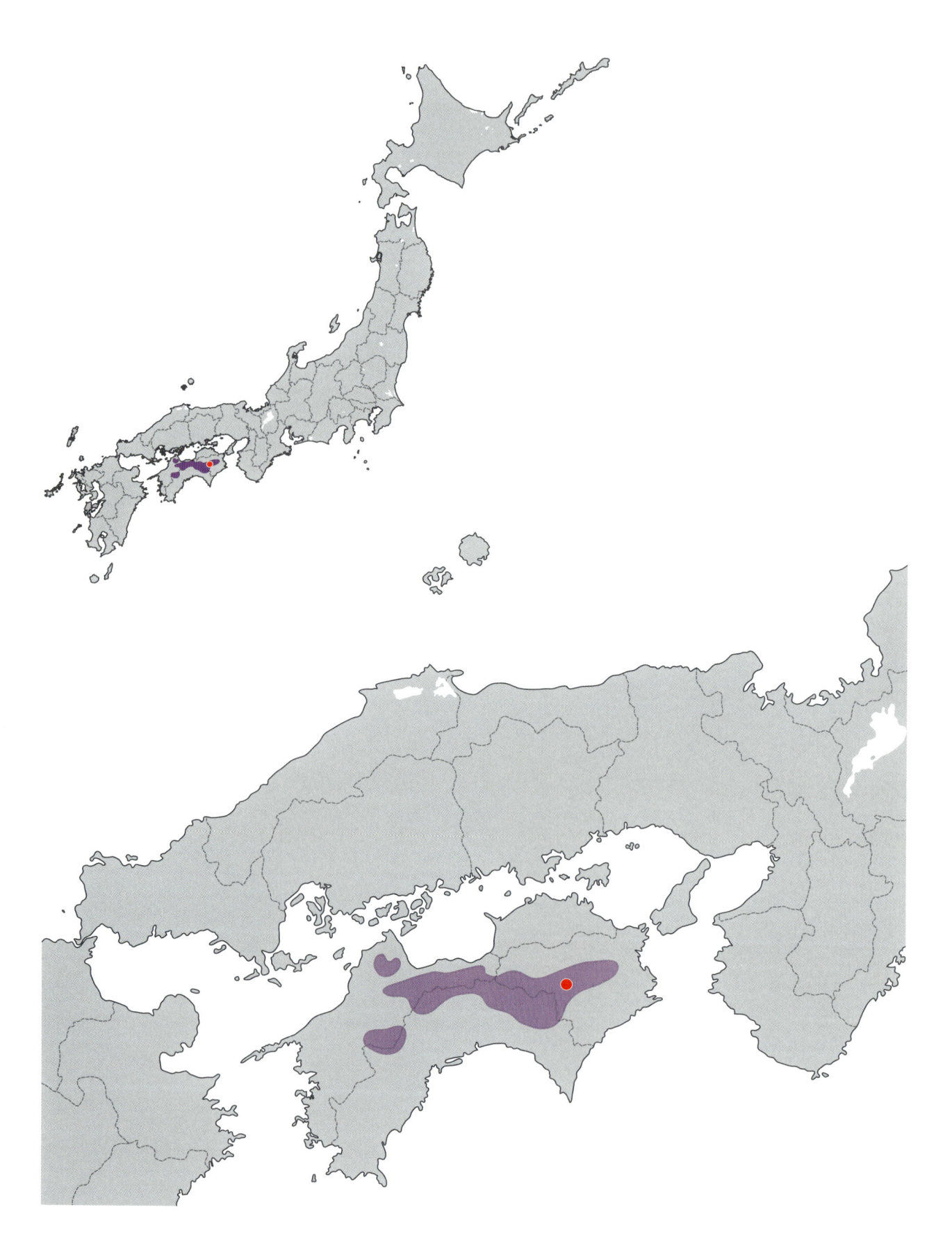

Fig. 26. Map showing the distributional range of ***Carabus*** (***Leptocarabus***) ***hiurai*** Kamiyoshi et O, Mizoguchi, 1960. ● : Type locality of the species (Mt. Tsurugi).

9. *Carabus* (*Leptocarabus*) *hiurai*

Fig. 27. Habitat of *Carabus* (*Leptocarabus*) *hiurai* (Mt. Tsurugi-san, Tokushima Pref.).
図27. シコククロナガオサムシの生息地(徳島県剣山).

Fig. 28. *Carabus* (*Leptocarabus*) *hiurai* passing the winter in rotten wood (Mt. Suiha-miné, Shikoku-chûô-shi, Ehime Pref.).
図28. 朽木中で越冬しているシコククロナガオサムシ(愛媛県四国中央市翠波峰).

Fig. 29. *Carabus* (*Leptocarabus*) *hiurai* passing the winter in the soil (Mt. Higashi-mitsumori-yama, Niihama-shi, Ehime Pref.).
図29. 土中で越冬しているシコククロナガオサムシ(愛媛県新居浜市東光森山).

10. *Carabus* (*Leptocarabus*) *kumagaii* (Kimura et Komiya, 1974)
オオクロナガオサムシ

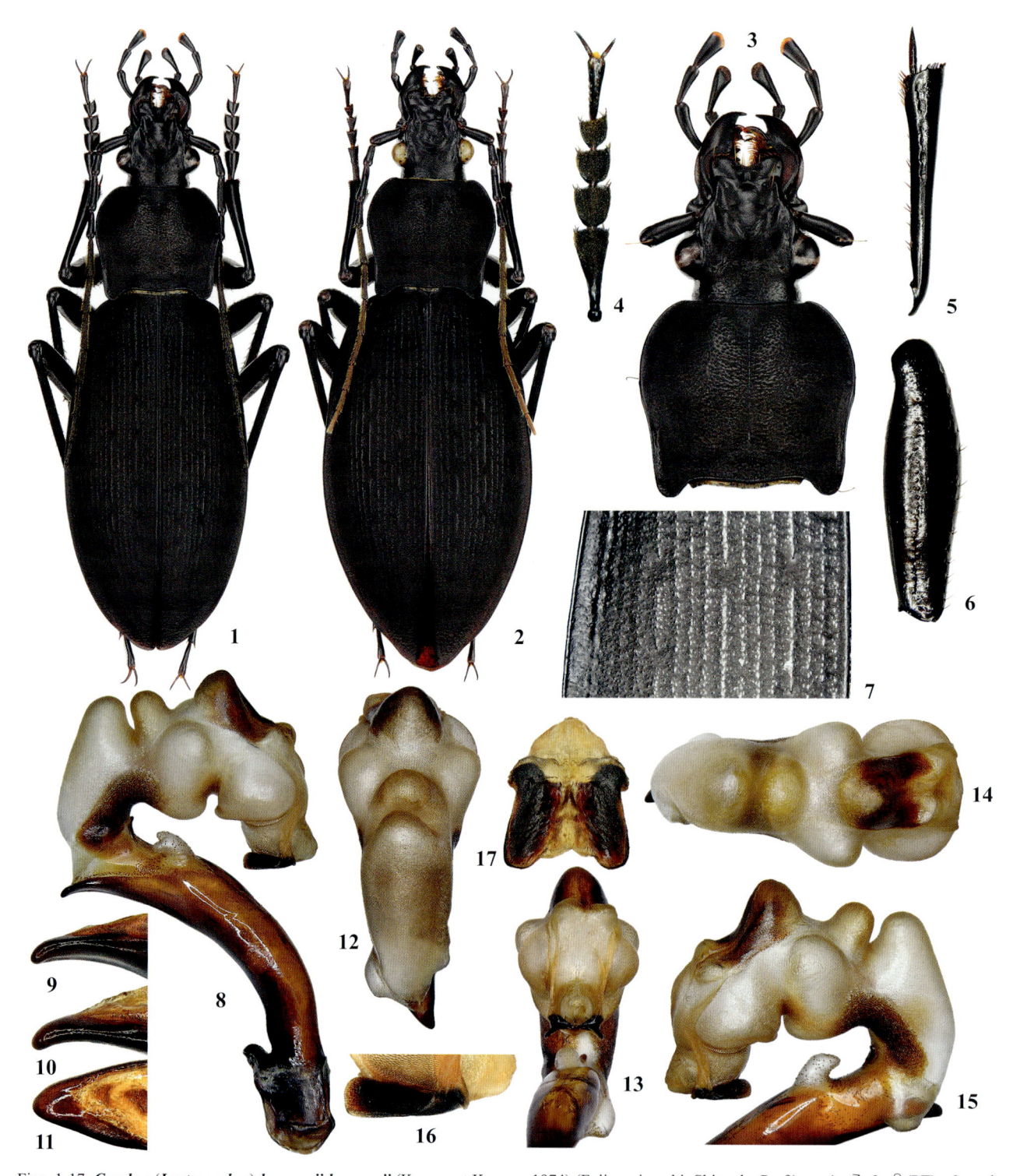

Figs. 1-17. *Carabus* (*Leptocarabus*) *kumagaii kumagaii* (Kimura et Komiya, 1974) (Fujinomiya-shi, Shizuoka Pref.). — 1, ♂; 2, ♀ (PT); 3, male head & pronotum; 4, male right protarsus in ventral view; 5, male right protibia in subdorsal view; 6, male right profemur in left lateral view; 7, male left elytron (median part); 8, penis & fully everted internal sac in right lateral view; 9, apical part of penis in right lateral view; 10, ditto in right subdorsal view; 11, ditto in dorsal view; 12, internal sac in caudal view; 13, ditto in frontal view; 14, ditto in dorsal view; 15, ditto in left sublateral view; 16, terminal plate of internal sac in frontal view; 17, ditto in right lateral view.

10. *Carabus* (*Leptocarabus*) *kumagaii*

Figs. 18-25. ***Carabus*** (***Leptocarabus***) ***kumagaii*** subspp. — 18-19, ***C***. (***L***.) ***k***. ***hida*** (18, ♂, 19, ♀, Uwano-machi, Takayama-shi, Gifu Pref.); 20-21, "**Tenryû-gawa population**" (20, ♂, 21, ♀, Ina-tajima, Nakagawa-mura, Nagano Pref.); 22-25, ***C***. (***L***.) ***k***. ***nishi*** (22, 24, ♂, 23, 25, ♀; 22-23, Mt. Kongô-san (= Kongô-zan), Gosé-shi, Nara Pref.; 24-25, Is. Tôshi-jima, Toba-shi, Mie Pref.).

図 18-25. オオクロナガオサムシ各亜種. — 18-19, 飛騨高山亜種(18, ♂, 19, ♀, 岐阜県高山市上野町); 20-21, **天竜川個体群**(20, ♂, 21, ♀, 長野県中川村伊那田島); 22-25, 近畿中部地方亜種(22, 24, ♂, 23, 25, ♀; 22-23, 奈良県御所市金剛山; 24-25, 三重県鳥羽市答志島).

10. *Carabus* (*Leptocarabus*) *kumagaii* (Kimura et Komiya, 1974) オオクロナガオサムシ

Original description *Leptocarabus kumagaii* Kimura et Komiya, 1974, Kontyû, 42 (4), p. 396, figs. 6a & 6b (p. 397).

History of research This species had been confused with *Carabus* (*Leptocarabus*) *procerulus* until it was described as a new species by Kimura and Komiya (1974) mainly based on the difference in shape of the terminal plate of internal sac of the male genitalia. It was revised by Kubota & Ishikawa (2004), and is currently classified into three subspecies.

Type locality Fujinomiya City, Shizuoka Pref.

Type depository Systematic Zoology Laboratory, Department of Biological Sciences, Tokyo Metropolitan University (Hachiôji) [Holotype, ♂]. Depository of the type specimen designated by Kimura & Komiya (1974) is the private collection of Kinji Kimura, but the holotype is now preserved in Tokyo Metropolitan University.

Morphology 26~39 mm. Body above black and mat. External features: Figs. 1~7. Description on details: penultimate segment of labial palpus bisetose; median tooth of mentum almost as long as, or a little longer than lateral lobes, narrowly extending anteriad, with apex not so sharply pointed in ventral view, protruded ventrad and tapered toward apex in lateral view; submentum smooth and asetose; male antennae with long and marked hairless ventral depressions from segment five to eight, though sometimes short and unclear in segment eight; pronotal margins basically bisetose; preapical emargination of elytra faintly recognized in female; elytral sculpture as in *C.* (*L.*) *kyushuensis* and *C.* (*L.*) *hiurai*, though primaries the widest and strongest, secondaries and tertiaries almost equally raised to form fine costae or rows of granules, areas between intervals coarsely scattered with small granules; umbilicate series usually recognized in basal half of elytra; propleura smooth; sides of sternites weakly rugoso-punctate; metacoxa basically bisetose, lacking anterior seta; sternal sulci clearly carved. Male genitalic features: Figs. 8~17. Penis less than half the elytral length; ostium lobe small and weakly bilobed at tip; internal sac almost as in *C.* (*L.*) *hiurai*, though presaccellar lobe is a little smaller, parasaccellar lobes are more weakly protruded, terminal plate is much longer and rounded at apex. Female genitalia with inner plate of ligular apophysis a little narrower than that of *C.* (*L.*) *procerulus*. From *C.* (*L.*) *procerulus*, this species is distinguished by larger body, broader pronotum, shorter umbilicate series, differently shaped terminal plate of internal sac, and narrower inner plate of ligular apophysis, etc.

Molecular phylogeny For details, see the same section of *C.* (*L.*) *kyushuensis*.

Distribution SC.~SW. Honshu / Iss. Tôshi-jima (Mie Pref.) & Oki-no-shima (Wakayama Pref.).

Habitat Hilly to mountainous region.

Bionomical notes Autumn breeder. Larvae feed on larvae of other insects.

Geographical variation Classified into three subspecies.

Etymology Named after Kômei Kumagai, a member of the research group on carabid beetles in Keihin Insect Lovers' Club, Tokyo.

研究史　本タクソンは長年，クロナガオサムシと混同されていたが，主として♂交尾器形態，とりわけ内嚢頂板の形態が異なるため，独立種として記載された．本種の記載はKONTYÛ（昆蟲）の42巻4号395~400ページに掲載された論文において行われたが，これに先だって発行された京浜昆虫同好会の会誌INSECT MAGAZINE 76号（奥付には1970年12月31日発行と記されているが，実際にこの雑誌が配布されたのは1971年に入ってかなり経ってからであったことが明らかなので，出版年は1971年とみなすべきである）の85~86ページには，ラテン語で記された新種名の提示と詳細な形態の記載が載っている．学名の後に「（未記載）」と記されているので，国際動物命名規約第4版条8.2「公表は棄権しうる」の条文に照らし，この論文は公表されたものにならないと解釈されるが，この一語がなければこの同好会誌上の報文が原記載と解釈される．記載から30年以上に亘り単型種として扱われてきたが，Kubota & Ishikawa (2004) により計3亜種に分類された．クロナガオサムシとの棲み分けについては近畿オサムシ研究グループにより詳しい調査がなされており，両種の分布境界線はP-K lineと呼ばれている.

タイプ産地　静岡県，富士宮市．

タイプ標本の所在　首都大学東京動物系統分類学研究室（八王子）［ホロタイプ，♂］　原記載論文ではホロタイプの保管先が「木村欣二個人コレクション」となっているが，現在は首都大学東京に収蔵されているようである．

形態　26~39 mm．背面は光沢を欠く黒色．外部形態：図1~7．細部の記載：下唇肢亜端節の剛毛は2本．下唇基節の中央歯は側葉とほぼ同長ないしやや長く，前方に向けて細長く伸び，先端はあまり鋭く尖らず，側面から見ると先端に向けて細まり，腹側に突出する．下唇亜基節は無毛で表面は平滑．♂触角第5節から第8節の腹面には，各節の全長近くに亘る顕著な無毛帯があるが，第8節のものは時にやや短く不鮮明となる場合がある．前胸背板の側縁剛毛は2対（中央と後角付近に各1対）．鞘翅端部は♀でごく僅かにえぐれる．鞘翅彫刻は前2種のそれに近いが，第一次間室の鎖線が通常，最も幅広く強く，第二次，第三次間室はほぼ同等に隆起した弱い隆条または顆粒列で，鞘翅全体が小顆粒により密に覆われる．丘孔点列は通常，鞘翅前半部に認められる．前胸側板は滑らかで，腹部腹板の側面には弱い皺と点刻を装う．後基節の剛毛は2本（前方剛毛を欠く）．腹部腹板の横溝は明瞭．♂交尾器形態：図8~17．陰茎は鞘翅長の1/2よりやや短い．葉片は小さく，先端は軽く分葉する．内嚢各部の基

10. *Carabus* (*Leptocarabus*) *kumagaii*

板はより幅広く，鞘翅の丘孔点列がより短く，♂交尾器内嚢先端の頂板の形が異なり，膣底部節片内板がより細いこと等により識別される．

分子系統 詳細はキュウシュウクロナガオサムシの分子系統の項を参照．

分布 本州中南部～中西部／答志島（三重県），沖ノ島（和歌山県）．クロナガオサムシとはほぼ側所的に棲み分けている（両種の分布状況についてはクロナガオサムシの同項を参照）が，分布を接する地域では両者の雑種と思われる個体が得られている（長野県下伊那郡，伊那市，三重県鈴鹿市，亀山市，津市など）．

生息環境 山地から丘陵，段丘上にかけて生息するが，大きい沖積平野では採集例が少ない．山地では深い森林にも見られるが，低地では乾燥した二次林に多く，この点においてオオオサムシ属各種と対照的であるとされる（近畿オサムシ研究グループ，1979）．

生態 秋繁殖・幼虫越冬型のオサムシで，幼虫は基本的に昆虫の幼虫食．

地理的変異 記載から30年以上に亘り単型種として扱われてきたが，Kubota & Ishikawa（2004）により計3亜種に分類された．

種小名の由来 京浜昆虫同好会オサムシグループの一員であった熊谷幸明に因む．

10-1. *C. (L.) k. kumagaii* (Kimura et Komiya, 1974) オオクロナガオサムシ基亜種

Original description, type locality & type depository See explanation of the species.

Diagnosis 29~35 mm. Pronotum deeply and densely punctate, with distinct transverse wrinkles. Granules in umbilicate series disappearing behind middle of elytra. Inner plate of ligular apophysis narrow.

Distribution Area between the Fuji-kawa drainage & E. bank of Riv. Ôi-gawa (C. Shizuoka Pref. & SW. Yamanashi Pref.), SC. Honshu.

原記載・タイプ産地・タイプ標本の所在 種の項を参照．

亜種の特徴 29~35 mm．前胸背板の点刻は深く密で，横皺が強い．丘孔点列は鞘翅中央よりも後方では消失する．♀交尾器膣底部節片内板の幅は狭い．

分布 富士川流域～大井川東岸（静岡県中部～山梨県南西部）．

10-2. *C. (L.) k. hida* (Kubota et Ishikawa, 2004) オオクロナガオサムシ飛騨高山亜種

Original description *Leptocarabus* (*Leptocarabus*) *kumagaii hida* Kubota et Ishikawa, 2004, Biogeography, 6, p. 33, figs. 2 (p. 28), 7 (p.30), 11 (p.31) & 16, 17 (p.32).

Type locality Uwano-chô [sic] (more precisely, Uwano-machi), 660 m, Takayama-shi, Gifu Pref.

Type depository Systematic Zoology Laboratory, Department of Biological Sciences, Tokyo Metropolitan University (Hachiôji) [Holotype, ♂].

Diagnosis 30~37 mm. Pronotum shallowly punctate and weakly rugulose. Elytra slenderer than in other two subspecies; granules in umbilicate series recognized continuously nearly to elytral apices. Female genitalia with inner plate of ligular apophysis apparently broader than in other two subspecies, though narrower than in *C. (L.) procerulus*.

Distribution NE. Takayama Basin (Gifu Pref.), C. Honshu.

Etymology The subspecific name, Hida, is a traditional regional name where the type locality of this subspecies is located.

タイプ産地 岐阜県，高山市，上野町（うわのまち；原記載では「うわのちょう」となっているが，「うわのまち」と読むべきもののようである），660 m．

タイプ標本の所在 首都大学東京動物系統分類学研究室（八王子）［ホロタイプ，♂］．

亜種の特徴 30~37 mm．前胸背板の点刻は浅く，横皺は弱い．鞘翅は他の2亜種よりも細長く，丘孔点列は鞘翅中央より後方まで連続し，断続しつつ鞘翅先端付近に達する．♀交尾器膣底部節片は他亜種より明らかに幅広い．

分布 高山盆地北東部（岐阜県高山市）．

亜種小名の由来と通称 亜種小名は飛騨地方に因む．原記載論文中で提唱された通称はヒダオオクロナガオサムシ．

10-3. *C. (L.) k. nishi* (Kubota et Ishikawa, 2004) オオクロナガオサムシ近畿・中部地方亜種

Original description *Leptocarabus* (*Leptocarabus*) *kumagaii nishi* Kubota et Ishikawa, 2004, Biogeography, 6, p. 29, figs. 6 (p. 30), 10 (p. 31) & 14, 15 (p. 32).

Type locality Mt Kongôzan, 500-1,100 m, Gose-shi, Nara Pref.

Type depository Systematic Zoology Laboratory, Department of Biological Sciences, Tokyo Metropolitan University (Hachiôji) [Holotype, ♂].

Diagnosis 26~39 mm. Pronotum shallowly punctate and weakly rugulose. Umbilicate series not extending beyond middle of elytra. Inner plate of ligular apophysis narrow.

Distribution Kinki Region (Kyoto Pref., Osaka Pref., Nara Pref., Wakayama Pref. & Mie Pref.), S. Chubu Region & Tokai

10. *Carabus* (*Leptocarabus*) *kumagaii*

Region (Shizuoka Pref. & Nagano Pref. by Riv. Tenryû-gawa), WC. Honshu / Iss. Tôshi-jima (Mie Pref.) & Oki-no-shima (Wakayama Pref.).

Habitat   In the Kinki Region, this subspecies is found from the plain up to the heights, at elevations where *C.* (*L.*) *procerulus* is not distributed. In the southern part of the Chubu Region, it inhabits hilly region by Riv. Tenryû-gawa in Shizuoka Prefecture extending northwards to the Ina-dani Valley in Nagano Prefecture, where the range appears to be roughly delimited by branches of the Katagirimatsu-kawa and Koshibu-gawa Rivers, parapatric with the Tenryû-gawa population.

Etymology   The subspecific name "Nishi" means West in Japanese, since this subspecies occupies the westernmost part of the distributional range of *C.* (*L.*) *kumagaii*.

タイプ産地　奈良県, 御所市, 金剛山, 500-1,100 m.

タイプ標本の所在　首都大学東京動物系統分類学研究室(八王子)[ホロタイプ, ♂].

亜種の特徴　26〜39 mm. 前胸背板の点刻は浅く, 横皺も基亜種に比べはるかに弱い. 鞘翅最広部は中央より後方にあり, 丘孔点列は鞘翅中央を大きく越えない. ♀交尾器膣底部節片は比較的幅の狭いものが多い.

分布　近畿地方中東部(京都, 大阪, 奈良, 和歌山, 三重各県)および中部地方南部(天竜川中下流域)／答志島(三重県), 沖ノ島(和歌山県). 近畿地方と天竜川中下流域に二つのかけ離れた分布域を持ち, 天竜川上流域の右岸(西岸)では大田切川(おおたぎりがわ)〜片桐松川(かたぎりまつかわ)間, 同左岸(東岸)では三峰川(みぶがわ)〜小渋川(こしぶがわ)間から記録されている. KUBOTA & ISHIKAWA(2004)では触れられていないが, この他に長野県南西部から岐阜県南東部にかけての木曽川上流域からも記録されている.

亜種小名の由来と通称　亜種小名は, 本亜種が本種の分布域の最西部を占める集団であることに因む. 原記載論文中で提唱された通称はニシオオクロナガオサムシ.

※ *C.* (*L.*) *kumagaii* "Tenryû-gawa population" オオクロナガオサムシ天竜川個体群

In the Ina-dani Basin, the upper reaches of the Tenryû-gawa River, a unique population of *C.* (*L.*) *kumagaii* is distributed, called "the Tenryû-gawa population" (KUBOTA & ISHIKAWA, 2004, p. 34). They are large-sized, with extended umbilicate series and broader inner plate of ligular apophysis of the female genitalia, though the length of umbilicate series is somewhat variable. This population is distributed in the area between Ôtagiri-gawa and Katagirimatsu-kawa Rivers on the western bank of the Tenryû-gawa River, and that between Mibu-gawa and Koshibu-gawa Rivers on the eastern bank. It is parapatric in the south with *C.* (*L.*) *k. nishi* and is intergraded with the latter.

天竜川上流域の伊那谷一帯には, やや大型で, 丘孔点列の長さは不規則ながらより後方へ伸張する傾向をもち, ♀交尾器膣底部節片がより幅広い, 飛騨高山亜種に似たオオクロナガオサムシの集団が分布しており, 天竜川個体群と名付けられている(KUBOTA & ISHIKAWA, 2004, p. 34). その分布範囲は狭く, 天竜川右岸(西岸)では大田切川〜片桐松川間, 同左岸(東岸)では三峰川〜小渋川間から知られており, 分布域南部では以南に分布するオオクロナガオサムシ近畿・中部亜種と分布を接し, 後者への移行を示す個体が得られている.

10. *Carabus* (*Leptocarabus*) *kumagaii*

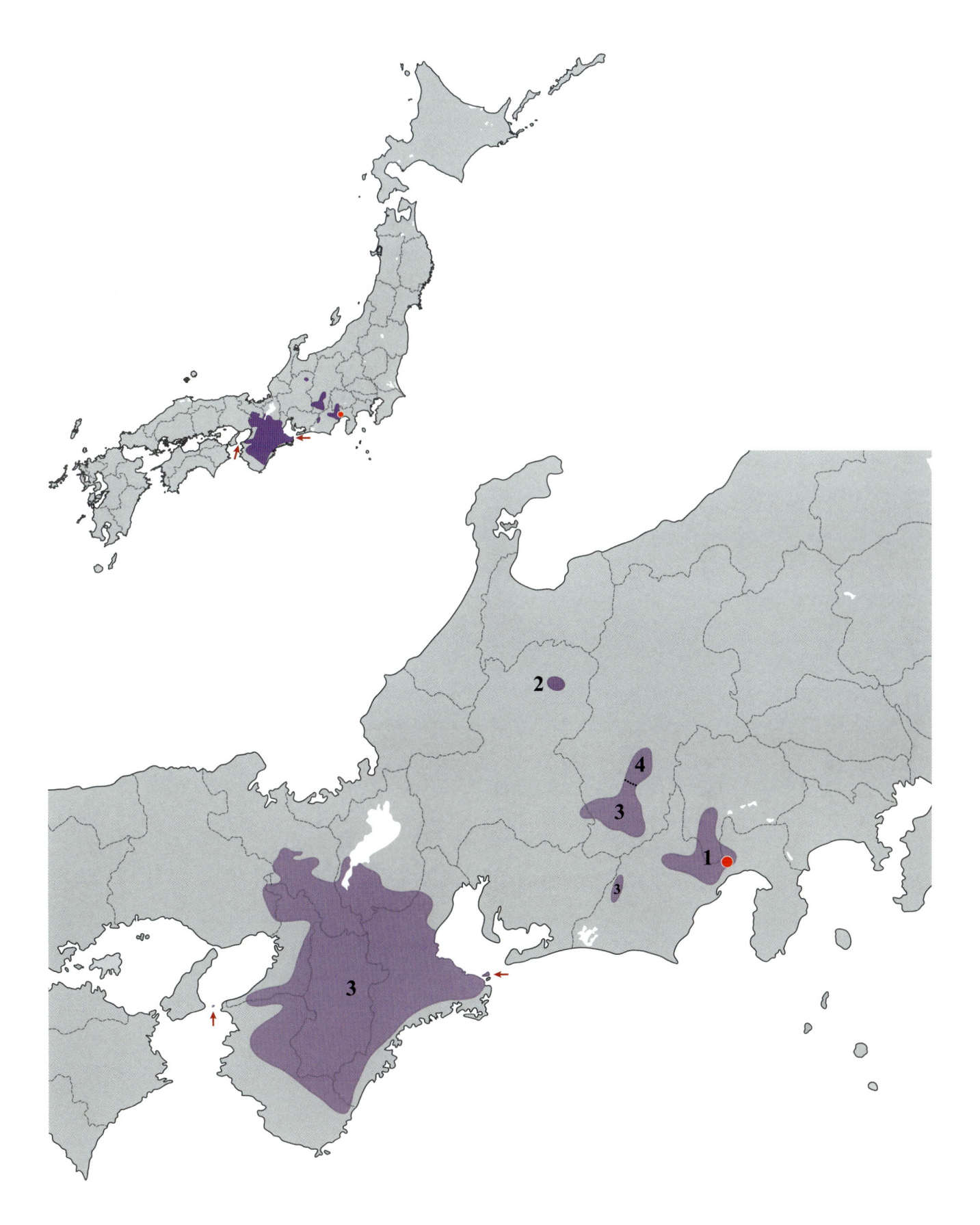

Fig. 26. Map showing the distributional range of ***Carabus*** (***Leptocarabus***) ***kumagaii*** (Kaimura et Komiya, 1971). — 1, Subsp. ***kumagaii***; 2, subsp. ***hida***; 3, subsp. ***nishi***; 4, Tenryû-gawa population. ●: Type locality of the species (Fujinomiya City).
図 26. **オオクロナガオサムシ**分布図. — 1, 基亜種; 2, 飛騨高山亜種; 3, 近畿・中部地方亜種; 4, 天竜川個体群. ●: 種のタイプ産地（富士宮市）.

11. *Carabus* (*Leptocarabus*) *procerulus* Chaudoir, 1862
クロナガオサムシ

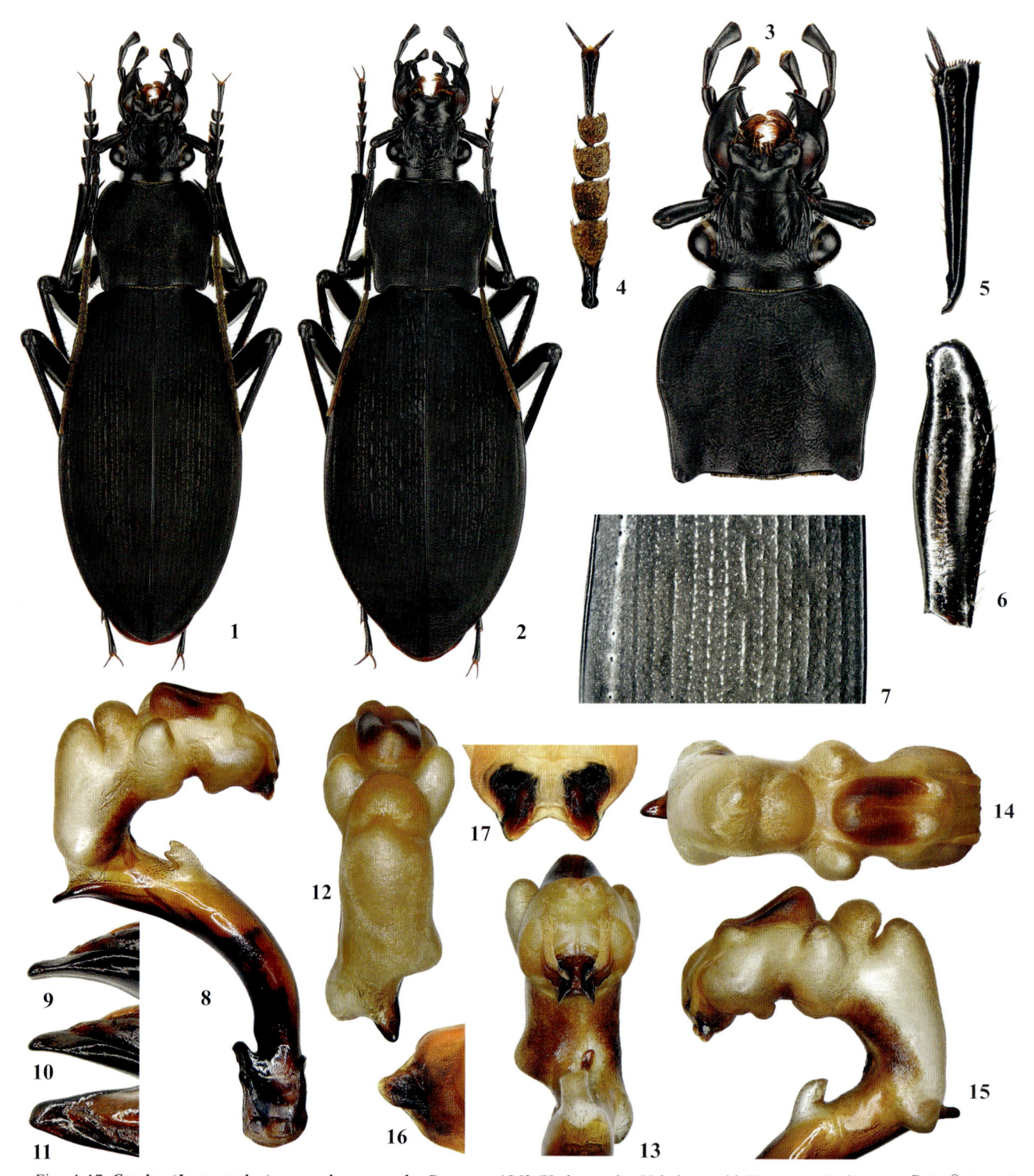

Figs. 1-17. ***Carabus*** (***Leptocarabus***) ***procerulus procerulus*** Chaudoir, 1862 (Hodogaya-ku, Yokohama-shi, Kanagawa Pref.). — 1, ♂; 2, ♀; 3, male head & pronotum; 4, male right protarsus in ventral view; 5, male right protibia in subdorsal view; 6, male right profemur in left lateral view; 7, male left elytron (median part); 8, penis & fully everted internal sac in right lateral view; 9, apical part of penis in right lateral view; 10, ditto in right subdorsal view; 11, ditto in dorsal view; 12, internal sac in caudal view; 13, ditto in frontal view; 14, ditto in dorsal view; 15, ditto in left sublateral view; 16, terminal plate of internal sac in frontal view; 17, ditto in right lateral view.

図 1-17. **クロナガオサムシ基亜種**(神奈川県横浜市保土ヶ谷区). — 1, ♂; 2, ♀; 3, ♂の頭部と前胸背板; 4, ♂右前跗節(腹面観); 5, ♂右前脛節(斜背面観); 6, ♂右前腿節(左側面観); 7, ♂左鞘翅中央部; 8, 陰基と完全反転下の内嚢(右側面観); 9, 陰基先端(右側面観); 10, 同(右斜背面観); 11, 同(背面観); 12, 内嚢(尾側面観); 13, 同(頭側面観); 14, 同(背面観); 15, 同(左側側面観); 16, 頂板(頭側面観); 17, 同(右側面観).

11. *Carabus* (*Leptocarabus*) *procerulus*

Figs. 18-29. ***Carabus*** (***Leptocarabus***) ***procerulus*** subspp. — 18-25, ***C***. (***L***.) ***p***. ***procerulus*** (18, 20, 22, 24, ♂, 19, 21, 23, 25, ♀; 18-19, Shikanai, Toyomanai, Gonohé-machi, Aomori Pref.; 20-21, Pass Daibosatsu-tôgé, Kôshû-shi, Yamanashi Pref.; 22-23, Azô-chô, Toyota-shi, Aichi Pref.; 24-25, Mt. Dai-sen, Daisen-chô, Tottori Pref.); 26-29, ***C***. (***L***.) ***p***. ***miyakei*** (26, 28, ♂, 27, 29, ♀; 26-27, Kita-senri-ga-hama, Kujû-san Mts., Taketa-shi, Oita Pref.; 28-29, Mt. Hiko-san, Soeda-machi, Fukuoka Pref.). — a, Penis in right lateral view; b, apical part of penis in dorsal view; c, terminal plate of internal sac in frontal view.

図 18-29. **クロナガオサムシ**各亜種. — 18-25, **基亜種**(18, 20, 22, 24, ♂, 19, 21, 23, 25, ♀; 18-19, 青森県五戸町豊間内鹿内; 20-21, 山梨県甲州市大菩薩峠; 22-23, 愛知県豊田市阿蔵町; 24-25, 鳥取県大山町大山); 26-29, **九州亜種**(26, 28, ♂, 27, 29, ♀; 26-27, 大分県竹田市九重山北千里ヶ浜; 28-29, 福岡県添田町英彦山). — a, 陰茎(右側面観); b, 陰茎先端(背面観); c, 内嚢頂板(頭側面観).

11. *Carabus* (*Leptocarabus*) *procerulus* Chaudoir, 1862 クロナガオサムシ

Original description　*Carabus procerulus* Chaudoir, 1862, Rev. Mag. Zool., 14, p. 486.
History of research　This species was described by Chaudoir (1862) from "Japon" without designation of detailed type locality. Nakane (1961) described subsp. *miyakei* from the Kujû Mountains of northern Kyushu.
Type locality　Japon.
Type depository　Muséum Nationale d'Histoire Naturelle, Paris [Lectotype, ♂].
Morphology　24~35 mm. Body above black and mat. External features: Figs. 1~7. Description on details: general features almost as in *C.* (*L.*) *kumagaii*; median tooth of mentum variable in length, but usually longer than lateral lobes and sharply pointed at tip in ventral view, above all in subsp. *miyakei*, and less strongly tapered toward apex than in *C.* (*L.*) *kumagaii* in lateral view; submentum smooth and asetose, though seldom setiferous in a certain population; male antennae with long and marked hairless ventral depressions from segment five to eight, though sometimes short and unclear in segment eight; pronotal margins basically bisetose, though often lacking anterior seta in subsp. *miyakei*; preapical emargination of elytra faintly recognized in female; elytral sculpture as in *C.* (*L.*) *kyushuensis* and *C.* (*L.*) *hiurai*, though primaries the widest and strongest, secondaries and tertiaries almost equally raised to form fine costae or rows of granules, areas between intervals a little less coarsely scattered with small granules; umbilicate series usually recognized from bases to tips of elytra; propleura smooth, at most vaguely punctate; sides of sternites weakly rugoso-punctate; metacoxa basically bisetose, lacking anterior seta; sternal sulci clearly carved. Male genitalic features: Figs. 8~17. Penis usually more than 0.4 times as long as elytra; ostium lobe small and weakly bilobed at tip; pigmentation, sclerotization and feature of ligulum of internal sac almost as in *C.* (*L.*) *hiurai*; basic structure of internal sac also similar to that of *C.* (*L.*) *hiurai*, though presaccellar lobe is a little smaller, parasaccellar lobes more weakly protruded; terminal plate similar to that of *C.* (*L.*) *kumagaii*, but a little shorter and acute-angled at outer corner.
Molecular phylogeny　For details, see the same section of *C.* (*L.*) *kyushuensis*.
Distribution　Honshu (from E. Aomori Pref. to NW. Hiroshima Pref.) & Kyushu (mountainous area) / Iss. Ô-shima (Kesennuma-shi, Miyagi Pref.), Noto-jima (Ishikawa Pref.) & Oki-no-shima [Lake Biwa-ko] (Shiga Pref.).
Habitat　Widely distributed from the plain to subalpine zone, and usually not found from the alluvial plain. Common in the edge of forests, either natural or planted, but uncommon in open lands or excessively deep, highly humid forest.
Bionomical notes　Autumn breeder. Larvae prefer to feed on larvae of other insects.
Geographical variation　Classified into two subspecies, though the geographical variation of so-called nominotypical subspecies has not been much researched as yet.
Etymology　The specific name means "somewhat elongated".

　研究史　フランスのChaudoir (ショドワー) (1862) によって記載された，邦産オサムシのなかでも最もなじみ深い種の一つであるが，タイプ産地の詳細は不明である．Nakane (1961) により九州亜種 *miyakei* が記載されており，同じNakaneにより同年，本種の亜種として *Carabus* (*Leptocarabus*) *procerulus kyushuensis* と *C.* (*L.*) *p. shikokuensis* (*C. hiurai* Kamiyoshi et Mizoguchi, 1960の異名) が記載されたが，現在これらはそれぞれ，独立種キュウシュウクロナガオサムシ，シコククロナガオサムシとして扱われている．また，長年本種と混同されてきたオオクロナガオサムシは1974年に独立種として記載された．本邦の特産種で，2亜種に分類されるが，いわゆる基亜種に相当するとされる本州産の集団の地理的変異や亜種分類に関する研究は不十分である．

　タイプ産地 (地域)　日本．
　タイプ標本の所在　パリ自然史博物館 [レクトタイプ，♂]．筆者 (井村) は，パリ自然史博物館に保管されている本種のレクトタイプ標本を借り受け，♂交尾器内嚢を解剖して頂板の形態を調べてみた．その結果，*C. procerulus* のレクトタイプの頂板形態は，オオクロナガオサムシのそれではなく，まさしく今日我々がクロナガオサムシと呼んでいる種のそれに該当するものであった．当然のように思われるかもしれないが，これは非常に大切な知見であるため，敢えて一言，触れておきたい．
　形態　24~35 mm．背面は光沢を欠く黒色．外部形態：図1~7．細部の記載：下唇肢亜端節の剛毛は2本．下唇基節の中央歯は側葉よりやや短いものからほぼ同長ないしやや長いものまで地域・個体により変異が見られ，先端はやや，ないし非常に鋭く尖り，側面から見るとオオクロナガオサムシほど先端に向けて細まることはない．下唇亜基節は無毛だが，有毛の個体が一定比率で混入する集団もある．♂触角第5節から第8節の腹面には顕著な無毛凹陥部がある．前胸背板の側縁剛毛は2対 (中央と後角付近に各1対) だが，分布域西部 (中部地方西部以西) のものと九州亜種では通常，前方剛毛を欠く．鞘翅端部はごく僅かにえぐれるが，♀においてより顕著で，♂ではしばしば不鮮明．鞘翅彫刻はオオクロナガオサムシのそれに近いが，鞘翅全体を覆う顆粒の密度がより低く，丘孔点列の顆粒は少なくとも鞘翅端近くまで認められる．前胸側板は平滑で，せいぜいごく弱い点刻を装う程度．腹部腹板の側面には弱い皺と点刻を装う．後基節の剛毛は2本 (前方剛毛を欠く)．腹部腹板の横溝は明瞭．♂交尾器形態：図8~17．陰茎はオオクロナガオサムシ同様，鞘翅長の0.4倍強のものが多い．葉片や内嚢の基本形態はシコククロナガオサムシのものに近い．頂板はオオクロナガオサムシ同様，強く硬化して突出するが，やや短く，先端外角は矢筈形に切れ込んで角をなす．
　分子系統　詳細はキュウシュウクロナガオサムシの分子系統の項を参照．
　分布　本州，九州，大島 (宮城県気仙沼市)，能登島 (石川県)，沖島 [琵琶湖] (滋賀県)．本州北端から中国地方の山地帯に

かけてほぼ連続的に分布するほか，九州中北部の山岳地帯にもかけ離れた分布圏を持つ．東北地方では日本海側の青森県西部〜秋田県〜山形県にかけての大半の地域から記録がなく，ごく数か所の孤立した小分布圏が知られるのみ．秋田県の日本海側では，県南部の由利本荘市(ゆりほんじょうし)鳥海町百宅(ちょうかいまちもももやけ)が唯一の産地として知られているに過ぎず(他に三ヶ所の記録があるが，いずれも青森，岩手，宮城県境に近い場所で，太平洋側の分布域の延長線上にあるとみなしうるもの)，同県のレッドデータブックでは絶滅危惧I類に指定されている．東北地方南部から北陸，東海地方にかけては，沿岸の平地を除き，ほぼ全域から記録があるが，伊勢湾以西の太平洋側では六甲山系西部の鈴蘭台から石楠花山(しゃくなげやま)にかけての地域に孤立した小分布圏が知られるのみで，基本的に太平洋側には分布しない．近畿地方における分布は北半部に限られ，南半部に分布するオオクロナガオサムシと棲み分けている．両者の境界(P-K line)は伊勢平野中央部〜安濃川(あのうがわ)〜野登山(ののぼりやま)東麓〜鈴鹿峠〜滋賀県日野町(ひのちょう)〜甲賀市(こうかし)水口町(みなくちちょう)〜雄琴(おごと)丘陵〜京都盆地北縁にあるとされる．オオクロナガオサムシのほうは亀岡盆地から北西へ旧和知町(わちちょう)(現京丹波町(きょうたんばちょう)の一部)の由良川(ゆらがわ)上流域まで進出している．両者の分布が接する地域では垂直的な棲み分けが見られ，本種が高所を，オオクロナガオサムシが低所を占めるが，鈴鹿山地の東側では本種が亀山市〜津市を結ぶ丘陵に進出していて上下関係が逆転している．東北地方から中部地方までは低地から山地，一部亜高山帯にかけて広く分布するが，西へ向かうほど山地性となり，中国地方では脊梁山地沿いに島根・広島県境の新造山(しんぞうじさん/しんぞうじやま)まで帯状に細長く分布を伸ばし，やや西にかけ離れた広島県西部，北広島町の聖山(ひじりやま)からも記録されている．この地域では高標高域に本種，より下部にキュウシュウクロナガオサムシが分布している場合が多いが，日本海側斜面ではかなりの低所まで本種が分布を拡げている場所もある．九州では英彦山(ひこさん)から国見岳を経て市房山(いちふさやま)にかけての九州山地高所に分布し，別亜種となる．

　生息環境　平地から丘陵，段丘，さらに山地に至る広い範囲に生息している．一般に大きい沖積平野には見られない場合が多いが，河川敷や湿地から得られたという記録もある．乾燥した二次林や，やや湿潤な森林を好む傾向にあり，開けた環境や極端に湿潤なうっ閉森林には少ない．

　生態　秋繁殖・幼虫越冬型のオサムシで，幼虫は基本的に節足動物(他種昆虫の幼虫など)食．土中・朽木中で越冬し，複数の個体が同一の越冬窩内に見られることもある．

　地理的変異　本種の地理的変異に関する検討はいまだ不十分である．木村(1971)は本種のタイプ産地(地域)を横浜付近と仮定したうえで，東北亜種(♂交尾器内嚢の頂板先端が矢筈形に切れ込み鋭角をなす，東北地方に分布．命名規約上適格な形での記載はなされていない)，基亜種原型，同中部型(上翅の彫刻が粗く陰茎先端部上面の窪みが明瞭，山梨県などのもの)，同名古屋型(前胸背板側縁剛毛は1対のみで胸側板の点刻を欠く，名古屋周辺〜京都北部に分布)，同西日本型(上翅彫刻が比較的滑らかで，陰茎はなだらかな弧を描いて内湾し，表面が滑らか，伯耆大山とその周辺地域のものが典型的)，九州亜種などに分類している．ここでは，あくまで暫定的ながら本州産の基亜種と九州亜種の2亜種に分類し，解説を加えた．

　種小名の由来　種小名は「高い，長い，伸びた」を意味する procerus に「小さい，やや〜な」を表す縮小辞 -ulus がついたもので，「やや長く伸びた」，「やや高い」を意味する．

11-1. *C. (L.) p. procerulus* Chaudoir, 1862　クロナガオサムシ基亜種

　Original description, type locality & type depository　See explanation of the species.
　Diagnosis　25~35 mm. See explanation of the species and the same section of the following subspecies, *miyakei*.
　Distribution　Honshu / Iss. Ô-shima (Kesennuma-shi, Miyagi Pref.), Noto-jima (Ishikawa Pref.) & Oki-no-shima [Lake Biwa-ko] (Shiga Pref.).

　原記載・タイプ産地(地域)・タイプ標本の所在　種の項を参照．
　亜種の特徴　25~35 mm．種の形態に関する解説および九州亜種の特徴の項を参照．
　分布　本州(北端から中国地方の山地帯にかけてほぼ連続的に分布)／大島(宮城県気仙沼市)，能登島(石川県)，沖島(滋賀県琵琶湖)．

11-2. *C. (L.) p. miyakei* Nakane, 1961　クロナガオサムシ九州亜種

　Original description　*Carabus* (*Leptocarabus*) *procerulus miyakei* Nakane, 1961, Fragm, coleopterol., Kyoto, (1), p. 2.
　Type locality　Kitasenrigahama, Bungo, Kyushu.
　Type depository　Hokkaido University Museum, Sapporo [Holotype, ♂].
　Diagnosis　24~30 mm. Body above black and mat. Differs from the nominotypical subspecies in the following points: 1) body a little smaller and slenderer; 2) median tooth of mentum more strongly protruded anteriad, and more sharply pointe at apex; 3) pronotum slenderer, with lateral margins less strongly sinuate, usually lacking anterior seta.
　Distribution　Kyushu (mountainous region).
　Etymology　Named after Yoshikazu Miyake.

11. *Carabus* (*Leptocarabus*) *procerulus*

　　タイプ産地　九州, 豊後, 北千里浜. 原記載に記されたタイプ産地名は "Kitasenrigahama" で, これは大分県の九重山群にある北千里浜(きたせんりがはま)のことを指すものと思われるが, ホロタイプに付されているラベルには, 表に "[KYUSHU]/ 豊後北千里ヶ浜/18.VIII.1943/Y. MIYAKE", 裏に "**Kusasenrigahama** (1300) near Mt. Kujû" と書かれている. "Kusasenrigahama" (草千里ヶ浜)は熊本県の阿蘇山にある地名で, 九重山群にはなく, ラベルまたは原記載の記述のいずれかに誤りがあるものと思われるが, ここでは一応, 原記載の記述に従って九重山群の北千里浜を本種のタイプ産地とみなしておく. また, 細かいことだが, 採集年月日も, 原記載では "18. VII. 1945" となっているのに対し, ラベルには "18. **VIII**. 194**3**" と記されているように読める. 著者の中根による誤読があったかもしれない.

　　タイプ標本の所在　北海道大学総合博物館(札幌)[ホロタイプ, ♂].

　　亜種の特徴　24~30 mm. 基亜種とは以下の点で異なる: 1)やや小型で体型は細い; 2)下唇基節の中央歯は側葉を越えてより長く前方に伸長し, 先端が非常に鋭く尖る; 3)前胸背板がより細く, 側縁の波曲は弱く, 通常, 前方剛毛を欠く.

　　分布　九州(九州山地高所).

　　亜種小名の由来と通称　亜種小名は三宅義一に因む. ホソムネクロナガオサムシという通称で呼ばれることが多いが, ミヤケクロナガオサムシという名が提唱されたこともある.

11. *Carabus* (*Leptocarabus*) *procerulus*

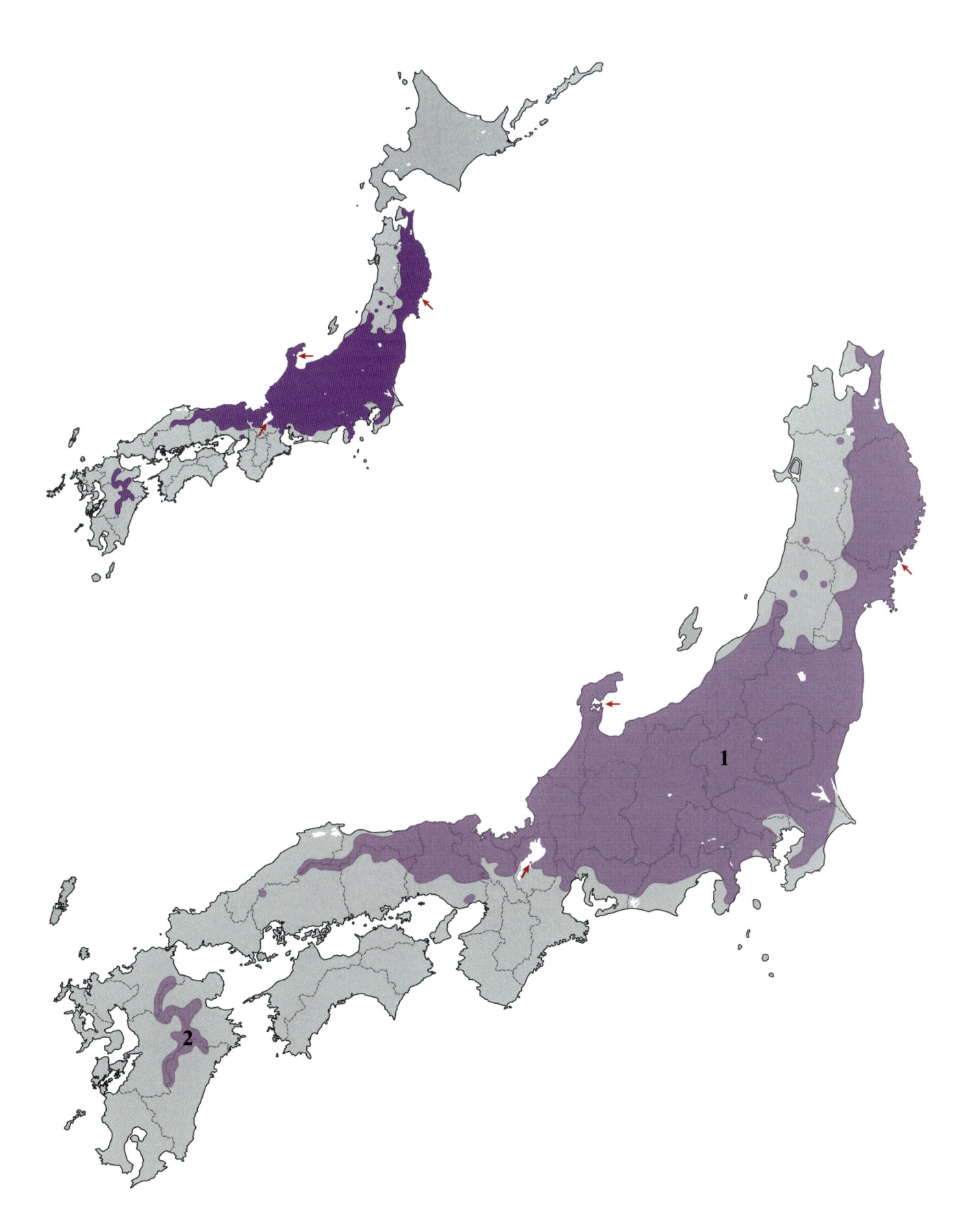

Fig. 30. Map showing the distributional range of ***Carabus*** (***Leptocarabus***) ***procerulus*** CHAUDOIR, 1862. — 1, Subsp. ***procerulus***; 2, subsp. ***miyakei***. Type locality of the species is "Japon", detailed locality is unknown.

図 30. **クロナガオサムシ**分布図. — 1, **基亜種**; 2, **九州亜種**. 種のタイプ産地は「日本」(詳しい場所は不明).

12. *Carabus* (*Leptocarabus*) *arboreus* Lewis, 1882
コクロナガオサムシ

Figs. 1-17. *Carabus* (*Leptocarabus*) *arboreus arboreus* Lewis, 1882 (Lake Junsai-numa, Nanaé-chô, Hokkaido). — 1, ♂; 2, ♀; 3, male head & pronotum; 4, male right protarsus in ventral view; 5, male right protibia in subdorsal view; 6, male right profemur in left lateral view; 7, male left elytron (median part); 8, penis & fully everted internal sac in right lateral view; 9, apical part of penis in right lateral view; 10, ditto in right subdorsal view; 11, ditto in dorsal view; 12, internal sac in caudal view; 13, ditto in frontal view; 14, ditto in dorsal view; 15, ditto in left sublateral view; 16, terminal plate of internal sac in frontal view; 17, ditto in right lateral view.

図 1-17. コクロナガオサムシ基亜種（北海道七飯町蓴菜沼）. — 1, ♂; 2, ♀; 3, ♂の頭部と前胸背板; 4, ♂の右前跗節（腹面観）; 5, ♂右前脛節（斜背面観）; 6, ♂右前腿節（左側面観）; 7, ♂左鞘翅中央部; 8, 陰茎と完全反転下の内嚢（右側面観）; 9, 陰茎先端（右側面観）; 10, 同（右斜背面観）; 11, 同（背面観）; 12, 内嚢（尾面観）; 13, 同（頭側面観）; 14, 同（背面観）; 15, 同（左斜側面観）; 16, 頂板（頭側面観）; 17, 同（右側面観）.

12. *Carabus* (*Leptocarabus*) *arboreus*

Figs. 18-29. *Carabus* (*Leptocarabus*) *arboreus* subspp. — 18-19, *C.* (*L.*) *a. pararboreus* (18, ♂, 19, ♀; Naka-ashoro, Ashoro-chô, Hokkaido); 20-21, *C.* (*L.*) *a. ishikarinus* (20, ♂, 21, ♀; Hitsuji-ga-oka, Sapporo-shi, Hokkaido); 22-23, *C.* (*L.*) *a. karibanus* (22, ♂, 23, ♀; Mt. Kariba-yama, Shimamaki-mura, Hokkaido); 24-25, *C.* (*L.*) *a. nepta* (24, ♂, 25, ♀; Mt. Osoré-zan, Mutsu-shi, Aomori Pref.); 26-27, *C.* (*L.*) *a. pronepta* (26, ♂, 27, ♀; southwestern Ninohé-shi (former Jôbôji-machi), Iwate Pref.); 28-29, *C.* (*L.*) *a. shimoheiensis* (28, ♂, 29, ♀; Ishinazawa, Miyako-shi, Iwate Pref.). — a, Penis in right lateral view; b, apical part of penis in right subdorsal view.

図 18-29. コクロナガオサムシ各亜種. — 18-19, 道央道東道北亜種(18, ♂, 19, ♀; 北海道足寄町中足寄); 20-21, 石狩亜種(20, ♂, 21, ♀; 北海道札幌市羊ヶ丘); 22-23, 渡島半島北部亜種(22, ♂, 23, ♀; 北海道島牧村狩場山); 24-25, 青森県亜種(24, ♂, 25, ♀; 青森県むつ市恐山); 26-27, 岩手県北部亜種(26, ♂, 27, ♀; 岩手県二戸市南西部(旧浄法寺町)); 28-29, 下閉伊亜種(28, ♂, 29, ♀; 岩手県宮古市石名沢). — a, 陰茎(右側面観); b, 陰茎先端部(右斜背面観).

12. *Carabus* (*Leptocarabus*) *arboreus*

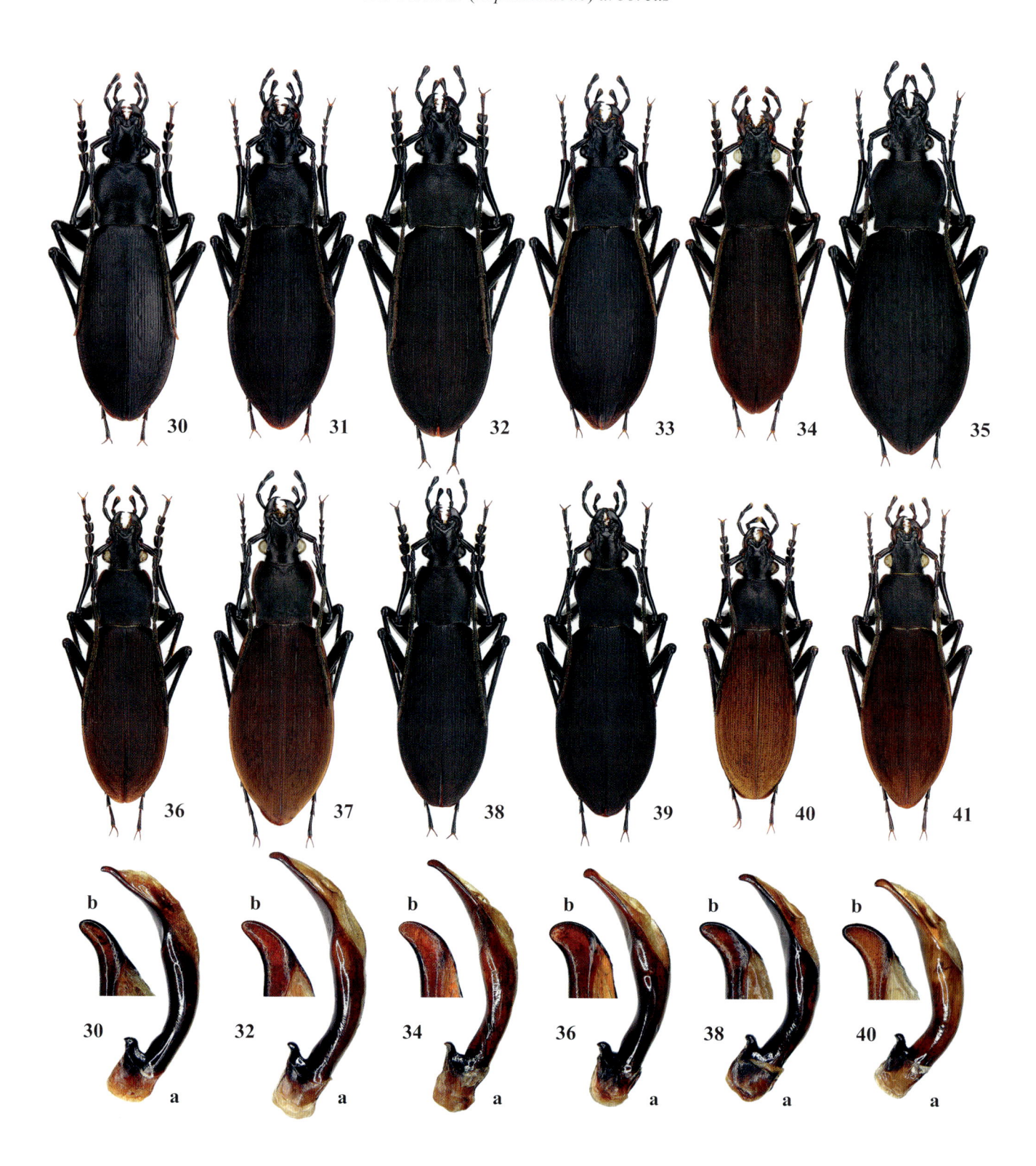

Figs. 30-41. *Carabus* (*Leptocarabus*) *arboreus* subspp. — 30-31, *C.* (*L.*) *a. kitakamisanus* (30, ♂, 31, ♀; Hikoroichi-chô, Ôfunato-shi, Iwate Pref.); 32-33, *C.* (*L.*) *a. akitanus* (32, ♂, 33, ♀; Kyôwa-funasawa, Daisen-shi, Akita Pref); 34-35, *C.* (*L.*) *a. parexilis* (34, ♂, 35, ♀; Aizu-wakamatsu, Fukushima Pref.); 36-37, *C.* (*L.*) *a. tenuiformis* (36, ♂, 37, ♀; Lake Chûzenji-ko, Nikkô-shi, Tochigi Pref.); 38-39, *C.* (*L.*) *a. babai* (38, ♂, 39, ♀; Mt. Hiuchi-yama, Myôkô-shi, Niigata Pref.); 40-41, *C.* (*L.*) *a. shinanensis* (40, ♂, 41, ♀; Shibu-no-yu, Mt. Yatsu-ga-také, Chino-shi, Nagano Pref.). — a, Penis in right lateral view; b, apical part of penis in right subdorsal view.

図 30-41. コクロナガオサムシ各亜種. — 30-31, 北上山地亜種 (30, ♂, 31, ♀; 岩手県大船渡市日頃市町); 32-33, 秋田県亜種 (32, ♂, 33, ♀; 秋田県大仙市協和船沢); 34-35, 東北地方南部亜種 (34, ♂, 35, ♀; 福島県会津若松); 36-37, 北関東上越亜種 (36, ♂, 37, ♀; 栃木県日光市中禅寺湖); 38-39, 妙高連峰亜種 (38, ♂, 39, ♀; 新潟県妙高市火打山); 40-41, 八ヶ岳亜種 (40, ♂, 41, ♀; 長野県茅野市八ヶ岳渋の湯). — a, 陰茎 (側面視); b, 陰茎先端部 (右斜背面視).

12. *Carabus* (*Leptocarabus*) *arboreus*

Figs. 42-53. *Carabus* (*Leptocarabus*) *arboreus* subspp. — 42-43, *C.* (*L.*) *a. ogurai* (42, ♂, 43, ♀; (Mt. Daibosatsu, Kôshû-shi, Yamanashi Pref.); 44-45, *C.* (*L.*) *a. fujisanus* (44, ♂, 45, ♀; Subashiri, Oyama-chô, Shizuoka Pref); 46-47, *C.* (*L.*) *a. horioi* (46, ♂, 47, ♀; Pass Sanpuku-tôgé, Ôshika-mura, Nagano Pref.); 48-49, *C.* (*L.*) *a. gracillimus* (48, ♂, 49, ♀; Mt. Ontaké-san, Kiso-machi, Nagano Pref.); 50-51, *C.* (*L.*) *a. hakusanus* (50, ♂, 51, ♀; Mt. Haku-san, Hakusan-shi, Ishikawa Pref.); 52-53, *C.* (*L.*) *a. ohminensis* (52, ♂, 53, ♀; Ôminé Mts., Tenkawa-mura, Nara Pref.). — a, Penis in right lateral view; b, apical part of penis in right subdorsal view.

図 42-53. コクロナガオサムシ各亜種. — 42-43, **奥秩父亜種**(42, ♂, 43, ♀; 山梨県甲州市大菩薩); 44-45, **富士箱根丹沢亜種**(44, ♂, 45, ♀; 静岡県小山町須走); 46-47, **赤石山脈亜種**(46, ♂, 47, ♀; 長野県大鹿村三伏峠); 48-49, **飛騨御嶽木曽亜種**(48, ♂, 49, ♀; 長野県木曽町御嶽山); 50-51, **白山亜種**(51, ♂, 52, ♀; 石川県白山市白山); 52-53, **大峰山脈亜種**(52, ♂, 53, ♀; 奈良県天川村大峰山脈). — a, 陰茎(右側面観); b, 陰茎先端部(右斜背面観).

12. *Carabus* (*Leptocarabus*) *arboreus* Lewis, 1882 コクロナガオサムシ

Original description *Carabus arboreus* Lewis, 1882, Trans. ent. Soc. London, 1882, p. 526.

History of research *Carabus* (*Leptocarabus*) *arboreus* in the present sense may be regarded as an aggregation of several taxa originally described as distinct species, that is, *Carabus arboreus* from Hokkaido (Lewis, 1882), *C. exilis* from Is. Sado-ga-shima (Bates, 1883), *C. tenuiformis* from Nikko, *C. gracillimus* from Mt. On-také and *C. Fujisanus* from Subashiri (Bates, 1883). Later, Born (1922) described *C. shinanensis* from Mt. Yatsu-ga-také of Nagano Prefecture in central Honshu. Nakane (1961) described three subspecies of *Carabus* (*Leptocarabus*) *exilis*, namely, *parexilis* from Aizu of Fukushima Prefecture, *horioi* from the Akaishi Mountains of central Honshu and *hakusanus* from Mt. Haku-san of Ishikawa Prefecture. Ishikawa described *C.* (*L.*) *e. babai* from the Myôkô Mountains on the borders between Niigata and Nagano Prefectures (1968b) and *C.* (*L.*) *e. ogurai* from the Daibosatsu Mountains of Yamanashi Prefecture (1969a). In 1978, a new isolated population of this group was unexpectedly found from the Ôminé Mountains of the Kii Peninsula, Nara Prefecture, and it was described by Harusawa as a new subspecies of *C.* (*Adelocarabus*) *arboreus* under the name of *ohminensis*. Ishikawa (1984b) described *Leptocarabus* (*A.*) *arboreus nepta* from Mt. Osore-zan of Aomori Prefecture. Ishikawa (1992) unified these lower taxa into a single polytypical species, *L.* (*A.*) *arboreus*, revised it morphologically and classified it into 20 subspecies with descriptions of seven new subspecies mainly from northern Japan under the names of *pararboreus, ishikarinus, karibanus, pronepta, shimoheiensis, kitakamisanus* and *akitanus*.

Type locality Junsai (near Lake Junsai-numa in Nanaé-chô, S. Hokkaido at present). This is the most probable type locality judging from the morphological characters of the lectotype specimen, though the type locality of *C. arboreus* designated by Lewis (1882) is "near Sapporo".

Type depository Natural History Museum, London [Lectotype, ♂].

Morphology 16~33 mm. Body above mat, with coloration black, dark brown or reddish brown. External features: Figs. 1~7. Description on details: penultimate segment of labial palpus bisetose; median tooth of mentum almost as long as, or a little shorter than lateral lobes, with margin narrowly rimmed by low ridge, apex pointed in ventral view and more or less protruded ventrad in lateral view; submentum bisetose; male antennae with hairless ventral depressions from segment five to eight, though that of segment five is not so clear; pronotal margins basically bisetose, though lacking apical pair in a certain subspecies; preapical emargination of elytra faintly recognized in female; elytral sculpture triploid; primaries indicated by chain striae and both secondaries and tertiaries indicated by contiguous low costae or rows of granules, though degree of development of each interval is variable according to subspecies; umbilicate series also variable in length according to subspecies, usually longer in northern populations, shorter in southern ones; sculpture of propleura and sides of sternites variable according to subspecies; metacoxa trisetose, though lacking anterior setae in a certain subspecies; sternal sulci sometimes unclear excluding that of segment seven. Male genitalic features: Figs. 8~17. Penis 0.4 to 0.5 times as long as elytra; ostium lobe and internal sac drastically variable in shape according to subspecies; terminal plate weakly developed. Female genitalia with outer plate of ligular apophysis broad oval, inner plate narrowly sclerotized and vestigial.

Molecular phylogeny For details, see the same section of *Carabus* (*Leptocarabus*) *kyushuensis*.

Distribution Hokkaido & Honshu / Iss. Okushiri-tô, Kunashiri-tô (Hokkaido), Izu-shima, Kinkasan, Aji-shima, Miyato-jima, Sabusawa-jima, Nono-shima (Miyagi Pref.) & Sado-ga-shima (Niigata Pref.). Outside Japan: Is. Sakhalin & S. Kurils.

Habitat In Hokkaido, this species is an inhabitant of the forest, and widely distributed from the plain to rather high mountains with the altitude more than 1,000 m. In the Tohoku Region, it is common from hills to high mountains, and is often predominant over other *Carabus* species. In the area south of central Honshu, it is usually subalpine or alpine species, inhabiting such environment as the deciduous broadleaved forest in the upper temperate zone, subalpine mixed forest, alpine meadow, or even in the barren summit area with the altitude more than 3,000 m.

Bionomical notes Autumn breeder. Larvae most probably prefer to feed on other insects' larvae.

Geographical variation This species shows remarkable geographical variation, and is classified into 20 subspecies (Ishikawa, 1992) as to the population distributed in Japan. The population in South Sakhalin has been described as subsp. *exarboreus* (Ishikawa, 1992).

Etymology Specific name means arborous (= of, relating to, or formed by trees).

研究史　今日，一般的にコクロナガオサムシとして1種にまとめられているものは，当初その多くが独立種として記載された複数のタクソンの集合体に近い．まず，Lewis（ルイス）(1882) により *Carabus arboreus* が記載され，次いで Bates（ベイツ）(1883) により *C. exilis, C. tenuiformis, C. gracillimus, C. Fujisanus* の4種が，更に40年近いブランクを経て Born（ボーン）(1922) により *C. shinanensis* が記載された．更にその約40年後，Nakane (1961) により *C.* (*Leptocarabus*) *exilis* の3新亜種, *parexilis*（会津），*horioi*（南アルプス），*hakusanus*（白山）が記載された．Ishikawa は同じく *C.* (*L.*) *exilis* の亜種として妙高連峰から *C.* (*L.*) *e. babai* (Ishikawa, 1968b) を，また翌年には大菩薩から *C.* (*L.*) *e. ogurai* (Ishikawa, 1969a) を記載した．春沢 (1978) は，それまでの本種の分布概念からはるか南方にかけはなれた大峰山脈の高所から *C.* (*Adelocarabus*) *arboreus* の亜種として *ohminensis* を記載した．その6年後には，青森の恐山から *Leptocarabus* (*A.*) *arboreus* の亜種として *nepta* が記載された (Ishikawa, 1984b)．Ishikawa (1992) は，外部形態および陰茎の形態に基づいて本種（種名としては *L.* (*A.*) *arboreus* を採用）の再検討を行い，北日本のものを細分し，北海道と東北地方から *pararboreus, ishikarinus, karibanus, pronepta, shimoheiensis, kitakamisanus, akitanus* の計7亜

12. *Carabus* (*Leptocarabus*) *arboreus*

種を新たに記載し，本種を計20亜種に分類した．しかし，この論文では本種の各集団間で大きく変化する♂交尾器内嚢の形態が考慮されておらず，また各「亜種」の移行帯や交雑帯とされる地域に分布する集団をどう取り扱うかについても検討を要する点が多い．将来より豊富な材料を用いて再検討を行う必要があるだろう．

タイプ産地 蓴菜（北海道亀田郡七飯町；レクトタイプの形態学的特長から推測される産地）．*Carabus arboreus*はLewis(1882)により「札幌近郊」から記載され，後にBates(1883)によって「札幌，美々，蓴菜」等の標本に基づき再記載された．現在，ロンドン自然史博物館にはこのうち1♂の標本しか残されておらず，この1頭がIshikawa(1992)によってレクトタイプに指定されている．この標本には「札幌」というラベルが付けられているが，外部形態，交尾器形態とも，その特徴は札幌近郊のものとは明らかに異なり，渡島半島方面に産する集団のそれに一致するため，ベイツによって示された本種の産地の一つである「蓴菜」をタイプ産地とみなすのが妥当であろうと考えられている．

タイプ標本の所在 ロンドン自然史博物館［レクトタイプ，♂］．原記載論文ではホロタイプが指定されておらず，Ishikawa(1992)により "Type / H. T. // *arboreus* / Lewis // Sapporo // Japan / G Lewis. / 1910-320" と記されたラベルを持つ♂がレクトタイプに指定されている．

形態 16～33 mm．背面，とりわけ鞘翅の色彩は黒色から褐色，さらに赤褐色まで変化し，光沢は鈍い．外部形態：図1～7．細部の記載：下唇肢亜端節の剛毛は2本．下唇基節の中央歯は側葉とほぼ同長か，やや短く，辺縁は低い稜により縁取られ，先端は尖り，腹側への突出の度合いは集団により異なる．下唇亜基節には2剛毛を具える．♂触角第5節から第8節の腹面には基本的に無毛部があり，第6節から第8節において顕著．前胸背板の側縁剛毛は基本的に2対（中央と後角付近に各1対）だが，亜種によっては後角付近の1対のみとなる．鞘翅端は♀でごく僅かにえぐれる．鞘翅彫刻は三元的で，基本的に第一次間室は鎖線，第二次，第三次間室は連続する低い隆条ないし顆粒列となり，各間室の発達の程度は集団により異なる．丘孔点列の長さも集団により異なるが，一般に北のものほど長く，南のものほど短い．胸側板と腹部腹板側面の彫刻も集団により変化する．後基節の剛毛は基本的に3本だが，亜種によっては前方剛毛を欠く．腹部腹板の横溝は第7節のものを除き不明瞭な場合がある．♂交尾器形態：図8～17．陰茎は鞘翅長の0.4～0.5倍程度で，葉片の発達度や内嚢形態は集団により大きく変化する．頂板の発達は弱い．♀交尾器膣底部節片外板は幅広い楕円形で，内板はごく狭く硬化し，退化的．

分子系統 キュウシュウクロナガオサムシの分子系統の項を参照．本種は大陸の*Adelocarabus*（アデロカラブス）亜属に分類されていた時期があったが，分子系統の上からは邦産の他の4種のレプトカラブス亜属と共に単系統群を形成することが判明している．

分布 北海道，本州／奥尻島，国後島（以上，北海道），出島，金華山，網地島，宮戸島，寒風沢島，野々島（以上，宮城県），佐渡島（新潟県）．国外：サハリン，千島列島南部．

生息環境 北海道では平地から標高500 mくらいまでの地域に多いが，標高1,000 mを越す地点からも記録がある．主として森林帯に生息し，草原や湿原からは殆ど発見されないが，沖積平野の河川敷からの採集記録はある．森林であればカラマツの単相林のような人工的かつ単調な環境にも生息している．東北地方では低地や丘陵から高山にかけて広く見られ，その場所におけるオサムシの最優先種となっていることも多い．本州中部以西では一般に生息域の下限がより高くなり，山地から亜高山帯，高山帯にかけて生息する．落葉広葉樹葉林，針広混交林，針葉樹林はもとより，高山帯の風衝地やお花畑にも生息している．

生態 秋繁殖型で，成虫・幼虫双方で越冬する．越冬成虫は土中の他，好んで朽木中でも越冬している．幼虫の基本的な食性は節足動物（昆虫の幼虫など）食であろう．

地理的変異 顕著な地理的変異を示し，わが国のものはIshikawa(1992)により目下，20亜種に分類されている．これらは基亜種群（下記1～6の6亜種，北海道と東北北部に分布），*exilis*亜種群（下記7～16の10亜種；東北南部と中部地方に分布）および*gracillimus*亜種群（下記17～20の4亜種；南北アルプス，木曽山脈，御嶽，白山，大峰山脈に分布）の3種群に分けられているが，♂交尾器内嚢や分子系統の結果を考慮すると，これとはまた異なる分類になる可能性もあるので，ここでは便宜上，命名記載されている20亜種を羅列するに留める．国外ではサハリン南部の集団が*exarboreus*という名で別亜種として記載されている．

種小名の由来 種小名は「樹木の」を意味する．キタクロナガオサムシ，エゾクロナガオサムシ（とくに，基亜種を含む北海道産の集団に対して）と呼ばれることもある．

12-1. *C.* (*L.*) *a. pararboreus* (Ishikawa, 1992) コクロナガオサムシ道央道東道北亜種

Original description *Leptocarabus* (*Adelocarabus*) *arboreus pararboreus* Ishikawa, 1992, TMU Bull. nat. Hist, Tokyo, (1), p. 10, figs. 26, 27 (pl. 10), 28~30 (pl. 11) & 32~35 (pl. 12).
Type locality Naka-ashoro, Ashoro-machi (more precisely, Ashoro-chô), Tokachi.
Type depository Systematic Zoology Laboratory, Department of Biological Sciences, Tokyo Metropolitan University (Hachiôji) [Holotype, ♂].
Diagnosis 23~33 mm. Body above entirely black and mat. Similar to subsp. *ishikarinus*, but distinguished from that race in the following points: 1) secondary and tertiary intervals of elytra, above all the latter, are weaker, and usually recognized as scattered granules; 2) umbilicate series behind middle of elytra are composed of a single plain row of separate granules; 3) penis more plainly cylindrical in median portion, with right margin not or barely swollen; 4) rudimentary ostium lobe is very rarely recognized.
Distribution C. & NE. Hokkaido / Is. Kunashiri-tô.

Etymology   The subspecific name means "allied to *arboreus*".

タイプ産地　十勝, 足寄町(あしょろちょう), 中足寄.
タイプ標本の所在　首都大学東京動物系統分類学研究室(八王子)［ホロタイプ, ♂］.
亜種の特徴　23~33 mm. 黒色で光沢は鈍い. 次に述べる石狩亜種に近いが, 以下の諸点により識別される: 1) 鞘翅の第二次, 第三次原線はより減弱する傾向にあり, とりわけ後者は通常ごく弱い顆粒列として認められるにすぎない; 2) 丘孔点列の後半部は一列の顆粒列からなる; 3) 陰茎中央部はより単純な円柱形に近く, 右側縁の膨隆は殆ど認められない; 4) ごく稀に痕跡的な葉片を持つ個体が見られる.
分布　石狩平野よりも北東方の北海道中・北・東部／国後島.
亜種小名の由来と通称　亜種小名は「*arboreus*に近いもの」を意味する. オクエゾクロナガオサムシという通称で呼ばれることもある.

12-2. *C. (L.) a. ishikarinus* (Ishikawa, 1992)　コクロナガオサムシ石狩亜種

Original description   *Leptocarabus* (*Adelocarabus*) *arboreus ishikarinus* Ishikawa, 1992, TMU Bull. nat. Hist, Tokyo, (1), p. 9, figs. 19 (pl. 8), 20~23 (pl. 9) & 24 (pl. 10).
Type locality   Hitsujigaoka, Sapporo-shi.
Type depository   Systematic Zoology Laboratory, Department of Biological Sciences, Tokyo Metropolitan University (Hachiôji) [Holotype, ♂].
Diagnosis   25~33 mm.   Body above entirely black and mat. This is an intermediate form between subspp. *pararboreus* and *arboreus*, showing a considerable individual variation, but may be characterized by the following points: 1) elytral sculpture variable, though intervals usually degenerated as in *pararboreus*; 2) umbilicate series similar to those of *arboreus* and *karibanus*, indicated by irregularly set double rows of granules at least partly behind middle of elytra extending to elytral apices; 3) apical part of penis variable in shape, though relatively short, robust and strongly bent ventrad as in *pararboreus*.
Distribution   Ishikari Area, SW. Hokkaido (approximately from east of Kuromatsunai to west of Yubari Mts.).
Etymology   Named after the Ishikari Area, the main distributional range of this subspecies.

タイプ産地　札幌市, 羊ヶ丘.
タイプ標本の所在　首都大学東京動物系統分類学研究室(八王子)［ホロタイプ, ♂］.
亜種の特徴　25~33 mm. 黒色で光沢は鈍い. 道央道東道北亜種と基亜種の中間的な形態を有する集団で, 以下の諸点により識別される: 1) 鞘翅彫刻のうち, 間室の形態は道央道東道北亜種のそれに近い; 2) 丘孔点列の状態は渡島半島北部亜種と基亜種のそれに近い; 3) 陰茎先端の形態は個体変異が大きいが, 一般に道央道東道北亜種のそれに似て比較的短く, 腹側への湾曲が強い.
分布　石狩地方とその周辺(黒松内低地帯以東, 夕張山地以西).
亜種小名の由来と通称　亜種小名は, 主な分布域である石狩地方に因む. イシカリクロナガオサムシと呼ばれることもある.

12-3. *C. (L.) a. karibanus* (Ishikawa, 1992)　コクロナガオサムシ渡島半島北部亜種

Original description   *Leptocarabus* (*Adelocarabus*) *arboreus karibanus* Ishikawa, 1992, TMU Bull. nat. Hist, Tokyo, (1), p. 8, figs. 15 (pl. 7) & 16~18 (pl. 8).
Type locality   Mt. Karibayama, 1000-1519 m.
Type depository   Systematic Zoology Laboratory, Department of Biological Sciences, Tokyo Metropolitan University (Hachiôji) [Holotype, ♂].
Diagnosis   22~28 mm.   Body above entirely black and mat. Externally similar to mountainous form of subsp. *arboreus*, but characterized by the following points: 1) elytral intervals less strongly costate and areas between them are indicated by rows of distinct granules which are slightly weaker than tertiaries; 2) most part of umbilicate series indicated by irregularly set double rows of granules; 3) propleura vaguely and almost evenly scattered with punctures, mesepisterna smooth and impunctate, metepisterna and sides of metasterna rugulose, sometimes scattered with punctures; 4) sides of sternites densely punctate, sternal sulci usually more clearly impressed in front; 5) penis peculiar in shape, with apical lobe narrowly elongated, strongly arcuate and bent ventrad in lateral view, ventral side of membraneous part weakly but apparently convex, widely and strongly concave and rugulose on its right side.
Distribution   N. Oshima Penins. (areas between Kuromatsunai lowland and Riv. Toshibetsu-gawa), SW. Hokkaido.
Etymology   Named after the type locality, Mt. Kariba-yama.

タイプ産地　狩場山, 1000-1519 m.
タイプ標本の所在　首都大学東京動物系統分類学研究室(八王子)［ホロタイプ, ♂］.
亜種の特徴　22~28 mm. 黒色で光沢は鈍い. 外見的には基亜種の山地型に似ているが, 以下の諸点により識別される: 1) 鞘

翅原線は基亜種のそれほど強い隆条とはならず，間室間には第三次間室よりも僅かに弱いがはっきりとした顆粒列を具える；2) 丘孔点列は大部分が不規則な二重の顆粒列となる；3) 前胸側板には痕跡的な点刻を均等に装い，中胸前側板は平滑で無点刻．後胸前側板と後胸腹板の側方は皺状で，さらに点刻を装う場合もある；4) 腹部腹板の側面は密に点刻され，腹板の横溝は通常側方において鮮明に刻まれる；5) 陰茎先端は細長く，腹側へ強く屈曲し，膜状開口部の腹側は弱いながら明らかに膨隆し，右側縁は強くえぐれる．

分布 渡島半島北部（黒松内低地帯以南，利別川以北）．

亜種小名の由来と通称 亜種小名は，タイプ産地の狩場山に因む．カリバクロナガオサムシと呼ばれることもある．

12-4. *C. (L.) a. arboreus* LEWIS, 1882 コクロナガオサムシ基亜種

Original description, type locality & type depository See explanation of the species.

Diagnosis 24~29 mm. Body above entirely black and mat. Characterized by relatively shorter penis with longer apical lobe and regularly raised elytral intervals with tertiaries well defined and never degenerated to scattered granules as in other races of Hokkaido.

Distribution C. & S. Oshima Penins., SW. Hokkaido / Is. Okushiri-tô.

Geographical variation Specimens from the mountainous regions are constantly smaller and slenderer, with the elytral intervals very regularly and strongly costate.

原記載・タイプ産地・タイプ標本の所在 種の項を参照．

亜種の特徴 24~29 mm. 黒色で光沢は鈍い．鞘翅原線が規則的に隆起し，第三次原線が顆粒列に減退せず，はっきりと認められること，ならびに陰茎が相対的に短く，その一方で先端部は比較的長く伸長することにより，北海道内に分布する他亜種から識別される．鞘翅彫刻は規則的かつ顕著に刻まれ，第三次原線は常に明瞭に認められる．間室間の顆粒は，認められる場合でも第三次間室のそれより弱い．丘孔点列は不規則な二列の顆粒列からなり，鞘翅端付近まで続く．前胸側板にはせいぜい痕跡的な点刻を装う程度．中胸前側板には点刻を欠き，表面は平滑で光沢がある．後胸前側板と後胸腹板の側面には浅いながら密な点刻を装う．腹板の横溝は不鮮明なものから鮮明なものまで変化がある．陰茎は比較的短いが，先端部は長く，僅かに湾曲するのみで，先端に向けて急激に屈曲することはない．

分布 渡島半島中~南部／奥尻島．

地理的変異 山地に産するものは一般に小型で細く，鞘翅の隆条が規則的かつ顕著に認められる．

12-5. *C. (L.) a. nepta* (ISHIKAWA, 1984) コクロナガオサムシ青森県亜種

Original description *Leptocarabus (Adelocarabus) arboreus nepta* ISHIKAWA, 1984, Akitu, Kyoto, (N. S.), (68), p. 1, figs. 1, 2 (p. 2).

Type locality Mt. Osorezan, Mutsu-shi, Aomori Pref.

Type depository Systematic Zoology Laboratory, Department of Biological Sciences, Tokyo Metropolitan University (Hachiôji) [Holotype, ♂].

Diagnosis 20~30 mm. Elytra and ventral side of body black or sometimes feebly brownish. Identified by the following characters: 1) male antennae reaching middle of elytra; 2) pronotum with basal foveae shallow and unclear, hind angles weakly protruded posteriad; 3) elytral margins with weak glossy stripe, granules of umbilicate series reaching elytral apices though sporadical in posterior halves; 4) episterna coarsely scattered with punctures; 5) sternal sulci unclear; 6) penis long and slender, hardly convex at middle, with membraneous part long and tapered toward base, ostium lobe small and vestigial. Specimens from the lowlands in the Tsugaru Peninsula are larger and broader than those from hills or mountainous areas, but there is no definitive difference between them in the characters of penis.

Distribution Aomori Pref. & N. Akita Pref. (N. of Riv. Yoneshiro-gawa), NE. Honshu. This subspecies is intergraded to subsp. *akitanus* in the southeastern part of its distributional range, and to subsp. *pronepta* in the southeast.

Etymology Nepta, or Nebuta, is the name of traditional summer festival held in every August in Aomori Prefecture where is the main distributional range of this subspecies.

タイプ産地 青森県，むつ市，恐山．

タイプ標本の所在 首都大学東京動物系統分類学研究室（八王子）［ホロタイプ，♂］．

亜種の特徴 20~30 mm. 鞘翅と体下面は黒色ないし僅かに茶色味を帯びる．以下の特徴を有する：1) ♂触角の先端は鞘翅中央に達する；2) 前胸背板の基部凹陥が浅く不明瞭で，後角の突出は弱い；3) 鞘翅側縁には弱い光沢帯があり，丘孔点列の顆粒は鞘翅端近くまで続くが，後半部では間隔が大きくなる；4) 胸側板には密に点刻を装う；5) 腹板横溝は不鮮明で，第4節から第6節において浅く窪む程度；6) 陰茎は細長く，中央部で殆ど膨らまず，膜状部は長く，基部に向けて狭まることが多く，葉片は小さく退化的．津軽半島の低地に産するものは，丘陵~山地に産するものに比し，大型で幅広い体型となるが，陰茎の基本的形態に殆ど違いは見られない．

分布　青森県のほぼ全域と秋田県北部(米代川以北)．分布域南西部の青森県旧碇ヶ関村(いかりがせきむら)(現在の平川市の一部)や秋田県小坂町(こさかまち)などのものは秋田県亜種との雑種とされ，分布域南東部では徐々に岩手県北部亜種へと移行する．

亜種小名の由来と通称　亜種小名は，ねぷた(もしくはねぶた)祭り(主として青森県内の各地で行われる伝統的な夏祭りの名)に因む．通称はムツクロナガオサムシ．

12-6. *C. (L.) a. pronepta* (Ishikawa, 1992) コクロナガオサムシ岩手県北部亜種

Original description　*Leptocarabus* (*Adelocarabus*) *arboreus pronepta* Ishikawa, 1992, TMU Bull. nat. Hist, Tokyo, (1), p. 13, figs. 52~59 (pls. 17~18).

Type locality　Pass, the east of Jyôbôji-machi (= present Ninohé-shi of Iwate Pref.), 400 m.

Type depository　Systematic Zoology Laboratory, Department of Biological Sciences, Tokyo Metropolitan University (Hachiôji) [Holotype, ♂].

Diagnosis　20~30 mm. Closely allied to subsp. *nepta*, but differs from that race in the following points: 1) penis shorter and more plainly arcuate in lateral view, with inner margin not convex but evenly arched; 2) membraneous part of penis broader, not narrowed toward base, area before internal sac wholly wrinkled with larger ostium lobe.

Distribution　SE. Aomori Pref. & N. Iwate Pref., NE. Honshu. Distributed in the areas surrounded by the ranges of subspp. *nepta*, *shimoheiensis* and *akitanus*.

Etymology　The subspecific name means "just as *nepta*".

タイプ産地　浄法寺町東方の峠，標高400 m(現在は岩手県二戸市(にのへし)に属する)．

タイプ標本の所在　首都大学東京動物系統分類学研究室(八王子)[ホロタイプ，♂]．

亜種の特徴　20~30 mm. 青森県亜種に近いが，以下の点で異なる：1)陰茎はより短く，側面から見て湾曲がより緩やかで，内側縁に膨隆部分は見られず，ほぼ均等に弧を描く；2)陰茎膜状部はより広く，基部に向けて狭まらず，葉片はより大きい．

分布　青森県南東部～岩手県北部．青森県亜種，下閉伊亜種，秋田県亜種の三者に囲まれた地域に分布する．

亜種小名の由来と通称　亜種小名は「*nepta*のような」を意味する．リクチュウクロナガオサムシという通称で呼ばれることもある．

12-7. *C. (L.) a. shimoheiensis* (Ishikawa, 1992) コクロナガオサムシ下閉伊(しもへい)亜種

Original description　*Leptocarabus* (*Adelocarabus*) *arboreus shimoheiensis* Ishikawa, 1992, TMU Bull. nat. Hist, Tokyo, (1), p. 14, figs. 65~68 (pl. 20).

Type locality　Ishinazawa, 200 m, Tarô-cho (= present Miyako-shi).

Type depository　Systematic Zoology Laboratory, Department of Biological Sciences, Tokyo Metropolitan University (Hachiôji) [Holotype, ♂].

Diagnosis　20~30 mm. According to the original description, this race is a transitional form connecting subspp. *pronepta* and *kitakamisanus* in the characters of the penis, not being distinguished definitely from either in other morphological features. Its status as a subspecies is therefore provisional, because it is uncertain whether it is an imperfectly isolated southern extension of subsp. *pronepta* or primary hybrids between subspp. *pronepta* and *kitakamisanus* (Ishikawa, 1992, p. 14).

Distribution　C. & E. Iwate Pref. (areas between Riv. Omoto-gawa & Riv. Hei-gawa), NE. Honshu.

Etymology　Named after the Shimohei Area occupying east-central part of Iwate Prefecture, where is the main distributional range of this subspecies.

タイプ産地　田老町，石名沢，200 m(田老町は合併により廃止され，石名沢は現在，岩手県宮古市に属する)．

タイプ標本の所在　原記載論文にはタイプ標本の保管場所が記されていないが，現在は首都大学東京動物系統分類学研究室(八王子)に保管されているようである[ホロタイプ，♂]．

亜種の特徴　20~30 mm. 本亜種の原記載(Ishikawa, 1992, p. 14)によれば，本集団は陰茎の特徴が青森県亜種と北上山地亜種との中間を示し，他の形態からは明確に識別できない．従って，亜種としての扱いはあくまで暫定的なものであり，青森県亜種の分布域の最南限を占める，隔離の不十分な集団，あるいは青森県亜種と北上山地亜種との一時的な交雑集団であるかもしれないという．

分布　岩手県中東部(小本川～閉伊川間)．北上山地北部の狭い範囲から知られ，小本川と閉伊川に挟まれた地域に分布する．

亜種小名の由来と通称　亜種小名は本亜種の主分布域である下閉伊地方(岩手県中東部)に因む．シモヘイクロナガオサムシという通称で呼ばれることもある．

12-8. *C. (L.) a. kitakamisanus* (Ishikawa, 1992) コクロナガオサムシ北上山地亜種

Original description   *Leptocarabus* (*Adelocarabus*) *arboreus kitakamisanus* Ishikawa, 1992, TMU Bull. nat. Hist, Tokyo, (1), p. 15, figs. 69~74 (pls. 21~22).
Type locality   Nakaitayô, 40 m, Hikoroichi, Ofunato-shi.
Type depository   Systematic Zoology Laboratory, Department of Biological Sciences, Tokyo Metropolitan University (Hachiôji) [Holotype, ♂].
Diagnosis   21~28 mm. This race, as well as subsp. *akitanus*, appears to be northern extension of wide ranging subsp. *parexilis*. Similar to subsp. *akitanus*, but differs from it in differently featured penis, which is narrowed toward base from widest part in dorsal view, its right margin more strongly angulated medially and always more deeply concavely emarginate therefrom as in subsp. *parexilis*, apical part longer and slenderer, not so abruptly narrowed toward tip.
Distribution   SC. Iwate ref. & NE. Miyagi Pref. (areas between Riv. Hei-gawa & Kesennuma), NE. Honshu.
Etymology   Named after the Kitakami Mountains, the main distributional range of this subspecies.

タイプ産地　大船渡市, 日頃市, 中板用, 40 m.
タイプ標本の所在　首都大学東京動物系統分類学研究室（八王子）［ホロタイプ, ♂］.
亜種の特徴　21~28 mm. 秋田県亜種に近いが, 主として陰茎形態の違いにより識別される. 陰茎は右側縁中央部でより顕著に角張り, 最広部から基部に向けて細くなる. 膜状開口部の右腹側はより強くくぼみ, 皺を有する. 陰茎先端部はより細長く, 先端に向けてそれほど急激に細まることはない.
分布　岩手県中南部〜宮城県北東端（閉伊川〜気仙沼）. 閉伊川以南の北上山地に分布し, 南方の気仙沼付近で東北地方南部亜種に移行する.
亜種小名の由来と通称　亜種小名は本亜種が分布する北上山地に因む. キタカミクロナガオサムシという通称で呼ばれることもある.

12-9. *C. (L.) a. akitanus* (Ishikawa, 1992) コクロナガオサムシ秋田県亜種

Original description   *Leptocarabus* (*Adelocarabus*) *arboreus akitanus* Ishikawa, 1992, TMU Bull. nat. Hist, Tokyo, (1), p. 15, figs. 77~80 (pl. 23).
Type locality   Funazawa (more precisely, Funasawa), 70 m, Kyôwa-machi, Akita Prefecture (= present Kyôwa-funasawa in Daisen-shi).
Type depository   Systematic Zoology Laboratory, Department of Biological Sciences, Tokyo Metropolitan University (Hachiôji) [Holotype, ♂].
Diagnosis   23~29 mm. Similar to subsp. *kitakamisanus*, but differs from it in shape of penis which is less strongly convex at right margin near middle and less deeply concave therefrom toward apex, with apical part shorter and abruptly narrowed though never so digitate as in subsp. *kitakamisanus*. Also allied to subsp. *parexilis*, but differs from that race by differently shaped apical part of penis and more ovate elytra, above all in male. Umbilicate series always completely recognized throughout elytral margins. This subspecies appears to hybridize with subsp. *nepta* at the northern borders of its range by the Yoneshiro-gawa River.
Distribution   Akita Pref. & WC. Iwate Pref., NE. Honshu (S. of Riv. Yoneshiro-gawa, N. of Riv. Omono-gawa & E. of Riv. Waga-gawa).
Etymology   Named after Akita Prefecture, the main distributional range of this subspecies.

タイプ産地　秋田県, 協和町, 船沢（ふなさわ; 原記載では「ふなざわ」と記されている）, 70 m（現在は秋田県大仙市協和船沢）.
タイプ標本の所在　首都大学東京動物系統分類学研究室（八王子）［ホロタイプ, ♂］.
亜種の特徴　23~29 mm. 陰茎は北上山地亜種のものに近いが, 右側縁中央部の膨らみがやや弱く, 膜状開口部右腹側面の窪みも浅く, 陰茎先端部が明らかにより短く, 先端はより鈍く丸まる点などによって区別される. 東北地方南部亜種にも近いが, 陰茎先端部の形状が異なり, 鞘翅は特に♂でより楕円形に近くなり, 東北地方南部亜種のように両側縁が平行に近くならない.
分布　秋田県〜岩手県中西部（米代川以南〜雄物川以北〜和賀川以東）.
亜種小名の由来と通称　亜種小名は, 本亜種の主分布域である秋田県に因む. デワクロナガオサムシという通称で呼ばれることもある.

12-10. *C. (L.) a. parexilis* Nakane, 1961 コクロナガオサムシ東北地方南部亜種

Original description   *Carabus* (*Leptocarabus*) *exilis parexilis* Nakane, 1961, Fragm, coleopterol., Kyoto, (1), p. 2
Type locality   Aizu, Fukushima Pref., Honshu.
Type depository   Hokkaido University Museum (Sapporo) [Holotype, ♂].
Diagnosis   21~33 mm.   Body above usually black; specimens from higher place are often a little brownish. Identified by the

following characters: 1) elytra with the shoulders more distinct, lateral margins less strongly arcuate in anterior two-thirds and more or less parallel-sided above all in male, so that they appear less oval, above all in large sized specimens; 2) umbilicate series usually short, but may be extending toward elytral apices as a row of separate granules; 3) penis with the basal portion long and cylindrical, widened medially, strongly narrowed therefrom toward apex, with apical lobe rather gradually bent ventrad though not so abruptly narrowed from membraneous ostium part as in subsp. *akitanus* nor attenuated as in subsp. *kitakamisanus*; 4) right inner margin of penis distinctly edged, angulated at widest part and evenly concavely arcuate to base of apical lobe, underside of angulated edge is distinctly excavated and wrinkled; 5) membraneous area wide, its basal margin extended toward base of penis, preostium narrow but fully wrinkled.

Distribution   SC. Tohoku Region & NC. Kanto Region (S. Akita Pref., Miyagi Pref., Yamagata Pref., Fukushima Pref., N. Niigata Pref., Tochigi Pref. (excepting south-central part), NE. Gunma Pref., NC. Ibaraki Pref. & NC. Chiba Pref.), NE. Honshu / Izu-shima, Kinkasan, Aji-shima, Miyato-jima, Sabusawa-jima, Nono-shima (Miyagi Pref.).

Geographical variation   The specimens from the plains or lower altitudes are larger, broader and wholly black, resembling *C. (L.) procerulus* with which this subspecies is often coexist, while those from the high altitudinal areas are smaller, narrower and more or less brownish in color, resembling *C. (L.) a. tenuiformis*.

Etymology   The subspecific name means "allied to *exilis*".

タイプ産地　本州, 福島県, 会津.
タイプ標本の所在　北海道大学総合博物館(札幌)[ホロタイプ, ♂].
亜種の特徴　21~33 mm. 背面は通常黒色だが, 高所に産するものは褐色味を帯びる. 以下の特徴を有する: 1)♂触角先端は鞘翅中央を越える; 2)前胸背板は密に点刻され, 基部凹陥は深く大きい; 3)鞘翅は側縁が直線的で, とりわけ♂では両側縁が平行に近くなるものが多い; 4)鞘翅の間室は細く, 隆起は弱く, 側縁部の光沢帯を欠き, 丘孔点列は基部1/3にのみ認められる; 5)陰茎は中央部で明らかに膨らみ, 膜状部の右腹側縁は強くえぐれ, 先端部は扁平で側面から見て腹側へ湾曲し, 先端は鈍く丸まる. 膜状部には葉片がある.
分布　秋田県南部~宮城・山形・福島各県~新潟県北部(南限は阿賀野川(あがのがわ))~中南部を除く栃木県~群馬県北東部~茨城県中北部~千葉県中北部/出島, 金華山, 網地島, 宮戸島, 寒風沢島, 野々島(以上, 宮城県).
地理的変異　平地や低標高地域のものは大型で幅広く黒色で, 同所的に生息することの多いクロナガオサムシに似る. 一方, 山地のものは小型で細く, 褐色がかる傾向にあり, 北関東上越亜種などに似た外観を有する.
亜種小名の由来と通称　亜種小名は「*exilis*に近いもの」を意味する. 通称としてよく用いられてきた名はトウホククロナガオサムシだが, 東北地方からは複数の本種の亜種が知られているので紛らわしい. アイヅクロナガオサムシという名も提唱されている.

12-11. *C. (L.) a. tenuiformis* BATES, 1883　コクロナガオサムシ北関東上越亜種

Original description   *Carabus tenuiformis* BATES, 1883, Trans. ent. Soc. Lond., 1883, p. 226.
Type locality   Niohozan and Chiuzenji (= Mt. Nyohô-zan and Chûzenji of Nikko in Tochigi Pref.).
Type depository   Natural History Museum, London [Lectotype, ♀].
Diagnosis   20~31 mm. Body above often brownish or rather reddish, though entirely black in the specimens inhabiting northwestern localities in Niigata Prefecture. Pronotum narrow, with lateral sides less strongly cordate in apical halves, hind angles sharply protruded, basal foveae shallow and unclear. Propleura smooth and impunctate, or at most very shallowly punctate. Umbilicate series shorter than pronotal length, not recognized in the posterior parts of elytra. Penis similar to that of subsp. *parexilis*, but apical part is shorter, flatter, more strongly bent ventrad and abruptly tapered toward tip.

Distribution   Jôetsu (= Echigo) Mountain Range (including Taishaku Mts. & Mikuni Mts.), Asama-yama Massif & Higashi-kubiki Hills (SW. Fukushima, W. Gunma, NW. Tochigi, E. Niigata, NE. Nagano), C. Honshu. Subalpine to alpine zone (in the southeastern part of the distributional range) or hills and plains (in the northwestern part of the distributional range).

Geographical variation   This subspecies is sporadically distributed in the southeastern part of its range inhabiting the altitudes where it is slender, smaller in size and usually more or less brownish, but in the northwestern localities in Niigata Prefecture, it inhabits from the altitudes down to hills and plains where it is larger or very large-sized and wholly black, just as in subsp. *parexilis* of the corresponding habitats though more elongate and pronotum is particularly narrower with less cordate sides. The southeastern and the northwestern forms thus differ from each other superficially, but they share some characteristic features of the aedeagal characters and smooth and impunctate or at most very shallowly punctate propleura. The specimens from Mt. Kurohimé-yama and Mt. Hishi-ga-také in the Higashi-kubiki Hills have narrower apical part of penis, with tendency of losing gular setae, suggesting a relationship with subsp. *babai* of the Myôkô Mountains. Hybrids between this subspecies and subsp. *parexilis* in Fukushima Prefecture were collected only at the altitudes where they are distinguishable only by the characters of penis and propleura (ISHIKAWA, 1992, p. 19).

Etymology   The subspecific name means "slender".

タイプ産地　ニオホザンとチウゼンジ(日光の女峰山と中禅寺を指すものと思われる).
タイプ標本の所在　ロンドン自然史博物館[レクトタイプ, ♀]. 原記載論文ではホロタイプが指定されておらず, Ishikawa (1992)により "Type / H. T. // tenuiformis / Bates // Chiuzenji. / 19.VIII.- 24.VIII.81 // Japan / G Lewis. / 1910-320 // 19.8.81" と記されたラベルを持つ♂がレクトタイプに指定されている.
亜種の特徴　20~31 mm. 背面の色彩は, 分布域南東部の高所に産するものでは赤褐色味を帯びるが, 同北西部の新潟県などに産するものでは黒色となる. 以下の特徴を有する: 1)前胸背板は細く, 側縁前方の丸みが弱く, 基部凹陥は浅く不明瞭で, 後角は鋭く突出する; 2)中胸前側板は滑らかで, 顕著な点刻はなく, せいぜい浅く不明瞭な圧痕が見られる程度; 3)鞘翅は細く(特に♂), 丘孔点列は後方で消失する; 4)陰茎は東北地方南部亜種のものに似るが, 先端部はより太短く, 側面から見てより幅広く, より急激に腹側へ曲がる.
分布　上越(越後)山地(帝釈山脈, 三国山地を含む)~浅間山から東頚城丘陵(福島県南西部~群馬県西部~栃木県北・西部~新潟県東部~長野県北東部).
地理的変異　分布域の南東部では主として高所に産し, 細く小型で背面が茶褐色を帯びるが, 分布域北西部の新潟県中東部では低標高域にも分布し, 大型で黒色となる(外見は東北地方南部亜種に似るが, 体形が異なり, 特に前胸背板がより細長く, 側縁はより直線的になる). このように, 分布域の両端において外見上かなりの違いが見られるが, 陰茎の形と前胸背板が滑らかで殆ど点刻されないことにより, 東北地方南部亜種から区別される. 東頚城丘陵の黒姫山や菱ヶ岳のものは, 陰茎先端が非常に細く, 咽頭剛毛が消失する傾向にあり, 妙高連峰亜種との関連をうかがわせる. 福島県南西端の高所では, 本亜種と東北地方南部亜種との雑種が見られる(Ishikawa, 1992, p. 19).
亜種小名の由来と通称　亜種小名は「細長い形の」を意味する. 通称はホソクロナガオサムシ.

12-12. *C. (L.) a. exilis* Bates, 1883　コクロナガオサムシ佐渡島亜種

Original description　*Carabus exilis* Bates, 1883, Trans. ent. Soc. Lond., 1883, p.226.
Type locality　Island of Sado.
Type depository　Natural History Museum, London [Lectotype, ♂ (Fig. 54)].
Diagnosis　20 mm (lectotype specimen). Head and pronotum brownish black, pronotum brown. Frons with fine longitudinal wrinkles which are coarser at sides; cervix mat, with distinct transverse wrinkles; submentum setiferous, with a pair of normal setae plus one additional seta in lectotype specimen; antennae with hairless areas on ventral side; pronotum subquadrate, a little wider than long, with sides weakly sinuate, very weakly convexly arcuate in front, and nearly parallel-sided behind, hind angles strongly protruded though blunt at tips, disk densely rugoso-punctate; elytra long oval and weakly convex above, with marginal areas densely granulate; granules of umbilicate series almost reaching before elytral apices; intervals evenly and narrowly raised; primaries strongly costate with subpolished tops, segmented by shallow foveoles; secondaries and tertiaries almost equally raised, their tops indicated by subpolished broken lines; propleura smooth though a little opaque, and vaguely punctate in front; prosterna impunctate; mesepisterna almost smooth and polished; mesepimera smooth; metepisterna weakly rugulose and sporadically scattered with shallow punctures; sides of sternites vaguely punctate; sternal sulci carved at sides on 4th and 5th sternites, completely so across center on 6th. Penis similar to that of subsp. *tenuiformis* or of subsp. *parexilis*, but maybe identified by the following characters or combination of them: 1) widest nearly at middle, with right inner margin sharply carinate around there; 2) area below the carina concave, with surface rather smooth, not remarkably rugulose nor aciculate; 3) apical part spatulate as in subsp. *parexilis* though shorter, and less strongly bent ventrad than in subsp. *tenuiformis*; 4) outer margin of membraneous ostium area strongly sinuous; 5) ostium lobe recognized.
Distribution　Is. Sado-ga-shima (Niigata Pref.). Known so far only from a single male specimen now preserved in the Natural History Museum, London.
Etymology　Subspecific name means "exiguous" or "poor".

タイプ産地　佐渡島(新潟県).
タイプ標本の所在　ロンドン自然史博物館[レクトタイプ, ♂(図54)]. 原記載論文ではホロタイプが指定されておらず, Ishikawa(1992)により "Type / H. T. // exilis / Bates // Sado / 6.10.81 // Japan / G Lewis. / 1910-320." と記されたラベルを持つ♂がレクトタイプに指定されている.
亜種の特徴　20 mm(レクトタイプ標本). 頭部と前胸背板は黒褐色, 鞘翅は茶褐色. 前頭には細かい線状の縦皺が刻まれ, 側方においてより密. 頚部は光沢を欠き, 顕著な横皺を有する. 咽頭には1対の剛毛に加え, もう1本の付随毛がある. 触角腹面には無毛部がある. 前胸背板は四角形に近く, 長さより僅かに幅広く, 側縁の波曲は弱く, 側縁前方はごく僅かに弧状に膨らみ, 後方では左右ほぼ平行. 後角は強く後方に突出するが, 先端は鈍い. 前胸背板表面には密な点刻と皺を有する. 鞘翅は長卵形で, 軽く膨隆する. 辺縁には密な顆粒を具える. 丘孔点列は鞘翅のほぼ全長に達する. 各間室はほぼ一様に隆起し, 第一次原線の隆起は強く, 頂部にはやや光沢を有し, 浅い第一次凹陥により分断される. 第二次および第三次原線はほぼ同程度に隆起し, 頂部にやや光沢をもつ破線となる. 前胸側板の表面は平滑ながらやや光沢を欠き, 前方部分に痕跡的な点刻を有する. 前胸

腹板には点刻を欠く．中胸前側板には光沢があり，表面は平坦で，点刻は殆ど目立たない．中胸後側板は平滑で無点刻．後胸前側板には弱い皺があり，ごく浅い点刻を疎に装う．腹部腹板の側面には不明瞭な点刻を有する．腹板横溝は第4，第5腹板では側方に，第6腹板では中央部まで完全に刻まれる．陰茎は北関東上越亜種あるいは東北地方南部亜種のものに近いが，ほぼ中央で最も幅広く，右腹側の凹陥部表面は滑らかで皺を欠き，先端部は東北地方南部亜種に似て扁平だがより短く，北関東上越亜種のものより腹面への湾曲がやや弱い．膜状開口部には葉片があり，開口部の外縁は顕著に波曲する．陰茎およびその他の外部形態上の特徴から，北関東上越亜種あるいは東北地方南部亜種に最も類縁が近いと考えられている．

分布　佐渡島．ただし，原記載以来確実な追加記録はなく，ロンドン自然史博物館に保管されている1♂（レクトタイプ）が現存する唯一の標本と思われる．Nakane（1962, pl. 2, fig. 28）には「Mt. Kinpoku, Sado」産の1♂（14. VIII. 1940, Miyazaki leg.）が図示されているが，真偽のほどは定かではない．

亜種小名の由来と通称　亜種小名は「痩せた，貧弱な」を意味する．サドクロナガオサムシもしくは狭義のコクロナガオサムシという通称で呼ばれる．

12-13. *C.* (*L.*) *a. babai* Ishikawa, 1968　コクロナガオサムシ妙高連峰亜種

Original description　*Carabus* (*Leptocarabus*) *exilis babai* Ishikawa, 1968, Bull. natn. Sci. Mus., Tokyo, 11 (3), p. 264, fig. 2 (p. 265).
Type locality　Mt. Hiuchiyama, 2100 m, in the Myôkô Mountains.
Type depository　National Museum of Nature and Science, Tokyo [Holotype, ♂].
Diagnosis　18~25 mm. Most closely allied to subsp. *tenuiformis*, but distinguished from that race in the following points: 1) dorsum constantly black without brownish tinge; 2) frons much smoother; 3) gular setae absent 4) penis shorter, with inner margin less strongly convex at middle where it is not at all carinate and barely concave below though rugose, apical part shorter and narrower, gradually tapered toward tip and strongly bent ventrad from base.
Distribution　Myôkô Mts. (Niigata Pref. & Nagano Pref.; subalpine to alpine zone, usually above 1,900 m), C. Honshu.
Etymology　Named after Kintarô Baba (Niigata), a medical doctor and a keen entomologist.

タイプ産地　妙高連峰，火打山，2100 m.
タイプ標本の所在　国立科学博物館［ホロタイプ，♂］．
亜種の特徴　18~25 mm. 北関東上越亜種に最も近いが，背面は完全な黒色で，褐色味を帯びることはなく，前頭表面が滑らかなことにより一見して識別される．咽頭剛毛を欠く点も本亜種の特徴である．前胸背板には細かい横皺が多く，点刻は不明瞭．中胸側板は点刻を欠き，平滑．丘孔点列は非常に短い．陰茎は中央の膨らみが弱く，先端部は基部から曲がり，側面から見て細い．

分布　妙高連峰（頸城（くびき）山塊）．妙高連峰の高所（標高およそ1,900 m以上）に産する．

亜種小名の由来と通称　亜種小名は，医師であり熱心な昆虫愛好家・研究者でもあった新潟の馬場 金太郎に因む．ミョウコウクロナガオサムシという通称が一般的だが，ミョウコウホソクロナガオサムシと呼ばれることもある．

12-14. *C.* (*L.*) *a. shinanensis* Born, 1922　コクロナガオサムシ八ヶ岳亜種

Original description　*Carabus shinanensis* Born, 1922, Ent. Mitt., Berlin, 11 (4), p. 173.
Type locality　Mt. Yatzuzadake [sic] (misspelling of Yatsugatake), Provinz Shinano, Insel Nippon
Type depository　Unknown.
Diagnosis　17~25 mm. Allied to subsp. *babai*, but elytra and venter are occasionally brownish, gular setae usually absent, umbilicate series a little longer, and apical part of penis is less slenderer though shorter than that of subsp. *tenuiformis*. From subsp. *ogurai* distributed in the adjacent regions, this subspecies is readily discriminated by smooth and impunctate propleura.
Distribution　Mt. Yatsu-ga-také & Chikuma Mts. (Nagano Pref. & Yamanashi Pref.; subalpine to alpine zone, usually above 1,500 m in the altitude), C. Honshu.
Etymology　Named after Shinano or Shinano-no-kuni, one of the ryôsei provinces of Japan used until the early Meiji period, which roughly corresponds to the present Nagano Prefecture.

タイプ産地　日本列島，信濃（の国），"ヤツザダケ"（=八ヶ岳）．
タイプ標本の所在　不明．Bornのオサムシ類のコレクションはチューリッヒ大学，ウィーン自然史博物館，ゾロトゥルン博物館（スイス）のいずれかに収められているはずであるが，本亜種のタイプ標本の所在は確認できていない．
亜種の特徴　17~25 mm. 妙高連峰亜種に似るが，鞘翅と腹面は時に褐色味を帯び，丘孔点列はやや長く，陰茎先端は妙高連峰亜種のものほど細くない（ただし，北関東上越亜種のそれに比し，やや短い）．隣接する地域から知られる奥秩父亜種からは，前胸側板が滑らかで点刻を欠く点により容易に識別される．

12. Carabus (Leptocarabus) arboreus

分布　八ヶ岳，千曲山地．八ヶ岳および千曲山地（長野～山梨県）の高所（標高1,500～1,600 m以上）に産する．

亜種小名の由来と通称　亜種小名は，かつての令制国（りょうせいこく）の一つで現在の長野県にほぼ相当する信濃国（しなののくに）に因む．シナノクロナガオサムシ，ヤツクロナガオサムシ，ヤツホソクロナガオサムシ等の通称がある．

12-15. *C. (L.) a. ogurai* Ishikawa, 1969　コクロナガオサムシ奥秩父亜種

Original description　*Carabus* (*Leptocarabus*) *exilis ogurai* Ishikawa, 1969, Bull. natn. Sci. Mus., Tokyo, 12 (1), p. 35, fig. 2 (p. 35).

Type locality　Mt. Daibosatsu, Yamanashi Pref.

Type depository　National Museum of Nature and Science, Tokyo [Holotype, ♂].

Diagnosis　18~25 mm. Elytra often brownish. Similar to subsp. *shinanensis*, but readily distinguished from that race by distinctly punctate propleura. Penis also as in subsp. *shinanensis*, though its inner margin more weakly convex medially at the widest part where it is neither carinate nor concave below, apical part shorter and more strongly narrowed from broader base.

Distribution　Kanto, Misaka & Tenshi (or Tenshu) Mts. (Tokyo Met., Saitama Pref., Gunma Pref., Yamanashi Pref. & Nagano Pref.; subalpine zone), SC. Honshu.

Etymology　Named after Akio Ogura who collected the first series of this subspecies from Daibosatsu.

タイプ産地　山梨県，大菩薩（嶺）．

タイプ標本の所在　国立科学博物館［ホロタイプ，♂］．

亜種の特徴　18~25 mm．鞘翅は茶褐色味を帯びるものが多い．八ヶ岳亜種に近く，咽頭剛毛を欠くが，腹面，特に前胸側板に顕著な点刻を具える点により容易に識別される．陰茎の形態も八ヶ岳亜種のものに似るが，内側縁中央腹側の膨らみは弱く，同部の右側縁は稜をなさず，腹面も殆どえぐれない．陰茎先端部は八ヶ岳亜種のそれよりも短く，基部はより幅広く，そこから先端に向けてより急激に細くなる．葉片の先端は基本的に分葉する．天子山地の毛無山（けなしやま）に産するものは，外部形態の特徴が富士箱根丹沢亜種のそれに近いが，陰茎の形態は本亜種のそれに一致する（ただし，葉片は単葉）．

分布　関東・御坂・天子（又は天守）山地（東京都及び埼玉，群馬，山梨，長野各県）．

亜種小名の由来と通称　亜種小名は，大菩薩で最も初期に本亜種を採集した小倉暁雄に因む．ダイボサツクロナガオサムシ，チチブクロナガオサムシ，チチブホソクロナガオサムシ等の通称がある．

12-16. *C. (L.) a. fujisanus* Bates, 1883　コクロナガオサムシ富士箱根丹沢亜種

Original description　*Carabus Fujisanus* Bates, 1883, Trans. ent. Soc. Lond., 1883, p. 227.

Type locality　Subashiri, near Fujisan.

Type depository　Natural History Museum, London [Lectotype, ♂].

Diagnosis　19~26 mm. Black or dark brown. Readily identified by the following characters: 1) gular setae absent; 2) 5th segment of male antenna without hairless ventral area; 3) pronotum only with a pair of marginal setae (anterior seta is absent); 4) metacoxa bisetose, anterior seta is absent; 5) ostium lobe completely lost; 6) elytral intervals much degenerated. Penis as in subsp. *ogurai*, but distinguished by having longer and subparallel-sided widest part and absence of ostium lobe.

Distribution　Mt. Fuji-san, Hakoné Mts. & Tanzawa Mts. (subalpine to alpine zone), SC. Honshu.

Etymology　Named after Mt. Fuji-san, the type locality of this subspecies.

タイプ産地　富士山近傍，須走．

タイプ標本の所在　ロンドン自然史博物館［レクトタイプ，♂］．原記載論文ではホロタイプが指定されておらず，Ishikawa (1992) により "Type / H. T. // Tsubashiri // Japan / G. Lewis / 1910-320 // *Fujisanus* // Fujiyama / Bates" と記されたラベルを持つ♂がレクトタイプに指定されている．

亜種の特徴　19~26 mm．黒色ないし暗褐色で，体型は細長い．咽頭剛毛と♂触角第5節腹面の無毛部を欠き，前胸背板の側縁剛毛も後角前方の一対のみである点で特異．前胸背板は細長く，側縁の波曲が弱く，後角は鋭く突出する．鞘翅間室の隆条は極めて弱く，全面に小顆粒を装い，丘孔点列は非常に短い．胸側板の点刻は顕著．後基節は前方剛毛を欠く．陰茎は膜状部より先端にかけて急激に狭まり，先端部は短い．膜状部基部には葉片の痕跡を欠く．本種中，特殊化の程度が最も高い亜種の一つで，咽頭剛毛や前胸背板前方剛毛を欠くこと，ならびに陰茎膜状開口部基部に葉片の痕跡が見られないことなど，いくつかの退化的な形質を有する点が特徴である．

分布　丹沢山地，箱根山地，富士山．

亜種小名の由来と通称　亜種小名は富士山に因む．通称はフジクロナガオサムシ．

12-17. *C. (L.) a. horioi* NAKANE, 1961 コクロナガオサムシ赤石山脈亜種

Original description　*Carabus* (*Leptocarabus*) *exilis horioi* NAKANE, 1961, Fragm, coleopterol., Kyoto, (1), p. 2.
Type locality　Sampuku Pass in S. Japan Alps, Honshu.
Type depository　Hokkaido University Museum (Sapporo) [Holotype, ♂].
Diagnosis　18~25 mm. Elytra black, brown or reddish brown. Characterized by relatively robuster and strongly cordate pronotum, rather smooth pro- and mesepisterna, shorter umbilicate series restricted to basal third of elytra and characteristically shaped apical part of penis, etc., though showing a considerable ecogeographical variation. From three allied subspecies such as *gracillimus*, *hakusanus* and *ohminensis*, this race is distinguished by strongly wrinkled pronotum, more distinctly bent apical part of penis and presence of ostium lobe though rudimentary.
Distribution　Akaishi Mts. (subalpine to alpine zone), C. Honshu.
Etymology　Named after T. HORIO who collected the type specimen from the Pass Sanpuku-tôgé (= Sampuku Pass).

　タイプ産地　本州，南アルプスの三伏峠．中根（1962, p. 58）は「三福峠」と記しているが，三伏峠の誤記と思われる．
　タイプ標本の所在　北海道大学総合博物館（札幌）［ホロタイプ，♂］．
　亜種の特徴　18~25 mm．鞘翅は黒色ないし褐色で，赤褐色に近い個体もある．前胸背板が比較的太短く，後方に向けて強く狭まる点，胸側板が平滑で点刻を欠く点，丘孔点列が短く，鞘翅前方1/3に限られる点，特徴的な陰茎先端部の形態（基部から腹側に湾曲しつつ先端に向けて比較的急に狭まる）等により特徴付けられる．本種の分布域の西部に分布する近縁の3亜種（飛騨御嶽木曽亜種，白山亜種，大峰山脈亜種）とは，陰茎中央部の膨らみが弱く，先端がほぼ円錐形に細まり，葉片が退化的である点で共通しているが，本亜種は前胸背板の皺が強く，陰茎先端の腹側への湾曲が強く，葉片が痕跡的ながら認められること等より識別される．
　分布　赤石山脈高所．
　地理的変異　標高によって形態に変化が見られ，一般に2,000 m以下の亜高山針葉樹林帯に生息するものは平均して大型で鞘翅と腹面は茶褐色を帯び，前胸背板の皺は密ながら弱く，鞘翅側縁光沢帯内の丘孔点列は前胸背板とほぼ同長で，その後方の顆粒列の後端は鞘翅中央を越えず，鞘翅間室の隆起は弱く，第一次原線は分断された鎖線状で，頂部には細い光沢部があり，第二次，第三次原線は顆粒列となる．これに対し，高山帯に生息するものでは，これらの形態的特徴が一般により顕著に発達する傾向にある．すなわち，前胸背板の皺は強く，丘孔点列の後端は鞘翅中央より後方に達し，鞘翅間室の隆起はより強く，第一次原線は深い第一次凹陥により頻繁に分断された結節列状となり，第二次原線は鎖線状ないし刻み目を持つ隆条，第三次間室は不規則な顆粒列となり，全体的な印象として網眼状に近い状態を呈する．分布域の北端に位置する入笠山のものでは，陰茎先端部の形態や葉片の大きさに変化が見られ，咽頭剛毛を欠く個体が高頻度で現れるなど，八ヶ岳亜種との交雑を思わせる個体が混入する．
　亜種小名の由来と通称　亜種小名は，タイプ産地の三伏峠で本亜種を採集した堀尾 貞太郎に因む．サンプククロナガオサムシという通称が一般的だが，アカイシクロナガオサムシという名が提唱されたこともある．

12-18. *C. (L.) a. gracillimus* BATES, 1883 コクロナガオサムシ飛騨御嶽木曽亜種

Original description　*Carabus gracillimus* BATES, 1883, Trans. ent. Soc. Lond., 1883, p. 227.
Type locality　The summit of Ontake.
Type depository　Natural History Museum, London [Lectotype, ♂].
Diagnosis　16~24 mm. Body above black, often brownish. Allied to subsp. *horioi*, but distinguished from it by weaker pronotal wrinkles, slenderer and less strongly bent apical part of penis, etc. *Carabus* (*Leptocarabus*) *exilis hidamontanus* described by ISHIKAWA (1968b, p. 266) from "Mt. Shiroumadake, 2600-2933m. alt." of the Hida Mountains is synonymized with the present subspecies according to ISHIKAWA (1992),
Distribution　Hida Mts., Kiso Mts. & Mt. Ontaké-san (subalpine to alpine zone), C. Honshu.
Etymology　Subspecific name means "extremely slender".

　タイプ産地　御嶽（山）山頂．
　タイプ標本の所在　ロンドン自然史博物館［レクトタイプ，♂］．原記載論文中ではホロタイプが指定されておらず，ISHIKAWA（1992）により"Type H. T. // Japan / G. LEWIS / 1910-320 // 18-8. 1 // *gracillimus* / BATES"と記されたラベルを持つ♂がレクトタイプに指定されている．）
　亜種の特徴　16~24 mm．鞘翅は黒色で，褐色味を帯びることが多く，稀に茶褐色．腹面は褐色味を帯びる．鞘翅側縁の丘孔点列は少なくとも前胸背板と同長か，より長いが，先端が鞘翅中央を越えて後方に達することは殆どない．赤石山脈亜種に近いが，陰茎中央部はより細く，同先端部は細長い円錐形で，腹側への湾曲はごく弱い．赤石山脈亜種と同様，体長や鞘翅彫刻にかなり変異が見られるが，前胸背板の皺がより弱く，陰茎の形態が異なることにより識別される．飛騨山脈北部の白馬岳より*Carabus* (*Leptocarabus*) *exilis hidamontanus*という名で記載された集団（ISHIKAWA, 1968b, p. 266）は，ISHIKAWA（1992）に倣い，本亜種のシノ

ニムとして処理しておく.
　分布　飛騨山脈, 御嶽山, 木曽山脈.
　亜種小名の由来と通称　亜種小名は「非常に細長い」を意味する. 御嶽産のものはオンタケクロナガオサムシという通称で呼ばれる. 飛騨山脈のものはヒダクロナガオサムシと呼ばれることもある.

12-19. *C. (L.) a. hakusanus* NAKANE, 1961　コクロナガオサムシ白山亜種

　Original description　*Carabus* (*Leptocarabus*) *exilis hakusanus* NAKANE, 1961, Fragm. coleopterol., Kyoto, (1), p. 3.
　Type locality　Mt. Hakusan, Kaga, Ishikawa Pref., Honshu.
　Type depository　Hokkaido University Museum (Sapporo) [Holotype, ♂].
　Diagnosis　19~25 mm. Body above usually black, rarely brownish. Characterized by frequently lacking anterior setae of pronotum. Penis as in subsp. *gracillimus*, though relatively thicker in median portion in right subdorsal view, with apical part a little robuster and blunt at tip.
　Distribution　Mt. Haku-san (subalpine to alpine zone), C. Honshu.
　Etymology　Named after the type locality, Mt. Haku-san, which means "white mountain" in Japanese.

　タイプ産地　本州, 石川県, 加賀, 白山.
　タイプ標本の所在　北海道大学総合博物館(札幌)［ホロタイプ, ♂］
　亜種の特徴　19~25 mm. 背面の色彩が通常, 褐色味を欠いた黒色で, 前胸背板側縁の前方剛毛を高頻度で欠き, 鞘翅彫刻が比較的粗く, 丘孔点列は短く, 肩部に限られる. 陰茎は斜背面から見ると中央部で比較的幅広く, 先端部は比較的太短く, 鈍く丸まって終わる.
　分布　白山(石川県~岐阜県~福井県).
　亜種小名の由来と通称　亜種小名は, タイプ産地の白山に因む. 通称はハクサンクロナガオサムシ.

12-20. *C. (L.) a. ohminensis* HARUSAWA, 1978　コクロナガオサムシ大峰山脈亜種

　Original description　*Carabus* (*Adelocarabus*) *arboreus ohminensis* HARUSAWA, 1978, Bull. Osaka Mus. nat. Hist., Osaka, (31), p. 43.
　Type locality　Ohmine Mountain Range, alt. 1,700 m, Kii Peninsula.
　Type depository　Osaka Museum of Natural History [Holotype, ♂].
　Diagnosis　19~22 mm. Elytra brownish black to brown. Similar to subsp. *gracillimus* from the forest zone of the southern localities of its range, but distinguishable from that race by characteristically featured pronotum, elytral sculpture and apical part of penis as follows: 1) pronotum weakly cordate, with hind angles weakly protruded, disk at most rudimentarily punctate and faintly rugulose, distinctly depressed laterally so that lateral margins sharply and strongly curved above throughout, basal foveae broad and deep; 2) elytral intervals weakly but regularly raised, which is the most noticeable characteristic of this subspecies; 3) penis as in subsp. *gracillimus*, though a little more strongly arcuate ventrad in apical portion and blunt at tip.
　Distribution　Ôminé Mts. (very narrowly restricted to the summit of a certain mountain, though detailed locality name has not yet announced), Nara Pref., SW. Honshu.
　Conservation　The distribution of this species is confined to the summit of a certain mountain in the Ôminé Mountains and the species has become much rarer during the past three decades probably due to excessive hunting by collectors. This species is designated as threatened species (NT; near threatened species) in the 4th edition of Red List (Threatened Wildlife of Japan) of Ministry of the Environment (2012).
　Etymology　Named after the Ôminé Mountains in the Kinki Region, on which the type locality of this taxon lies.

　タイプ産地　紀伊半島, 大峰山脈, 標高 1,700 m.
　タイプ標本の所在　大阪市立自然史博物館［ホロタイプ, ♂］.
　亜種の特徴　19~22 mm. 鞘翅は黒褐色ないし褐色. 形態学的には, 白山亜種よりも飛騨御嶽木曽亜種, とくに分布域南部の森林地帯に産するものに近いが, 前胸背板, 鞘翅彫刻および陰茎先端の形態等により識別される. 1)前胸背板は後方に向けて軽く狭まった心臓形で, 後角の突出は弱く, 背面の点刻は痕跡的で皺もごく軽微. 側縁内側は著しく押圧され, 辺縁は強く反り, 基部凹陥は広く深い; 2)鞘翅間室の隆起は弱く, ほぼ均等(本亜種の最も顕著な特徴とされる); 3)陰茎は飛騨御嶽木曽亜種のそれに近いが, 先端部の腹側への湾曲がやや強く, あまり鋭く尖らない.
　分布　大峰山脈にある某山の山頂付近.
　保全　本亜種は大峰山脈にある某山(場所の詳細については公表されていない)の山頂付近から知られているにすぎず, コクロナガオサムシの亜種の中ではとびぬけてかけ離れた孤立分布を示すため, 稀種として珍重されている. 生息地では比較的多くの個体が見られるようだが, 生息範囲が狭いため, 複数の採集者により多数のトラップが波状的にかけ続けられるといった過度の

12. *Carabus* (*Leptocarabus*) *arboreus*

採集圧がかかれば，その将来は必ずしも楽観できない．環境省第4次レッドリスト（2012）では，準絶滅危惧種（NT）となり，それまでより評価の下がったダウンリスト種となっている．

亜種小名の由来と通称　亜種小名は大峰山脈に因む．通称はオオミネクロナガオサムシ．

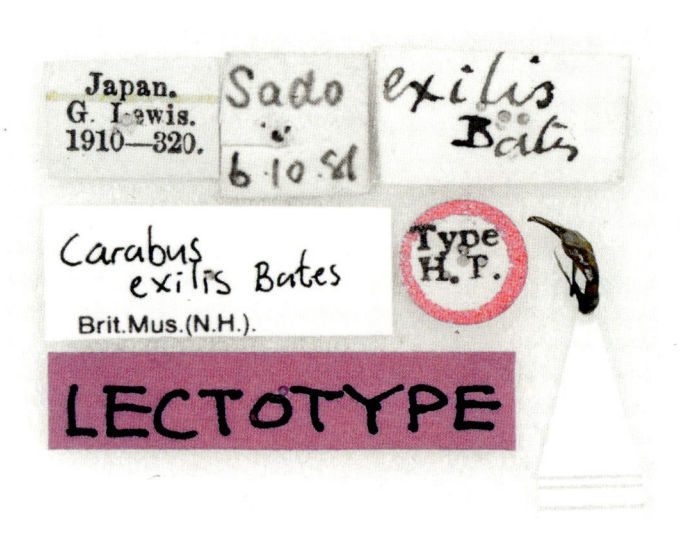

Fig. 54. Lectotype (♂) of *Carabus* (*Leptocarabus*) *arboreus exilis* BATES, 1883 (preserved in the collection of the Natural History Museum, London).
図 54. コクロナガオサムシ佐渡島亜種レクトタイプ（♂，ロンドン自然史博物館所蔵）．

12. *Carabus* (*Leptocarabus*) *arboreus*

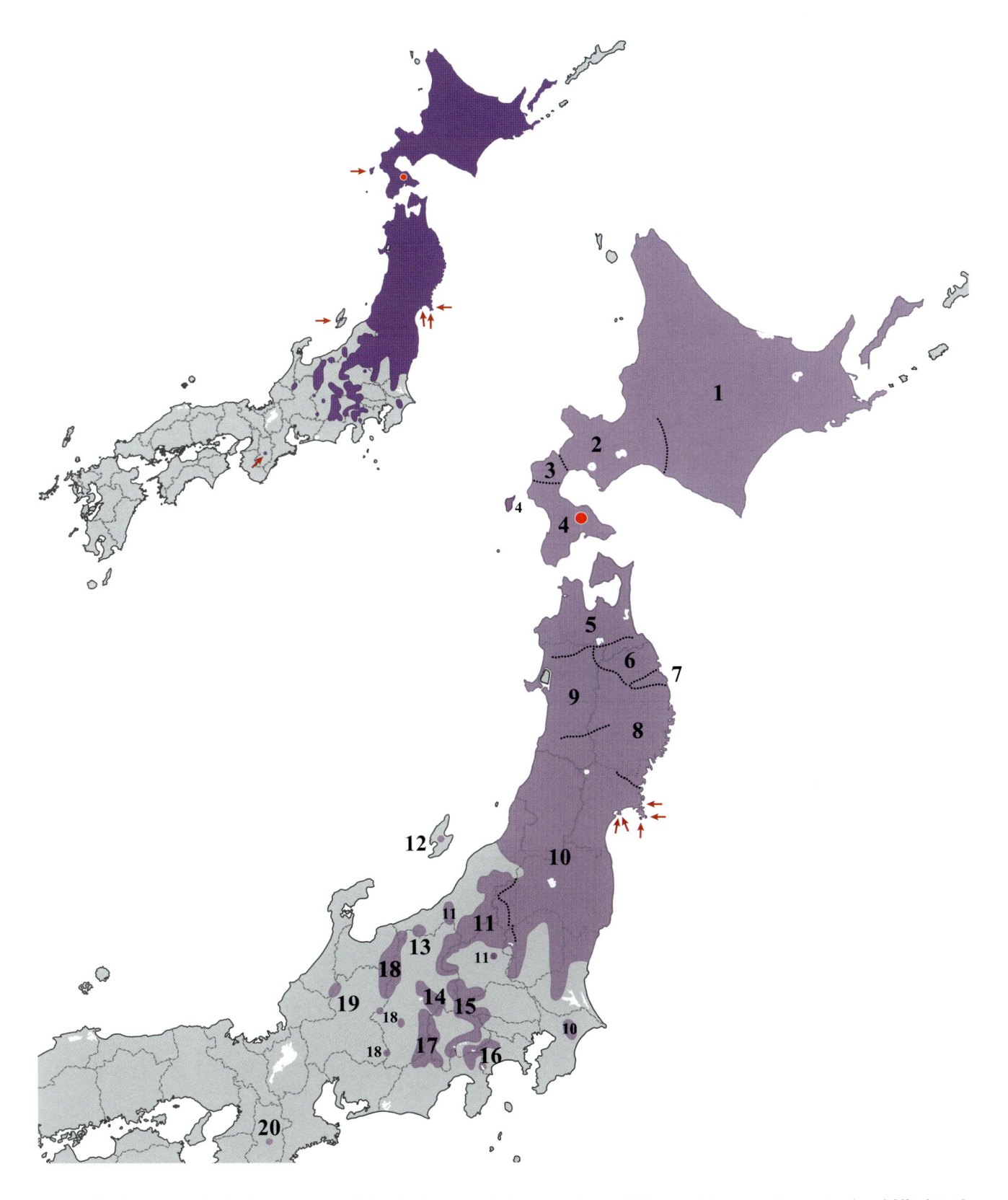

Fig. 55. Map showing the distributional range of **Carabus** (**Leptocarabus**) **arboreus** Lewis, 1882. — 1, Subsp. *pararboreus*; 2, subsp. *ishikarinus*; 3, subsp. *karibanus*; 4, subsp. **arboreus**; 5, subsp. *nepta*; 6, subsp. *pronepta*; 7, subsp. *shimoheiensis*; 8, subsp. *kitakamisanus*; 9, subsp. *akitanus*; 10, subsp. *parexilis*; 11, subsp. *tenuiformis*; 12, subsp. *exilis*; 13, subsp. *babai*; 14, subsp. *shinanensis*; 15, subsp. *ogurai*; 16, subsp. *fujisanus*; 17, subsp. *horioi*; 18, subsp. *gracillimus*; 19, subsp. *hakusanus*; 20, subsp. *ohminensis*. ●: Type locality of the species (Junsai = Lake Junsai-numa).

図 55. コクロナガオサムシ分布図. — 1, 道央道東道北亜種; 2, 石狩亜種; 3, 渡島半島北部亜種; 4, 基亜種; 5, 青森県亜種; 6, 岩手県北部亜種; 7, 下閉伊亜種; 8, 北上山地亜種; 9, 秋田県亜種; 10, 東北地方南部亜種; 11, 北関東上越亜種; 12, 佐渡島亜種; 13, 妙高連峰亜種; 14, 八ヶ岳亜種; 15, 奥秩父亜種; 16, 富士箱根丹沢亜種; 17, 赤石山脈亜種; 18, 飛騨御嶽木曽亜種; 19, 白山亜種; 20, 大峰山脈亜種. ●: 種のタイプ産地（蓴菜沼）.

13. *Carabus* (*Asthenocarabus*) *opaculus* Putzeys, 1875
ヒメクロオサムシ

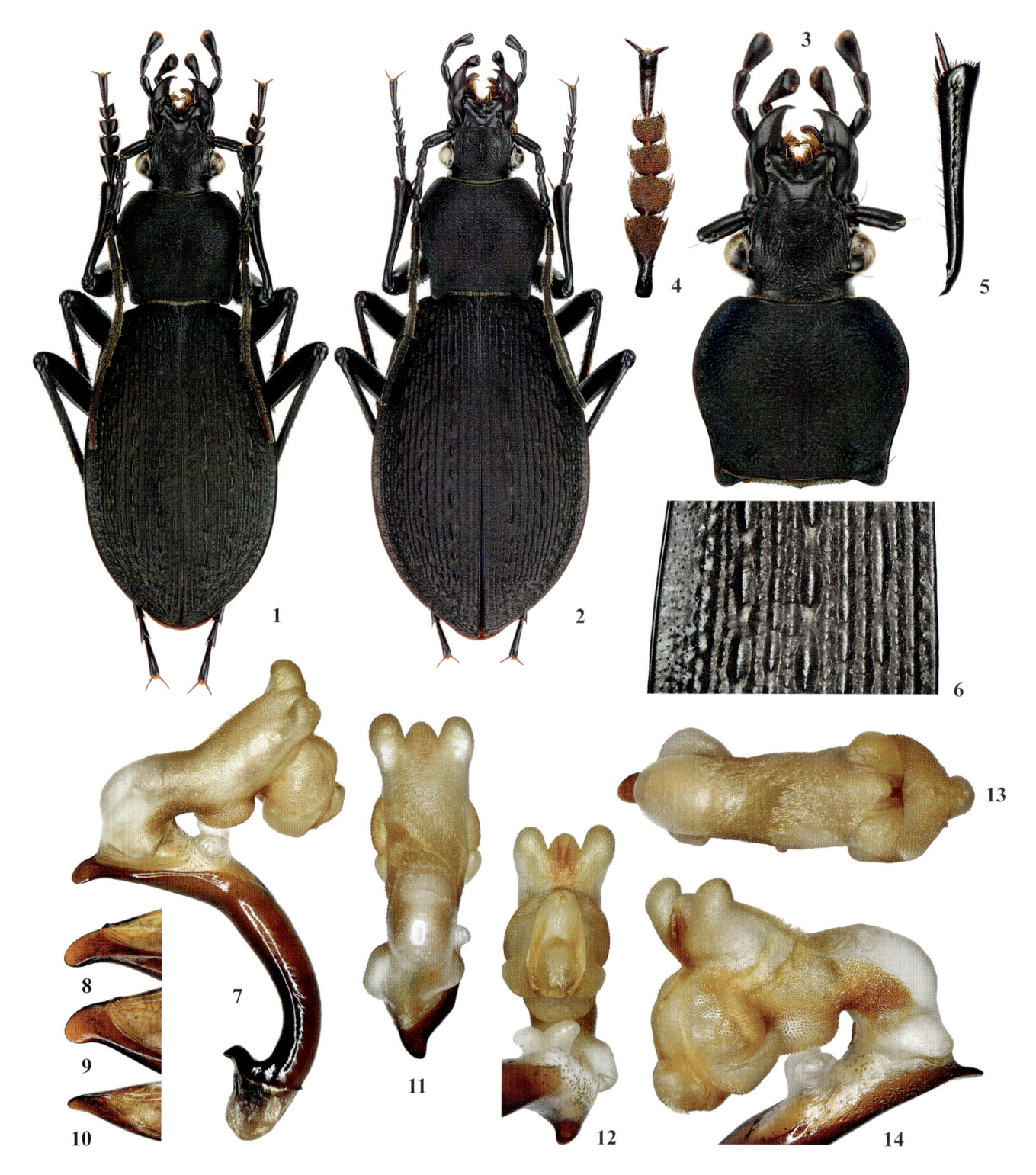

Figs. 1-14. *Carabus* (*Leptocarabus*) *opaculus opaculus* Putzeys, 1875 (Mt. Yokotsu-daké, Nanaé-chô, Hokkaido). — 1, ♂; 2, ♀; 3, male head & pronotum; 4, male right protarsus in ventral view; 5, male right protibia in subdorsal view; 6, male left elytron (median part); 7, penis & fully everted internal sac in right lateral view; 8, apical part of penis in right lateral view; 9, ditto in right subdorsal view; 10, ditto in dorsal view; 11, internal sac in caudal view; 12, ditto in frontal view; 13, ditto in dorsal view; 14, ditto in left sublateral view.

図 1-14. ヒメクロオサムシ異曲種（北海道七飯町横津岳）．— 1, ♂; 2, ♀; 3, ♂の頭部と前胸背板; 4, ♂右前跗節（腹面観）; 5, ♂右前脛節（斜背面観）; 6, ♂左鞘翅中央部; 7, 陰茎と完全反布下の内嚢（右側面観）; 8, 陰茎先端（右側面観）; 9, 同（右斜背面観）; 10, 同（背面観）; 11, 内嚢（尾側面観）; 12, 同（頭側面観）; 13, 同（背面観）; 14, 同（左斜側面観）.

13. *Carabus* (*Asthenocarabus*) *opaculus*

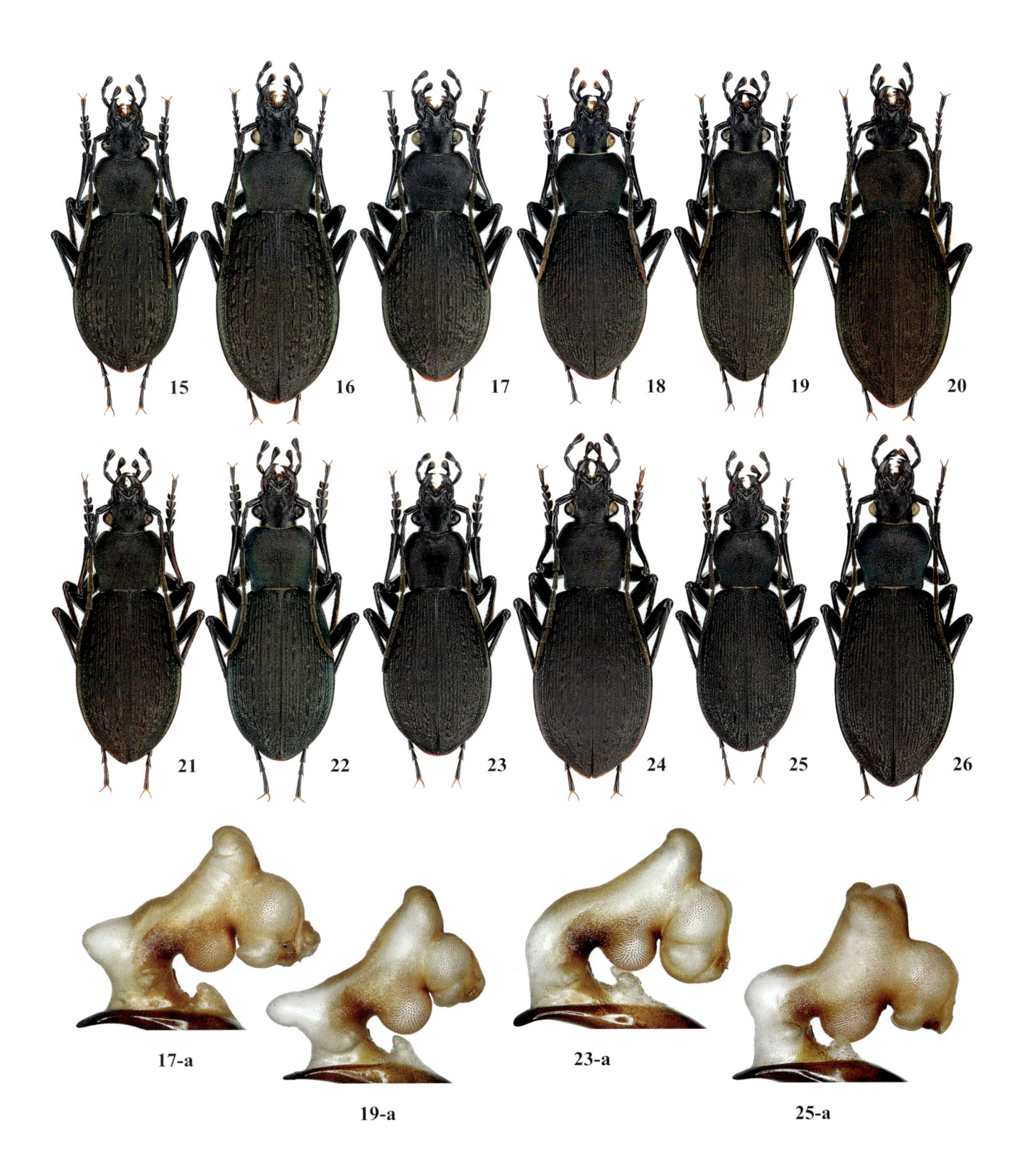

Figs. 15-26. *Carabus* (*Leptocarabus*) *opaculus* subspp. — 15-22, *C.* (*A.*) *o. kurosawai* (15, 17-19, 21, 22, ♂, 16, 20, ♀; 15-16, Is. Rishiri-tô; 17, Yukomanbetsu, Daisetsu-zan Mts.; 18, Pass Asahi-tôgé, Engaru-chô; 19-20, Futatsu-yama, Shibecha-chô; 21, Lake Chôboshi-ko, Nemuro-shi; 22, Mt. Yûbari-daké, Yûbari-shi); 23-24, *C.* (*A.*) *o. opaculus* (23, ♂, Pass Nakayama-tôgé, Sapporo-shi; 24, ♂, Is. Okushiri-tô); 25-26, *C.* (*A.*) *o. apoi* (25, ♂, PT, Mt. Apoi-daké, Samani-chô; 26, ♀, PT, Horoman, Samani-chô). — a, Fully everted internal sac in right lateral view. All from Hokkaido.

図 15-26. ヒメクロオサムシ各亜種. — 15-22, **道央道東道北亜種**(18, 17-19, 21, 22, ♂, 16, 20, ♀; 15-16, 利尻島; 17, 大雪山勇駒別; 18, 遠軽町旭峠; 19-20, 標茶町二ツ山; 21, 根室市長節湖; 22, 夕張市夕張岳); 23-24, **基亜種**(23, ♂, 札幌市中山峠; 24, ♂, 奥尻島); 25-26, **日高山脈南端部亜種**(25, ♂, PT, 様似町アポイ岳; 26, ♀, PT, 様似町幌満). — a, 完全反転下の内囊(右側面観). すべて北海道産.

13. *Carabus* (*Asthenocarabus*) *opaculus*

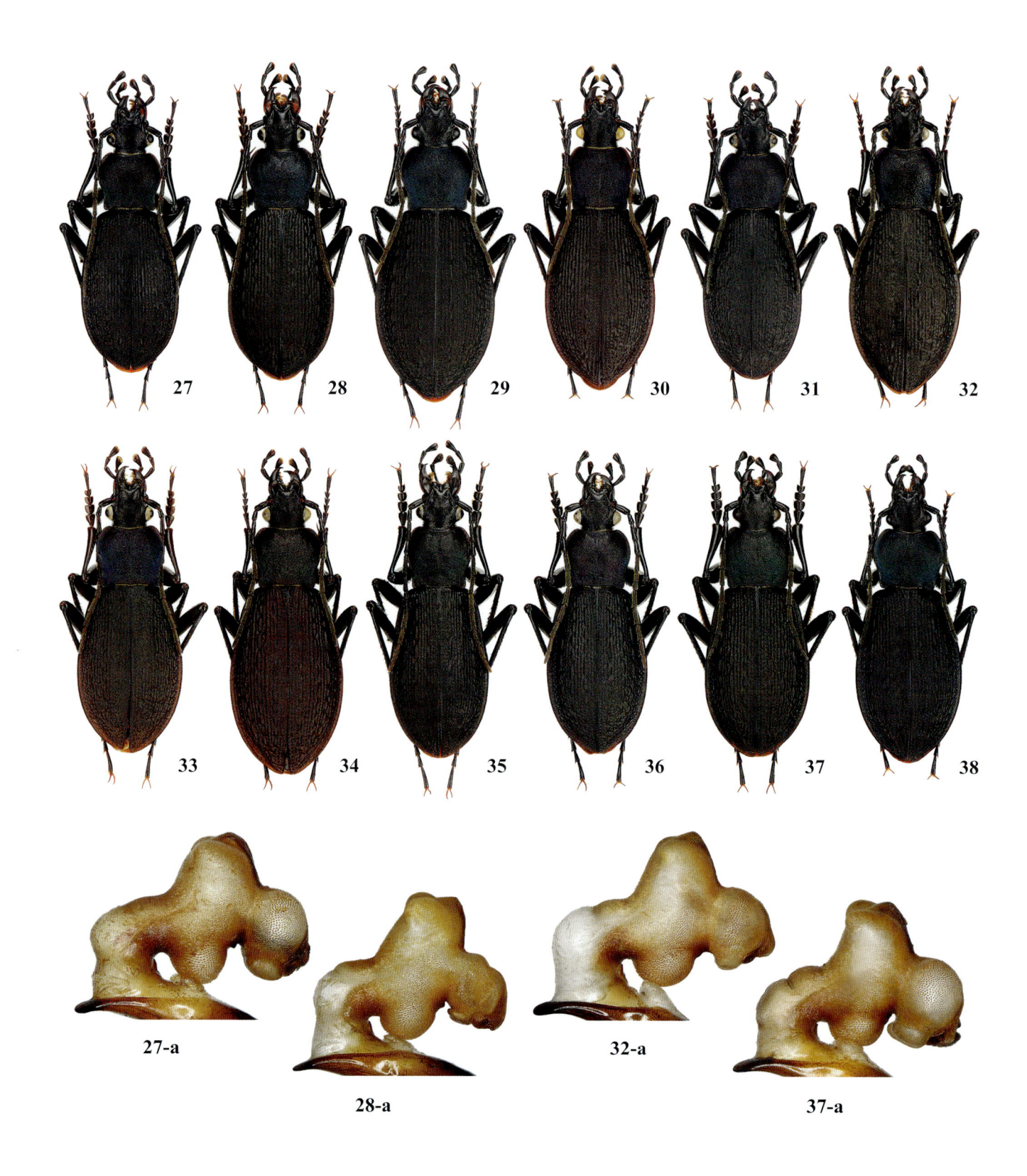

Figs. 27-38. ***Carabus*** (***Leptocarabus***) ***opaculus shirahatai***. — 27, 28, 31, 33, 35-37, ♂, 29, 30, 32, 34, 38, ♀; 27, Mt. Iwaki-san, Aomori Pref.; 28-29, Mt. Ô-daké, Hakkôda-san Mts., Aomori Pref.; 30, Mt. Hayachiné-san, Iwate Pref.; 31-32, Mt. Chôkai-san, Akita Pref.; 33-34, Mt. Gassan, Yamagata Pref.; 35, Mt. Zaô-san, Miyagi Pref.; 36, Mt. Azuma-yama, Fukushima Pref.; 37-38, Mt. Iidé-san, Niigata Pref. — a, Fully everted internal sac in right lateral view.

13. *Carabus* (*Asthenocarabus*) *opaculus* Putzeys, 1875 ヒメクロオサムシ

Original description *Carabus opaculus* Putzeys, 1875, Annls. Soc. ent. Belg., Bruxelles, 18, p. 48.

History of research This species was originally described by Putzeys (1875) from Jesso which is an old name of Hokkaido. Although Lapouge (1930) established for this species a new subgenus *Asthenocarabus* of the genus *Leptocarabus*, adopted genus and subgenus for Putzeys' species have been different according to authors, i.e., *Carabus* (*Eucarabus*) *opaculus* (Breuning, 1932), *C.* (*Asthenocarabus*) *opaculus* (Nakane, 1962), *Leptocarabus* (*Asthenocarabus*) *opaculus* (Ishikawa, 1972 etc.), *C.* (*Tomocarabus*) *opaculus* (Imura & Mizusawa, 1996), etc. Judging from peculiarly shaped internal sac of the male genitalia and the result of molecular phylogenetical analyses, it is regarded as a member of the subgenus *Asthenocarabus* in the *Tomocarabus* subgenus-group of the grand genus *Carabus* in this book. Breuning (1957) described subsp. *kurosawai* from the Daisetsu-zan Mountains of Hokkaido, which has often been regarded as a mere synonym of the nominotypical *opaculus*. Nakane (1961) described subsp. *shirahatai* from Mt. Chôkai-san of Tohoku Region. A brief comment was given by Okumura (1971) on the geographical variation of this species, and he suggested that the population distributed in southern Hidaka Region and Shibecha of eastern Hokkaido are characteristic. Imura (2011) focused on the taxonomical importance of the male genitalic features of this species, above all on those of internal sac in fully everted condition. He recognized *kurosawai* as a good subspecies, and described another subspecies from the southernmost part of the Hidaka Mountains of Hokkaido under the name of *apoi*.

Type locality Jesso (= ancient name of Hokkaido). Most probably Hakodaté or its nearby regions.

Type depository Institut Royal des Sciences Naturelles de Belgique (Brussels) [Holotype, ♂].

Morphology 15~22 mm. Entirely black and mat, often with purplish, bluish or rarely with greenish tinge on head, pronotum and elytral margins. External features: Figs. 1~6. Description on details: penultimate segment of labial palpus with two setae near base; median tooth of mentum almost as lateral lobes in length, narrowly rimmed by low ridge, with apex sharply pointed in ventral view, forming vertical plate in lateral view; submentum bisetose; ventral side of male antennae without hairless areas; pronotal margins basically bisetose on each side; preapical emargination of elytra only faintly recognized in both sexes, though usually unclear; elytral sculpture triploid; primaries usually the strongest, indicated by rows of costae of long ovoid-shaped tubercles irregularly segmented by small and shallow primary foveoles; both secondaries and tertiaries indicated by contiguous low costae or rows of granules; areas between intervals sporadically scattered with fine, small granules; umbilicate series variable in length according to population; propleura smooth; sides of sternites weakly rugoso-punctate; metacoxa basically bisetose, lacking inner seta; sternal sulci clearly carved. Male genitalic features: Figs. 7~14; penis about 0.4 times as long as elytra; ostium lobe large and markedly bilobed at tip; ligulum indicated by assemblage of large granules; internal sac with weakly inflated basal lateral lobes on both sides; median lobe variable in size and shape according to population; both parasaccellar lobes and saccellus strongly protruded dorsad; apical portion of internal sac short and robust, with moderately developed podian lobes; inflexed side of internal sac with a large, hemispherical inflation near middle; peripheral rim of gonopore very weakly sclerotized and pigmented. Female genitalia with outer plate of ligular apophysis elongate oval, inner plate wedge-shaped and sclerotized.

Molecular phylogeny In the molecular phylogenetic tree of the world Carabina constructed by mitochondrial ND5 gene sequences, this species belongs to an independent lineage close to those of *Tomocarabus* spp., and no close affinity is recognized to the species belonging to the subgenus *Leptocarabus*. The tree suggests that there are two major lineages within the cluster of *C.* (*A.*) *opaculus*: one is composed of the specimens from Hokkaido excluding those of southernmost Hidaka, and another is composed of the specimens from southernmost Hidaka and the Tohoku Region (Osawa et al., 2004).

Distribution Hokkaido & NE. Honshu / Iss. Rebun-tô, Rishiri-tô, Daikoku-jima, Okushiri-tô, Shikotan-tô, Kunashiri-tô & Etorofu-tô (Hokkaido). Outside Japan: Is. Sakhalin (southern part) & S. Kurils.

Habitat In Hokkaido, this species is eurytopic, widely distributed from the forest to grassland, and is often predominant over other *Carabus* species. The range of vertical distribution is also wide, recorded from nearly at sea level to high mountains with the altitude around 2,000 m. In Honshu, it is subalpine to alpine species, and is found locally in ridges, barren slopes or alpine meadows higher than 800 to 1,000 m.

Bionomical notes Autumn breeder. Adult has its main activity peak in June (low altitudinal area) to July (higher or colder area). Newly emerged adult appears from mid-summer to early autumn. Larvae prefer to feed on other insets or their larvae (Inoue, 1953c).

Etymology The specific name *opaculus* is composed of opacus and -ulus, meaning "a little dark".

研究史　本タクソンはPutzeys (プッツェイス) により明治時代初頭の1875年, イェッソ (蝦夷＝北海道) において得られた1♂に基づき, カラブス属の新種として記載された. Lapouge (ラプージュ) は本種を基準種とするアステノカラブス亜属 (レプトカラブス属) を設立し, Ishikawa (1972等) もこれに倣ったが, カラブス属 (狭義) のアステノカラブス亜属 (中根, 1962), カラブス属 (広義) のエウカラブス亜属 (Breuning (ブロイニング), 1932), カラブス属 (広義) のトモカラブス亜属 (井村・水沢, 1996) 等, 著者によりその扱いは様々であった. 本書では, ♂交尾器内嚢形態, 分子系統解析の結果などから, 本種を広義のカラブス属の中のトモカラブス亜属群, アステノカラブス亜属の一員として扱う. Breuning (1957) によって大雪山から記載された亜種 *kurosawai* は, 基亜種の異名として処理されることが多かったが, 本書ではImura (2011) に従い, 独立した亜種とみなす. Nakane (1961) は山形県の鳥海山から亜種 *shirahatai* を記載した. 奥村 (1971) は本種の地理的変異に触れ, 大雪山 (山頂付近及び山麓), 札幌周辺, 日高「方面」(奥村の

13. *Carabus* (*Asthenocarabus*) *opaculus*

表現による）（上歌別，アポイ岳，西舎等），知床，標茶，利尻・礼文両島，渡島半島の集団の特徴を述べ，日高と標茶のものにそれぞれ顕著な特徴があり，他地域のものから識別できるとしている．2011年になって，これまで顧みられることのなかった本種の♂交尾器内嚢形態を分類形質として重視した IMURA は，本種の亜種の再定義を行い，*kurosawai* を亜種と認めたうえで，日高山脈南端部から亜種 *apoi* を記載した．

タイプ産地　イェッソ（蝦夷＝北海道）．原記載には場所の詳細に関する記述がないが，ホロタイプ標本の形態ならびに当時外国と交流があった地域という点から勘案し，函館ないしその周辺地域から得られたものである可能性が高い．

タイプ標本の所在　ベルギー王立自然史博物館（ブリュッセル）［ホロタイプ，♂］．

形態　15～22 mm．黒色で光沢は鈍く，前胸背板と鞘翅側縁はしばしば青～紫，時に緑色を帯びる．外部形態：図1～6．細部の記載：下唇肢亜端節の剛毛は基部よりに2本．下唇基節中央歯は側葉と同長ないしやや短く，狭い縁取りを有し，先端は腹面から見て鋭く尖り，側面から見て垂直版を形成する．下唇亜基節には2本の長毛を有する．♂触角腹面に無毛部はない．前胸背板の側縁剛毛は2対で，中央と後角付近に各1本．鞘翅端は♂♀共にごく僅かにえぐれる場合があるが一般に不明瞭，鞘翅彫刻は三元的で，第一次間室は通常最も強く，小さく浅い第一次凹陥により分断された鎖線ないし長楕円形の瘤状隆起列となり，第二次，第三次間室は不規則に分断された低い隆条ないし顆粒列．間室間の基面にはごく小さい顆粒を疎に装う．丘孔点列は鞘翅基半部のみに見られるものから鞘翅のほぼ全長に亘るものまで，集団ないし個体により変化がある．前胸側板は滑らかで，腹部腹板の側面には弱い皺と点刻を装う．後基節の剛毛は2本で，内側剛毛を欠く．腹部腹板の横溝は明瞭．♂交尾器形態：図7～14．陰茎は鞘翅長の0.4倍前後．葉片は大きく，先端は顕著に分葉する．内嚢基部の基斑は粗大な顆粒の集簇．内嚢は両側にそれほど顕著でない基部側葉を具え，中央葉はこれを欠くものから大きく三角形に発達するものまで，集団により変化が著しい．傍盤葉と内嚢盤は共に強く背側に突出し，内嚢先端は太短く，脚葉が目立つ．内嚢屈側中央に半球形の大きな膨隆部がある．射精孔縁膜の硬化と色素沈着はごく軽度．♀交尾器腟底部節片は比較的大きく，外板は長楕円形，内板は楔形で硬化する．

分子系統　ミトコンドリアDNAによる世界のオサムシ亜族の分子系統樹上で，本種はトモカラブス亜属各種に比較的近い位置を占める単系統の枝を構成し（OSAWA *et al.*, 2004），系統的には独立した位置を占めているように思われる．種内では大きく2系統に分かれ，両者の分岐年代は古い．第一の系統は日高地方南端部（様似など）産を除くすべての北海道産からなり，第二の系統には北海道日高地方南端部と本州産のものが含まれる．分子系統上，日高南端のものと東北地方の集団が近縁である点は非常に興味深いが，鞘翅彫刻や内嚢形態からも両者の類縁の近さが示唆される．

分布　北海道，本州（東北地方）／礼文島，利尻島，大黒島，奥尻島，色丹島，国後島，択捉島（以上，北海道）．国外：サハリン南部，千島列島南部．北海道のほぼ全域と本州東北地方の山岳地帯に分布し，国外ではサハリン南部と千島列島南部から記録されている．北海道では平地・山地を問わず広く分布するが，東北地方では山地性で，分布は断続する．周辺島嶼では利尻，礼文，奥尻の各島のほかに大黒島（北海道厚岸町）から記録があり，北方領土の島々にかけて分布する．

生息環境　北海道では平地からかなりの高所まで広く生息し，各地で優占種となっている場合が多いが，生息密度並びに同所的に産するオサムシの中で本種の占める比率は，標高を増すにつれて高くなる傾向にある．寒帯～亜寒帯の適湿森林から準乾燥疎林に至るまで，主として森林帯に多く，特に下部針広混交林において個体密度が高い．高山帯では，いわゆるお花畑から岩礫，砂礫地，さらに雪渓の周辺や渓流に近い湿潤地などにも比較的広く見られる．東北地方では主として亜高山帯～高山帯に生息するが，鳥海山（分布下限標高約800 m）や早池峰山（はやちねさん）（同約1,000 m）などのように中山帯の上部まで下りてきている場所もある．

生態　秋繁殖・幼虫越冬型のオサムシで，井上（1953c）によれば幼虫は各種の鱗翅目幼虫やその他の昆虫を捕食するというから，基本的には節足動物食（他の昆虫の幼虫等）であろう．野外における成虫の摂食活動の観察記録は少ないが，小型鱗翅類の幼虫やフキバッタの一種を食していたという記録がある（山谷・草刈，1989）．冬期における成虫の採集記録も極めて少なく，「やや砂がかった土質のガケ」（奥村，1972）や高山帯の石下（草刈，1984）等から数例の報告があるにすぎない．北海道では平地で6月，高所では7月頃から成虫が活動を始め，盛夏から初秋の頃にかけて新成虫が混入し，個体数を増す．夏場の短い高山帯においては，2年1化のような生活環を有している可能性もある．

地理的変異　本種は従来，基亜種と東北地方亜種の2亜種に分類され，大雪山から記載された亜種 *kurosawai* は基亜種のシノニムとみなされることが多かったが，ここでは IMURA（2011）に倣ってこれを独立した亜種と認め，更に同じ IMURA（2011）により日高山脈南端部から記載された亜種 *apoi* を加えて計4亜種に分類した．道内の分布はほぼ連続しているうえ，同一集団内における個体変異も著しいため，本種の地理的変異に関しては更なる検討が必要であろう．

種小名の由来　「暗い」，「陰の」を意味するラテン語の opacus に縮小辞の -ulus がついたもので，「やや暗い」を意味する．

13-1. *C. (A.) o. kurosawai* BREUNING, 1957　ヒメクロオサムシ道央道東道北亜種

Original description　*Carabus* (*Eucarabus*) *opaculus* PUTZ. ssp. *Kurosawai* BREUNING, 1957, Ent. Arb. Mus. Frey, Tutzing, 8, p. 275.

Type locality　Insel Yesso, Mount Daisetusu [sic] (misspelling of Daisetsu), alpine Region.

Type depository　Unknown. The type specimen of this taxon is not found in the Breuning collection now preserved in the Zoölogisch Museum Amsterdam.

Diagnosis 15~22 mm. External features variable according to localities or individuals, though this subspecies is defined by peculiarly featured internal sac of male genitalia as follows: 1) median lobe well-developed; 2) dorsal margin of internal sac from median lobe to saccellus gently curved in lateral view; 3) a large hemispherical lobe on inflexed side not adhered to ostium lobe in fully everted condition.

Distribution C.~N.~E. Hokkaido / Iss. Rishiri-tô, Rebun-tô, Daikoku-jima, Shikotan-tô, Kunashiri-tô, Etorofu-tô.

Etymology Named after Yoshihiko KUROSAWA, a former director of the Department of Zoology, National Museum of Nature and Science, Tokyo.

タイプ産地　蝦夷(島),大雪山,高山帯.

タイプ標本の所在　不明［タイプ標本は3頭だが,ホロタイプを含め,性別指定はなされていない］.BREUNING(1957)は本亜種の原記載論文でタイプ標本の保管先を自身のコレクションと指定しているが,彼のコレクションの大半が納入されているアムステルダム動物学博物館には,井村が調査した範囲では,本亜種のタイプ標本は保管されていない.

亜種の特徴　15~22 mm. 地理的変異の項で述べるように,本亜種は体型や鞘翅彫刻の変化に富み,外部形態のみで亜種の定義付けを行うことは困難であるが,♂交尾器内嚢基部伸側に顕著な中央葉を有する点で他亜種から識別される.中央葉から内嚢盤にかけての曲線は側面から見て基亜種同様なだらかで,内嚢盤と傍盤葉を含む背面膨隆部が内嚢の基軸に対し急な角度をもって立ち上がることはない.内嚢腹面中央にある半球形の膨隆部は,内嚢を完全に反転させた状態で葉片と接することはない.

分布　北海道中,東,北部／礼文島,利尻島,大黒島,色丹島,国後島,択捉島.

地理的変異　本書では,♂交尾器内嚢に明らかな中央葉を認めるものに対し,一括して本亜種名を適用したが,外部形態の上からは亜種内においても明らかな地理的変異が見られる.利尻・礼文両島や道北の宗谷地方の集団は,一般に鞘翅第一次間室が発達し,瘤状に強く隆起した隆条片となる一方で,第三次間室は殆ど隆起せず,平坦な基面の上に小顆粒列が並んだように見えるものが多い.このような個体は道央や道南地方にも出現するので,必ずしもこの地域の集団に特有の特徴であるとは限らないが,三つの間室がほぼ均等に隆起することの多い日高山脈南端部亜種や道東の一部(北見等,オホーツク海沿岸に近い地域のもの)に産する集団からは比較的容易に識別できる.根室など道東のものは第一次間室の隆起部分が太く,瘤状に近くなるが,道北のものより一般に第三次間室がより強く発達する.同じ道東でも標茶一帯に産するものは細型で,第一次間室の隆条の分断される頻度が低いものが多く,特異である.

亜種小名の由来と通称　亜種小名は黒澤良彦(元国立科学博物館動物研究部長)に因む.通称ダイセツヒメクロオサムシ.

13-2. *C. (A.) o. opaculus* PUTZEYS, 1875 ヒメクロオサムシ基亜種

Original description, type locality & type depository　See explanation of the species.

Diagnosis 15~22 mm. External features variable according to localities and individuals, though this subspecies is readily discriminated by characteristically featured internal sac of male genitalia as follows: 1) median lobe not developed; 2) dorsal margin of internal sac from basal third of extended side to saccellus gently curved in lateral view.

Distribution SW. Hokkaido / Is. Okushiri-tô.

原記載・タイプ産地・タイプ標本の所在　種の項を参照.

亜種の特徴　15~22 mm. 本亜種も外部形態には変化があるが,一般に鞘翅の三間室は比較的均等に隆起し,第一次間室の隆起部分が瘤状に発達するものは少ない.♂交尾器内嚢は中央葉を欠き,基部屈側1/3から内嚢盤にかけての背側縁が側方から見てなだらかなので,一見して他亜種から識別される.

分布　北海道南西部(渡島半島から石狩地方,更に南端部を除く日高山脈西半部にかけての地域)／奥尻島.

13-3. *C. (A.) o. apoi* (IMURA, 2011) ヒメクロオサムシ日高山脈南端部亜種

Original description　*Asthenocarabus opaculus apoi* IMURA, 2011, Coléoptères, Paris, 17 (3), p. 24, figs. 3-a, b (p. 23).

Type locality　Japan, south-central Hokkaido, Samani-chô of Samani-gun, western foot of Mt. Apoi-daké, 50-100 m in altitude

Type depository　Muséum National d'Histoire Naturelle, Paris [Holotype, ♂].

Diagnosis 17~22 mm. Dorsal surface black, often with dark bluish or blue-purplish tinge on head and pronotum. Distinguished from other races of *C. (A.) opaculus* by the following points: 1) body slenderer, above all in pronotum and elytra, more remarkably so in male; 2) elytral sculpture with primaries narrow but always more strongly costate than other intervals, rather irregularly interrupted by small, shallow foveae into segments with various lengths; both secondaries and tertiaries almost equally raised, though a little weaker than primaries, to form catenulate ridges or rows of granules; 3) internal sac of male genitalia rather robust, without median lobe and almost rectangularly inflexed at basal third; median portion strongly inflated, with extended side constructed by parasaccellar lobes and saccellus nearly rectangularly protruding dorsad, inflexed side with a large hemispherical lobe closely adhered to ostium lobe in fully everted condition; apical portion strongly inflated, podian lobes also well recognized.

Molecular phylogeny On the ND5 molecular phylogenetic tree, *C. (A.) opaculus* is divided into two different lineages; one is composed of the specimens from the greater part of Hokkaido excluding those from Samani-chô in the southernmost part of the Hidaka Mountains, which is one of the type locality of the present new subspecies, and another is composed of those from Samani-chô and Honshu, suggesting that the Samani-chô population is phylogenetically closer to those inhabiting Honshu than to those inhabiting other parts of Hokkaido (Osawa *et al.*, 2004). The present morphological classification does not consistent with the molecular phylogenetical view.

Distribution Narrowly localized in the southernmost part of the Hidaka Mountains near the Cape Erimo-misaki (Samani-chô & Erimo-chô) in south-central Hokkaido, Northeast Japan.

Etymology Named after its type locality, Mt. Apoi-daké.

タイプ産地　北海道中南部, 様似郡様似町, アポイ岳西麓, 標高 50-100 m.

タイプ標本の所在　パリ自然史博物館［ホロタイプ, ♂］.

亜種の特徴　17~22 mm. 他亜種に比し, 1) 体形はやや細長く, この傾向は♂においてより顕著; 2) 鞘翅第二次, 第三次間室は第一次間室よりもやや弱めながら, 顕著に隆起した隆条あるいは連続した顆粒列となり, 第一次間室の間にほぼ均等に隆起した3本の隆条を具えるものが多い; 3) ♂交尾器内嚢は基部1/3付近でほぼ直角に近く屈曲し, 中央葉を欠き, 内嚢盤と傍盤葉を含む背面膨隆部分は内嚢の基軸に対しほぼ直角に立ちあがる. 傍盤葉は東北地方亜種のそれに比し, より強く突出する. 内嚢腹面中央にある半球形の大きな膨隆部は非常に大きく, 内嚢を完全に反転させた状態では葉状片と密着する (他亜種では通常, 両者が密着することはない). 分子系統樹の上では, 本集団は本州産の集団との類縁が近く, 他の北海道各地産の集団とは異なる系統に属し, ♂交尾器内嚢形態から見た分類と矛盾しない.

分布　日高山脈南端部の様似町からえりも町にかけてのごく狭い範囲から知られている. 分布域の西部では浦河町西部から新ひだか町南東部にかけての地域で基亜種へと移行しているようである. 東部では広尾町南部あたりまで本亜種に相当するものが分布しているようだが, 大樹町以北では道央道東道北亜種の特徴を有する集団になるようである. 日高山脈主稜線に沿った高所における変異の実態はまだよくわかっていない.

亜種小名の由来　亜種小名はタイプ産地のアポイ岳に因む.

13-4. *C. (A.) o. shirahatai* Nakane, 1961 ヒメクロオサムシ東北地方亜種

Original description *Carabus (Asthenocarabus) opaculus shirahatai* Nakane, 1961, Fragm, coleopterol., Kyoto, (1), p. 1.

Type locality Mt. Chokai, Yamagata Pref., Honshu.

Type depository Unknown. The holotype specimen of this taxon is not found in the Nakane collection now preserved in the Hokkaido University Museum (Sapporo).

Diagnosis 15~20 mm. Differs from other subspecies in the following points: 1) body, above all elytra, usually a little slenderer; 2) body above usually black, and less strongly or hardly bearing purplish- or bluish tinge; 3) secondary and tertiary intervals of elytra almost similarly elevated; 4) glossy stripe of umbilicate series not continuous excepting shoulders, with granules smaller in number and more sporadically set; 5) internal sac similar to that of subsp. *apoi*, though more gently inflexed at basal third, median portion a little less strongly protruded dorsad.

Distribution NE. Honshu (mountainous region) (Aomori Pref., Iwate Pref., Akita Pref., Yamagata Pref., Miyagi Pref., Fukushima Pref. & Niigata Pref.).

Etymology Named after Kôtarô Shirahata, the famous amateur entomologist who lived in Yamagata Prefecture.

タイプ産地　本州, 山形県, 鳥海山.

タイプ標本の所在　不明. ホロタイプは1957年8月20日白畑 孝太郎採集の♂であるが, 北海道大学総合博物館 (札幌) の中根コレクション中にある本亜種の標本は, 白畑の採集品ではあるものの, 1952年8月6日に得られた1♂のみで, タイプ標本であることを示す証拠 (「タイプ」の文字や赤ラベル等) も付されておらず, ホロタイプの所在は不明である.

亜種の特徴　15~20 mm. 以下の諸点により他亜種から識別される: 1) 体型, とりわけ鞘翅がより細長い; 2) 背面は黒色に近いものが多く, 青, 紫, 緑等を強く帯びた個体は見られない; 3) 鞘翅第二次, 第三次間室の隆起はほぼ同高となるものが多い; 4) 丘孔点列の光沢帯は肩部を除き連続せず, 顆粒も数が少なく, 間隔が大きい; 5) ♂前跗節基部4節の幅が他亜種よりやや狭いものが多い; 6) ♂交尾器内嚢は日高山脈南端部亜種のものに最も近いが, 基部1/3付近における屈曲の程度はやや弱く, 内嚢盤と傍盤葉を含む背面膨隆部分の背側への突出がやや弱い.

分布　東北地方の主として高所 (青森, 岩手, 秋田, 山形, 宮城, 福島, 新潟の各県).

亜種小名の由来と通称　亜種小名は, 山形県酒田市のアマチュア昆虫研究家 (本職は警察官), 白畑 孝太郎に因む. チョウカイヒメクロオサムシという通称が一般的だが, トウホクヒメクロオサムシという名が提唱されたこともある.

13. *Carabus* (*Asthenocarabus*) *opaculus*

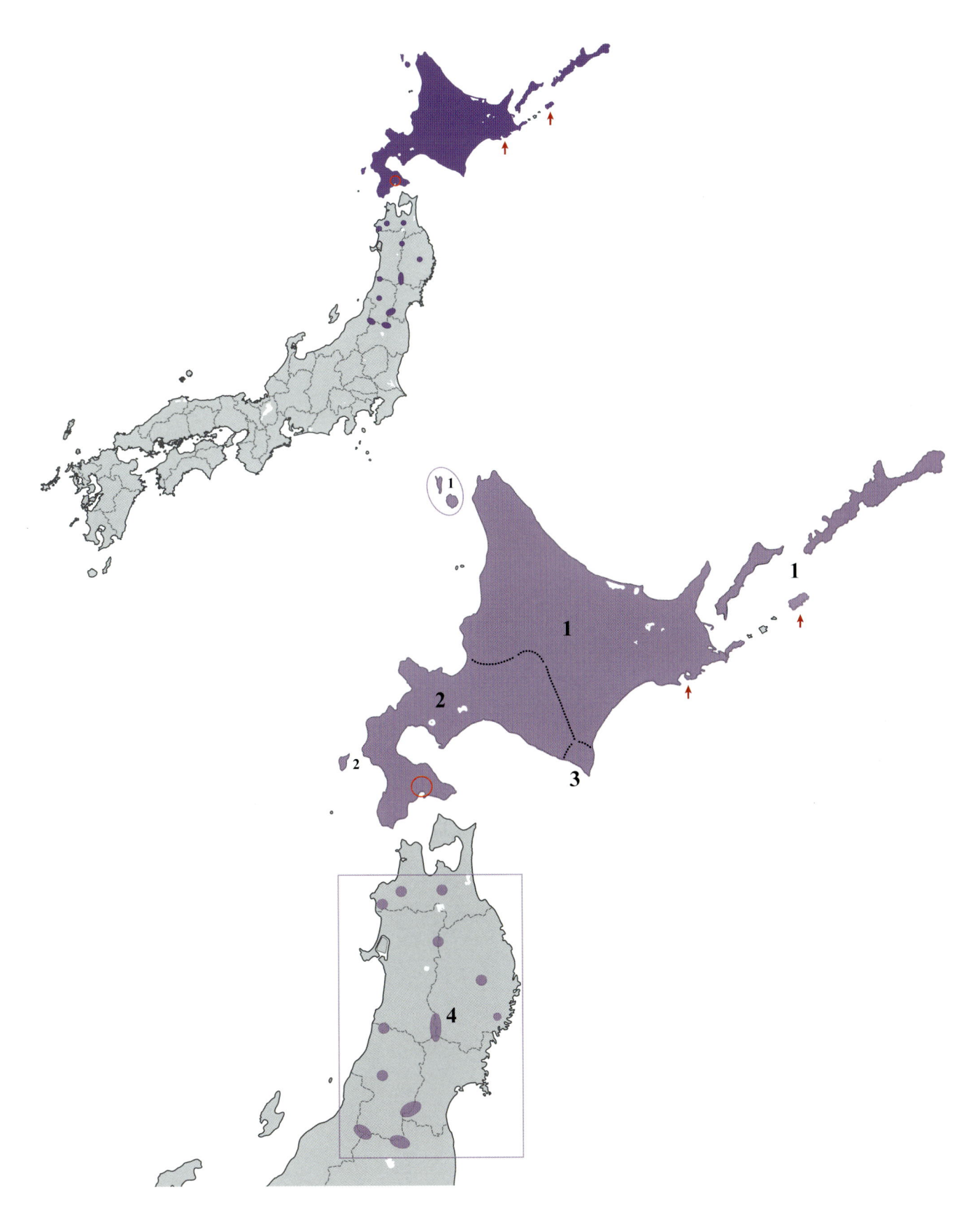

Fig. 39. Map showing the distributional range of ***Carabus*** (***Asthenocarabus***) ***opaculus*** PUTZEYS, 1875. — 1, Subsp. ***kurosawai***; 2, subsp. ***opaculus***; 3, subsp. ***apoi***; 4, subsp. ***shirahatai***. ●: Most likely type area of the species (Hakodaté or its nearby regions).
　図39. ヒメクロオサムシ分布図. — 1, 道央道東道北亜種; 2, 基亜種; 3, 日高山脈南端部亜種; 2, 東北地方亜種. ●: 最も可能性の高い種のタイプ地域（函館ないしその周辺）.

14. *Carabus (Carabus) arvensis* HERBST, 1784
コブスジアカガネオサムシ

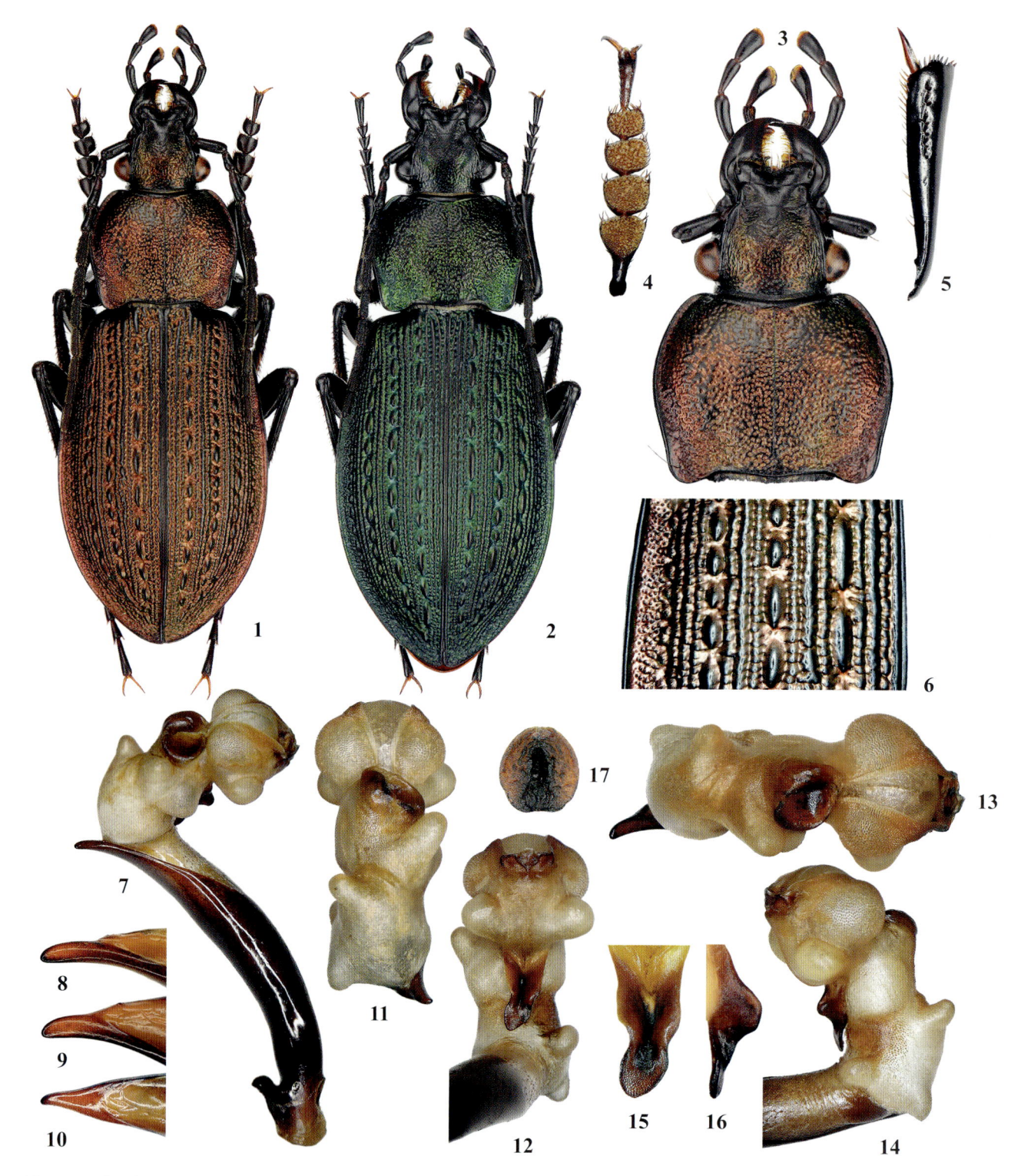

Figs. 1-17. *Carabus (Carabus) arvensis hokkaidensis* LAPOUGE, 1924 (Teshio-chô, Hokkaido). — 1, ♂; 2, ♀; 3, male head & pronotum; 4, male right protarsus in ventral view; 5, male right protibia in subdorsal view; 6, male left elytron (median part); 7, penis & fully everted internal sac in right lateral view; 8, apical part of penis in right lateral view; 9, ditto in right subdorsal view; 10, ditto in dorsal view; 11, internal sac in caudal view; 12, ditto in frontal view; 13, ditto in dorsal view; 14, ditto in left sublateral view; 15, digitulus in frontal view; 16, ditto in right lateral view; 17, inner plate of ligular apophysis.

14. *Carabus* (*Carabus*) *arvensis*

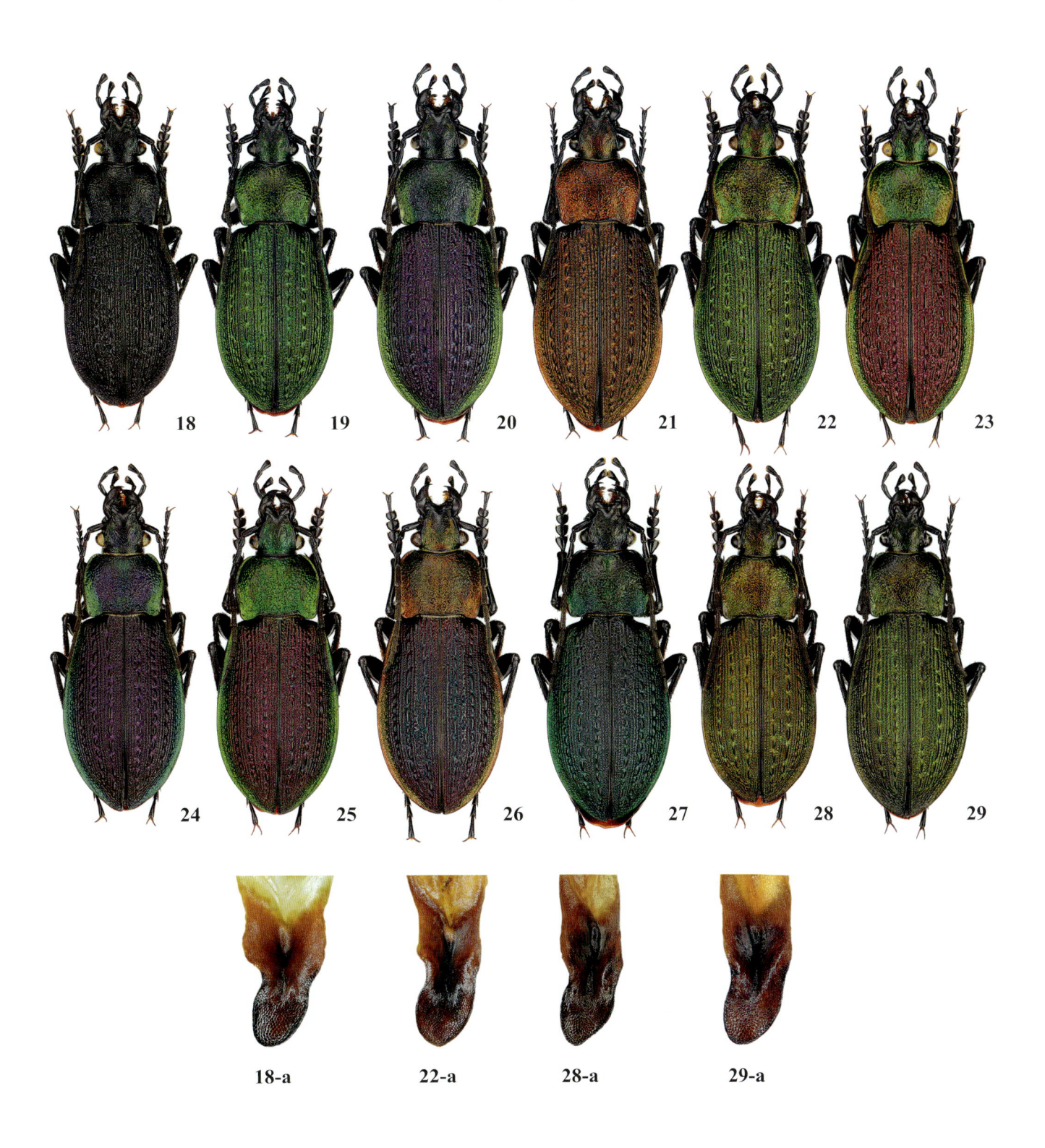

Figs. 18-29. ***Carabus* (*Carabus*) *arvensis hokkaidensis***. — 18-20, 22, 23, 25-28, ♂, 21, 24, 29, ♀; 18-21, Is. Rishiri-tô; 22-24, Pass Asahi-tôgé, Engaru-chô; 25, Pass Rukushi-tôgé, Kitami-shi; 26, Shiretoko; 27, Mt. É-san, Hakodaté-shi; 28, Is. Yagishiri-tô; 29, Is. Teuri-tô. — a, Digitulus in frontal view. All from Hokkaido.

図 18-29. コブスジアカガネオサムシ北海道亜種. — 18-20, 22, 23, 25-28, ♂, 21, 24, 29, ♀; 18-21, 利尻島; 22-24, 遠軽町旭峠; 25, 北見市ルクシ峠; 26, 知床; 27, 函館市恵山; 28, 焼尻島; 29, 天売島. — a, 指状片（頭側面観）. すべて北海道産.

14. *Carabus* (*Carabus*) *arvensis* Herbst, 1784 コブスジアカガネオサムシ

Original description *Carabus arcensis* Herbst, 1784, Fueßly, Arch. Insekteng., 5, p. 132.

History of research This taxon was described by Herbst (1784) from Pommern (= Pomerania) of Northwest Poland (originally given spelling of the specific name is *arcensis*, however, *arvensis* has been sanctioned by usage; on this matter, see Turin *et al*., 2003, p. 182). It is widely distributed in the Eurasian Continent and its attached islands, from Great Britain, Central France to Far East including Japan, and is divided into many subspecies. Representative geographical races are as follows: *C*. (*C*.) *arvensis sylvaticus* Dejean, 1826 of western Europe, *C*. (*C*.) *a*. *carpathus* Born, 1902 of the Carpathian Mountains, *C*. (*C*.) *a*. *conciliator* Fischer, 1822 of central to eastern Eurasia, etc.

Type locality Pommern (= Pomerania).

Type depository Unknown. (Museum für Naturkunde der Humboldt-Universität zu Berlin ?)

Morphology 13~25 mm. Body above variable in color, coppery, green, purple, black, etc., or two-toned (e.g., head, pronotum and elytral margins are green, elytra excepting margins are purple, etc.). External features: Figs. 1~6. Description on details: penultimate segment of labial palpus bisetose; median tooth of mentum shorter than lateral lobes, triangular in ventral view, with apex hardly protruding ventrad in lateral view; submentum bisetose; male antennae with hairless ventral depression from segment five to nine, though that of segment five is often unclear; pronotal margins basically trisetose; preapical emargination of elytra only faintly recognized in female, though usually unclear; elytral sculpture triploid; primaries usually the widest, indicated by chain striae frequently and irregularly segmented by clearly recognized primary foveoles; both secondaries and tertiaries indicated by contiguous low costae or rows of granules, with dorsal surface crenulated; striae between intervals irregularly and sporadically scattered with fine punctures; propleura sporadically scattered with fine punctures; sides of sternites weakly rugoso-punctate; metacoxa trisetose; sternal sulci clearly carved. Genitalic features: Figs. 7~17. Penis less than half the length of elytra; ostium lobe absent; ligulum small, indicated by longitudinally elongated assemblage of large granules bearing many long hairs on its surface; internal sac with right basal lateral lobe weakly inflated, left one much larger, with apex triangularly protruded, paraligula distinct; digitulus tongue-shaped, asymmetrical, constricted a little before middle in frontal view, median longitudinal part near base elevated to form vertical plate; peripheral rim of gonopore weakly sclerotized and pigmented, though not forming terminal plate. Female genitalia with inner plate of ligular apophysis (Fig. 17) nearly circular in dorsal view, its margin apparently elevated in apical half, almost flat in basal half, disk irregularly rugulose.

Molecular phylogeny On the molecular phylogenetic tree constructed by mitochondrial ND5 gene sequences, this species belongs to the same cluster as *Carabus* (*Carabus*) *deyrollei*, *C*. (*C*.) *sculpturatus*, *C*. (*C*.) *granulatus* and *C*. (*C*.) *vanvolxemi*. The specimens of *C*. (*C*.) *arvensis* from Hokkaido reveal closer affinity with those from Kamchatka than those from the Eurasian Continent, but anyway the sequence difference is not so large among the specimens from various localities of Japan and the Eurasian Continent. The Japanese population of *C*. (*C*.) *arvensis* is considered to be established by immigration through the Kurils or Sakhalin, probably over land bridges in the glacial era (Osawa & Su, 2001a; Osawa *et al*., 2002, 2004; Su *et al*., 2004).

Distribution Hokkaido / Iss. Rebun-tô, Rishiri-tô, Teuri-tô, Yagishiri-tô, Yururi-tô, Kunashiri-tô & Etorofu-tô (Hokkaido). Outside Japan: Eurasian Continent (E. France~Far East) / Iss. Great Britain, Sakhalin & S. Kurils.

Habitat This species is widely distributed from the plain to mountain, from the meadow to forest, and is one of the most popular species in Hokkaido excluding southwestern part, though usually not found in the area higher than the subalpine zone.

Bionomical notes Spring breeder. According to Turin *et al*. (2003), both the adult and larva of *C*. (*C*.) *a*. *arvensis* prefer to feed on earthworm, larvae of Lepidoptera and snails in the cage. Larva of this species is basically considered an earthworm feeder. According to Sugie & Fujiwara (1981), however, that of *C*. (*C*.) *a*. *hokkaidensis* was fed larvae of Lepidoptera and imagoes of Carabidae, and emerged successfully in captivity.

Geographical variation See "history of research" section of this species. In Japan, it is endemic to Hokkaido and represented by subsp. *hokkaidensis*.

研究史　Herbst（ヘアブスト／ヘルプスト）(1784)により記載された古くから知られているオサムシで，アカガネオサムシと並び，ユーラシア大陸北部に広く生息する広域分布種．種小名に関しては，原記載で用いられた綴り*arcensis*を採用する著者もいるが，学名の安定性保持の観点から，長年慣用されてきた*arvensis*が用いられる場合が多い（Turin（トゥリン）ら，2003, p. 182）．多くの亜種に分けられ，代表的なものに西ヨーロッパの *C*. (*C*.) *arvensis sylvaticus* Dejean, 1826, カルパチア山脈の *C*. (*C*.) *a*. *carpathus* Born, 1902, 中央〜東ユーラシアの *C*. (*C*.) *a*. *conciliator* Fischer, 1822などがある．我が国にはLapouge（ラプージュ）(1924)により天塩から記載された北海道固有の亜種 *C*. (*C*.) *a*. *hokkaidensis* を産する．

タイプ産地　ポンメルン（＝ポメラニア）（ポーランド北西部）．

タイプ標本の所在　不明．（フンボルト大学自然博物館（ベルリン）？）

形態　13~25 mm. 色彩の変異が著しく，背面は銅色，緑色，紫，黒或いはそれらの中間的色彩を持つものから，頭部・前胸背板，鞘翅側縁と，側縁を除く鞘翅の大部分の色彩が異なる「ツートン」のものまで，変化の幅が広い．外部形態：図1~6．細部の記載：下唇肢亜端節の剛毛は2本．下唇基節中央歯は側葉より短く，腹面から見て三角形，側面から見て腹側へ殆ど突出し

ない．下唇亜基節には2本の長毛を有する．♂触角第5節から第9節の腹面には顕著な無毛凹陥部があるが，第5節のものはやや不明瞭．前胸背板の側縁剛毛は片側に3~4本．鞘翅端のえぐれは♀でごく僅かに認められる場合があるが，一般に不明瞭．鞘翅彫刻は三元的．第一次間室は通常最も幅広く，比較的明瞭な第一次凹陥により頻繁に分断された鎖線．第二次，第三次間室はより低い隆条ないし粗大顆粒列で，背面は小鈍鋸歯状となる．間室間の線条には不規則な弱い点刻を装う．前胸側板には小点刻が散在し，腹部腹板の側面には弱い皺と点刻を装う．後基節の剛毛は3本．腹部腹板の横溝は明瞭．交尾器形態：図7~17．陰茎長は鞘翅長の1/2弱．膜状部に葉片を欠く．基斑は内嚢の縦軸に沿って伸長した小顆粒の集簇で，表面に多数の長毛を有する．右基部側葉は目立たず，左基部側葉は小さいが先端は顕著に突出し，側舌は明瞭．指状片は左右非対称の舌形で，背面から見ると基部でくびれ，先端は鈍く丸まり，基部中央に縦軸に沿った垂直の板状隆起がある．射精孔縁膜の硬化と色素沈着は軽度で，頂板は形成しない．♀交尾器膣底部節片内板はほぼ円形に近く，側縁は前半部分で明らかに隆起するが，後半部分はほぼ平坦で，壁面には不規則な皺を装う．

分子系統 ミトコンドリアND5遺伝子を用いた分子系統樹上で，コブスジアカガネオサムシ北海道亜種はユーラシア大陸各地のものよりカムチャツカの集団に近縁で，遥か遠く離れたユーラシア西部の集団からの分岐開始時期も新しい．本種は，氷期の海退により生じた陸橋を経由して大陸から日本列島へと渡ってきた侵入種の一つであろうと考えられている（大澤・蘇，2001a；大澤ら，2002；OSAWA et al., 2004; SU et al., 2004）．

分布 北海道／礼文島，利尻島，天売島，焼尻島，ユルリ島，国後島，択捉島（以上，北海道）．国外：ユーラシア大陸（フランス東部~極東地方）／グレート・ブリテン島，サハリン，千島列島．文献上は奥尻島からも記録されているが，近年における確実な記録はなく，また数回に亘り実際に同島でトラップ採集を行われた方のお話を伺っても，全く採集される気配がないという．奥尻島に本種が分布すると記した文献はいくつか見られるが，実状を鑑み，本書では分布域から省いた．

生息環境 平地の草原から森林地帯まで幅広い環境に生息し，個体数も多いが，あまり高所には生息していない場合が多い．

生態 春繁殖型のオサムシで，TURINら（2003）によれば，基亜種の成虫は飼育下でミミズ，鱗翅類の幼虫，カタツムリなどを食し，幼虫の食性もこれに準じたものであるというから，幼虫は基本的に環形動物食（ミミズなど）と考えられるが，杉江・藤原（1981）によれば，北海道亜種の幼虫は飼育下で鱗翅目の幼虫とゴミムシの成虫を食し，蛹化に至ったという．

地理的変異 基亜種はポーランド北西部ポメラニア地方から記載されたもので，亜種以下の分類階級に対して数多くの名称が与えられており，西部ヨーロッパの *C. (C.) a. sylvaticus* DEJEAN, 1826，カルパティア山脈の *C. (C.) a. carpathus* BORN, 1902，中央~東ユーラシアの *C. (C.) a. conciliator* FISCHER, 1822などが代表的なものである．我が国には北海道亜種 *C. (C.) a. hokkaidensis* LAPOUGE のみを産するが，これはユーラシア東部の *C. (C.) a. conciliator* に近縁で，後者を独立種とみなし，その一亜種とされることもある．

14-1. *C. (C.) a. hokkaidensis* LAPOUGE, 1924 コブスジアカガネオサムシ北海道亜種

Original description *Carabus arvensis (conciliator) Hokkaidensis* LAPOUGE, 1924, Misc. ent., p.186.
Type locality Hokkaido: Teshio.
Type depository The type series once preserved in the LAPOUGE's collection seem to have been lost.
Diagnosis 17~24 mm. Body above variable in color, reddish coppery, green, black, or two-toned as shown in Figs. 20, 23, 25 and 26. Differs from the nominotypical subspecies and some other races of the Eurasian Continent in the following points: 1) size larger on an average; 2) head and pronotum more remarkably rugulose; 3) pronotum more transverse, more strongly convex above, with posterior parts of lateral margins more strongly curved above; 4) elytral sculpture more strongly uneven, with primaries indicated by rows of tubercles, secondaries and tertiaries more remarkably notched; 5) apical part of penis shorter.
Distribution Hokkaido; Iss. Rebun-tô, Rishiri-tô, Teuri-tô, Yagishiri-tô, Yururi-tô, Shikotan-tô, Kunashiri-tô & Etorofu-tô.
Etymology Named after Hokkaido.

タイプ産地 北海道：天塩．
タイプ標本の所在 本亜種のタイプ標本は著者LAPOUGEのコレクションに保管されていたと思われるが，ホソヒメクロオサムシのタイプ標本の項で述べた通り，劣悪な環境下で管理されている間に失われてしまった可能性が高いという．
亜種の特徴 17~24 mm．赤銅色系と緑色系のものを中心に，中間的色彩のもの，全体黒色に近いもの，あるいは鞘翅の色彩が中央部と側縁部で異なりツートンカラーとなるタイプまで，さまざまな型が出現し，地域により各色彩型の出現頻度が異なる．道東・道北など，気温の低い地域では，前胸背板や鞘翅側縁が濃緑色ないし紫赤色に輝く美しい個体が出現する．基亜種を初めとするユーラシア大陸西部の諸集団とは以下の点で異なる：1) 平均して大型；2) 頭部と前胸背板の皺がより顕著；3) 前胸背板はより横長で，より強く膨らみ，側縁後方の反りが強い；4) 鞘翅彫刻の凹凸がより顕著で，第一次間室はより明瞭な瘤状隆起列，第二次，第三次間室は横の刻み目がより顕著；5) 陰茎先端部がより短い．
分布 北海道／礼文島，利尻島，天売島，焼尻島，ユルリ島，色丹島，国後島，択捉島．種の解説の項でも述べた通り，奥尻島は分布域から省いた．道南の渡島半島からは記録が少なく，島牧村の一部に分布するほか，南方にかけ離れた横津岳の山頂付近や函館市恵山（えさん）付近等に孤立した分布域を持つ．

14. *Carabus (Carabus) arvensis*

生息環境　原野や湿原から森林地帯まで幅広く生息する．産地における個体数も一般に多く，その地域の優占種となっている場合が多い．西島(1989)は本種の個体数が著しく多い地域として，道北の礼文島，利尻島，サロベツ原野，道東の根釧原野，道央の滝里・富良野付近，ニセコ周辺，岩内の目国内岳(めくんないだけ)，道南の恵山，海向山(かいこうざん)などを挙げているが，これら以外にも本種の多産地は多い．

生態　春繁殖・成虫越冬型で，基本的に環形動物(ミミズ)食のオサムシと考えられる．夏季にはトラップで大量に採集されるが，冬期における採集記録は非常に少ない．自然状態下でエゾアカガネオサムシと交尾中の個体が見られることもあるという．後翅はよく発達しており，飼育中の本種が鞘翅を開き，後翅を伸張させ，羽音を立てながら震わせていたという観察記録がある(荒井, 1987b)．

亜種小名の由来　亜種小名は北海道に由来する．我が国には本亜種のみを産するため，コブスジアカガネオサムシという種和名がそのまま北海道亜種に対する亜種和名として慣用的に用いられており，和名だけでは種，亜種いずれを指すのかを識別できず，混乱をきたす一因となっている．これは，マークオサムシ，セスジアカガネオサムシ，アカガネオサムシ等でも同様で，慣用されてきた固有の亜種和名を用いる大きな弊害の一つである．

14. *Carabus* (*Carabus*) *arvensis*

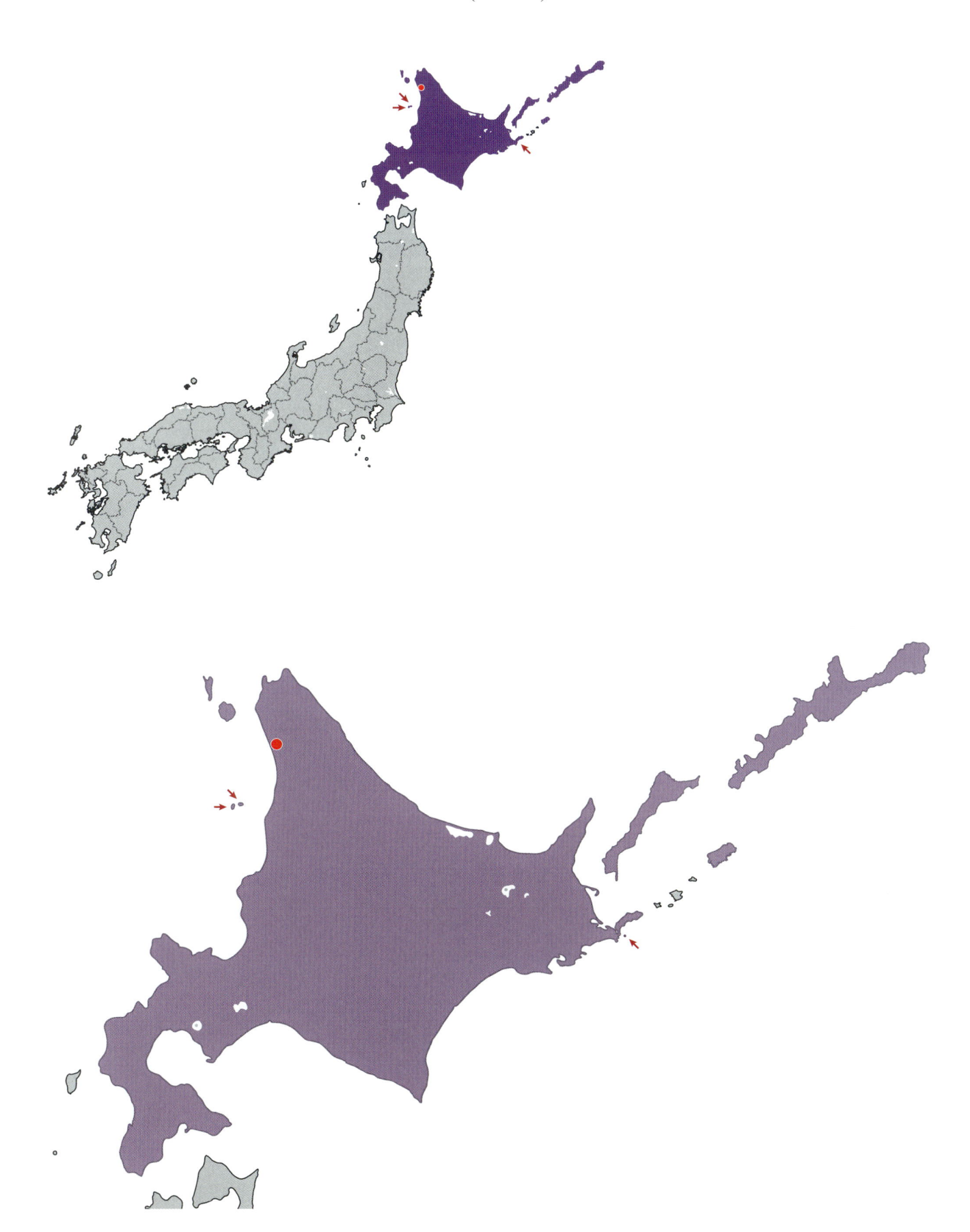

Fig. 30. Map showing the distributional range of ***Carabus*** (***Carabus***) ***arvensis hokkaidensis*** Lapouge, 1924. ●: Type locality of the subspecies (Teshio).
図 30. **コブスジアカガネオサムシ北海道亜種**分布図. ●: 亜種のタイプ産地（天塩）.

15. *Carabus* (*Carabus*) *granulatus* Linnaeus, 1758
アカガネオサムシ

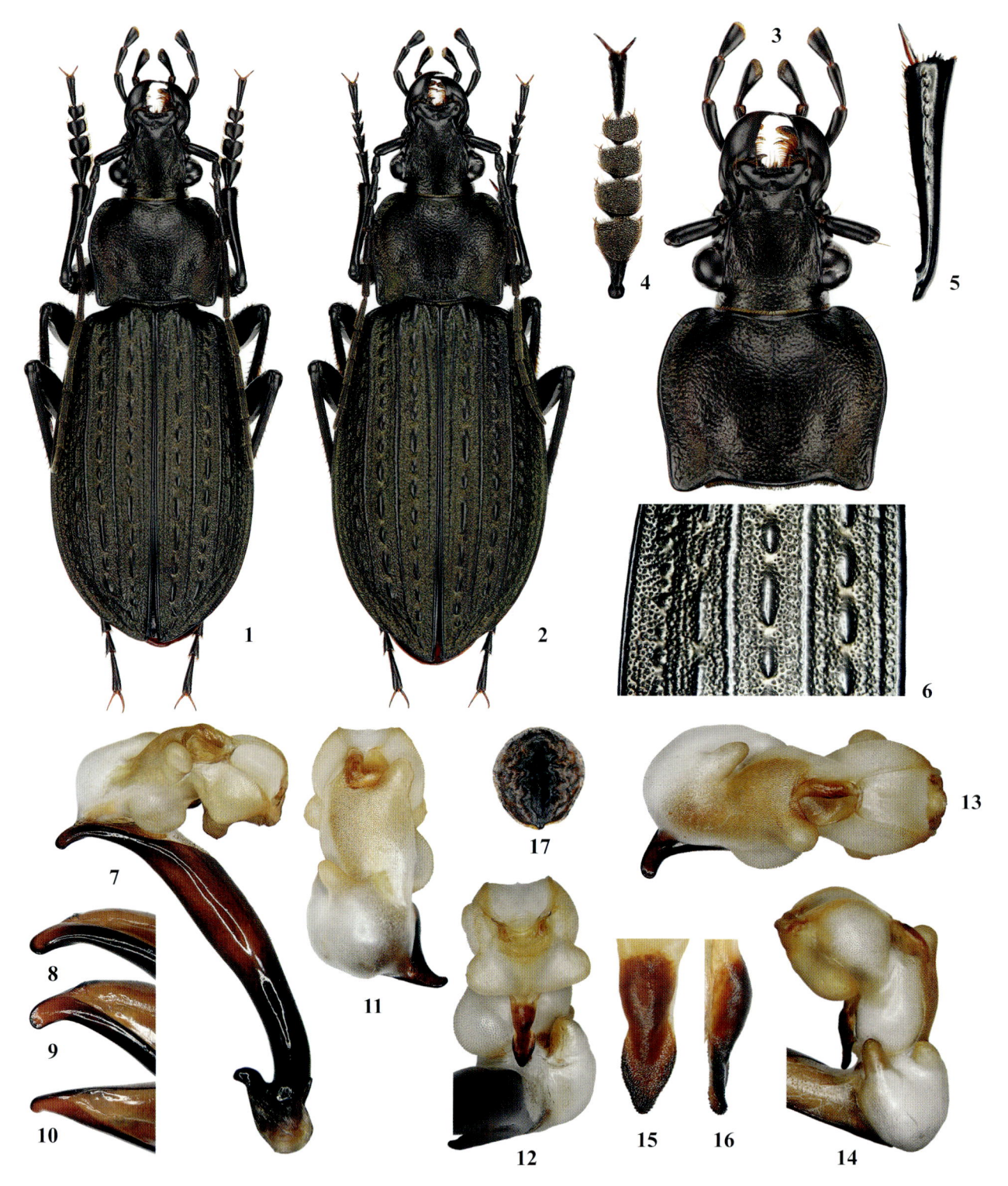

Figs. 1-17. *Carabus* (*Carabus*) *granulatus telluris* Lewis, 1882 (Nogi-machi, Tochigi Pref.). — 1, ♂; 2, ♀; 3, male head & pronotum; 4, male right protarsus in ventral view; 5, male right protibia in subdorsal view; 6, male left elytron (median part); 7, penis & fully everted internal sac in right lateral view; 8, apical part of penis in right lateral view; 9, ditto in right subdorsal view; 10, ditto in dorsal view; 11, internal sac in caudal view; 12, ditto in frontal view; 13, ditto in dorsal view; 14, ditto in left sublateral view; 15, digitulus in frontal view; 16, ditto in right lateral view; 17, inner plate of ligular apophysis.

図 1-17. アカガネオサムシ：本州亜種 (栃木県野木町). — 1, ♂; 2, ♀; 3, ♂の頭部と前胸背板; 4, ♂右前跗節 (腹面観); 5, ♂右前脛節 (斜背面観); 6, ♂左鞘翅中央部; 7, 陰茎と完全反転下の内嚢 (右側面観); 8, 陰茎先端 (右側面観); 9, 同 (右斜背面観); 10, 同 (背面観); 11, 内嚢 (尾側面観); 12, 同 (頭側面観); 13, 同 (背面観); 14, 同 (左斜側面観); 15, 指状片 (頭側面観); 16, 同 (右側面観); 17, 膣底部節片内板.

15. *Carabus* (*Carabus*) *granulatus*

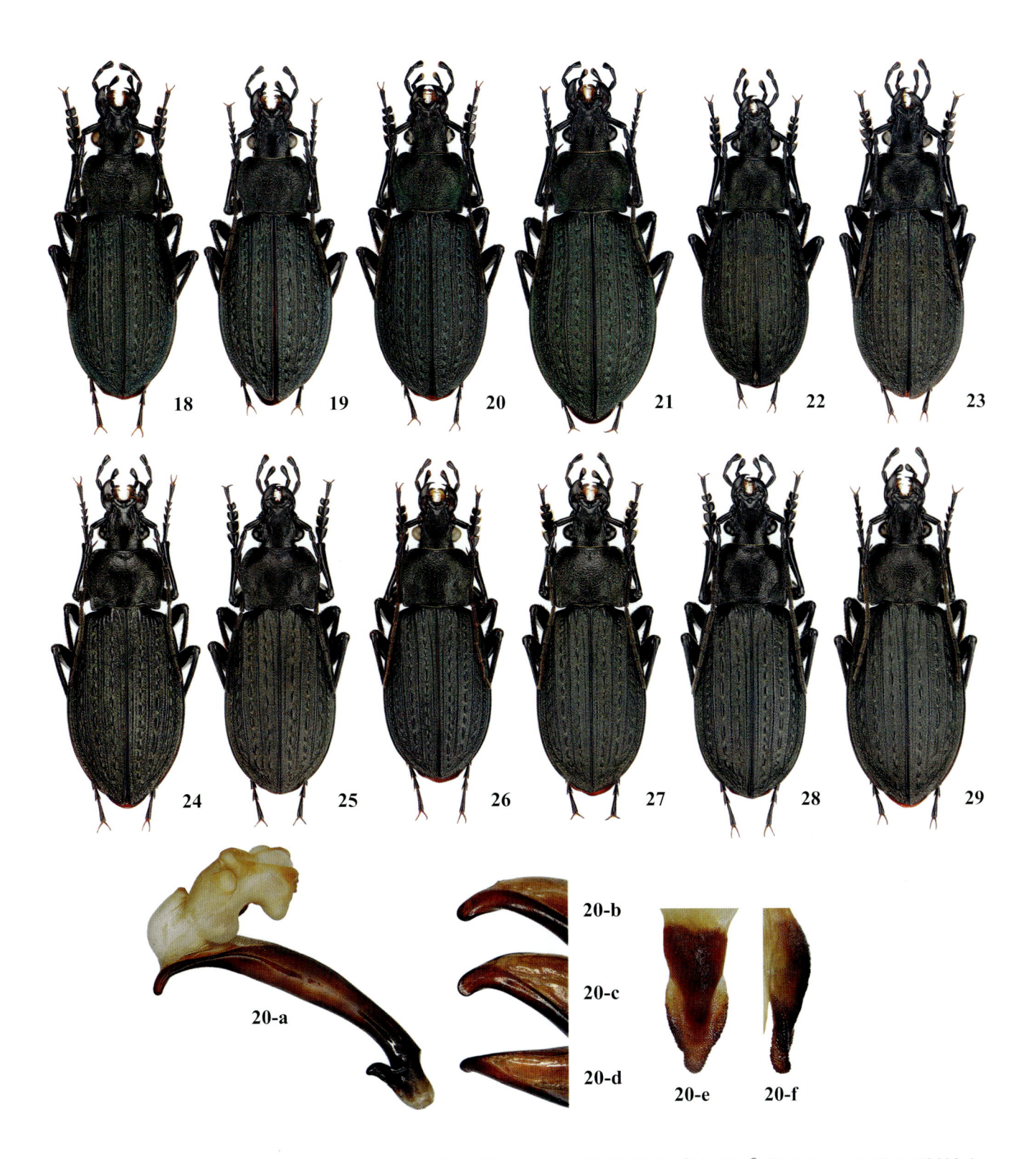

Figs. 18-29. ***Carabus*** (***Carabus***) ***granulatus*** subspp. — 18-23, ***C***. (***C***.) ***g. yezoensis*** (18, 20, 22, 23, ♂, 19, 21, ♀; 18, Is. Rebun-tô; 19, Is. Rishiri-tô; 20-21, Mt. Moiwa-yama, Sapporo-shi; 22, Is. Yagishiri-tô; 23, Is. Teuri-tô; all from Hokkaido); 24-29, ***C***. (***C***.) ***g. telluris*** (25-28, ♂, 24, 29, ♀; 24, Nakadomari-machi, Aomori Pref.; 25, Anetai-chô, Mizusawa-shi, Iwate Pref.; 26, N. bank of Riv. Kaji-kawa, Shibata-shi, Niigata Pref.; 27, Tazawa, Azumino-shi, Nagano Pref.; 28, Ô-aota, Kashiwa-shi, Chiba Pref.; 29, Nippa-bashi, Riv. Tsurumi-gawa, Yokohama-shi, Kanagawa Pref.). — a, Penis & fully everted internal sac in right lateral view; b, apical part of penis in right lateral view; c, ditto in right subdorsal view; d, ditto in dorsal view; e, digitulus in frontal view; f, ditto in right lateral view.

図 18-29. **アカガネオサムシ**各亜種. — 18-23, **北海道亜種**(18, 20, 22, 23, ♂, 19, 21, ♀; 18, 礼文島; 19, 利尻島; 20-21, 札幌市藻岩山; 22, 焼尻島; 23, 天売島; 以上すべて北海道産); 24-29, **本州亜種**(25-28, ♂, 24, 29, ♀; 24, 青森県中泊町; 25, 岩手県水沢市姉体町; 26, 新潟県新発田市加治川北岸; 27, 長野県安曇野市田沢; 28, 千葉県柏市大青田; 29, 神奈川県横浜市鶴見川新羽橋). — a, 陰茎と完全反転下の内嚢(右側面観); b, 陰茎先端(右側面観); c, 同(右斜背面観); d, 同(背面観); e, 指状片(頭側面観); f, 同(右側面観).

15. *Carabus* (*Carabus*) *granulatus* Linnaeus, 1758 アカガネオサムシ

Original description *Carabus granulates* Linnaeus, 1758, Syst. Nat. ed. 10, p. 413.

History of research This is one of the oldest *Carabus* species described by Linnaeus in 1758. It is widely distributed in the northern part of the Eurasian Continent and its attached islands, from Is. Ireland and northern Iberian Peninsula to Japan, and is divided into several subspecies. Although its type locality is not fixed, the population distributed in north-central Europe is regarded to be the nominotypical subspecies. Representative geographical races are as follows: *C.* (*C.*) *granulatus daghestanicus* Lapouge, 1924 of N. Caucasus, *C.* (*C.*) *g. duarius* Fischer, 1844 of Dzungaria, *C.* (*C.*) *g. telluris* Bates, 1883 of Far East and Honshu, Japan, and *C.* (*C.*) *g. yezoensis* Bates, 1883 of Hokkaido, Japan.

Type locality Not fixed.

Type depository Unknown.

Morphology 16~30 mm. Body above variable in color, coppery brown, dark brownish black, black with violet tinge, or dark greenish coppery. External features: Figs. 1~6. Description on details: penultimate segment of labial palpus bisetose; median tooth of mentum much shorter than lateral lobes, not strongly protruding both anteriad and ventrad; submentum bisetose; ventral side of male antennae without hairless area; pronotal margins multisetose, with two to six setae on each side; preapical emargination of elytra weakly but rather remarkably recognized in female; elytral sculpture triploid; primaries indicated by chain striae frequently segmented by small but rather deep primary foveoles; secondaries usually indicated by contiguous low costae; tertiaries much reduced, either at most forming rows of vague granules or vestigial; areas between intervals coarsely scattered with small granules; propleura smooth, at most weakly rugulose; sides of sternites either smooth or weakly rugoso-punctate; metacoxa bisetose, lacking inner seta; sternal sulci carved, though often unclear above all in median portion. Genitalic features: Figs. 7-17. Penis a little shorter than half the elytral length; membraneous area very narrow, without ostium lobe; ligulum small, resembling a dot, bearing many short hairs on its surface; left basal lateral lobe fused with median lobe to form a finger-like protrusion directed toward apex of internal sac; paraligula large and finger-like; parasaccellar lobe asymmetric, right lobe strongly protruded dorsal, left one vestigial; apical part of internal sac with well-recognized podian lobes; peripheral rim of gonopore very weakly sclerotized and hardly pigmented; digitulus elongated spade-shaped and not flexed dorsad in lateral view. Female genitalia with inner plate of ligular apophysis nearly circular in dorsal view, with inner wall irregularly wrinkled.

Molecular phylogeny Molecular phylogenetic tree constructed by mitochondrial ND5 gene sequence shows that the genetic distance between the specimens of *C.* (*C.*) *granulatus* from the Eurasian continent and those from Sakhalin and several localities of Japan including Hokkaido and Honshu are very close, though the specimens from Japan reveal closer affinity with those from Sakhalin than those from the continent. This suggests the recent immigration of the Eurasian component to Japan, presumably via Sakhalin (Osawa & Su, 2001a; Osawa *et al*., 2002, 2004; Su *et al*., 2004).

Distribution Hokkaido & C.~NE. Honshu / Iss. Rebun-tô, Rishiri-tô, Teuri-tô, Yagishiri-tô, Daikoku-jima, Shikotan-tô & Kunashiri-tô (Hokkaido). Outside Japan: Eurasian Continent (N. Iberian Penins.~Far East) / Iss. Great Britain, Ireland & Sakhalin. NE. America (introduced species).

Habitat In Europe, this is a very eurytopic species of moist to wet habitats (Turin *et al*., 2003). In Hokkaido, it is adapted to various environment and widely distributed from the woodland to grassland or wetland. In Honshu, however, it is stenotopic and hygrophilous, strictly inhabitant of wetland, inhabiting marshy environment near a river or lake.

Bionomical notes Spring breeder, earthworm feeder.

Etymology The specific name means "having many granules".

研究史　Linnaeus(リンナエウス)により「自然の体系」第10版(1758)の中で命名された，オサムシの中でも最も初期に記載された種の一つ．タイプ産地は定められていないが，ヨーロッパ中北部に分布するものが基亜種とみなされている．ユーラシア大陸と周辺島嶼に広く分布し，亜種以下の分類階級に対して与えられた名称は，不適格名も含め，これまでに50を越える．我が国からはLewis(ルイス)(1882)により本州亜種(原記載では独立種*Carabus telluris*として記載)が，さらにBates(ベイツ)(1883)により北海道亜種(同じく独立種*Carabus Yezoensis*として記載)が記載されている．

タイプ産地　未確定．

タイプ標本の所在　不明．

形態　16~30 mm. 背面は多彩で，銅色，暗褐色，緑褐色，紫色，黒色等のものが出現する．外部形態：図1~6．細部の記載：下唇肢亜端節の剛毛は2本．下唇基節中央歯は側葉よりはるかに短く，前方と腹側への突出は弱い．下唇亜基節は有毛(2剛毛を具える)．♂触角腹面に無毛部はない．前胸背板の側縁には片側にそれぞれ2~6本の剛毛がある．♀鞘翅端は弱いながら明らかにえぐれる．鞘翅彫刻は三元的で，第一次間室は小さく深い第一次凹陥により頻繁に分断されて鎖線となり，第二次間室は通常，連続する低い隆条，第三次間室はこれらよりはるかに減退し，殆ど認められないか，せいぜいごく弱い顆粒列となる程度．間室間の鞘翅基面には小顆粒を密に装う．前胸側板は滑らかで，せいぜい弱い皺が認められるにすぎず，腹部腹板も平滑で，せいぜい弱い皺と点刻を有する程度．後基節の剛毛は2本で，内側剛毛を欠く．腹板の先端部側には横溝があるが，中央部ではしばしば不明瞭．交尾器形態：図7~17．陰茎は鞘翅長の半分よりやや短い．膜状部は極めて小さく，葉片を欠く．基斑は小さく，点

15. *Carabus* (*Carabus*) *granulatus*

状で，表面には多数の短毛を具える．左基部側葉は大きく膨らみ，中央葉及び顕著に突出する指状の側舌と融合する．傍盤葉は左右非対称で，右葉のみ突出する．内嚢先端部には脚葉が明らかに認められ，射精孔縁膜の硬化と色素沈着は殆ど見られない．指状片は細長いスペード形で，側面から見て背側に屈曲しない．♀交尾器膣底部節片内板はほぼ円形で，内側壁には多数の不規則な皺を有する．

分子系統 ミトコンドリアDNAの分子系統樹上で，日本産の2亜種はサハリン産の集団と共にまとまったクラスターを形成し，ユーラシア大陸各地に産する本種の他亜種から分岐した時期もかなり新しい．コブスジアカガネオサムシと同様，おそらくは氷期の海退により生じた陸橋を経由して大陸から日本列島へ侵入してきた種であろうと考えられる（大澤・蘇，2001a；大澤ら，2002；Osawa *et al*., 2004；Su *et al*., 2004）．

分布 北海道，本州中～北東部／礼文島，利尻島，天売島，焼尻島，大黒島，色丹島，国後島（以上，北海道）．海外：ユーラシア大陸（イベリア半島北部～極東地方）／グレート・ブリテン島，アイルランド島，サハリン．移入種として北米大陸北東部からも記録がある．

生息環境 ヨーロッパではやや湿潤な環境を好む広生性の種とされるが，我が国では北海道と本州で生息環境と分布状況が著しく異なる．北海道では原野や湿原から森林地帯まで様々な環境に適応しており，垂直分布も平地から標高1,600 m以上までと幅広く，個体数も一般に多い．これに対し，本州では湿地に強く依存した局地的な分布を示し，主として河川流域や湖沼の周辺から記録されている．

生態 春繁殖・成虫越冬型で基本的に環形動物（ミミズ）食のオサムシと考えられている．冬期には土中，朽木中，石下などから越冬成虫の採集記録がある．

地理的変異 地理的変異に富み，亜種以下の分類階級に対して与えられた名称はこれまでに50以上あるが，比較的顕著な形態学的特徴を持つ亜種として，北カフカスの *C.* (*C.*) *g. daghestanicus* Lapouge, 1924，ジュンガリア（新疆ウィグル自治区北東部）の *C.* (*C.*) *g. duarius* Fischer, 1844，北海道の *C.* (*C.*) *g. yezoensis* Bates, 1883，そして日本の本州から記載された *C.* (*C.*) *g. telluris* Bates, 1883などがある．中国北東部（北京，山西省から東北地方），朝鮮半島北部，ロシア沿海地方の一部などに産する集団にも，いわゆる本州亜種 *C.* (*C.*) *g. telluris* の学名が適用されることが多いが，大陸産と本州産の集団の比較研究は不十分である．

種小名の由来 種小名は「多くの小粒のある」，「ざらざらした」の意．

15-1. *C.* (*C.*) *g. yezoensis* Bates, 1883　アカガネオサムシ北海道亜種

Original description　*Carabus Yezoensis* Bates, 1883, Trans ent. Soc. Lond., 1883, p. 223.
Type locality　Sapporo, and across to Junsai Lake.
Type depository　Museum Nationale d'Histoire Naturelle, Paris [Lectotype, ♂].
Diagnosis　18~27 mm. Body above black, usually bearing dark greenish or coppery greenish tinge, and less strongly shiny. Differs from the nominotypical *granulatus* and other continental subspecies in the following points: 1) dorsum less strongly shiny; 2) punctures on head and pronotum fused with one another to form irregular wrinkles; 3) lateral portions of pronotum including basal foveae usually more remarkably granulate; 4) primary intervals of elytra more frequently segmented to form rows of tubercles, tertiaries usually well-recognized as irregularly set rows of granules; 5) apical lobe of penis a little robuster; 6) apical part of digitulus longer and slenderer.
Distribution　Hokkaido / Iss. Rebun-tô, Rishiri-tô, Teuri-tô, Yagishiri-tô, Daikoku-jima, Shikotan-tô & Kunashiri-tô.
Habitat　Unlike that of subsp. *telluris* which is rather hygrophilous, the habitat of the present subspecies is not restricted to wetlands but widely adapted to various environment such as forests, grasslands, meadows and marshy places.
Etymology　Named after Yezo or Ezo, an old name of Hokkaido.

タイプ産地 札幌および蓴菜沼にかけての地域．ただし，レクトタイプに付されているラベルには "Yezo Island" としか記されていない（Toulgoët（トゥルゴエットゥ），1975a）．

タイプ標本の所在 パリ自然史博物館［レクトタイプ，♂］．

亜種の特徴 18~27 mm．背面は通常，暗緑色ないし銅緑色を帯びた黒色で，色彩の変異は少なく，光沢は鈍い．基亜種を始めとする大陸産の各亜種とは以下の点で異なる：1）背面の光沢がより鈍い；2）頭部と前胸背板の点刻は互いに融合し，不規則な皺を形成する；3）前胸背板側縁部から基部凹陥にかけて，より顕著な顆粒を具える；4）鞘翅第一次間室はより頻繁に分断され，瘤状隆起列を形成し，第三次間室は通常，顆粒列として明らかに認められる；5）陰茎先端部がやや太い；6）指状片の先端がより細長く伸長する．

分布 北海道／礼文島，利尻島，天売島，焼尻島，大黒島，色丹島，国後島．

生息環境 道内各地の原野や湿原から森林地帯まで幅広い環境に生息し，湿地やそれに準じた環境に強く依存している本州亜種とはやや異なっている．垂直分布域も幅広く，西島（1989）によれば，知床の遠音別岳（おんねべつだけ）では平地から標高300 m付近まで，十勝川源流部では1,000 m付近まで，夕張岳では1,100 m付近まで，道南の横津岳では1,167 mの山頂付近でもトラップされているという．本亜種の個体数が特に多い地域として，西島（1989）は，礼文島桃岩付近の草原，日高沙流川（さるがわ）の二風谷（にぶたに）付近の河川敷，道南の恵山（えさん），川汲峠（かっくみとうげ）などを挙げている．

16. *Carabus (Carabus) vanvolxemi* Putzeys, 1875
ホソアカガネオサムシ

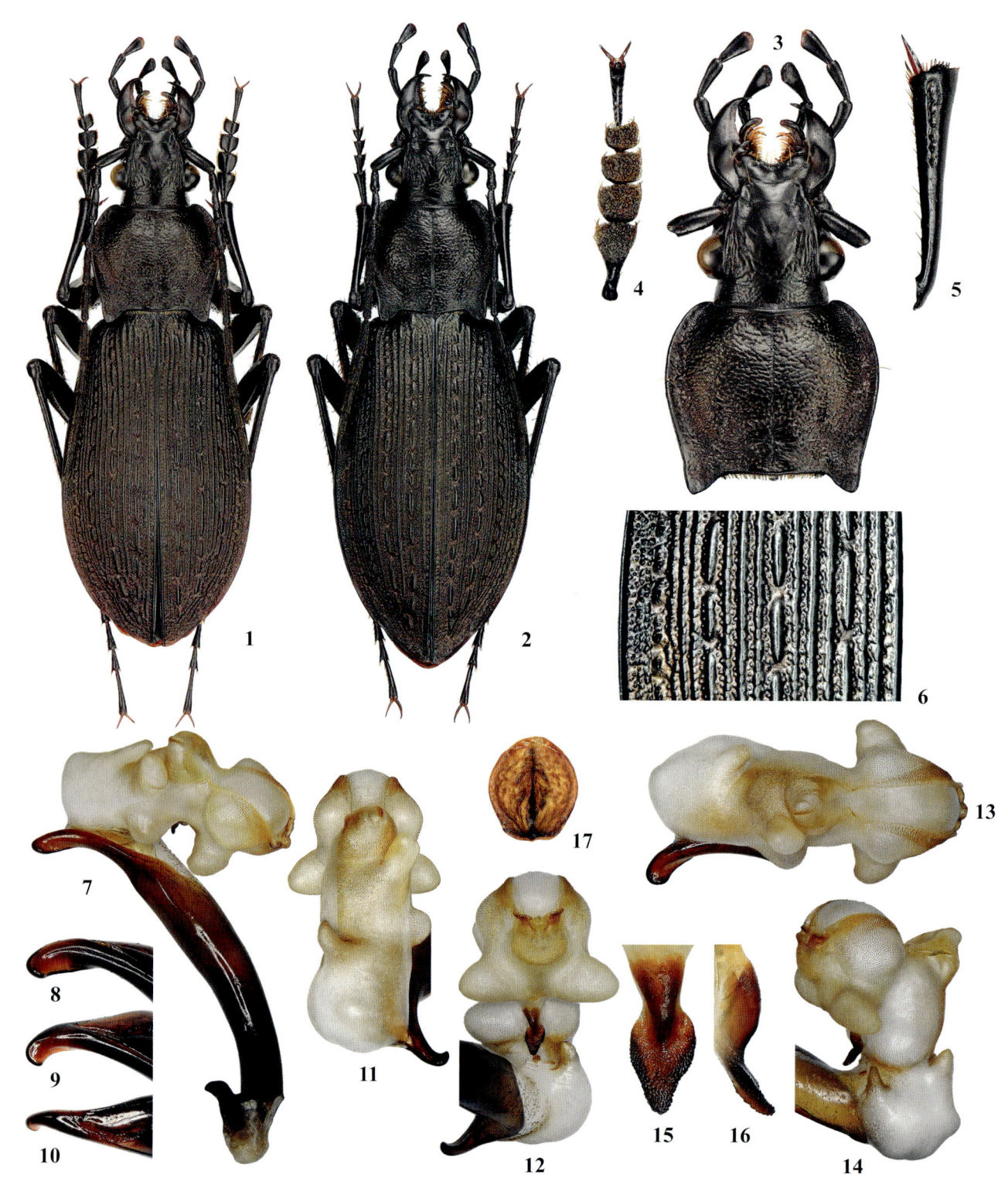

Figs. 1-17. *Carabus (Carabus) vanvolxemi vanvolxemi* Putzeys, 1875 (Lake Chûzenji-ko, Nikkô-shi, Tochigi Pref.). — 1, ♂; 2, ♀; 3, male head & pronotum; 4, male right protarsus in ventral view; 5, male right protibia in subdorsal view; 6, male left elytron (median part); 7, penis & fully everted internal sac in right lateral view; 8, apical part of penis in right lateral view; 9, ditto in right subdorsal view; 10, ditto in dorsal view; 11, internal sac in caudal view; 12, ditto in frontal view; 13, ditto in dorsal view; 14, ditto in left sublateral view; 15, digitulus in frontal view; 16, ditto in right lateral view; 17, inner plate of ligular apophysis.

図 1-17. ホソアカガネオサムシ基亜種（栃木県日光市中禅寺湖畔）．― 1, ♂, 2, ♀; 3, ♂の頭部と前胸背板; 4, ♂右前跗節（腹面観）; 5, ♂右前脛節（斜背面観）; 6, ♂左鞘翅中央部; 7, 陰茎と完全反転下の内嚢（右側面観）; 8, 陰茎先端（右側面観）; 9, 同（右斜背面観）; 10, 同（背面観）; 11, 内嚢（尾側面観）; 12, 同（頭側面観）; 13, 同（背面観）; 14, 同（左斜側面観）; 15, 指状片（頭側面観）; 16, 同（右側面観）; 17, 膣底部節片内板．

16. *Carabus* (*Carabus*) *vanvolxemi*

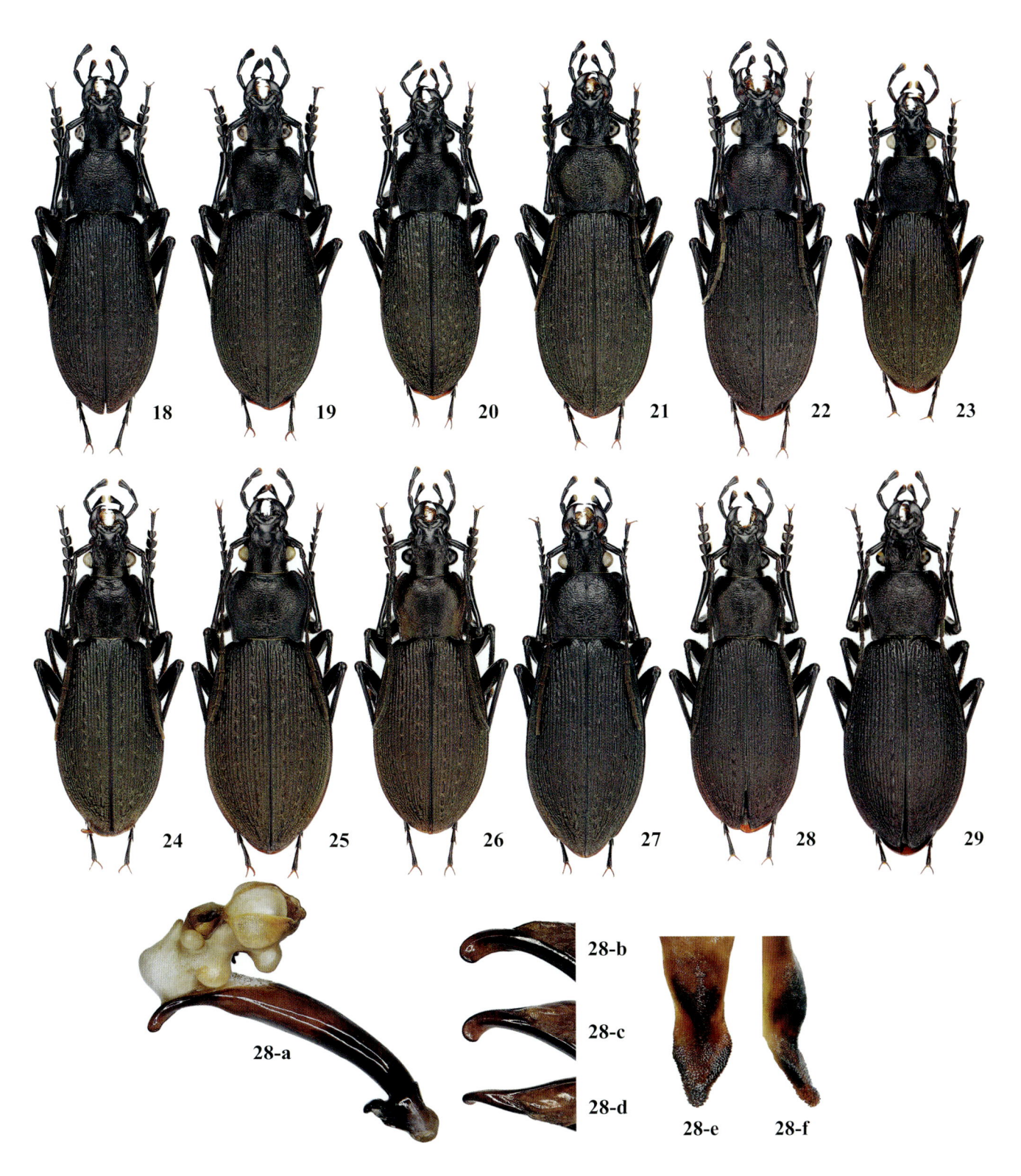

Figs. 18-29. ***Carabus*** (***Carabus***) ***vanvolxemi*** subspp. — 18-27, ***C.*** (***C.***) ***v. vanvolxemi*** (18-24, 26, ♂, 25, 27, ♀; 18, Gôzawa, Yomogita-mura, Aomori Pref.; 19, Nagano, Yonaizawa, Kita-akita-shi, Akita Pref.; 20, Pass Nizawa-tôgé, Sumita-chô, Iwate Pref.; 21, Mt. Ômori-yama, Shinjô-shi, Yamagata Pref.; 22, Taiwa-chô, Miyagi Pref.; 23, Mt. Futamata-yama, Ten'ei-mura, Fukushima Pref.; 24, Mt. Sasa-ga-miné, Myôkô-shi, Niigata Pref.; 25, Kiyosato, Hokuto-shi, Yamanashi Pref.; 26, Jûrigi, Susono-shi, Shizuoka Pref.; 27, Hiwada-kôgen, Takayama-shi, Gifu Pref.; 28-29, ***C.*** (***C.***) ***v. noesskei*** (28, ♂, 29, ♀; Mt. Myôken-san, Is. Sado-ga-shima, Niigata Pref.). — a, Penis & fully everted internal sac in right lateral view; b, apical part of penis in right lateral view; c, ditto in right subdorsal view; d, ditto in dorsal view; e, digitulus in frontal view; f, ditto in right lateral view.

図 18-29. **ホソアカガネオサムシ**各亜種. — 18-27, **基亜種**(18-24, 26, ♂, 25, 27, ♀; 18, 青森県蓬田村郷沢; 19, 秋田県北秋田市米内沢長野; 20, 岩手県住田町荷沢峠; 21, 山形県新庄市大森山; 22, 宮城県大和町; 23, 福島県天栄村二岐山; 24, 新潟県妙高市笹ヶ峰; 25, 山梨県北杜市清里; 26, 静岡県裾野市十里木; 27, 岐阜県高山市日和田高原; 28-29, **佐渡島亜種**(28, ♂, 29, ♀; 新潟県佐渡島妙見山). — a, 陰茎と完全反転下の内嚢(右側面観); b, 陰茎先端(右側面観); c, 同(右斜背面観); d, 同(背面観); e, 指状片(頭側面観); f, 同(右側面観).

16. *Carabus* (*Carabus*) *vanvolxemi* PUTZEYS, 1875 ホソアカガネオサムシ

Original description　*Carabus Van Volxemi* PUTZEYS, 1875, Annls. Soc. ent. Belg., Bruxelles, 18, p. 46.

History of research　This taxon was described by PUTZEYS (1875) as an independent species based on the specimens collected by Jean VAN VOLXEM from Nikkô of Tochigi Prefecture without designation of the holotype. The type series had been left untouched for many years until IMURA (2004a) examined all the seven (2♂5♀) syntypes now preserved in the Institut Royal des Sciences Naturelles de Belgique (Brussels) and designated one of them as the lectotype. *Carabus* (*Leptinocarabus*) *Noesskei* described by VAN EMDEN (1932) from Sado (= Is. Sado-ga-shima) is regarded as a subspecies of *C.* (*C.*) *vanvolxemi*.

Type locality　N. Nipon [sic] (error of Nippon = Japan), dans la forêt entre Niko [sic] (error of Nikkkô) et le temple de Fiu-Sendji [sic] (indicating Chûzenji, most probably), au bord du lac Takaï (the meaning of this word is not clear).

Type depository　Institut Royal des Sciences Naturelles de Belgique (Brussels) [Lectotype, ♂].

Morphology　19~26 mm. Body above black to dark coppery brown, often with faint purplish or blue-greenish tinge. External features: Figs. 1~6. Description on details: penultimate segment of labial palpus bisetose; median tooth of mentum much shorter than lateral lobes, with apex triangular in ventral view and weakly protruding ventrad in lateral view; submentum bisetose; ventral side of male antennae without hairless area; pronotal margins multisetose, with two to six setae on each side; preapical emargination of elytra faintly recognized in female; elytral sculpture triploid, with all three intervals almost equally raised; primaries indicated by chain striae frequently segmented by not so deep primary foveoles; secondaries and tertiaries indicated by contiguous low costae, or partly by rows of large granules with dorsal surface crenulated; areas between intervals scattered with small granules forming longitudinally arranged single row; propleura smooth, at most weakly rugulose; sides of sternites either smooth or weakly rugoso-punctate; metacoxa bisetose, inner seta absent; sternal sulci carved, though often unclear in median part. Genitalic features: Figs. 7~17. Closely allied to those of *C.* (*C.*) *granulatus*, but differs from that species in the following points: 1) apical lobe of penis longer, less strongly bent ventrad but more remarkably hooked at apex which is spatulate ventrad and weakly pointed at tip; 2) paraligula narrow and sharply pointed at tip; 3) podian lobes of internal sac strongly protruded laterad; 4) digitulus distinctly constricted near base in frontal view, and apparently flexed dorsad in lateral view; 5) inner plate of ligular apophysis of female genitalia more elongated longitudinally, with edge of median groove less strongly wrinkled.

Molecular phylogeny　In the ND5 phylogenetic tree, *C. vanvolxemi* belongs to the same cluster as that consisting of *C. granulatus*, and both the species are considered to have been derived from the common ancestor. The ND5 sequences of *C. vanvolxemi* from various localities of Japan including Is. Sado-ga-shima (the locality of subsp. *noesskei*) reveal that the diversification within the species started rather recently (OSAWA & SU, 2000; OSAWA et al., 2002, 2004).

Distribution　C.~NE. Honshu / Is. Sado-ga-shima (Niigata Pref.).

Habitat　Inhabiting temperate to cool temperate forests composed of beech, Mongolian oak, etc.

Bionomical notes　Spring breeder, earthworm feeder.

Etymology　Named after Jean VAN VOLXEM who visited Japan in 1874 and collected this species from Nikkkô.

　研究史　本タクソンはPUTZEYS（プッツェイス）（1875）により日光産の複数の個体に基づいて新種として記載された．海外からは記録されておらず，本邦特産種とみなされている．タイプシリーズは長年，調査されずに放置されてきたが，2004年になってIMURAによりブリュッセルのベルギー王立自然科学博物館に保管されている7頭（2♂5♀）のシンタイプが詳しく調査され，このうちの1♂がレクトタイプに指定されている．VAN EMDEN（ファン・エムデン）（1932）により佐渡島から独立種*Carabus* (*Leptinocarabus*) *noesskei*として記載されたタクソンは現在，ホソアカガネオサムシの一亜種として扱われている．

　タイプ産地　北ニポン，ニコとフィウセンジ寺間の森，タカイ湖畔．「ニポン」は日本，「ニコ」は日光，「フィウセンジ」は中禅寺のことであろう．「タカイ湖」がどこを指すのか不詳だが，原記載に記されたこれらの内容から推測すると，タイプ産地は日光の中禅寺湖畔である可能性が極めて高い（「高い所にある湖」といったニュアンスの現地民の日本語を聞きなしてローマ字表記されたものかもしれない）．

　タイプ標本の所在　ベルギー王立自然史博物館（ブリュッセル）［レクトタイプ，♂］．

　形態　19~26 mm. 背面は黒色ないし暗銅色で，しばしば暗紫色の，時に暗青紫色の光沢を帯びる．外部形態；図1~6. 細部の記載：下唇肢亜端節の剛毛は2本．下唇基節中央歯は側葉より明らかに短く，三角形で腹側への突出は弱い．下唇亜基節には2本の長毛を具える．♂触角腹面に無毛部はない．前胸背板側縁は多毛で，左右にそれぞれ2ないし6本の剛毛がある．鞘翅端のえぐれは♀で軽度に認められる．鞘翅彫刻は三元的で，各間室はほぼ均等に隆起する．第一次間室はそれほど深くない第一次凹陥によって頻繁に分断された鎖線．第二次，第三次間室は連続する隆条で，部分的に粗大顆粒列となり，両者の中間的状態を呈する部域においては表面が小鈍鋸歯状となる．間室間の鞘翅基面には通常，一列の小顆粒列を装う．前胸側板は滑らかで，せいぜい不規則な弱い皺を装う程度．腹部腹板の側面には弱い皺と点刻がある．後基節の剛毛は2本で，内側剛毛を欠く．腹板には横溝があるが，中央部ではしばしば不明瞭．交尾器形態；図7~17. アゲハオサムシ類のものによく似ているが，以下の点で異なる：1) 陰茎先端部はやや長く，腹側への湾曲はやや弱いが，先端部の屈曲はより強く，同部の腹側はやや匙状で，中央で軽く尖る；2) 側舌はより細く，先端はより鋭く尖る；3) 内嚢先端部の脚葉がより顕著に側方へ突出する；4) 指状片は頭側か

ら見て基部付近でより強くくびれ，側面から見て明らかに背側へ屈曲する；5) ♀交尾器膣底部節片内板は縦方向にやや細長く，中央溝側縁部の皺が弱い．

分子系統　本種は形態学的に見てアカガネオサムシに最も近いが，分子系統樹の上でも同種と組み，同種が古日本列島域の一部に隔離された後に同種から種分化を起こしたものであろうと考えられている．いっぽう，種内の各地域集団間における分化の歴史は浅く，一旦幾つかの小地域に隔離されたものが，比較的新しい時期に分布を拡大したのではないかと推測されている（大澤・蘇，2000; 大澤ら，2002; OSAWA *et al.*, 2004）．

分布　本州（長野県西部以東の中〜北東部）／佐渡島．本州では中部地方以東・以北に分布し，その西限は糸魚川−静岡構造線にほぼ一致する．太平洋岸における西限は富士山付近．飛騨山脈では一部の地域に分布が認められる．

生息環境　ブナ，ミズナラを主体とする温帯林帯から冷温帯林にかけて生息し，東北地方では海岸に近い低所にも生息しているが，本州中部では山地性となる．

生態　春繁殖型で夏に幼虫期を持ち，成虫で越冬するタイプのオサムシ．幼虫は環形動物食で，フトミミズ科のミミズ類を好む．幼虫は飼育下で容器内を頻繁に歩き回る徘徊型の捕食者で，比較的おとなしく待ち伏せ型の捕食を行うオホモプテルス亜属の幼虫と好対照を示すとされる（船山ら，1998）．冬季には朽木中，土中いずれにおいても越冬し，同一の越冬窩に複数の個体が入っている場合もある．

地理的変異　本州の基亜種と佐渡島の別亜種 *noesskei* に分けられている．本州内における分布域は必ずしも連続しておらず，各集団間に若干の形態学的相違が見られるが，地理的変異に関する検討は不十分である．

種小名の由来　種小名は，1874年に日本を訪れ，日光で本種を採集したベルギーのJean VAN VOLXEMに献名されたもの．我が国では種小名を英語式に「バンボルクセミ」と読み，それをもじって「バンボレ」という略称で呼ばれることが多いが，オランダ系の人間であれば正しい発音はファン・フォルクセンに近いものであろう．

16-1. *C. (C.) v. vanvolxemi* PUTZEYS, 1875　ホソアカガネオサムシ基亜種

Original description, type locality & type depository　See explanation of the species.
Diagnosis　19~26 mm. For diagnosis, see the "Morphology" section of the species and explanation of the following subspecies.
Distribution　C.~NE. Honshu.

原記載・タイプ産地・タイプ標本の所在　種の項を参照．
亜種の特徴　19~26 mm. 種の解説及び佐渡島亜種の特徴の項を参照．
分布　本州（長野県西部以東の中〜北東部）．

16-2. *C. (C.) v. noesskei* VAN EMDEN, 1932　ホソアカガネオサムシ佐渡島亜種

Original description　*Carabus* (*Leptinocarabus*) *Noesskei* VAN EMDEN, 1932, Neue Beiträge zur system. Insektenkunde, 5 (4/5), p. 63, fig. 1 (tab. 1).
Type locality　Japan: Sado.
Type depository　Unknown. (Staatliches Museum für Tierkunde, Dresden ?)
Diagnosis　19~26 mm. Differs from the nominotypical subspecies in having narrower and shallower lateral furrows of pronotum, less strongly curved pronotal margins, a little deeper preapical emargination of elytra, less strongly constricted basal part of digitulus, less roundly arcuate lateral margins and apical tip of digitulus, though all of these differences are not so large.
Distribution　Is. Sado-ga-shima (Niigata Pref.)
Etymology　Named after Kurt Hermann Gustav Otto NOESSKE (1873~1946).

タイプ産地　日本：佐渡（新潟県佐渡島）．
タイプ標本の所在　不明．著者VAN EMDENのコレクションは国立動物学博物館（ドレスデン）に保管されている可能性が高いが，本亜種のタイプ標本の所在は確認できていない．
亜種の特徴　19~26 mm. 基亜種に比し，前胸背板の側縁に沿った溝がより狭く浅く，同側縁部は背面への反りが弱く，鞘翅端のえぐれがやや深いものが多いとされるが，いずれの差異も比較的軽微なものである．指状片は基部のくびれがやや弱く，両側縁と先端部の丸みが乏しいものが多いようだが，これも個体変異の幅が大きい．
分布　佐渡島．
亜種小名の由来と通称　亜種小名はドイツ人のKurt Hermann Gustav Otto NOESSKE（クルト・ヘルマン・グスタフ・オットー・ネスケ）（NOESSKEの"OE"はö（オー・ウムラウト）を表すので，「ノエスケ」とは発音しない）（1873~1946）に因む．ドレスデンで活動していた外科医で，ドイツ昆虫学協会の会員でもあった．メクラチビゴミムシ等，洞窟性の甲虫に興味があったようで，論文も何編か著している．亜種の通称としてサドオサムシ，サドホソアカガネオサムシ等の名が用いられるが，前者はクロオサムシ佐渡島亜種に対しても用いられたことがあるため，通称による呼称は紛らわしく，かつ分類群を正確に特定できない危険性を孕んでいる．

16. *Carabus* (*Carabus*) *vanvolxemi*

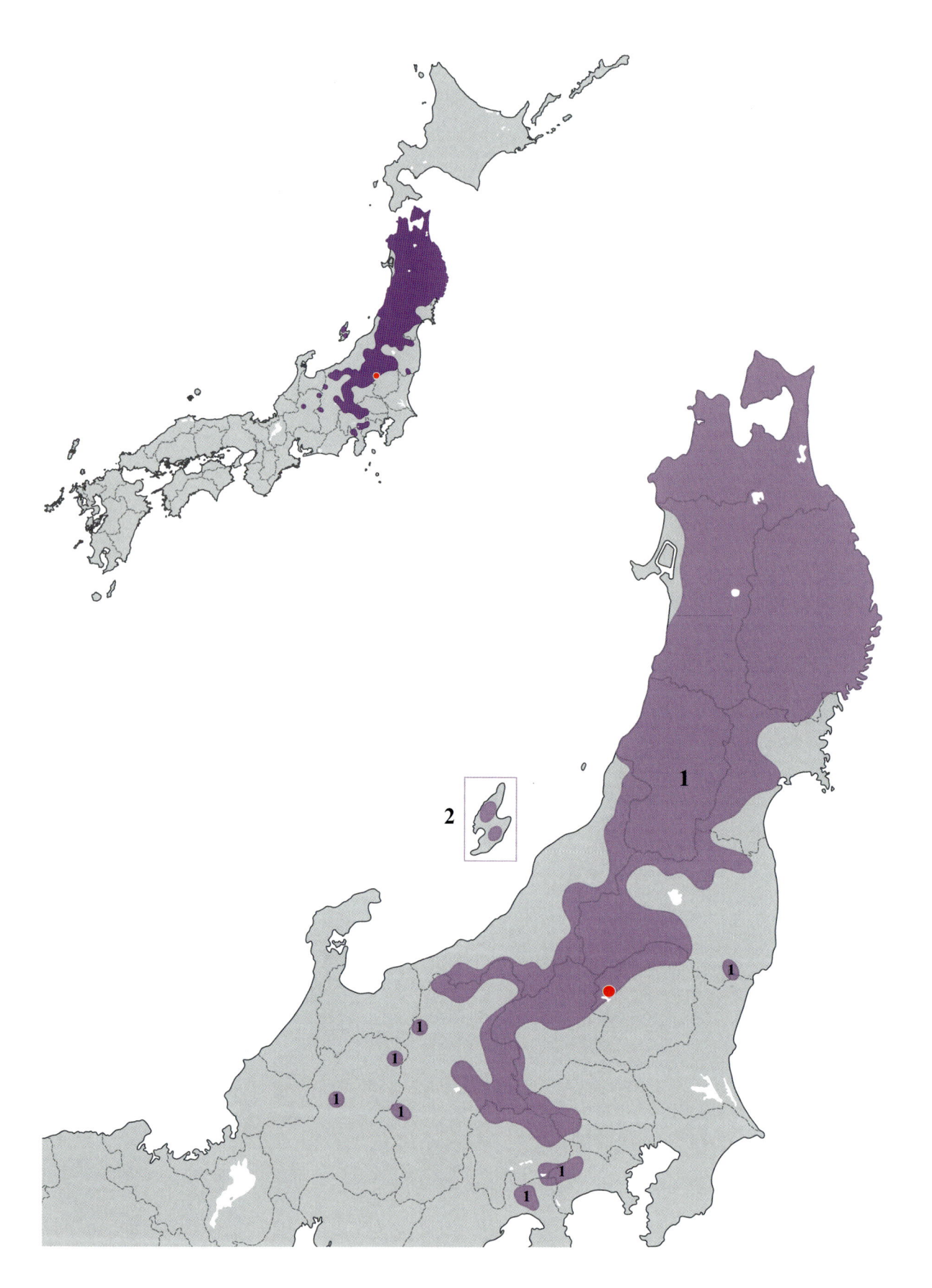

Fig. 30. Map showing the distributional range of ***Carabus*** (***Carabus***) ***vanvolxemi*** PUTZEYS, 1875. — 1, Subsp. ***vanvolxemi***; 2, subsp. ***noesskei***. ● : Type locality of the species (dans la forêt entre Nikko et le temple de Fin-Senji, au bord du lac Takai (most probably Lake Chûzenji ln.r (Nikkô)).

図 30. ホソアカガネオサムシ分布図. — 1, 基亜種; 2, 佐渡島亜種. ● : 種のタイプ産地 (おそらく日光中禅寺湖)

16. *Carabus* (*Carabus*) *vanvolxemi*

Fig. 31. Habitat of *Carabus* (*Carabus*) *vanvolxemi* (Yamanakako-mura, Yamanashi Pref.).

図31. ホソアカガネオサムシの生息地(山梨県山中湖村).

Fig. 32. *Carabus* (*Carabus*) *vanvolxemi* passing the winter in rotten wood (Suyama, Susono-shi, Shizuoka Pref.).

図32. 朽木中で越冬しているホソアカガネオサムシ(静岡県裾野市須山).

Fig. 33. *Carabus* (*Carabus*) *vanvolxemi* passing the winter in rotten wood (Suyama, Susono-shi, Shizuoka Pref.).

図33. 朽木中で越冬しているホソアカガネオサムシ(静岡県裾野市須山).

17. *Carabus* (*Ohomopterus*) *yamato* (Nakane, 1953)
ヤマトオサムシ

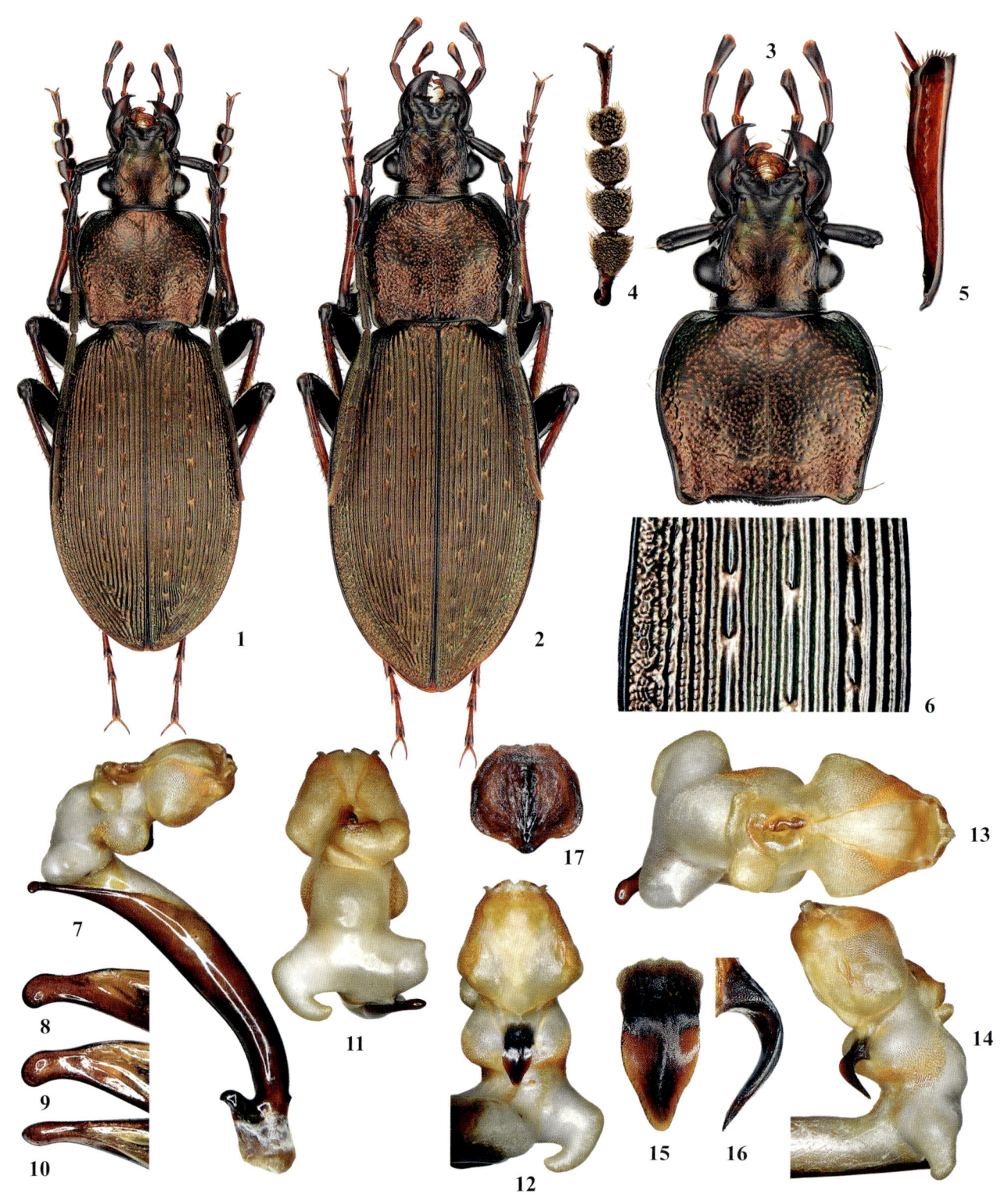

Figs. 1-17. *Carabus* (*Ohomopterus*) *yamato yamato* (Nakane, 1953) (Mt. Hiei-zan, Kyoto-shi, Kyoto Pref.). — 1, ♂; 2, ♀; 3, male head & pronotum; 4, male right protarsus in ventral view; 5, male right protibia in subdorsal view; 6, male left elytron (median part); 7, penis & fully everted internal sac in right lateral view; 8, apical part of penis in right lateral view; 9, ditto in right subdorsal view; 10, ditto in dorsal view; 11, internal sac in caudal view; 12, ditto in frontal view; 13, ditto in dorsal view; 14, ditto in left sublateral view; 15, digitulus in frontal view; 16, ditto in right lateral view; 17, inner plate of ligular apophysis.

17. *Carabus* (*Ohomopterus*) *yamato*

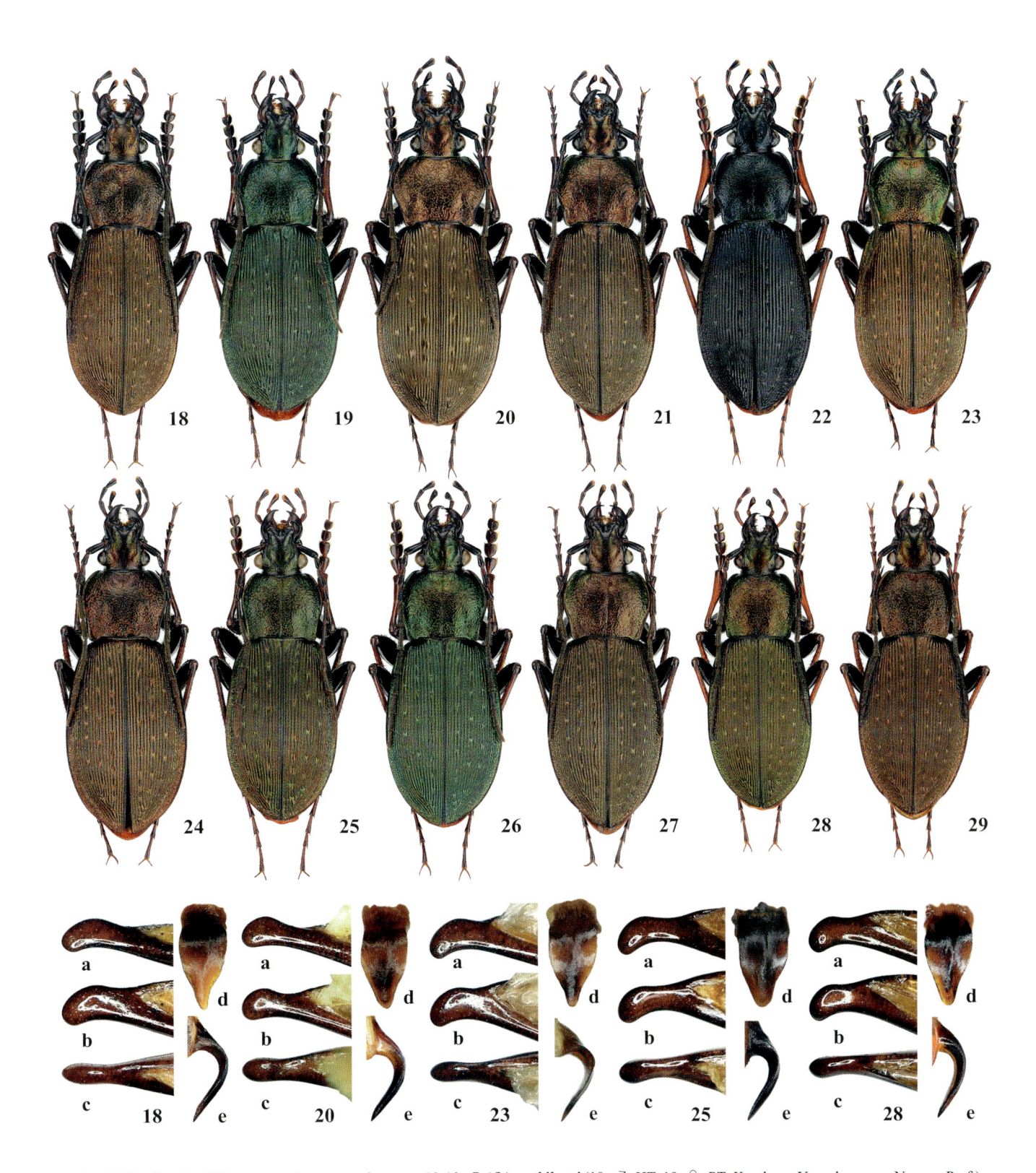

Figs. 18-29. *Carabus* (*Ohomopterus*) *yamato* subspp. — 18-19, *C.* (*O.*) *y. shikatai* (18, ♂, HT, 19, ♀, PT; Karakasa, Yasuoka-mura, Nagano Pref.); 20-22, *C.* (*O.*) *y. ojikai* (20 (HT), 22, ♂, 21, ♀, PT; 20, Tsukudé-sugidaira, Shinshiro-shi, Aichi Pref.; 21, ruins of Iwamura Castle, Iwamura-chô, Ena-shi, Gifu Pref.; 22, Lake Ontaké-ko, Ôtaki-mura, Nagano Pref.); 23-24, *C.* (*O.*) *y. takanonis* (23, ♂, HT, 24, ♀, PT; San-no-kuma, Toyama-shi, Toyama Pref.); 25-27, *C.* (*O.*) *y. kinkimontanus* (25 (HT), 26 (PT), ♂, 27, ♀, PT; Mt. Yamato-katsuragi-san, Gosé-shi, Nara Pref.); 28-29, *C.* (*O.*) *y. kitai* (28, ♂, HT, 29, ♀, PT; Mt. Nagao-yama, Kumano-shi, Mie Pref.). — a, Apical part of penis in right lateral view; b, ditto in right subdorsal view; c, ditto in dorsal view; d, digitulus in frontal view; e, ditto in right lateral view.

図 18-29. ヤマトオサムシ各亜種. — 18-19, 下伊那亜種(18, ♂, HT, 19, ♀, PT; 長野県泰阜村唐笠); 20-22, 三河亜種(20 (HT), 22, ♂, 21, ♀, PT; 20, 愛知県新城市作手杉平; 21, 岐阜県恵那市岩村町岩村城址; 22, 長野県王滝村御岳湖); 23-24, 北陸地方亜種(23, ♂, HT, 24, ♀, PT; 富山県富山市三熊); 25-27, 近畿地方中東部亜種(25 (HT), 26 (PT), ♂, 27, ♀, PT; 奈良県御所市大和葛城山); 28-29, 熊野亜種(28, ♂, HT, 29, ♀, PT; 三重県熊野市長尾山). — a, 陰茎先端(右側面観); b, 同(右斜背面観); c, 同(背面観); d, 指状片(頭側面観); e, 同(右側面観).

147

17. *Carabus* (*Ohomopterus*) *yamato* (NAKANE, 1953) ヤマトオサムシ

Original description *Apotomopterus albrechti yamato* NAKANE, 1953, Scient. Rept. Saikyo Univ., Kyoto, (Nat. Sci. & Liv. Sci.), 1, p. 96, figs. 16 D-i, i', 12, D-j, 13 & D-k, 14 (p. 102).

History of research This taxon was originally described by NAKANE (1953) as a subspecies of "*Apotomopterus albrechti*" (= *Carabus* (*Ohomopterus*) *albrechti* in the present sense) based on 42 specimens from 10 different localities, without designation of the holotype. In current classification, it is regarded as a full species mainly because of peculiarities of male genitalic features (ISHIKAWA, 1969b), though adopted genus has been different according to the authors, that is, *Apotomopterus* (HIURA, 1965, etc.), *Ohomopterus* (Kinki Research Group of Carabid Beetles, 1979), *Carabus* s. str. (ISHIKAWA, 1985, etc.) and *Carabus* s. lat. (IMURA & MIZUSAWA, 1996). Lectotype designation of this taxon was made by IMURA & MIZUSAWA (2002b), and four new subspecies were described at the same time under the names of *shikatai*, *ojikai*, *takanonis* & *kinkimontanus*. One more subspecies was described by IMURA (2004b) from Kumano-shi of southern Mie Prefecture under the name of *kitai*.

Type locality Mt. Hiei (Kioto) (designated by IMURA & MIZUSAWA, 2002b) (originally given type locality (NAKANE, 1953) is "Kinki and Chûbu (Tôkai) Regions of central Honshu, Japan").

Type depository Hokkaido University Museum (Sapporo) [Lectotype, ♂].

Morphology 18~24 mm. Body above reddish coppery, sometimes with greenish tinge on head, pronotum and elytral margins, rarely entirely greenish, very rarely entirely black with faint blue-greenish tinge. External features: Figs. 1~6. Description on details: penultimate segment of labial palpus bisetose; median tooth of mentum a little shorter than lateral lobes, triangular in shape, not strongly protruding both anteriad and ventrad; submentum bisetose; male antennae with hairless ventral depression from segment five to eight; pronotum with three to four marginal setae on each side; preapical emargination of elytra faintly recognized in female; elytral sculpture triploid homodyname; primaries indicated by chain striae segmented by small and shallow primary foveoles; secondaries and tertiaries indicated by contiguous low costae, with surface often crenulated near base and apex; striae between intervals vaguely scattered with fine punctures; propleura smooth; sides of sternites weakly punctato-rugulose; metacoxa usually trisetose; sternal sulci clearly carved; inner margin of male protibia subangulate a little behind middle. Genitalic features: Figs. 7~17. Penis about half the length of elytra, with characteristically shaped apex; membraneous area narrow, lacking ostium lobe; right basal lateral lobe short and robust, with two vaguely inflated humps at apex; left basal lateral lobe with apex conspicuously hooked ventrad; para-saccellar lobes asymmetric, right lobe markedly protruded dorsad, left one vestigial; apical and podian lobes not developed; peripheral rim of gonopore neither sclerotized nor strongly pigmented; digitulus as shown in Figs. 15, 16, 18, 20, 23, 25 and 28. Female genitalia with inner plate of ligular apophysis subquadrate in profile in dorsal view, deeply concave and cup-like, with hind angles strongly and nearly vertically protruded dorsal, hind edge not protruded dorsad.

Molecular phylogeny A molecular phylogenetic study of the carabid beetles was initiated in 1993 by the project team of the Biohistory Research Hall, Takatsuki. They analyzed over 90% of supraspecific categories and more than 40% of hitherto known species of the subtribe Carabina, using mainly mitochondrial gene encoding NADH dehydrogenase subunit 5 (ND5). Their works are compiled into two books (OSAWA et al., 2002, 2004). As to the species belonging to the subgenus *Ohomopterus*, an initial molecular phylogenetic analysis using a mitochondrial gene revealed serious discordance of mitochondrial gene genealogies with morphologically defined species. To explain this incongruence, a hypothesis called the type-switching was advocated (SU, TOMINAGA et al., 1996). However, a subsequent analysis of mitochondrial and nuclear gene genealogies suggested that this might be the result of repeated introgression of mitochondria through hybridization (SOTA, 2000a; SOTA & VOGLER, 2001; SOTA et al., 2001). Further analyses were therefore made using mitochondrial and nuclear gene sequences to clarify the relationship between the complicated history of introgressive hybridization and the specific differentiation of this subgenus (SOTA, 2000c, 2002a, b, 2003; SOTA & VOGLER, 2001, 2003; TOMINAGA, OKAMOTO et al., 2005; Research group of molecular phylogeny on the *Ohomopterus* ground beetles, 2005, 2006a; SU et al., 2006; OKAMOTO et al., 2007, etc.). According to a simultaneous analysis of six nuclear DNA sequences (4164 bp), the subgenus *Ohomopterus* consists of three species-groups, that is, the *albrechti* species-group (*albrechti*, *lewisianus*, *kimurai* and *yamato*), the *daisen* species-group (*daisen*, *yaconinus*, *kawanoi*, *tosanus*, *chugokuensis*, *japonicus* and *dehaanii*) and the *iwawakianus-insulicola* species-group (*iwawakianus*, *maiyasanus*, *uenoi*, *arrowianus*, *komiyai*, *esakii* and *insulicola*) (classification of the species is partly modified following that of this book) (SOTA & NAGATA, 2008). For details, refer to individual references shown above and the "Molecular phylogeny" section of each species. As to the molecular genealogy of the present species, see SU, TOMINAGA et al. (1996), OYAMA et al. (2000), OSAWA et al. (2002, 2004) and NAGATA et al. (2007).

Distribution WC. Honshu (Chubu, Hokuriku & Kinki Regions).

Habitat Basically distributed in hilly to mountainous region consisting of the basement rock, and not found from the Quaternary strata. Inhabitant of woodland in low to middle altitudinal area, but sometimes found from higher place with the altitude more than 1,000 m. Common in the forest edge of broadleaved trees or planted cedar / cypress trees.

Bionomical notes Presumed to be a spring breeder, earthworm feeder.

Geographical variation Classified into six subspecies.

Etymology "Yamato" is an old regional name for the central part of the Kinki Region, mainly for the northern part of Nara Prefecture.

17. *Carabus* (*Ohomopterus*) *yamato*

研究史 本タクソンはNakane（1953）により*Apotomopterus albrechti yamato*（クロオサムシの一亜種）として記載されたが，陰茎先端が独特の形態を呈することから，Ishikawa（1969b）以後，独立種として扱われている．原記載は10地点から得られた42頭の標本に基づいてなされたが，ホロタイプは指定されておらず，地理的変異に関する検討も不十分であった．2002年になって，Imura & Mizusawa（2002b）により，レクトタイプの指定と形態学的再検討に基づく4新亜種（下伊那亜種，三河亜種，北陸地方亜種，近畿地方中東部亜種）の記載がなされた．その後，Imura（2004b）によりさらに熊野亜種が記載されている．

タイプ産地 比叡山（京都）．Nakane（1953）により示されたタイプ産地（地域）は「日本，本州中部，近畿および中部（東海）地方」で，シンタイプの産地として"Mt. Hiei", "Kurama in Kyoto", "Mt. Kongo", "Iwawaki", "Amami in Kawachi", "Horaiji in Mikawa", "Mt. Ryuso", "Hamaishi", "Ashitaka in Suruga", "Mt. Katsuragi in Kawachi"といった10箇所が挙げられており，一部に別種スルガオサムシが含まれていた．このうち，最初に図示された比叡山産の♂がImura & Mizusawa（2002b）によってレクトタイプに指定されている．

タイプ標本の所在 北海道大学総合博物館（札幌）［レクトタイプ，♂］．

形態 18~24 mm．背面は赤銅色で，頭部から鞘翅側縁にかけてしばしば弱い緑色光沢を帯び，時に緑色型，稀に全身が黒化し，弱い青緑色光沢を帯びた型が出現する．外部形態：図1~6．細部の記載：下唇肢亜端節の剛毛は2本．下唇亜基節中央歯は側葉よりやや短く，三角形で前方及び腹側への突出は弱い．下唇亜基節には2本の長毛を有する．♂触角第5節から第8節の腹面に無毛凹陥部がある．前胸背板側縁には左右にそれぞれ3~4本の剛毛を有する．鞘翅端は♀でごく軽くえぐれる．鞘翅彫刻は三元同規的で，第一次間室は小さく浅い第一次凹陥により分断された鎖線，第二次，第三次間室は共に連続する隆条で，基部と翅端部ではしばしば表面が小鈍鋸歯状となる．間室間の線条はごく弱く点刻される．前胸側板は平滑で，腹部腹板側面は弱い点刻と皺を有し，後基節の剛毛は通常3本．腹板の横溝は明瞭．♂前脛節内側縁は中央やや前方で鈍く角ばる．交尾器形態：図7~17．陰茎長は鞘翅長の約半分で，先端は側方から見て亜三角形の特徴的な形に膨らみ，膜状部は小さく，葉片を欠く．右基部側葉は太短く，先端にはごく弱い二つの瘤状隆起があり，左基部側葉の先端は腹側に向けて鉤状に強く屈曲する．傍盤葉は左右非対称で，右葉は顕著に突出し，左葉は退化的．内嚢先端部の脚葉，先端葉は共に発達せず，射精孔縁膜の硬化と色素沈着は殆ど見られない．指状片は図15, 16, 18, 20, 23, 25, 28に示す通り．♀交尾器膣底部節片内板は角の丸い四角形に近く，後角が強くほぼ垂直に背側に向けて突出するため深いカップ状を呈するが，後縁は隆起しない．本種以下，ヒメオサムシまでの8種は，比較的小型で，各鞘翅に3条の第一次間室を有し，指状片は小さく，長三角形を呈するという形質を共有し，形態学的にはヒメオサムシ種群としてまとめられる．

分子系統 オサムシの分子系統学的研究に先鞭を付けたのは1993年にスタートした生命誌研究館のプロジェクトチームで，最終的に世界のオサムシ亜族の上位分類群の9割以上，日本産の全種を含む当時の既知種の4割以上について，主としてミトコンドリアDNA（mtDNA）のND5（NADH dehydrogenase subunit 5）遺伝子を用いた分析が行われた．一連の研究の成果は2冊の著書に集大成されている（大澤ら，2002；Osawa *et al.*, 2004）．オホモプテルス亜属については，レプトカラブス亜属と同様，形態分類とmtDNAの遺伝子系統樹に示される系統関係の間に著しい不一致が生じていたため，この現象を説明するために当初，タイプスィッチング（形態転換機構）による平行放散進化説が唱えられた（Su, Tominaga *et al.*, 1996等）が，複数の核遺伝子を用いた分子系統解析の結果は従来の形態分類と概ね一致し，また特に交雑帯が見られる近縁種間においてmtDNAの系統の共有が頻繁に見られることなどから，種間の交雑によるミトコンドリアの浸透（浸透性交雑）が大きく関係していると考えられるに至った（曽田，2000a; Sota & Vogler, 2001；Sota *et al.*, 2001）．そこで，本亜属についてはmtDNAに加え，核遺伝子を用いた詳細な検討が加えられ（曽田，2000c, 2002a, 2003; Sota, 2002b; Sota & Vogler, 2001, 2003；冨永・岡本ら，2005；オオサムシ属の分子系統研究グループ，2005, 2006a; Su *et al.*, 2006；岡本ら，2007等），複雑な交雑の歴史が本亜属の種分化に大きな影響を与えたことが明らかになっている．複数の核遺伝子配列を用いて構築された系統樹から，本亜属は3種群（クロオサムシ種群（ヤマト，スルガ，ルイス，クロ），ダイセンオサムシ種群（ダイセン，ヤコン，アワ，トサ，ヒメ，アキ，オオ），イワキーアオオサムシ種群（イワキ，カケガワ，シズオカ，アオ，ドウキョウ，ミカワ，マヤサン）；各種名末尾の「オサムシ」は省き，種の分類は本書のそれに倣って一部改編）によって構成されるという結果が出ている（Sota & Nagata, 2008）．しかし，本亜属の分子系統に関しては研究者（グループ）によってその見解が異なり，種間の分岐関係や種・亜種の分類に関してもなお完全にコンセンサスが得られているとは言い難いので，詳細は個々の文献を参照されたい．本種（ヤマトオサムシ）の場合，上記諸研究の結果から，スルガオサムシ，ルイスオサムシ，クロオサムシの3種と共にオホモプテルス亜属の中でも初期に分化した一種群を形成することはほぼ明らかなようで，同じく小型で三角形の指状片を持ち，形態学的には共にヒメオサムシ種群として纏められてきたダイセンオサムシ，ヒメオサムシ，アキオサムシ（未分析ながら，恐らくホウオウオサムシもそうであろう）等との類縁はかなり遠い．ミトコンドリアND5およびCOI遺伝子を用いた分子系統樹上で，ヤマトオサムシの種内では基亜種，近畿地方中東部亜種，三河亜種の三者が一つのクラスターにまとまるが，三河亜種と他の2亜種との分岐は比較的古い．北陸地方亜種に分類される中部地方の個体は，これとは別に本州北東部日本海沿岸沿いに分布するクロオサムシ各亜種のクラスター中に混在して出現し，ヤマトオサムシとクロオサムシとの交雑に由来する集団であろうと推測されている（Su, Tominaga *et al.*, 1996; Oyamaら，2000；大澤ら，2002; Osawa *et al.*, 2004）．同じくmtDNAを用いたNagata *et al.*（2007）による解析では，クロオサムシとヤマトオサムシの分布境界に近い地域に産する集団では，前者から後者へのmtDNAの浸透が頻繁に見られるが，スルガオサムシーヤマトオサムシ間に於けるmtDNAの共有はごく稀で，アキオサムシとの間には共有が見られなかったという．

17. *Carabus* (*Ohomopterus*) *yamato*

分布 本州(中部～近畿地方). 分布域北西部の3亜種(基亜種・近畿地方中東部亜種・北陸地方亜種)と南東部の2亜種(下伊那亜種・三河亜種)の間には, 飛騨山脈と濃尾平野に代表される大きな分布空白域が認められ, 紀伊半島南東部に孤立分布する熊野亜種と他亜種との間にも比較的大きな分布域のギャップがある.

生息環境 基本的には基盤山地に見られ, 第四紀層には生息せず, 垂直分布の下限はオオオサムシのそれよりやや高所にある(近畿オサムシ研究グループ, 1979). 低～中標高域の比較的乾燥した二次林において個体数が多いが, 1,000 mを越す高所からも記録されている.

生態 他の同亜属各種と同様, 春繁殖型で, 幼虫の基本的食餌は環形動物(フトミミズ等)であろうと思われる.

地理的変異 形態学的には以下に述べる6亜種に分類される.

種小名の由来 大和(やまと)地方ないし大和の国に因む.

17-1. *C.* (*O.*) *y. shikatai* (Imura & Mizusawa, 2002) ヤマトオサムシ下伊那亜種

Original description *Ohomopterus yamato shikatai* Imura & Mizusawa, 2002, Elytra, Tokyo, 30, p. 376, figs. 6 (p. 368), 20~23 (p. 369), 37~39 (p. 371) & 52~54 (p. 372).

Type locality Karakasa on the eastern bank of the Tenryû-gawa River of Yasuoka-mura in Shimo-ina Co., Nagano Pref.

Type depository National Museum of Nature and Science, Tokyo [Holotype, ♂].

Diagnosis 18~23 mm. Body above light reddish coppery, rarely with strong greenish tinge as a whole. 1) Basal foveae of pronotum rather deeply concave and not elongate longitudinally; 2) elytra relatively short and robust, with lateral margins often nearly parallel-sided, above all in male; 3) inner margin of male protibia moderately subangulate; 4) apical part of penis long, less strongly bent ventrad, its dorsal margin gently arcuate and not strongly emarginate at base in lateral view; 5) digitulus slender in frontal view, with apex often more sharply pointed than in other subspecies; viewed laterally, its outer margin not subangulate but gently rounded at basal third; 6) inner plate of ligular apophysis roundish in shape, moderately concave, with dorsal surface not remarkably uneven.

Distribution Mountainous regions along the middle course of Riv. Tenryû-gawa and its tributaries.

Etymology Named after Keiichiro Shikata of Iida City Museum, who collected the greater part of the type series of this subspecies.

タイプ産地 長野県, 下伊那郡, 泰阜村, 天竜川東岸, 唐笠.

タイプ標本の所在 国立科学博物館 [ホロタイプ, ♂].

亜種の特徴 18~23 mm. 背面は比較的明るい赤銅褐色で, ごく稀に緑色型が出現する. 1)前胸背板の基部凹陥は深く, 縦方向に伸長しない; 2)鞘翅は太短い箱型で, この傾向は♂においてより顕著; 3)♂前脛節内側縁の角ばりはそれほど強くない; 4)陰茎先端は長く, 腹側への湾曲は弱く, 背側縁は緩やかに弧を描き, 基部のくびれは弱い; 5)指状片は細く, 先端は鋭く尖り, 側面から見て基部1/3付近の外側縁は角ばらず, 比較的滑らかに弧を描くものが多い; 6)♀交尾器膣底部節片内板は丸く, 中等度に窪み, 表面の凹凸はそれほど顕著ではない.

分布 天竜川中流域.

亜種小名の由来と通称 亜種小名は, 飯田市博物館の四方 圭一郎に因む. 原記載論文中で提唱された通称はシモイナヤマトオサムシ.

17-2. *C.* (*O.*) *y. ojikai* (Imura & Mizusawa, 2002) ヤマトオサムシ三河亜種

Original description *Ohomopterus yamato ojikai* Imura & Mizusawa, 2002, Elytra, Tokyo, 30, p. 378, figs. 7 (p. 368), 24~27 (p. 369), 40~42 (p. 371) & 55~57 (p. 372).

Type locality Above Sugidaira, 400-500 m alt., of Tsukudé-mura in Minami-shitara Co., Aichi Pref.

Type depository National Museum of Nature and Science, Tokyo [Holotype, ♂].

Diagnosis 18~24 mm. Body above reddish coppery, partly with greenish tinge, and rarely entirely blackish. 1) Size largest on an average of all the known subspecies of *C.* (*O.*) *yamato*; 2) elytra relatively long, with secondary and tertiary intervals less remarkably notched; 3) apical part of penis less strongly bent ventrad and less strongly arcuate toward apex, with dorsal margin more strongly inflated in lateral view; 4) digitulus narrow and slender, with lateral margins nearly parallel-sided and not sharply pointed at tip in frontal view; viewed laterally, its outer margin not strongly subangulate but gently arcuate; 5) inner plate of ligular apophysis subquadrate in dorsal view, with disk moderately concave.

Distribution Mikawa Region (E. Aichi Pref.) & middle to upper parts of Kiso-dani Valley.

Etymology Named after Tôru Ojika of Anjô-shi in Aichi Prefecture.

タイプ産地 愛知県, 南設楽郡, 作手村, 杉平上方, 標高400-500 m (合併により, 現在は新城市(しんしろし)作手杉平(つくですぎだいら)となっている).

タイプ標本の所在 国立科学博物館 [ホロタイプ, ♂].

亜種の特徴 18~24 mm. 背面は比較的明るい赤銅褐色で，時に緑色味を帯び，ごく稀に黒色型が出現する．1）平均して大型（ヤマトオサムシの全亜種中，最も大型の集団）；2）鞘翅は長く，第二次，第三次間室の横の刻み目は弱い；3）陰茎先端の腹側への湾曲は弱く，背面の膨らみは強い；4）指状片は細長く，頭側から見て両側縁はほぼ平行で，先端はあまり鋭く尖らず，側方から見て外側縁の湾曲部は角ばらず，緩やかに弧を描く；5）膣底部節片内板は背面から見て角の丸い四角形に近く，中等度に窪む．

分布 三河地方～木曽谷．

亜種小名の由来と通称 亜種小名は愛知県安城市の小鹿 亨に因む．原記載論文中で提唱された通称はミカワヤマトオサムシ．

17-3. *C. (O.) y. takanonis* (Imura & Mizusawa, 2002) ヤマトオサムシ北陸地方亜種

Original description *Ohomopterus yamato takanonis* Imura & Mizusawa, 2002, Elytra, Tokyo, 30, p. 373, figs. 5 (p. 368), 16~19 (p. 369), 34~36 (p. 371) & 49~51 (p. 372).
Type locality San-no-kuma in Toyama City, Toyama Pref.
Type depository National Museum of Nature and Science, Tokyo [Holotype, ♂].
Diagnosis 19~24 mm. Body above reddish coppery, sometimes bearing strong greenish tinge, very rarely entirely greenish. Closely allied to the nominotypical subspecies, but distinguished from that race in the following points: 1) apical part of penis shorter, hardly narrowed at base, and mostly bent left laterad in dorsal view; 2) digitulus more acutely tapered toward apex in frontal view, with outer margin angulate at basal third in lateral view; 3) inner plate of ligular apophysis flatter, less strongly wrinkled and more narrowly rimmed along lateral margins.
Distribution Hokuriku Region (excepting the Noto Peninsula and most part of the alluvial plains on the Japan Sea side, with the northeastern edge sharply defined by Riv. Kurobé-gawa), Hida Highlands, Ryôhaku Mts., Etsumi Mts., approaching in the south to Ibuki Mts., Yôrô Mts. and northernmost part of Suzuka Mts. represented by Mt. Fujiwara-daké.
Etymology Named after Toshiaki Tatano (Toyama) who collected the holotype and most part of the paratypes of this subspecies from Toyama Prefecture.

タイプ産地 富山県，富山市，三熊．
タイプ標本の所在 国立科学博物館［ホロタイプ，♂］．
亜種の特徴 19~24 mm. 赤銅色で時に緑色光沢を強く帯びるが，緑色型は極めて稀．基亜種に近いが，以下の諸点で異なる：1）陰茎先端部は短く，基部で殆どくびれず，背面から見て左側に曲がる場合が多い；2）指状片は先端に向けてより急に狭まり，側方から見て外側縁は基部1/3で角ばるものが多い；3）膣底部節片は背側からのえぐれが弱く，より平坦な板状で，表面の皺が弱く，側縁の縁取りは狭い．
分布 北陸地方（能登半島および日本海側の沖積平野部を除く；北東限は黒部川），飛騨高地，両白山地，越美山地，伊吹・養老山地，および藤原岳を中心とする鈴鹿山脈北端．
亜種小名の由来と通称 亜種小名は富山市の高野敏明に因む．原記載論文中で提唱された通称はキタヤマトオサムシ．

17-4. *C. (O.) y. yamato* (Nakane, 1953) ヤマトオサムシ基亜種

Original description, type locality & type depository See explanation of the species.
Diagnosis 18~23 mm. Body above reddish coppery, often with faint greenish tinge; strongly greenish form was not found so far as concerned with the specimens examined by the authors. 1) Pronotum with basal foveae shallow and longitudinally elongate, lateral margins often with four setae on each side; 2) inner margin of male protibia gently rounded; 3) apical part of penis not so elongate, weakly bent ventrad, faintly constricted near base and gently rounded at tip in lateral view; viewed dorsally, it is almost straight or somewhat curved to right; 4) digitulus elongate pentagonal or lingulate in frontal view, with outer margin gently rounded and not subangulate at basal third in lateral view; 5) inner plate of ligular apophysis variable in shape according to individuals, either almost circular or subquadrate, with disk moderately concave.
Distribution Kinki Region (hilly to mountainous area around the Biwa-ko Lake; Mt. Hiei-zan, SW. Tanba Highland, Hira Mts., etc.).

原記載・タイプ産地・タイプ標本の所在 種の項を参照．
亜種の特徴 18~23 mm. 赤銅色でしばしば弱い緑色光沢を帯びる．筆者らの検しえた範囲では，緑色型は出現しない．1）前胸背板は基部回陥が浅く縦長で，側縁にはしばしば4本の剛毛を装う；2）♂前脛節内側縁は鈍く丸まる；3）陰茎先端部はあまり強く伸長せず，腹側への湾曲は弱く，基部で軽くくびれ，先端は丸く，背面から見るとまっすぐ伸長するか僅かに右に曲がる場合が多い；4）指状片は細長い五角形か舌形で，側方から見て滑らかに弧を描き，外側縁は基部1/3で角ばらない；5）膣底部節片内板の形態は個体による差が大きく，ほぼ円形のものから四角形に近いものまであり，カップ状に比較的強く窪む．

分布　琵琶湖周囲の丘陵〜山地（比叡山，丹波高地南西部，比良山地の一部等）．

17-5. *C. (O.) y. kinkimontanus* (Imura et Mizusawa, 2002)　ヤマトオサムシ近畿地方中東部亜種

Original description　*Ohomopterus yamato kinkimontanus* Imura & Mizusawa, 2002, Elytra, Tokyo, 30, p. 370, figs. 4 (p. 368), figs. 12~15 (p. 369), 31~33 (p. 371) & 46~48 (p. 372).
Type locality　Northern slope of Mt. Yamato-katsuragi-san, 700-900 m alt., in Gosé City, Nara Pref.
Type depository　National Museum of Nature and Science, Tokyo [Holotype, ♂].
Diagnosis　19~24 mm. Body above reddish coppery; entirely greenish individuals more frequently appears than in other subspecies of *C. (O.) yamato*, above all in the specimens distributed on the Izumi and Kongô Mountains. Allied to the nominotypical subspecies, but differs from that race in the following points: 1) coloration of tibiae and tarsi a little darker; 2) size a little larger on an average; 3) pronotum a little more transverse, with disk less strongly rugulose in basal portion; 4) primary foveoles of elytra a little larger on an average; 5) inner margin of male protibia more evidently subangulate at middle; 6) apical part of penis longer and more acutely curved ventrad, its basal portion more evidently constricted in lateral view; 7) digitulus usually a little robuster, with tip a little less sharply pointed in frontal view; 8) inner plate of ligular apophysis deeply concave and broadly margined except for distal portion.
Distribution　C.~E. Kinki Region (Izumi Mts., Kongô Mts., Kasagi Mts., Minakuchi Hills, Takami Mts., Nunobiki Mts., greater part of Suzuka Mts., etc.).
Etymology　The subspecific name means "of the mountains of the Kinki Region".

タイプ産地　奈良県，御所市，大和葛城山北麓，標高700-900 m.
タイプ標本の所在　国立科学博物館［ホロタイプ，♂］.
亜種の特徴　19~24 mm. 全亜種のなかで緑色型の出現比率が最も高く，この傾向は和泉山脈と金剛山地において特に著しい．基亜種に近いが，1）脛節と跗節はやや暗色；2）平均してやや大型のものが多い；3）前胸背板はやや横長で，基部の皺が弱い；4）鞘翅第一次凹陥は平均してやや大きい；5）♂前脛節の内縁は中央で顕著に角ばる；6）陰茎先端部は長く，腹側に向けて強く曲がり，基部のくびれが強い；7）指状片は幅広く，先端はあまり鋭く尖らないものが多い；8）膣底部節片内板はカップ状に深く窪み，近位端と側縁は幅広く縁取られる．
分布　近畿地方中〜東部（和泉山脈，金剛山地，笠置山地，水口丘陵，高見・布引山地とその東方山麓，及び鈴鹿山脈の大部分）．分布域西部では紀ノ川が分布の南縁となっているが，南東部では宮川(みやがわ)を越えて南方に分布を広げている．
亜種小名の由来と通称　亜種小名は「近畿地方の山地の」を意味する．原記載論文中で提唱された通称はミナミヤマトオサムシ．

17-6. *C. (O.) y. kitai* (Imura, 2004)　ヤマトオサムシ熊野亜種

Original description　*Ohomopterus yamato kitai* Imura, 2004, Elytra, Tokyo, 32, p. 4, fig. 2.
Type locality　Southwestern slope of Mt. Nagao-yama, 400-780 m in altitude, in Kumano City, southern part of Mié Prefecture, Central Japan.
Type depository　National Museum of Nature and Science, Tokyo [Holotype, ♂].
Diagnosis　18~22 mm. Body above reddish coppery. Most closely allied to subsp. *kinkimontanus*, but differs from it in the following points: 1) tibiae a little more strongly reddish; 2) size smaller on an average, with relatively shorter elytra and relatively longer antennae; 3) areas around basal foveae of pronotum more remarkably uneven; 4) apical part of penis robuster and more gently arcuate ventrad.
Distribution　Mt. Nagao-yama of Kumano-shi in S. Mie Pref.
Etymology　Named after Takaharu Kita (Nara) who discovered this subspecies.

タイプ産地　中部日本，三重県南部，熊野市，長尾山南西麓，標高400-780 m.
タイプ標本の所在　国立科学博物館［ホロタイプ，♂］.
亜種の特徴　18~22 mm. 近畿地方中東部亜種に近いが，1）脛節はより強く赤みを帯びる；1）より小型（全亜種中，平均して最も小型の集団）で，相対的に短い鞘翅と長い触角を持つ；3）前胸背板基部凹陥の周辺は基面の凹凸がより顕著；4）陰茎先端はより太短く，腹側への湾曲がより緩やかである．
分布　三重県熊野市長尾山．2003年の秋に発見された亜種で，熊野市西方にある長尾山の中腹より上に分布している．同所は近畿地方中東部亜種の分布南限から南へ50 km以上離れており，孤立した分布圏を形成しているように見える点は極めて興味深い．ただし，今後調査が進めば，これまで本種の南限と考えられていた櫛田川(くしだがわ)以南の台高山脈〜大峰山脈及びその周辺地域から更に本種の生息地が発見される可能性はあるだろう．
亜種小名の由来と通称　亜種小名は本亜種を発見した喜多孝治（奈良）に因む．ヤマトオサムシという通称がある．

17. *Carabus* (*Ohomopterus*) *yamato*

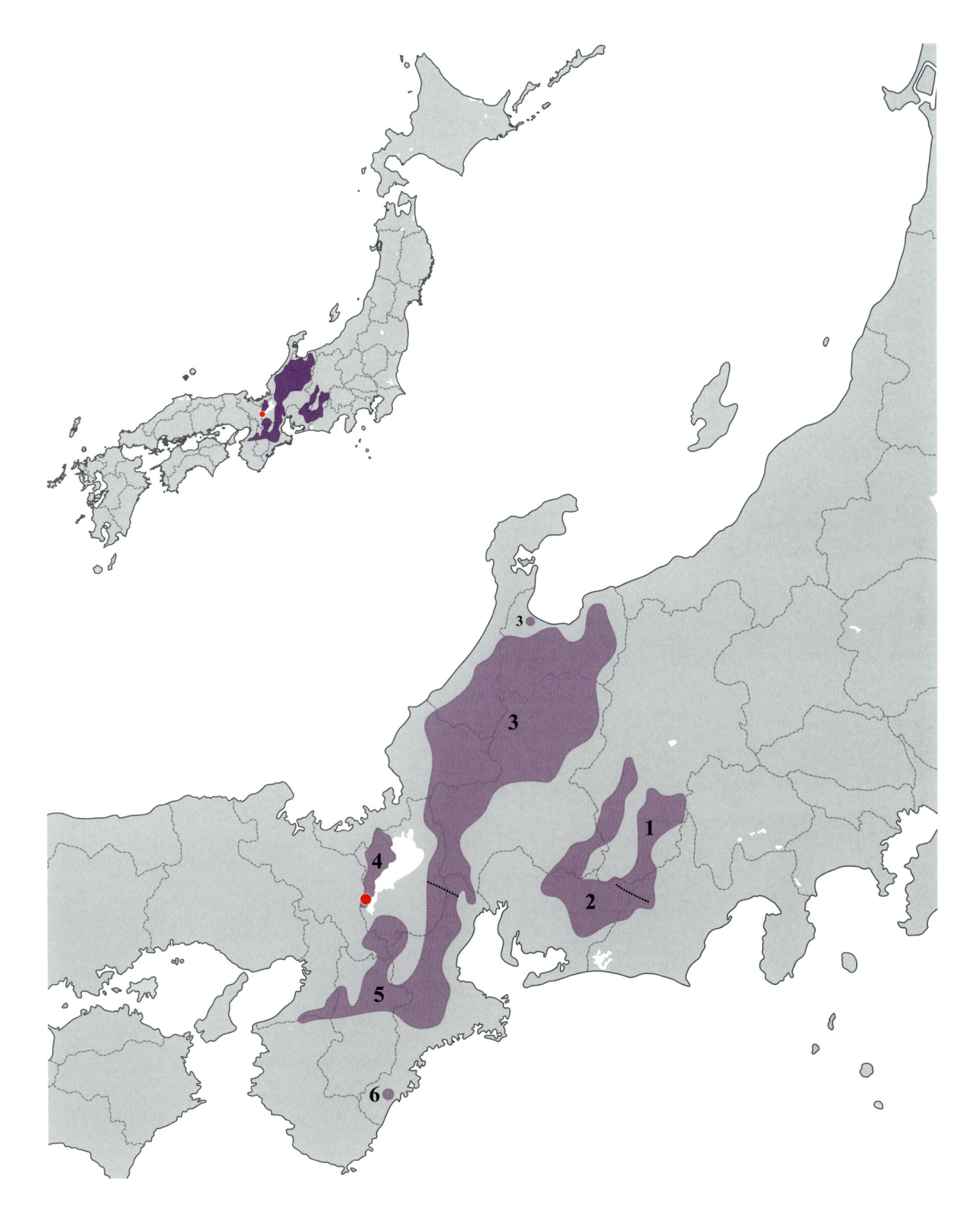

Fig. 30. Map showing the distributional range of ***Carabus*** (***Ohomopterus***) ***yamato*** NAKANE, 1953. — 1, Subsp. ***shikatai***; 2, subsp. ***ojikai***; 3, subsp. ***takanonis***; 4, subsp. ***yamato***; 5, subsp. ***kinkimontanus***; 6, subsp. ***kitai***. ●: Type locality of the species (Mt. Hiei).

図 30. ヤマトオサムシ分布図. — 1, 下伊那亜種; 2, 三河亜種; 3, 北陸地方亜種; 4, 基亜種; 5, 近畿地方中東部亜種; 6, 熊野亜種. ●: 種のタイプ産地(比叡山).

18. *Carabus* (*Ohomopterus*) *kimurai* (Ishikawa, 1966)
スルガオサムシ

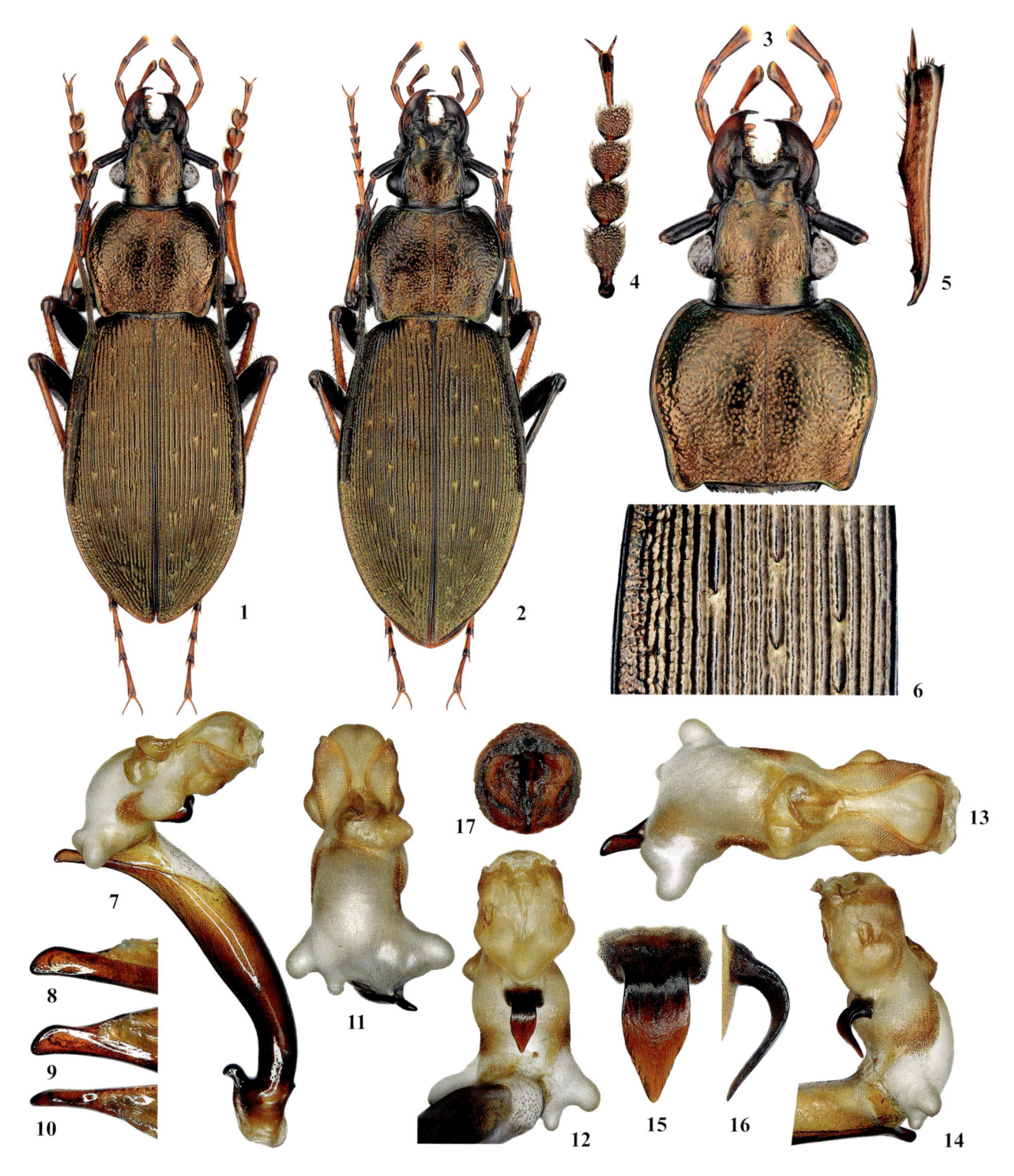

Figs. 1-17. *Carabus* (*Ohomopterus*) *kimurai* (Ishikawa, 1966) (Mt. Ryûsô-zan, Shizuoka-shi, Shizuoka Pref.). — 1, ♂; 2, ♀; 3, male head & pronotum; 4, male right protarsus in ventral view; 5, male right protibia in subdorsal view; 6, male left elytron (median part); 7, penis & fully everted internal sac in right lateral view; 8, apical part of penis in right lateral view; 9, ditto in right subdorsal view; 10, ditto in dorsal view; 11, internal sac in caudal view; 12, ditto in frontal view; 13, ditto in dorsal view; 14, ditto in left sublateral view; 15, digitulus in frontal view; 16, ditto in right lateral view; 17, inner plate of ligular apophysis.

18. *Carabus* (*Ohomopterus*) *kimurai* (Ishikawa, 1966) スルガオサムシ

Original description *Apotomopterus albrechti kimurai* Ishikawa, 1966, Bull. natn. Sci. Mus., Tokyo, 9 (1), p. 19, figs. C~E (p. 20).

History of research This taxon was originally described by Ishikawa (1966a) as a subspecies of *Apotomopterus albrechti* (= *Carabus* (*Ohomopterus*) *albrechti* in the present sense) from Mt. Ryûsôzan of Shizuoka Prefecture. It was later raised its taxonomic rank to a full species also by Ishikawa (1969b), mainly because of uniquely featured antenna, protibia and genitalia of the male.

Type locality Mt. Ryûsôzan, nr. Shizuoka City.

Type depository National Museum of Nature and Science, Tokyo [Holotype, ♂].

Morphology 16~22 mm. Body above coppery or reddish coppery, seldom strongly greenish; tibiae, tarsi and palpi wholly rufous excluding distal end of each segment which is a little darker. External features: Figs. 1~6. Allied to *C.* (*O.*) *yamato*, but differs from it in the following points: 1) appendages more strongly reddish; 2) male antennae without hairless ventral depressions; 3) elytra usually less strongly convex above, with primary foveoles larger, invading adjacent tertiaries, striae between intervals more strongly notched, so that elytral surface appears to be more uneven; 4) inner margin of male protibia more remarkably angulate at middle; 5) apical part of penis different in shape as shown in Figs. 8~10; 6) basal lateral lobes on both sides of internal sac also different in shape as shown in Figs. 11~14, above all in left one which is bilobed and not strongly hooked as in *C.* (*O.*) *yamato*; 7) digitulus with lateral margins more straight in apical halves, apex more sharply pointed; 8) inner margin of ligular apophysis more circular in profile in dorsal view.

Molecular phylogeny For the molecular phylogeny of the subgenus *Ohomopterus* including this species, see the same section of *C.* (*O.*) *yamato*. On the molecular phylogenetic tree of nuclear gene sequences, this species belongs to the same group as that consisting of three allied species, *C.* (*O.*) *yamato*, *C.* (*O.*) *lewisianus* and *C.* (*O.*) *albrechti* (Okamoto et al., 2007; Sota & Nagata, 2008, etc.).

Distribution Areas between Fuji-kawa & Tenryû-gawa Rivers in Shizuoka Pref., SC. Honshu.

Habitat Basically distributed in hilly to mountainous region consisting of the basement rock. Inhabitant of woodland in low to middle altitudinal area, and common in the edge of secondary forest of broadleaved trees or of planted trees such as cedar or cypress.

Bionomical notes Presumed to be spring breeder, earthworm feeder.

Geographical variation Monotypical species.

Etymology Named after Kinji Kimura, a member of the research group on carabid beetles in Keihin Insect Lovers' Club, Tokyo, that was active in the 1960s.

タイプ産地　静岡市近くの竜爪山（静岡県静岡市）．

研究史　Ishikawa（1966a）により当初，クロオサムシの亜種として記載されたタクソンで，後に同じくIshikawa（1969b）により，♂の触角，前脛節内側縁，陰茎等の形態が異なることから，独立種へと昇格された．

タイプ標本の所在　国立科学博物館［ホロタイプ，♂］．

形態　16~22 mm．背面は金銅色ないし金赤銅色で，時に緑色の強い個体が出現する．脛節，附節，口肢は赤褐色だが，各節の遠位端はより暗色．ヤマトオサムシに近いが，以下の諸点で異なる：1）口肢と脛節，附節の赤みがより強い；2）♂触角腹面に明瞭な無毛部がない；3）鞘翅の膨隆がやや弱く，第一凹陥はやや大きく，隣接する第三次間室を侵し，間室間の条溝の刻み目も強いため，鞘翅彫刻の凹凸がより目立つ；4）♂前脛節内側縁はほぼ中央で強く角ばる；5）陰茎先端の形態が図8~10に示すように異なる．すなわち，陰茎の腹側縁は僅かに波曲するものの殆ど直線状で，背面は中央付近で最も強く膨らみ，先端に向けて急激に狭まる；6）♂交尾器内嚢の左右基部側葉の形態が異なる（図11~14）．とりわけ，左基部側葉は大きく二分し，単葉で先端が鉤状に強く曲がるヤマトオサムシのそれとは大きく異なる；7）指状片の側縁先端半分がより直線的で，先端はより鋭く尖る；8）♀交尾器膣底部節片内板はより円形に近い．

分子系統　本種を含むオホモプテルス亜属の分子系統については，ヤマトオサムシの同項を参照．核遺伝子の分子系統樹によれば，本種はヤマトオサムシ，ルイスオサムシ，クロオサムシの3種と類縁が近く，オホモプテルス亜属の中でも初期に分化した種群に属すると考えられている（岡本ら，2007; Sota & Nagata, 2008等）．ヤマトオサムシとの分布域接点付近では，同種との間に少ない頻度ながらmtDNAの共有が見られるという（Nagata et al., 2007）．

分布　静岡県（富士川流域以西～天竜川以東）．分布域の東方では富士川東方の白糸の滝～田貫湖付近まで分布を伸ばしており，一部でルイスオサムシと分布域を接している．西方は秋葉山付近まで記録があり（平山，1981），一部でヤマトオサムシと分布域を接している．

生息環境　基本的には低～中標高域の基盤山地に生息し，落葉広葉樹の二次林やスギ・ヒノキ植林地の林縁等に多い．

生態　他の同亜属各種と同様，春繁殖型・環形動物（ミミズ）食と思われる．

地理的変異　亜種分類はなされておらず，単型種として扱われている．

種小名の由来　種小名は，京浜昆虫同好会オサムシグループの一員として1960年代を中心に精力的に活動した木村欣二に因む．キムラクロオサムシという名で呼ばれていたこともある．

18. *Carabus* (*Ohomopterus*) *kimurai*

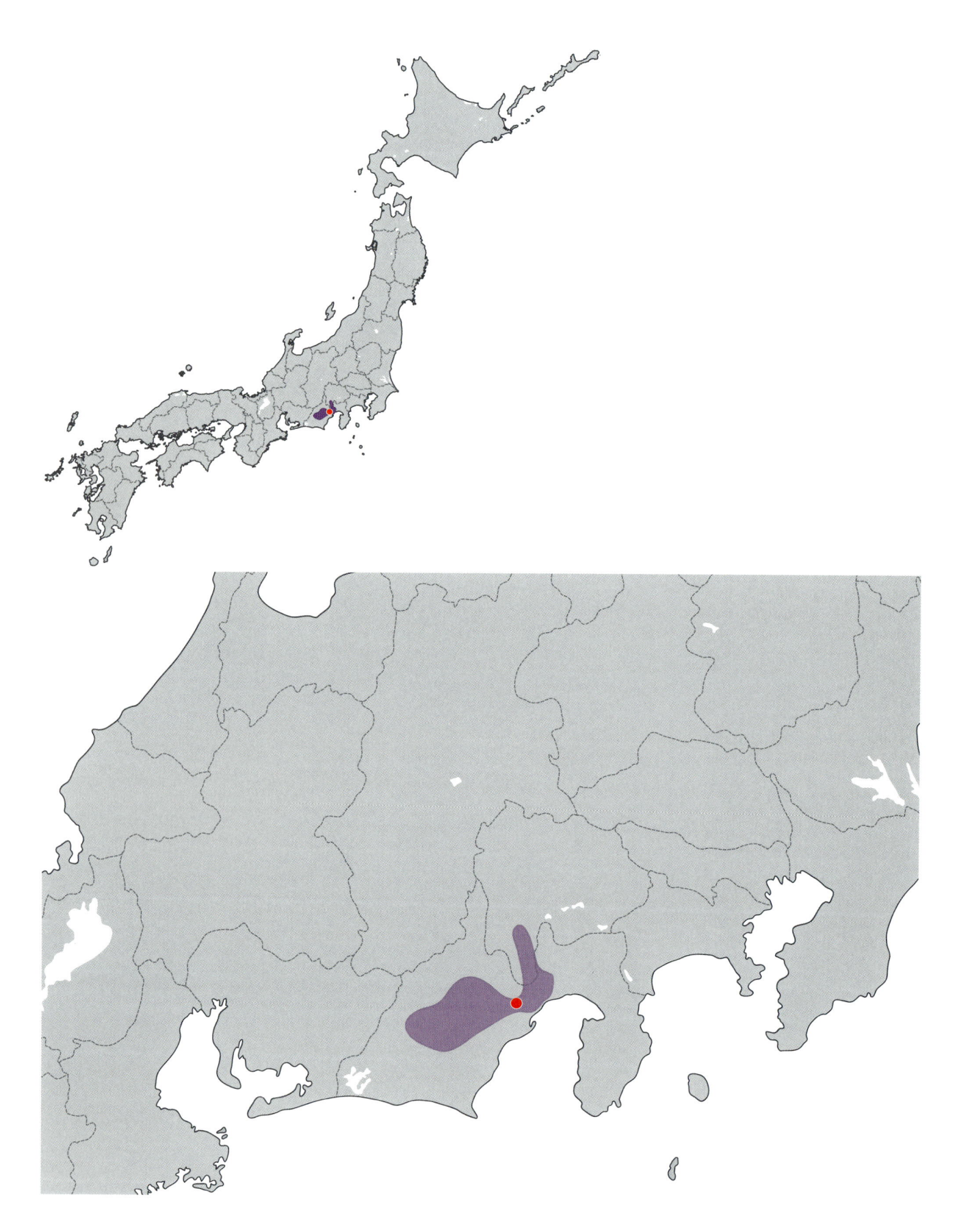

Fig. 18. Map showing the distributional range of *Carabus* (*Ohomopterus*) *kimurai* (ISHIKAWA, 1960). ●: Type locality of the species (Mt. Ryūsōzan).
図 18. **スルガオサムシ**分布図. ●: 種のタイプ産地（竜爪山）.

18. *Carabus* (*Ohomopterus*) *kimurai*

Fig. 19. Habitat of *Carabus* (*Ohomopterus*) *kimurai* (Mt. Ryûsô-zan, Shizuoka-shi, Shizuoka Pref., the type locality of the species). The adult usually pass the winter in the soil in such an environment.

図19. スルガオサムシの越冬環境（静岡県静岡市竜爪山）.

Fig. 20. *Carabus* (*Ohomopterus*) *kimurai* passing the winter in the soil (Mt. Ryûsô-zan, Shizuoka-shi, Shizuoka Pref.).

図20. 土中で越冬しているスルガオサムシ（静岡県静岡市竜爪山）.

Fig. 21. *Carabus* (*Ohomopterus*) *kimurai* passing the winter in the soil (Mt. Ryûsô-zan, Shizuoka-shi, Shizuoka Pref.).

図21. 土中で越冬しているスルガオサムシ（静岡県静岡市竜爪山）.

19. *Carabus* (*Ohomopterus*) *lewisianus* Breuning, 1932
ルイスオサムシ

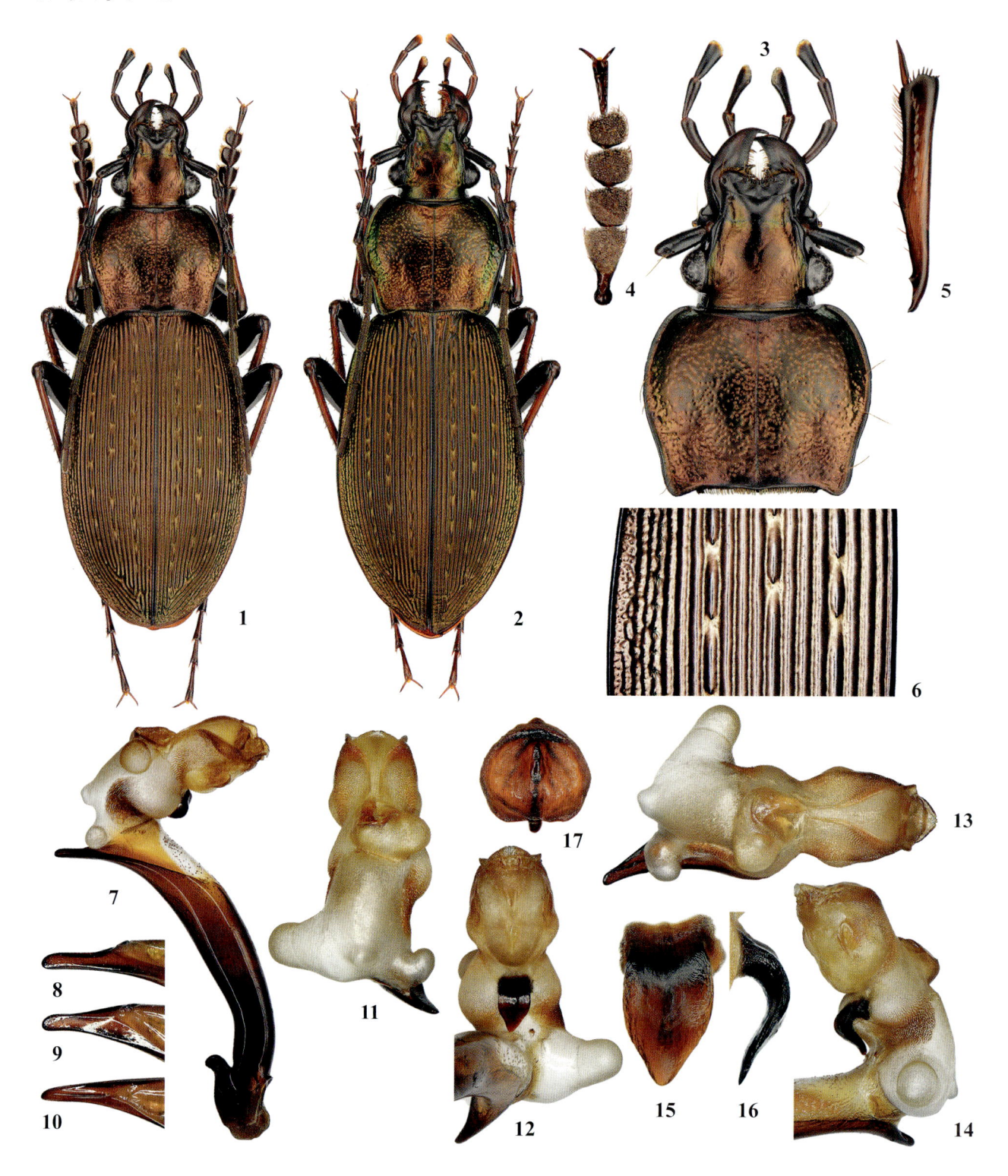

Figs. 1-17. *Carabus* (*Ohomopterus*) *lewisianus lewisianus* Breuning, 1932 (Hakoné-machi, Kanagawa Pref.). — 1, ♂; 2, ♀; 3, male head & pronotum; 4, male right protarsus in ventral view; 5, male right protibia in subdorsal view; 6, male left elytron (median part); 7, penis & fully everted internal sac in right lateral view; 8, apical part of penis in right lateral view; 9, ditto in right subdorsal view; 10, ditto in dorsal view; 11, internal sac in caudal view; 12, ditto in frontal view; 13, ditto in dorsal view; 14, ditto in left sublateral view; 15, digitulus in frontal view; 16, ditto in right lateral view; 17, inner plate of ligular apophysis.

19. *Carabus* (*Ohomopterus*) *lewisianus*

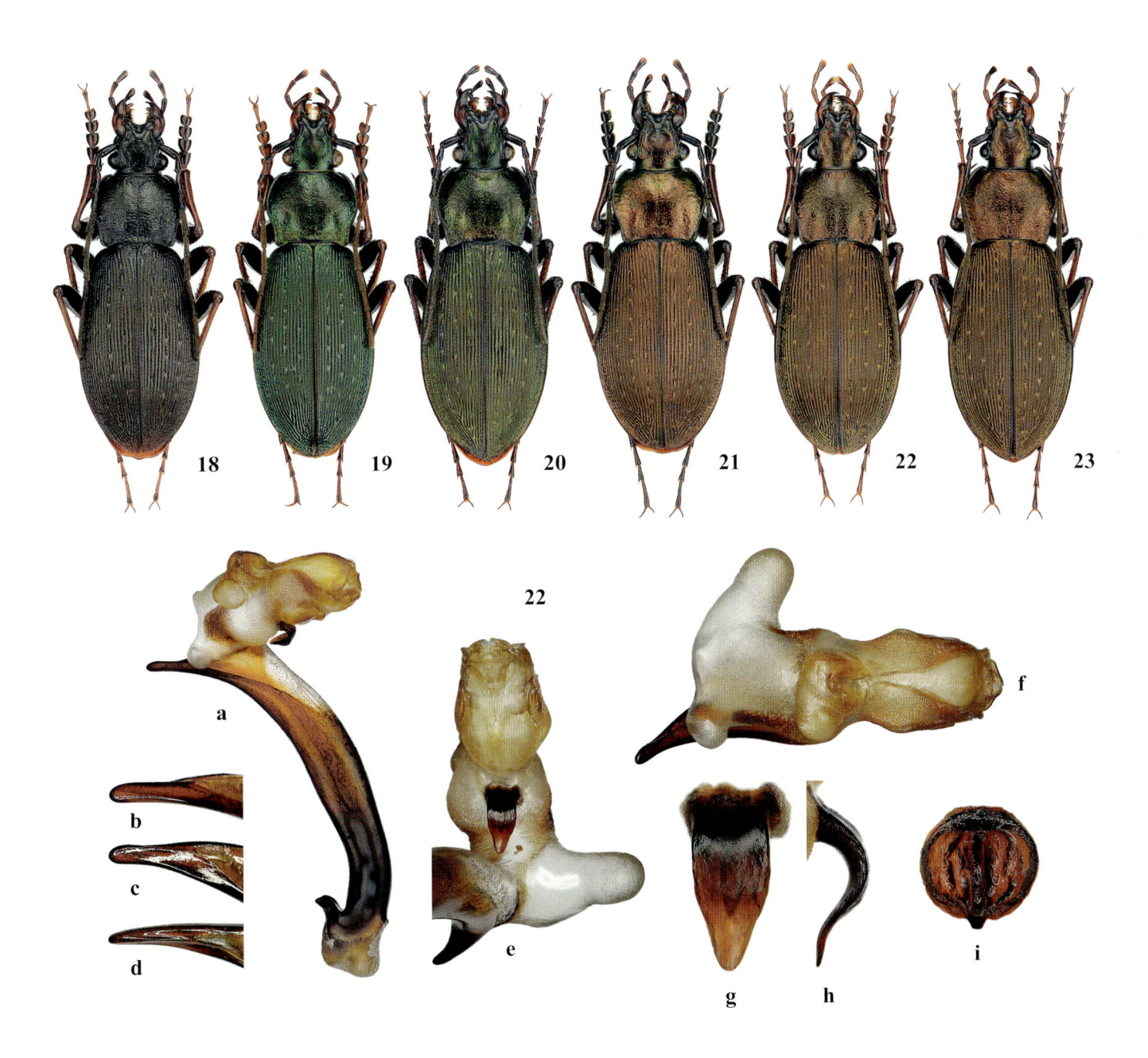

Figs. 18-23. *Carabus* (*Ohomopterus*) *lewisianus* subspp. — 18-21, *C*. (*O*.) *l. lewisianus* (18, 19, 21, ♂, 20, ♀; 18, Mt. Hinokibora-maru, Tanzawa Mts., Kanagawa Pref.; 19, Yaga, Yamakita-machi, Kanagawa Pref.; 20, Pass Otomé-tôgé, Hakoné-machi, Kanagawa Pref.; 21, Yamanakako-mura, Yamanashi Pref.); 22-23, *C*. (*O*.) *l. awakazusanus* (22, ♂, 23, ♀; Mt. Kiyosumi-yama, Kamogawa-shi, Chiba Pref.). — a, Penis & fully everted internal sac in right lateral view; b, apical part of penis in right lateral view; c, ditto in right subdorsal view; d, ditto in dorsal view; e, internal sac in frontal view; f, ditto in dorsal view; g, digitulus in frontal view; h, ditto in right lateral view; i, inner plate of ligular apophysis.

図 18-23. ルイスオサムシ各亜種. — 18-21, 基亜種(18, 19, 21, ♂, 20, ♀; 18, 神奈川県丹沢檜洞丸; 19, 神奈川県山北町谷峨; 20, 神奈川県箱根町乙女峠; 21, 山梨県山中湖村); 22-23, 房総半島南部亜種(22, ♂, 23, ♀; 千葉県鴨川市清澄山). — a, 陰茎と完全反転下の内嚢(右側面観); b, 陰茎先端(右側面観); c, 同(右斜背面観); d, 同(背面観); e, 内嚢(頭側面観); f, 同(背面観); g, 指状片(頭側面観); h, 同(右側面観); i, 膣底部節片内板.

19. *Carabus* (*Ohomopterus*) *lewisianus* BREUNING, 1932 ルイスオサムシ

Original description   *Carabus* (*Apotomopterus*) *japonicus* m. *lewisiana* BREUNING, 1932, Best.-Tab. eur. Coleopt., 104, p. 235.

History of research   This taxon was originally described by BREUNING (1932) as a morpha of *Carabus* (*Apotomopterus*) *japonicus* (= *C.* (*Ohomopterus*) *japonicus* in the present sense) based on the male collected from "Hakone" in Kanagawa Prefecture of central Honshu. NAKANE (1952a, b, 1953) regarded it as a subspecies of *Apotomopterus albrechti* and later combined it with *A. japonicus* (NAKANE, 1960). In his revisional study on *Apotomopterus japonicus* and its allied species, ISHIKAWA (1969b) regarded this taxon as a distinct species as well as some other lower taxa such as *A. albrechti*, *A. kimurai* and *A. yamato*. Since the redefinition of the higher taxa of the subtribe Carabina by ISHIKAWA (1973), it is treated as *Carabus* (*Ohomopterus*) *lewisianus*. In 1981, a new subspecies of this species was described by ISHIKAWA from the Bôsô Peninsula of Chiba Prefecture under the name of *awakazusanus*.

Type locality   Hakone.

Type depository   Zoölogisch Museum Amsterdam [Holotype, ♂].

Morphology   18~23 mm. Body above golden coppery or reddish coppery, seldom strongly greenish or entirely black. External features: Figs. 1~6. Genitalic features: Figs. 7~17. Allied to *C.* (*O.*) *kimurai*, but differs from it in the following points: 1) appendages less strongly reddish, above all in palpi which are dark brown partly or entirely blackish; 2) elytra usually a little more strongly convex above, with primary foveoles smaller and hardly invading adjacent tertiaries; 3) inner margin of male protibia angulate a little behind middle; 4) apical part of penis different in shape as shown in Figs. 8~10; 5) basal lateral lobes on both sides of internal sac also much different in shape as shown in Figs. 11~14; right lobe much smaller, with apex curved dorsad, left lobe much larger, uni-lobed and strongly protruded laterad; 6) digitulus much wider and robuster, with apex not sharply pointed but obtusely rounded; 7) inner plate of ligular apophysis less broadly rimmed, with surface less strongly uneven.

Molecular phylogeny   For an overview of the molecular phylogeny of *Ohomopterus*, see the same section of *C.* (*O.*) *yamato*. This species belongs to the same group as that consisting of three species, *C.* (*O.*) *yamato*, *C.* (*O.*) *kimurai* and *C.* (*O.*) *albrechti*. Analysis of mitochondrial ND5 gene reveals that haplotypes originating from *C.* (*O.*) *lewisianus* introgressed extensively into *C.* (*O.*) *albrechti* in their hybrid zone on the Tanzawa and Kantô Mountains in central Honshu (TAKAMI & SUZUKI, 2005).

Distribution   SC. Honshu (E. Shizuoka Pref., SE. Yamanashi Pref., W. Kanagawa Pref. & W. Tokyo Met.) & S. Bôsô Penins. (S. Chiba Pref.).

Habitat   Basically distributed in hilly to mountainous region consisting of the basement rock. Inhabitant of woodland in low to rather high altitudinal area, and common in the edge of the forest consisting of broadleaved trees or planted cedar- or cypress trees.

Bionomical notes   Spring breeder, earthworm feeder.

Geographical variation   Classified into two subspecies.

Etymology   Named after the famous British coleopterologist, George LEWIS.

研究史　本タクソンはBREUNING (ブロイニング) (1932) により *Carabus* (*Apotomopterus*) *japonicus* m. *lewisiana*という名で記載された．著者のBREUNINGは*lewisiana*をm. (= morpha モルファ) という階級 (序文の定義を見ると，恐らく亜種よりも低位の階級とみなしている) で記載しているので，国際動物命名規約第4版条45.6.2及び45.6.4の但し書きを適用するなら，この名は適格ではないことになる．しかし，BREUNINGはドイツ語の本文中でこのモルファという階級にVarianteあるいはFormといった用語を充てており，さらにこれまでの本学名の使用状況を勘案すれば，条45.6.4および条45.6.4.1の但し書き (1985年よりも前に種や亜種の有効名として使用されたか，または，古参同名として扱われた場合，条45.6.4の下では亜種よりも低位のある学名は，そうであるにもかかわらずその学名の原公表から亜種の階級であるとみなす)，とりわけ後者を適用し，原公表から亜種の階級であるとみなすほうが合理的であろう (ただし，属名と性を一致させるため，語尾を-usに換えて*lewisianus*とする必要がある)．中根 (1952a, b, 1953) はこれをクロオサムシの亜種として扱い，後にヒメオサムシの亜種に移している (NAKANE, 1960)．ISHIKAWA (1969b) は*A. albrechti*, *A. kimurai*, *A. yamato*と共に本タクソンを独立種と認め，これら小型の邦産アポトモプテルス属 (現在ではカラブス属のオホモプテルス亜属) 各種に対し，ヒメオサムシ種群という名を提唱した．1981年にはISHIKAWAにより房総半島南部産の集団が別亜種*awakazusanus*として記載されている．

タイプ産地　箱根．

タイプ標本の所在　アムステルダム動物学博物館 [ホロタイプ，♂].

形態　18~23 mm. 背面は金銅色ないし赤銅褐色で，頭部と前胸背板には緑色光沢を帯びるものが多い．基亜種では時に背面全体が緑色，汚緑色あるいは黒色のものが出現する．外部形態：図1~6．交尾器形態：図7~17．スルガオサムシに近く，♂触角腹面に無毛部を欠く点も両者に共通した特徴だが，以下の諸点により識別される：1) 口肢，脛節，跗節の赤みがやや弱く，とりわけ口肢は暗褐色で部分的あるいは全体的に黒褐色となる；2) 鞘翅は通常，より強く膨隆し，第一次凹陷はより小さく，隣接する第三次間室を侵さない；3) ♂前脛節内側縁は中央よりやや基部よりで角張る；4) 陰茎先端の形態が異なる (図8~10)；5) ♂交尾器内囊基部側葉の形態が大きく異なる (図11~14)．すなわち，右基部側葉ははるかに小さく，先端は背側に向けて湾曲し，左基部側葉ははるかに大きく，単葉で，側方に向けて顕著に突出する；6) 指状片はより太短く，先端は鋭く尖らず，鈍く丸まる；7) ♀交尾器腟底部節片内板の縁取りがより狭く，表面は凹凸に乏しい．

分子系統　オホモプテルス亜属内における分子系統の概要についてはヤマトオサムシの同項を参照．分子系統解析の結果から，本種はヤマトオサムシ，スルガオサムシ，クロオサムシの3種と共にクロオサムシ種群に入る．TAKAMI & SUZUKI (2005) によれ

ば，丹沢山地東部から関東山地南部にかけての本種とクロオサムシ関東地方北西部亜種の交雑帯においては，両者の浸透性交雑の結果，本種からクロオサムシへのミトコンドリアの移入・置換が広範に見られるという．

分布 本州中南部(静岡県東部，山梨県南西部，神奈川県西部，東京都西部)および房総半島南部(千葉県南部)．

生息環境 基本的には基盤岩により形成された丘陵～山地上に生息し，富士山周辺等ではかなりの高所まで生息している．落葉広葉樹やスギ・ヒノキ植林地の林縁等に多く見られ，産地における個体数は一般に多い．

生態 春繁殖，成虫越冬型で，幼虫は環形動物食(フトミミズ等)．

地理的変異 伊豆半島とその北方地域に分布する基亜種と房総半島南部亜種の2亜種に分類される．

種小名の由来 イギリスの著名な甲虫研究家，George Lewis(ジョージ・ルイス)に因む．彼は1880年3月17日から5月11日の間と同年12月20日に何度か箱根(宮ノ下，塔之澤，元箱根，宮城野等)を訪れ，その際採集された標本(1♂)をもとに後年，Breuningによって記載されたのが本種である．ハコネオサムシという和名が提唱されたこともあるが，殆ど使用されていない．

19-1. *C. (O.) l. lewisianus* (Nakane, 1952) ルイスオサムシ基亜種

Original description, type locality, type depository See explanation of the species.

Diagnosis 18~22 mm. Body above coppery with faint greenish tinge on head and pronotum, sometimes entirely greenish or blackish. Differs from subsp. *awakazusanus* mainly in more strongly protruded hind angles of pronotum, less strongly notched elytral striae, and differently featured genital organ of both sexes (see diagnosis of subsp. *awakazusanus*).

Distribution Izu Penins. & its northern area (E. Shizuoka Pref.~SE. Yamanashi Pref.~W. Kanagawa Pref.~W. Tokyo Met.), C. Honshu.

原記載・タイプ産地・タイプ標本の所在 種の項を参照．

亜種の特徴 18~22 mm. 背面の色彩は多彩で，標準的な金銅色，赤銅褐色の他，緑化型や黒化型も出現する．房総半島南部亜種とは，前胸背板後角の後方への突出がより強く，鞘翅間室間条溝の刻み目がはるかに弱く，雌雄交尾器の形態が異なる(房総半島南部亜種の項を参照)こと等により識別される．

分布 伊豆半島とその北方地域(静岡県東部～神奈川県西部～山梨県南東部～東京都西部)．

19-2. *C. (O.) l. awakazusanus* Ishikawa, 1981 ルイスオサムシ房総半島南部亜種

Original description *Carabus (Ohomopterus) lewisianus awakazusanus* Ishikawa, 1981, Kontyû, Tokyo, 49 (3), p. 498, figs. 2, 4, 6, 8 & 10 (p. 499).

Type locality Mt. Kiyosumiyama, ca 350m alt., Amatsu-kominato-chô, Awa-gun.

Type depository Systematic Zoology Laboratory, Department of Biological Sciences, Tokyo Metropolitan University (Hachiôji). [Holotype, ♂]

Diagnosis 19~23 mm. Body above coppery. Green or black individual has not been known. Externally similar to the nominotypical *lewisianus*, but readily discriminated from it in shape of genital organ as follows: 1) apical lobe of penis a little broader in lateral view, a little slenderer in dorsal view, with median longitudinal groove narrower and shallower; 2) internal sac with left basal lateral lobe more strongly protruded laterad, median lobe smaller; 3) digitulus much slenderer; 4) inner plate of ligular apophysis not transverse but circular in outline, with inner wall more strongly sclerotized and irregularly ridged; 5) outer plate of ligular apophysis broader, with front margin a little raised; 6) epivaginal sclerite narrower.

Distribution S. Bôsô Penins. (south of the line between Mt. Kanô-zan and Mt. Kiyosumi-yama except for south of Tateyama), S. Chiba Pref., C. Honshu.

Etymology "Awakazusa" is a compound word composed of two ryôsei provinces, AWA-no-kuni and KAZUSA-no-kuni ("no-kuni" means province), both existed in the southern part of the Bôsô Peninsula until the early Meiji period.

タイプ産地 安房郡，天津小湊町，清澄山，標高約350 m(天津小湊町は合併により消滅し，現在は鴨川市(かもがわし)の一部)．

タイプ標本の所在 首都大学東京動物系統分類学研究室(八王子)[ホロタイプ，♂]．

亜種の特徴 19~23 mm. 背面は銅色で，黒色型や緑色型は知られていない．外見上は基亜種に酷似するが，交尾器形態の違いにより明確に識別される．1)陰茎先端部は側面から見てより幅広く，背面から見てやや細く，背側縁中央の溝はより細く浅い；2)内嚢の左基部側葉はより強く側方へ突出し，中央葉はより小さい．右基部側葉は，原記載では「欠くか，あっても痕跡的」と記されているが，完全反転下で観察すると基亜種と同程度に膨隆するものが多く，明確な識別点とはならない；3)指状片は明らかにより幅が狭い；4)膣底部節片内板はより正円形に近く，壁面の硬化が顕著で，稜状の皺をより多く見える；5)同外版はより幅広く，前縁は隆起する；6)膣壁背面にある硬化・色素沈着した部域は線状で，基亜種より前後方向の幅が狭い．

分布 房総半島南部(鹿野山と清澄山を結ぶ線を北限とし，館山以南の半島先端部からも記録がない)．

亜種小名の由来と通称 亜種小名は，かつて房総半島南部地域にあった二つの令制国(りょうせいこく)，安房国(あわのくに)と上総国(かずさのくに)を組み合わせたもの．アワカズサオサムシという通称がある．

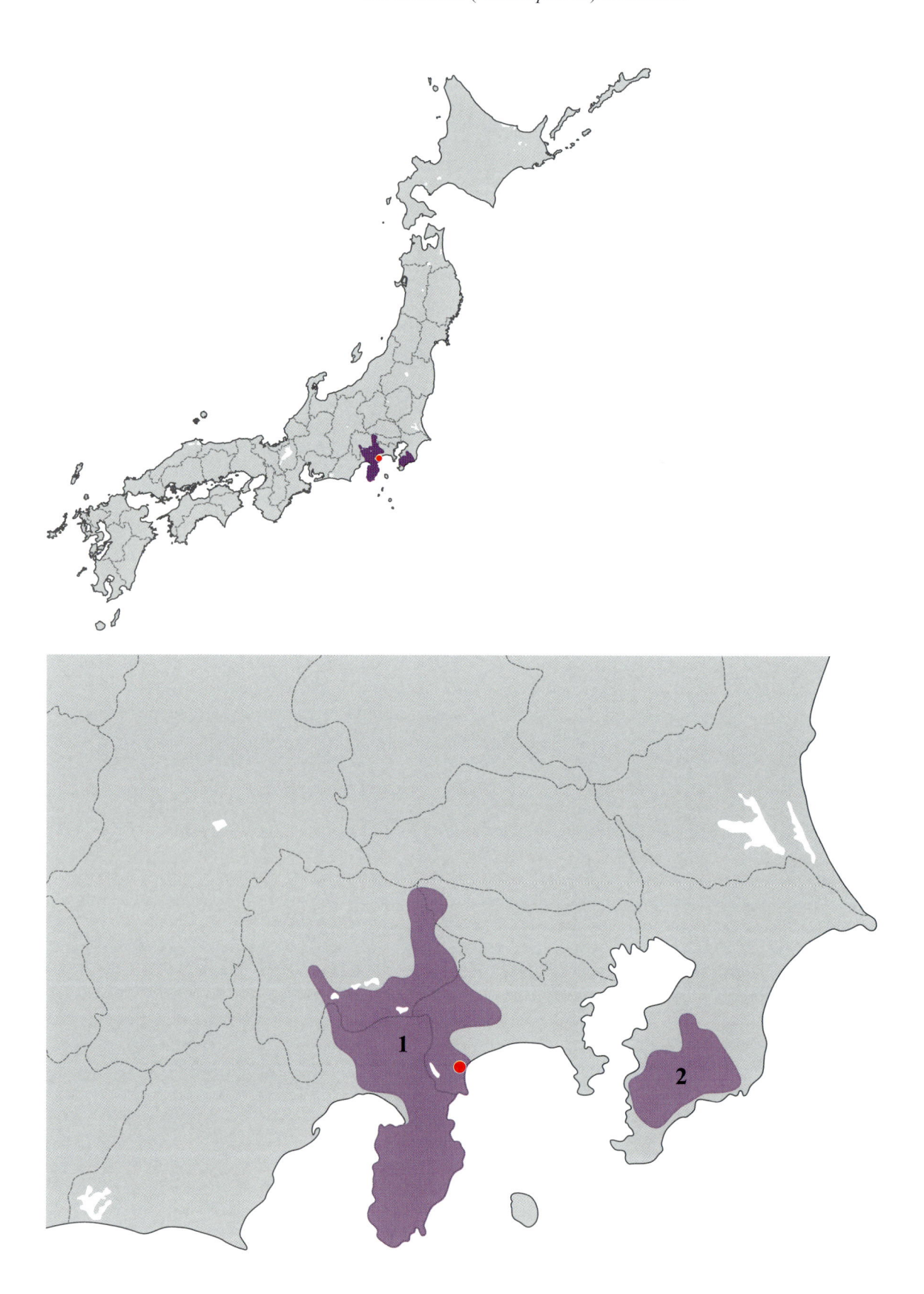

19. *Carabus* (*Ohomopterus*) *lewisianus*

Fig. 24. Map showing the distributional ranges of *Carabus* (*Ohomopterus*) *lewisianus* (Breuning, 1932). — 1, Subsp. *lewisianus*; 2, subsp. *awakazusanus*. ●: Type locality of the species (Hakone).

図 24. ルイスオサムシ分布図. — 1, 基亜種; 2, 房総半島南部亜種. ●: 種のタイプ産地（箱根）.

19. *Carabus* (*Ohomopterus*) *lewisianus*

Fig. 25. Habitat of *Carabus* (*Ohomopterus*) *lewisianus lewisianus*, *C.* (*O.*) *esakii*, *C.* (*Euleptocarabus*) *porrecticollis pacificus* and *C.* (*Damaster*) *blaptoides oxuroides* (artificial forest of planted cedar and cypress trees in Ôbuchi, Fuji-shi, Shizuoka Pref.).

図25. ルイスオサムシ基亜種, シズオカオサムシ, アキタクロナガオサムシ富士亜種, マイマイカブリ関東・中部地方亜種の産地(静岡県富士市大渕のスギ, ヒノキ人工林).

Fig. 26. *Carabus* (*Ohomopterus*) *lewisianus lewisianus* passing the winter in rotten wood (Suyama, Susono-shi, Shizuoka Pref.).

図26. 朽木中で越冬しているルイスオサムシ基亜種(静岡県裾野市須山).

Fig. 27. *Carabus* (*Ohomopterus*) *lewisianus awakazusanus* passing the winter in the soil (Mt. Kiyosumi-yama, Kamogawa-shi, Chiba Pref.).

図27. 土中で越冬しているルイスオサムシ房総半島南部亜種(千葉県鴨川市清澄山).

20. *Carabus* (*Ohomopterus*) *albrechti* Morawitz, 1862
クロオサムシ

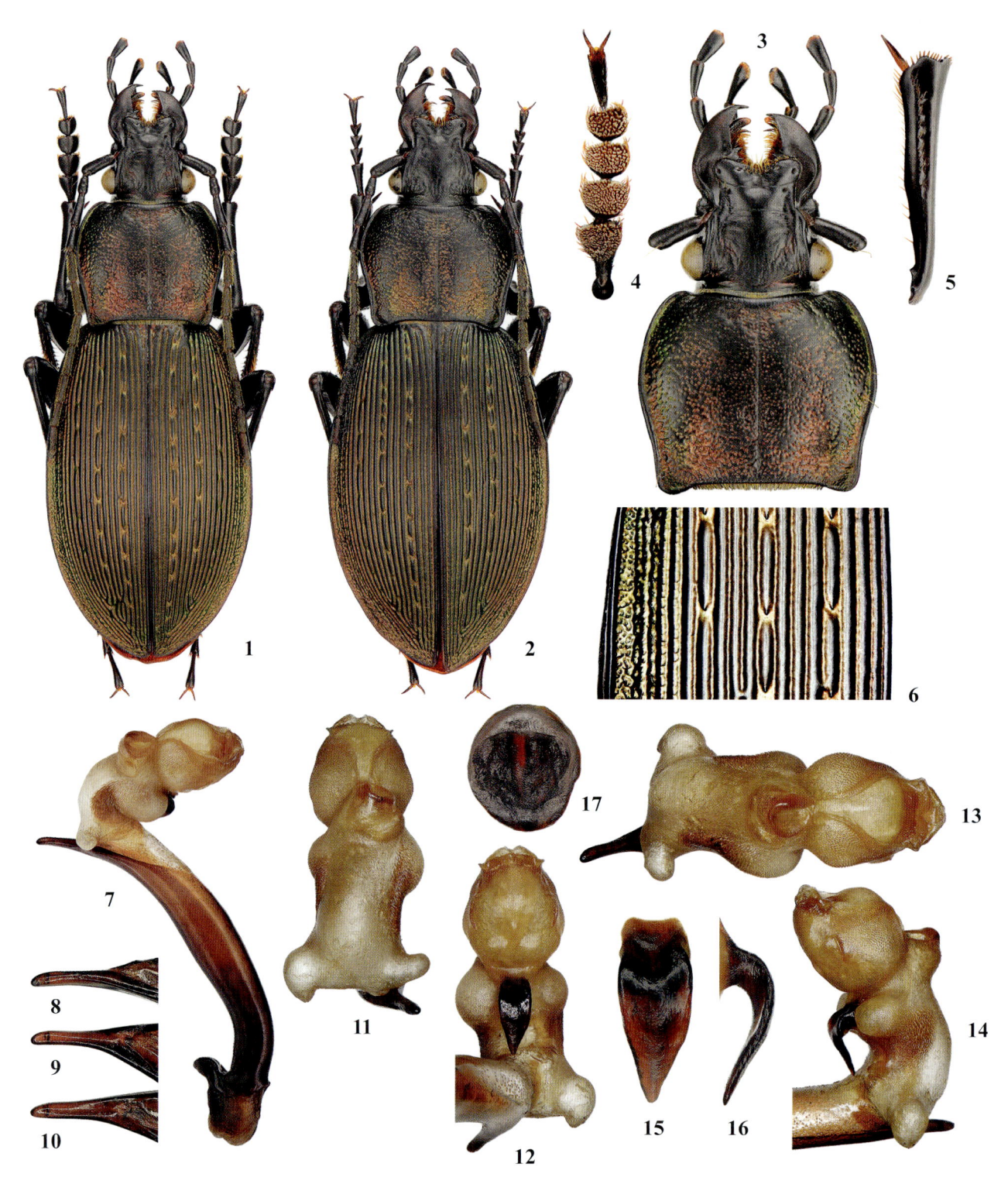

Figs. 1-17. *Carabus* (*Ohomopterus*) *albrechti albrechti* Morawitz, 1862 (Jinkawa-chô, Hakodaté-shi, Hokkaido). — 1, ♂; 2, ♀; 3, male head & pronotum; 4, male right protarsus in ventral view; 5, male right protibia in subdorsal view; 6, male left elytron (median part); 7, penis & fully everted internal sac in right lateral view; 8, apical part of penis in right lateral view; 9, ditto in right subdorsal view; 10, ditto in dorsal view; 11, internal sac in caudal view; 12, ditto in frontal view; 13, ditto in dorsal view; 14, ditto in left sublateral view; 15, digitulus in frontal view; 16, ditto in right lateral view; 17, inner plate of ligular apophysis.

図 1-17. クロオサムシ 基亜種 (北海道 函館市 神山町). — 1, ♂; 2, ♀; 3, ♂の頭胸部; 4, ♂の前跗節 (腹面観); 5, ♂の前脛節 (斜背面観); 6, ♂の左鞘翅中央部 (背面観); 7, 陰茎と完全に翻転した内嚢 (右側面観); 8, 陰茎先端 (右側面観); 9, 同 (右斜背面観); 10, 同 (背面観); 11, 内嚢 (尾側面観); 12, 同 (頭側面観); 13, 同 (背面観); 14, 同 (左斜側面観); 15, 指状片 (頭側面観); 16, 同 (右側面観); 17, 膣底部節片内板.

20. *Carabus* (*Ohomopterus*) *albrechti*

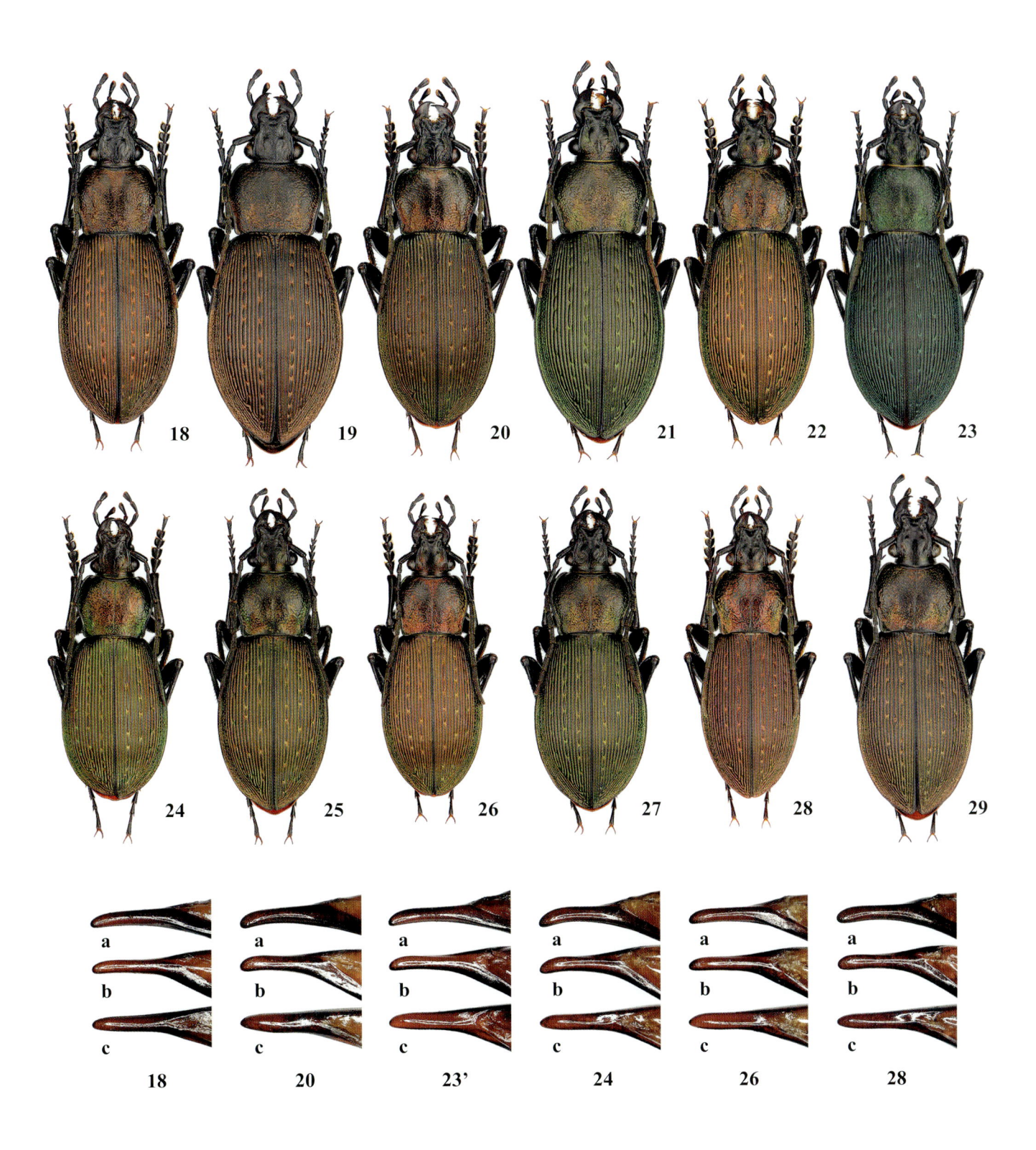

Figs. 18-29. ***Carabus*** (***Ohomopterus***) ***albrechti*** subspp. — 18-19, **C**. (***O***.) ***a***. ***itoi*** (18, ♂, 19, ♀; Rushin, Urahoro-chô, Hokkaido); 20-21, **C**. (***O***.) ***a***. ***hidakanus*** (20, ♂, 21, ♀; Asahi, Biratori-chô, Hokkaido); 22-23, 23', **C**. (***O***.) ***a***. ***albrechti*** (22, ♂; Is. Okushiri-tô, Hokkaido; 23 (♀), 23'(♂); Pass Unseki-tôgé, Yakumo-chô, Hokkaido); 24-25, **C**. (***O***.) ***a***. ***tohokuensis*** (24, ♂, 25, ♀; Matsuzono, Morioka-shi, Iwate Pref.); 26-27, **C**. (***O***.) ***a***. ***yamauchii*** (26, ♂, 27, ♀; Mt. Shikari-ga-daké, Ajigasawa-machi, Aomori Pref.); 28-29, **C**. (***O***.) ***a***. ***hagai*** (28, ♂, 29, ♀; Shin-chô, Yuzawa-shi, Akita Pref.). — a, Apical part of penis in right lateral view; b, ditto in right subdorsal view; c, ditto in dorsal view.

図 18-29. **クロオサムシ**各亜種. — 18-19, **道東亜種**(18, ♂, 19, ♀; 北海道浦幌町留真); 20-21, **日高亜種**(20, ♂, 21, ♀; 北海道平取町旭); 22-23, 23', **基亜種**(22, ♂; 北海道奥尻島; 23 (♀), 23'(♂); 北海道八雲町雲石峠); 24-25, **東北地方東部亜種**(24, ♂, 25, ♀; 岩手県盛岡市松園); 26-27, **白神山地亜種**(26, ♂, 27, ♀; 青森県鰺ヶ沢町然ヶ岳); 28-29, **東北地方中部亜種**(28, ♂, 29, ♀; 秋田県湯沢市新町). — a, 陰茎先端(右側面観); b, 同(右斜背面観); c, 同(背面観).

20. *Carabus* (*Ohomopterus*) *albrechti*

Figs. 30-41. ***Carabus*** (***Ohomopterus***) ***albrechti*** subspp. — 30-31, ***C***. (***O***.) ***a***. ***echigo*** (30, ♂, 31, ♀, Mugura-sawa, Uonuma-shi, Niigata Pref.); 32-33, ***C***. (***O***.) ***a***. ***awashimae*** (32, ♂, 33, ♀, Uchiura, Is. Awa-shima, Niigata Pref.); 34-35, ***C***. (***O***.) ***a***. ***freyi*** (34, ♂, Niibo; 35, ♀, Mt. Donden-yama; both in Is. Sado-ga-shima, Niigata Pref.); 36-37, ***C***. (***O***.) ***a***. ***esakianus*** (36, ♂, 37, ♀, Mt. Takao-san, Hachiôji-shi, Tokyo); 38-39, ***C***. (***O***.) ***a***. ***tsukubanus*** (38, ♂, Yatsuzaku, Ono-machi, Fukushima Pref.; 39, ♀, Mt. Tsukuba-san, Tsukuba-shi, Ibaraki Pref.); 40-41, ***C***. (***O***.) ***a***. ***okumurai*** (40, ♂, Mt. Daibosatsu-rei, Kôshû-shi, Yamanashi Pref.; 41, ♀, Kiyosato, Hokuto-shi, Yamanashi Pref.). — a, Apical part of penis in right lateral view; b, ditto in right subdorsal view; c, ditto in dorsal view.

図30-41. クロオサムシ各亜種. — 30-31, **越後亜種**(30, ♂, 31, ♀, 新潟県魚沼市葎沢); 32-33, **粟島亜種**(32, ♂, 33, ♀, 新潟県粟島内浦); 34-35, 佐渡島亜種(34, ♂, 小木潟; 35, ♀, ドンデン山; 新潟県佐渡島); 36-37, 関東中央山岳部亜種(36, ♂, 37, ♀, 東京都八王子市高尾山); 38-39, 関東地方北東部亜種(38, ♂, 福島県小野町谷津作; 39, ♀, 茨城県つくば市筑波山); 40-41, 山梨長野亜種(40, ♂, 山梨県甲州市大菩薩嶺; 41, ♀, 山梨県北杜市清里). — a, 陰茎先端(右側面観); b, 同(右斜背面観); c, 同(背面観).

20. *Carabus* (*Ohomopterus*) *albrechti* MORAWITZ, 1862 クロオサムシ

Original description *Carabus Albrechti* MORAWITZ, 1862, Bull. Acad. imp. Sci. St.-Petersb., 4, p. 321.

History of research This species was originally described by MORAWITZ (1862b) based on the specimens from Hakodade [sic] (misspelling of Hakodate = Hakodaté) near the southwestern tip of Hokkaido in northern Japan. BREUNING (1932) placed it in the subgenus *Apotomopterus* of the grand genus *Carabus*. VAN EMDEN (1932) regarded it as a member of the subgenus *Ohomopterus* of the genus *Carabus*, and described a new subspecies *freyi* from Is. Sado-ga-shima. NAKANE (1952a, b) regarded *Apotomopterus* as a distinct genus, and recognized three subspecies (excluding the nominotypical one) in *A. albrechti*, namely, *esakii*, *lewisianus* and *yamato* (NAKANE, 1953). Of these, *esakii* is a junior secondary homonym of *Carabus albrechti esakii* CSIKI, 1927, and a replacement name *esakianus* was given by NAKANE (1961). Later, he regarded *albrechti*, together with *lewisianus*, *yamato* and *daisen*, as subspecies of *A. japonicus* (NAKANE, 1962). ISHIKAWA (1966b) described a new subspecies of *A. albrechti* under the name *okumurai*, and later classified this species into six subspecies, nominotypical *albrechti*, *freyi*, *esakianus*, *okumurai*, *daisen* and *okianus* (ISHIKAWA, 1969b). Of these, the latter two are currently regarded an independent species (*Carabus* (*Ohomopterus*) *daisen*) and its subspecies (*C. (O.) d. okianus*). ISHIKAWA (1973) reconsidered a higher classification of the subtribe Carabina based on the male genitalic morphology, and moved all the Japanese species assigned until that time to *Apotomopterus* to the subgenus *Ohomopterus* of the genus *Carabus* in a narrow sense, and described later a new subspecies *C. (O.) a. tohokuensis* (ISHIKAWA, 1984b). ISHIKAWA & TAKAMI (1996) described four new subspecies of *C. (O.) albrechti* under the names of *itoi*, *hidakanus*, *echigo* and *awashimae*. TAKAMI & ISHIKAWA (1997) revised *C. (O.) albrechti* morphologically, and described three more subspecies, *yamauchii*, *hagai* and *tsukubanus*. Thus, *C. (O.) albrechti* is currently classified into 12 subspecies.

Type locality Hakodade [sic] (misspelling of Hakodaté).

Type depository Zoological Institute of Academy of Sciences (St. Petersburg) [Lectotype, ♂].

Morphology 17~26 mm. Body above usually coppery, often black, not rarely greenish. External features: Figs. 1~6. Genitalic features: Figs. 7~17. Allied to the former three species (*C. (O.) yamato*, *C. (O.) kimurai* and *C. (O.) lewisianus*), but differs from them in the following points: 1) male antennae with hairless ventral depressions from segment five to seven, though degree or extent of each depression differs according to subspecies; 2) elytral sculpture rather smooth, with intervals almost uniformly raised, primary foveoles not invading adjacent tertiaries; 3) inner margin of male protibia weakly subangulate a little before middle; 4) apical part of penis longer and slenderer than in *C. (O.) lewisianus* (Figs. 8~10); 5) internal sac with right basal lateral lobe small and narrow, left one larger, with tip weakly protruded ventrad (Figs. 11~14); 6) digitulus slender, more gradually narrowed toward apex which is pointed; 7) inner plate of ligular apophysis of female genitalia thickly rimmed, with inner wall less strongly uneven.

Molecular phylogeny For the molecular phylogeny of *Ohomopterus*, see the same section of *C. (O.) yamato*. This species belongs to the same group as that consisting of three species, *C. (O.) yamato*, *C. (O.) kimurai* and *C. (O.) lewisianus*, and forms the *albrechti* species-group on the molecular genealogical tree (SOTA & NAGATA, 2008). For details on the phylogenetic relations with the allied species, refer to OYAMA et al. (2000), TAKAMI & SUZUKI (2005) and OKAMOTO et al. (2007).

Distribution Hokkaido (discontinuously distributed in Shiretoko Penins., Shiranuka Hills, Pacific coast of Hidaka Region and Oshima Penins.) & NE. Honshu / Iss. Okushiri-tô (Hokkaido), Izu-shima, Kinkasan (Miyagi Pref.), Awa-shima, Sado-ga-shima (Niigata Pref.).

Habitat Basically distributed in hilly to mountainous regions consisting of the basement rock, and not found from the Quaternary strata. Inhabitant of woodland in low to middle altitudinal area, though rarely found from subalpine meadow with the altitude more than 2,000 m. Common in the edge of secondary forest of broadleaved trees or of planted trees such as cedar or cypress.

Bionomical notes Spring breeder, earthworm (mainly those belonging to the family Megascolecoidea) feeder. For the ecology and life cycle of this species, see SHIMIZU et al. (1989).

Geographical variation Classified into 12 subspecies.

Etymology The present species is named after Michael P. ALBRECHT, the Russian physician who stayed in Hakodaté as Russian consulate medical officer for several years beginning in 1858, and collected many insects and plants in response to the entomologists and botanists of his native country.

研究史　本タクソンは，函館産の複数の標本に基づいてMORAWITZ（モラウィッツ）（1862b）により独立種*Carabus Albrechti*として記載された．BREUNING（ブロイニング）（1932）はこれを広義のカラブス属のアポトモプテルス亜属に置いたが，VAN EMDEN（ファン・エムデン）（1932）はオホモプテルス亜属に移し，佐渡島から亜種*freyi*を記載した．中根（1952a, b）はアポトモプテルスを独立属とみなし，MORAWITZ（1862b）の記載した上記の種をその一員*A. albrechti*として扱い，基亜種の他に*esakii*, *lewisianus*, *yamato*の3亜種を認めた（NAKANE, 1953）．このうち，最初の亜種の亜種小名*esakii*は*Carabus albrechti esakii* CSIKI, 1927（シズオカオサムシ）の新参二次同名となるため，NAKANE（1961）により*esakianus*という置換名が与えられている（本書におけるクロオサムシ関東地方北西部亜種）．その後，中根（1962）は*A. albrechti*とそのすべての亜種をヒメオサムシの亜種として扱っている．ISHIKAWA（1966b）は*A. a. okumurai*（クロオサムシ山梨長野亜種）を記載し，*A. albrechti*の下に基亜種以外に*freyi*, *esakianus*, *okumurai*, *daisen*, *okianus*の5亜種を認めた（ISHIKAWA, 1969b）．現在では最後の二つ（*daisen*と*okianus*）は種レベルで異なるもの（ダイセンオサムシとその隠岐亜種）とみなされている．ISHIKAWA（1973）は，♂交尾器内囊の基本形態に基づくオサムシ亜族内の属・亜属の再検討を行い，そ

20. *Carabus* (*Ohomopterus*) *albrechti*

れまでアポトモプテルス属に置かれてきた日本産の全ての種をカラブス属(狭義)のオホモプテルス亜属に移し,1984年には *Carabus* (*Ohomopterus*) *albrechti* の種名の下に亜種 *tohokuensis* を記載した. Ishikawa & Takami (1996)はクロオサムシの4亜種 (*itoi, hidakanus, echigo, awashimae*)を記載し,さらに翌年,同種の再検討を行って *yamauchii, hagai, tsukubanus* の3亜種を追加記載した(Takami & Ishikawa, 1997). かくして本種は目下,形態学的には計12亜種に分類されている.

タイプ産地 ハコダデ(=函館).

タイプ標本の所在 科学アカデミー動物学研究所(サンクトペテルブルク)[レクトタイプ,♂]. 同研究所には現在,1♂5♀の本種のシンタイプが保管されており,そのうちの1♂が井村・水沢(1996, pl. 6, fig. 44-1)によりレクトタイプに指定されている.

形態 17〜26 mm. 背面は銅色系のものが多いが,黒色のものもしばしば出現し,時に全身緑色. 外部形態:図1〜6. 交尾器形態:図7〜17. 基本的形態は前3種に似るが,以下の諸点により識別される:1)♂触角第5節から第7節の腹面に無毛部があるが,その範囲と程度は亜種により異なる;2)鞘翅の間室は均一かつ平滑で,第一次凹陥は隣接する第三次間室を侵さない;3)♂前脛節内側縁は中央よりも基部よりで弱く角ばる;4)陰茎先端の形態が異なり,ルイスオサムシのそれよりも更に細長い(図8〜10);5)♂交尾器内嚢基部側葉は,右が小さく細く,左は大きく太く,先端が腹側に小さく突出する(図11〜14);6)指状片は細長く,先端に向けてより緩やかに狭まり,先端は尖る;7)♀交尾器膣底部節片内板の縁取りは厚く,内壁の表面は凹凸に乏しい.

分子系統 オホモプテルス亜属内における分子系統の概要についてはヤマトオサムシの同項を参照されたい. 本種はヤマトオサムシ,スルガオサムシ,ルイスオサムシの3種と近縁で,分子系統学的観点からはクロオサムシ種群に入り,形態学的にヒメオサムシ種群としてまとめられていたヒメオサムシ,アキオサムシ,ダイセンオサムシ等との類縁は遠い. 本種と関連種の分子系統に関してはOyamaら(2000), Takami & Suzuki (2005),岡本ら(2007)等に詳しい.

分布 北海道(知床半島,白糠丘陵,日高地方の沿岸部,道南などに不連続な分布を示す), 本州(中部以北)/奥尻島(北海道), 出島,金華山(以上,宮城県), 粟島,佐渡島(以上,新潟県). 北海道の分布について,清水ら(1989)は「大雪南東(中略)にも局地的に分布する」と述べ(p. 84),分布図(p. 28)にも道央部に複数のプロットを記しているが,すべて誤記録と思われる.

生息環境 丘陵〜山地に生息し,一般に沖積平野には見られない. 明るい広葉樹林を主な生活の場としているが,成熟したスギ・ヒノキ等の植林にも見られる. 北海道では低地から亜高山帯まで生息し,原生林・二次林を問わず落葉広葉樹林に多い. 東北地方では低地〜低山帯を中心に分布し,標高1,000 mを越す高地には少ない. 関東〜中部地方でも丘陵地から低山帯によく見られ,二次林や比較的乾燥した低山の尾根などに多いが,一部の地域においては低地の疎林や高標高のブナ帯,針葉樹林帯にも進出している. 長野県東御市(とうみし)の烏帽子岳(えぼしだけ)では,標高2,000 mを越す亜高山性の草原からアオオサムシと共に得られた記録がある(井村, 1971).

生態 春繁殖・成虫越冬型. 成虫は雑食性で,各種の肉類や昆虫類,果実を好んで食し,野外でも路上のミミズや果実を摂食している個体が目撃されることがある. 幼虫は基本的に環形動物食で,フトミミズ科(Megascolecidae)のミミズ類を主食としているものと思われ,飼育下ではドバミミズと呼ばれるものを好み,ツリミミズ科のシマミミズは餌として不適当であるという. 野外では5月頃から交尾・産卵を開始し,卵は約10日間で孵化,幼虫令数は3令までで,終令幼虫は土中に蛹室を作り,その中で仰向けになって前蛹となり,やがて蛹化する. 蛹は10日から2週間で羽化し,東日本では8月の前半に羽化のピークが見られることが多く,この時期には多くのテネラル個体が観察される. 越冬に入る時期は関東地方の早い所で10月上旬頃であり,同所的に産するコクロナガオサムシやマイマイカブリに比べ,やや早い場合が多い. 土中越冬が主体だが,高地では朽木や倒木の樹皮下で越冬している個体も見られる. 東北〜関東地方では冬期採集で得られる個体数が多いため,冬虫夏草による寄生例もしばしば観察されるが,実際の報告は稀である(以上,清水ら(1989)より抜粋).

地理的変異 形態学的には北海道に3亜種,本州に9亜種の計12亜種に分類されている.

種小名の由来 本種は,函館のロシア領事館で1858年からの数年間,医官を務めていたロシア人医師, Michael P. Albrecht (ミハエルP. アルブレヒト)に因んで命名された. 彼は妻と共に多くの昆虫や植物を採集し,本国の専門家に送ったという.

20-1. *C.* (*O.*) *a. itoi* Ishikawa et Takami, 1996 クロオサムシ道東亜種

Original description *Carabus* (*Ohomopterus*) *albrechti itoi* Ishikawa & Takami, 1996, Spec. Div., 1, p. 43, figs. 7, 8 (p. 40) & 15~18 (p. 42).

Type locality Rushin, Urahoro-chô, Tokachi, Hokkaido

Type depository Systematic Zoology Laboratory, Department of Biological Sciences, Tokyo Metropolitan University (Hachiôji) [Holotype, ♂].

Diagnosis 22~25 mm. Body above dark coppery; legs black, though sometimes reddish. 1) Antennae short, barely extending basal third of elytra in male, and hairless ventral areas not so distinct; 2) frons not punctate; 3) pronotum broad, with two to four (rarely five) marginal setae on each side, its lateral margins well convex in front, hind angles shorter than those of subsp. *hidakanus*; 4) elytra elongated oval, rather strongly convex above, striae between intervals at most very weakly notched; 5) inner margin of male protibia weakly convex and not angulate; 6) apical part of penis long and thick, not grooved dorsally, barely outlined by rudimentary ridge; 7) lateral margin of digimulus raised near base; 8) inner plate of ligular apophysis of female genitalia almost as in subsp. *hidakanus*, though outer plate is a little broader. Specimens from the Shiretoko Peninsula differ from the topotypical specimens as follows: 1) head and pronotum more strongly wrinkled; 2) apical part of penis more cylindrical, feebly bent beyond middle; 3) lateral margins of digitulus more distinctly raised; 4) outer plate of ligular apophysis shorter, inner plate with a shallow median longitudinal groove.

Distribution   Shiranuka Hills & Shiretoko Penins., E. Hokkaido.
Etymology   Named after Katsuhiko Ito (Nemuro).

タイプ産地　北海道, 十勝, 浦幌町, 留真.
タイプ標本の所在　首都大学東京動物系統分類学研究室(八王子)［ホロタイプ, ♂］.
亜種の特徴　22~25 mm. 背面の色彩は暗銅色. 脚は黒色で, 時にやや赤みを帯びる. 1) 触角は短く, ♂でも先端が鞘翅基部1/3をやや越える程度. ♂触角腹面の無毛部はあまり顕著でない; 2) 前頭部は点刻を欠く; 3) 前胸背板は幅広く, 側縁剛毛は2ないし4本(稀に5本)で, 側縁前半は外側に膨らみ, 後角は日高亜種のそれより短く, 後方に向けての突出は殆ど見られない; 4) 鞘翅は長卵形で, 膨らみは強く, 間室間線条の刻み目はごく弱い; 5) ♂前脛節内側縁中央は軽く膨らむ程度で角張らない; 6) 陰茎先端は長く, 比較的分厚く, 背面に明らかな溝はなく, 痕跡的な陵によってごく弱く縁取られるにすぎない; 7) 指状片の側縁は基部付近で隆起する; 8) ♀交尾器膣底部節片は日高亜種のそれに似るが, 外板はやや幅広い. 知床半島の集団は, 頭部と前胸背板の皺がより顕著で, 陰茎先端は筒状を呈し, 中央から先端にかけて僅かに腹側に湾曲し, 指状片側縁はより強く隆起し, 膣底部節片外板はより短く, 内板には浅い中央溝が認められる.
分布　白糠丘陵, 知床半島(北海道).
亜種小名の由来と通称　亜種小名は伊藤勝彦(根室)に因む. オクエゾクロオサムシという通称で呼ばれることもある.

20-2. *C. (O.) a. hidakanus* Ishikawa et Takami, 1996 クロオサムシ日高亜種

Original description   *Carabus* (*Ohomopterus*) *albrechti hidakanus* Ishikawa & Takami, 1996, Spec. Div., 1, p. 41, figs. 4~6 (p. 40) & 9~14 (p. 42).
Type locality   Biratori, 60 m, Biratori-chô, Hidaka, Hokkaido.
Type depository   Systematic Zoology Laboratory, Department of Biological Sciences, Tokyo Metropolitan University (Hachiôji) [Holotype, ♂].
Diagnosis   20~26 mm. Coloration of upper surface variable, dark coppery, greenish coppery, green, or black; legs black, though tibiae are sometimes dark rufous. 1) Antennae short, not reaching middle of elytra in male, with short and rudimentary hairless ventral areas from segment five to seven; 2) frons not or at most very weakly punctate; 3) pronotum broad and not so strongly convex above, with three to five marginal setae on each side, hind angles short and broad; 4) elytra long oval, strongly convex above, with striae very weakly notched; 5) inner margin of male protibia weakly convex, though not angulate; 6) apical part of penis very long and slender, without dorsal groove, weakly outlined by rudimentary ridge; 7) digitulus rather long; 8) outer plate of ligular apophysis long and narrow, anterior rim of inner plate broad but narrower than in subsp. *albrechti*, without median groove on disk.
Distribution   Hidaka Region and its surrounding area (from Sapporo area to Cape Erimo-misaki), SC. Hokkaido.
Etymology   Named after the Hidaka Region, the type area of this subspecies.

タイプ産地　北海道, 日高, 平取町, 平取, 60 m.
タイプ標本の所在　首都大学東京動物系統分類学研究室(八王子)［ホロタイプ, ♂］.
亜種の特徴　20~26 mm. 背面の色彩は多様で, 暗銅色・緑銅色・緑色・黒色などの変化が見られる. 脚は黒色だが, 脛節は時にやや赤みを帯びる. 1) 触角は短く, ♂でも先端は鞘翅中央に達しない. 無毛部は第5節から第7節に認められるが, 痕跡的で, 長さも各節の1/3~1/2程度; 2) 頭部には点刻を欠くか, あってもごく弱く刻まれるにすぎない; 3) 前胸背板は幅広く, 比較的平坦で, 3ないし5本の側縁剛毛を有し, 後角は幅広く短い; 4) 鞘翅は長卵形で膨隆は強く, 間室間線条の刻み目はごく弱い; 5) ♂前脛節内側縁中央は軽く膨らむが, 角張らない; 6) 陰茎先端はひじょうに細長く, 背側縁の溝を欠き, 痕跡的な陵によってごく弱く縁取られるにすぎない; 7) 指状片は細長い; 8) ♀交尾器膣底部節片外板は長三角形で, 内板の前縁は, 基亜種のそれよりやや狭いものの, 幅広く縁取られ, 中央溝は認められない.
分布　日高地方を中心とする北海道の太平洋沿岸地域(札幌付近から襟裳岬付近).
亜種小名の由来と通称　日高地方に因む. ヒダカクロオサムシという通称で呼ばれることもある.

20-3. *C. (O.) a. albrechti* Morawitz, 1862 クロオサムシ基亜種

Original description, type locality & type depository   See explanation of the species.
Diagnosis   20~26 mm. Body above coppery or dark coppery, sometimes dark green; legs black, though tibiae and tarsi are sometimes somewhat rufous. Identified by the following characters: 1) antennae short, not extending middle of elytra even in male, hairless ventral areas vaguely recognized from segment five to seven, each with less than half the length of each segment; 2) head weakly punctate and coarsely rugulose from frons to neck; 3) pronotum relatively small and cordiform. with three to five (rarely seven to eight) marginal setae, hind angles rather acute, weakly protruded postero-laterally; 4) elytra a little elongated oval, moderately

convex above, with striae very weakly notched; 5) inner margin of male protibia weakly protruded though not angulate; 6) apical part of penis long and slender, weakly constricted at middle in lateral view, weakly bent ventrad toward tip, almost flat or weakly grooved dorsad and weakly outlined by rudimentary ridge; 7) digitulus short and robust; 8) inner plate of ligular apophysis longer than wide, about 1.5 times as long as wide, lacking median groove, frontal margin widely rimmed.
Distribution   Oshima Peninsula & Is. Okushiri-tô, SW. Hokkaido.

原記載・タイプ産地・タイプ標本の所在　種の項を参照.
亜種の特徴　20~26 mm. 背面の色彩は銅色ないし暗銅色で, 時に緑色の個体も見られる. 脚は黒色だが, 脛節と跗節は時にやや赤みを帯びる. 以下の諸点により他亜種から識別される: 1) ♂触角は短く, 先端は鞘翅中央を越えず, 第5節から第7節の腹面には各節の1/2~1/3の長さの弱い無毛部がある; 2) 頭頂部は弱く点刻され, 前頭から頚部にかけて密に皺を装う; 3) 前胸背板は小さく, 側縁は後方で波曲して心臓形を呈し, 3~5本(稀に7~8本)の側縁剛毛を有する. 後角は鋭角で外側後方へ軽く突出する; 4) 鞘翅はやや長い卵形で膨隆は中程度. 間室間線条の刻み目はごく弱い; 5) ♂前脛節内側縁中央は軽く膨らむが, 角張らない; 6) 陰茎先端は細長く, 側方から見て中央部で僅かにくびれ, 先端に向けてごく弱く湾曲する. 背面はほぼ平坦ないし弱い溝を有し, 痕跡的な稜によって縁取られる; 7) 指状片は太短い; 8) ♀交尾器膣底部節片外板は長三角形で, 長幅比は約3対2. 内板には中央溝を認めず, 前縁は幅広く縁取られる. 大千軒岳等, 高所のものでは背面が明銅色を呈するものが多く, 低地や奥尻島の集団には暗銅色のものが多い. また, 奥尻島産の個体は平均してやや大型である.
分布　渡島半島, 奥尻島(北海道).
亜種の通称　クロオサムシの基亜種ではあるが, キタクロオサムシという通称で呼ばれることが多い. ただし, この名は北海道産の本種が3亜種に分割されるまでは道東亜種, 日高亜種に対しても用いられていたので, 注意が必要である.

20-4. *C. (O.) a. tohokuensis* Ishikawa, 1984　クロオサムシ東北地方東部亜種

Original description   *Carabus* (*Ohomopterus*) *albrechti tohokuensis* Ishikawa, 1984, Akitu, Kyoto, (N. S.), (68), p. 3.
Type locality   Morioka-shi, Matsuzono, alt. 200 m.
Type depository   Systematic Zoology Laboratory, Department of Biological Sciences, Tokyo Metropolitan University (Hachiôji) [Holotype, ♂].
Diagnosis   18~25 mm. Body above usually coppery; legs black. 1) Male antennae very short, barely extending basal third of elytra, hairless ventral areas vestigial; 2) head not punctate, only weakly wrinkled partly; 3) pronotum strongly convex above, its lateral margins weakly sinuate in posterior halves, with three to five marginal setae on each side, hind angles robust and very weakly protruded posteriad; 4) elytra robust and strongly convex above, striae between intervals weakly notched; 5) inner margin of male protibia faintly convex at middle; 6) apical part of penis cylindrical and slender, lacking groove on dorsal margin; 7) digitulus long; 8) outer plate of ligular apophysis of female genitalia weakly narrowed toward base, inner plate without median longitudinal groove and broadly rimmed in front.
Distribution   Pacific side of the Tohoku Region (Iwate Pref., E. Aomori Pref., E. Akita Pref. & NE. Miyagi Pref.), NE. Honshu.
Etymology   Named after the Tohoku Region of NE. Honshu.

タイプ産地　盛岡市, 松園, 標高200 m (岩手県盛岡市).
タイプ標本の所在　首都大学東京動物系統分類学研究室(八王子)[ホロタイプ, ♂].
亜種の特徴　18~25 mm. 背面の色彩は通常銅色で, 緑色や黒色の個体は見られない. 脚は黒色. 1) ♂触角は非常に短く, 先端は鞘翅基部1/3をかろうじて越し, 腹面の無毛部は痕跡的で各節の1/3~1/2の長さに亘りごく弱く認められるにすぎない; 2) 頭部には点刻を欠き, 部分的に弱い皺を装うのみ; 3) 前胸背板の膨隆は強く, 側縁は全体的に丸みを帯び, 後方へ向けての波曲は弱く, 3~5本の剛毛を有する. 後角は太短く, 後方への突出はごく弱い; 4) 鞘翅は短く, 側縁は丸みを帯び, 膨隆が強い. 間室間線条には軽微な刻み目を装う; 5) ♂前脛節の内側縁中央部はごく僅かに膨隆する; 6) 陰茎先端は細長い円筒形で, 背側縁には溝を欠く; 7) 指状片は長い; 8) ♀交尾器膣底部節片外板は後方へ向けての狭まりが弱く, 内板は中央溝を欠き, 前縁は幅広く縁取られる.
分布　東北地方太平洋側(岩手県, 青森県東部, 秋田県東部, 宮城県北東部).
亜種小名の由来と通称　東北地方に因む. トウホククロオサムシという通称があるが, 東北地方には複数の本種の亜種を産するので紛らわしいうえ, 東北地方産の本種が複数の亜種に分割されるまでは以下の数亜種に対してもこの通称が用いられていた可能性があるため, 注意が必要である.

20-5. *C. (O.) a. yamauchii* Takami et Ishikawa, 1997　クロオサムシ白神山地亜種

Original description   *Carabus* (*Ohomopterus*) *albrechti yamauchii* Takami & Ishikawa, 1997, TMU Bull. nat. Hist, Tokyo, (3), p. 65, figs. 22, 23 (p. 65) & 27, 30 (p. 66).
Type locality   Mt. Shikarigadake, Ajigasawa-machi, Aomori Prefecture.

Type depository   Systematic Zoology Laboratory, Department of Biological Sciences, Tokyo Metropolitan University (Hachiôji) [Holotype, ♂].

Diagnosis   20~23 mm. Body above coppery, legs black. 1) Male antennae barely reaching middle of elytra, with hairless ventral areas from segment five to seven, though that of segment five is rudimentary; 2) frons sometimes punctate; 3) pronotum narrow, its lateral margins arcuate in front and gently sinuate toward base, with three to five marginal setae, hind angles acute, rather distinctly protruded postero-laterad; 4) elytra relatively slender and less strongly convex above, striae between intervals weakly notched; inner margin of male protibia gently protruded at middle; 5) apical part of penis slender, its dorsal margin sometimes grooved and weakly outlined by rudimentary ridge; 6) digitulus slender; 7) outer plate of ligular apophysis gently narrowed toward base, inner plate with shallow median longitudinal groove and widely rimmed in front.

Distribution   Shirakami Mts. on the borders between Aomori Pref. and Akita Pref., NE. Honshu.

Etymology   Named after Satoshi YAMAUCHI who collected a part of the type series.

タイプ産地　青森県, 鰺ヶ沢町, 然ヶ岳.

タイプ標本の所在　首都大学東京動物系統分類学研究室(八王子)[ホロタイプ, ♂].

亜種の特徴　20~23 mm. 背面は銅色で, 脚は黒色. 1) ♂触角は辛うじて鞘翅中央に達する程度の長さで, 第5節から第7節にかけて無毛部があるが, 第5節のものは痕跡的; 2) 頭頂部は時に点刻を装う; 3) 前胸背板の幅は狭く, 側縁は前方で弧を描き, 後方に向けて緩やかに波曲し, 側縁剛毛は3~5本. 後角は鋭角で, 比較的顕著に突出する; 4) 鞘翅は細めで膨隆は弱く, 間室間の線条には弱い刻み目を装う; 5) ♂前脛節内側縁は中央部で軽く膨らむ; 6) 陰茎先端は細長く, 背側縁には時に溝を有し, 痕跡的な陵によって縁取られる; 6) 指状片は細長い; 7) ♀交尾器膣底部節片の外板は後方に向けて緩やかに狭まり, 内板は浅い中央溝を有し, 前縁の縁取りは幅広い.

分布　白神山地.

亜種小名の由来と通称　タイプシリーズの一部を採集した山内 智に因む. シラカミクロオサムシ, ヤマウチクロオサムシ等の通称で呼ばれることもある.

20-6. *C. (O.) a. hagai* TAKAMI et ISHIKAWA, 1997　クロオサムシ東北地方中部亜種

Original description   *Carabus* (*Ohomopterus*) *albrechti hagai* TAKAMI & ISHIKAWA, 1997, TMU Bull. nat. Hist, Tokyo, (3), p. 67, figs. 31~34 (p. 68) & 39, 40, 43, 44 (p. 69).

Type locality   Shinmachi (more precisely, Shinchô), Yuzawa-shi, Akita Prefecture.

Type depository   Systematic Zoology Laboratory, Department of Biological Sciences, Tokyo Metropolitan University (Hachiôji) [Holotype, ♂].

Diagnosis   19~24 mm. Body above dark coppery; legs black, though tibiae are a little reddish. 1) Male antennae barely reaching middle of elytra, with hairless ventral areas from segment five to seven; 2) head weakly punctate; 3) pronotum cordate, with three to five marginal setae, hind angles rather distinctly protruded and rounded at tips; 4) elytra slender, weakly convex above, striae between intervals strongly notched; 5) inner margin of male protibia protruded at middle; 6) apical part of penis not so elongated, with narrow flat part outlined by rudimentary ridges on its outer margin; 7) digitulus long; 8) outer plate of ligular apophysis triangular in shape, inner plate with shallow median groove and widely rimmed in front.

Distribution   C. Tohoku Region (S. Akita Pref., Yamagata Pref., SC. Miyagi Pref., N. Fukushima Pref. & N. Niigata Pref.), NE. Honshu / Iss. Izu-shima & Kinkasan (Miyagi Pref.) ? The main distributional range of this subspecies is Yamagata Pref. and Miyagi Pref., with the southwestern margin defined by Riv. Agano-gawa of Niigata Pref., intergrading to subsp. *tohokuensis* in C. Akita Pref., and to subsp. *tsukubanus* in NE. Fukushima Pref. Subspecific account of the Izu-shima and Kinkasan populations is still tentative.

Etymology   Named after Kaoru HAGA who collected a part of the type series.

タイプ産地　秋田県, 湯沢市, 新町. 原記載論文ではタイプ産地名が "Shinmachi" と綴られているが, 正しくは「しんちょう」と読むべきもののようである.

タイプ標本の所在　首都大学東京動物系統分類学研究室(八王子)[ホロタイプ, ♂].

亜種の特徴　19~24 mm. 背面の色彩は暗銅色で, 脚は黒色だが脛節はやや赤みを帯びる. 1) ♂触角の先端はかろうじて鞘翅中央に達する程度で, 第5節から第7節の腹面には各節の1/2~1/3の長さの無毛部を有する; 2) 頭部の点刻は弱い; 3) 前胸背板側縁は後方に向けて波曲し, 3~5本の側縁剛毛を有する. 後角の突出は比較的顕著で, 先端は丸い; 4) 鞘翅は細長く, 膨隆は弱く, 間室間の線条には強い刻み目を具える; 5) ♂前脛節内側縁は中央部で膨隆する; 6) 陰茎先端はそれほど細長くなく, 背面外側基部には弱い陵によって縁取られた狭い平坦部分がある; 7) 指状片は長い; 8) ♀交尾器膣底部節片外板は三角形で, 内板には浅い中央溝があり, 前縁の縁取りは幅広い.

分布　東北地方中部(秋田県南部~山形県~宮城県中南部~福島・新潟両県北部(阿賀野川以北)/出島, 金華山(以上, 宮城県)？ 山形・宮城両県を中心とする東北地方中部に分布し, 南西限は新潟県の阿賀野川. 秋田県中部で東北地方東部亜

種に，また福島県北東部で関東地方北東部亜種に移行する．出島と金華山に分布する集団の亜種分類は暫定的である．
　亜種小名の由来と通称　タイプシリーズの一部を採集した芳賀 馨に因む．ハガクロオサムシという通称で呼ばれることもある．

20-7. *C. (O.) a. echigo* Ishikawa et Takami, 1996 クロオサムシ越後亜種

Original description　*Carabus* (*Ohomopterus*) *albrechti echigo* Ishikawa & Takami, 1996, Spec. Div., 1, p. 44, figs. 19~24, 27~29 (p. 45).
　Type locality　Mugurazawa (more precisely, Mugurasawa), 190 m, Yunotani-mura (= present Uonuma-shi), Niigata Prefecture.
　Type depository　Systematic Zoology Laboratory, Department of Biological Sciences, Tokyo Metropolitan University (Hachiôji) [Holotype, ♂].
　Diagnosis　19~23 mm. Body above variable in color, light-, dark-, greenish coppery or black; legs and palpi black. 1) Male antennae extending middle of elytra, with hairless areas from segment five to seven on ventral side, length of each area about half to two-thirds of each antennal segment; 2) inner margin of male protibia produced and angulate; 3) pronotum with three to four (rarely five) marginal setae, posterior parts of lateral margins distinctly emarginate posteriorly behind middle, hind angles strongly protruded with rounded tips; 4) elytra slender, widest behind middle, with shoulders effaced, sides nearly straight before widest part above all in male, and disc weakly convex above; striae between intervals very weakly notched; 5) apical part of penis rather broad in lateral view, with distinct groove on dorsal margin; 6) digitulus sharply pointed at tip; 7) outer plate of ligular apophysis triangular though rather narrow, inner plate with broader outer rim and shallow median groove.
　Distribution　Niigata Pref. and its nearby regions, NC. Honshu. Northern margin is defined by Riv. Agano-gawa, southern limit may be northern Nagano Pref. (from Hakuba-mura to Iiyama-shi).
　Etymology　Named after Echigo or Echigo-no-kuni, one of the ryôsei provinces of Japan used until the early Meiji period, which corresponds to the present Niigata Prefecture except for island areas.

　タイプ産地　新潟県，湯之谷村，葎沢，190 m（合併により現在は魚沼市の一部）．原記載ではタイプ産地名が "Mugurazawa" と記されているが，正式には「むぐらさわ」と濁らずに読むべきもののようである．
　タイプ標本の所在　首都大学東京動物系統分類学研究室（八王子）［ホロタイプ，♂］．
　亜種の特徴　19~23 mm．背面の色彩には明ないし暗銅色，緑銅色，黒色等の変化があり，口肢と脚は黒色．1)♂触角の先端は鞘翅中央よりも後方に達し，第5節から第7節の腹面に浅い無毛部を有し，無毛部の長さは各節の1/2ないし2/3；2)♂前脛節内側縁は明瞭に角張る；3)前胸背板には3ないし4本（稀に5本）の側縁剛毛を有し，側縁は最広部より後方で顕著にえぐれ，後角は強く後方に突出し，先端は丸まる；4)鞘翅の幅は狭く，最広部は中央より後方にあり，膨隆は弱い．肩部の張り出しは弱く，側縁は前方部分でほぼ直線的で，この傾向は♂において著しい．間室間の線条にはごく弱い刻み目を有する；5)陰茎先端は側方から見て比較的幅広く，背側縁には顕著な溝を有する；6)指状片先端は鋭く尖る；7)♀交尾器膣底部節片外板は長三角形，同内板は円形で幅広く縁取られ，中央の溝は浅い．
　分布　新潟県とその周辺地域（北限は阿賀野川，南限は長野県北部の白馬村〜犀川（さいがわ）〜飯山市にかけての一帯）．
　亜種小名の由来と通称　亜種小名は，かつての令制国（りょうせいこく）の一つで，現在の新潟県の本州部分に相当する越後国（えちごのくに）に因む．エチゴクロオサムシという通称で呼ばれることもある．

20-8. *C. (O.) a. awashimae* Ishikawa et Takami, 1996 クロオサムシ粟島亜種

Original description　*Carabus* (*Ohomopterus*) *albrechti awashimae* Ishikawa & Takami, 1996, Spec. Div., 1, p. 46, figs. 25, 26, 30 (p. 45).
　Type locality　Awashima Is., Niigata Prefecture.
　Type depository　Systematic Zoology Laboratory, Department of Biological Sciences, Tokyo Metropolitan University (Hachiôji) [Holotype, ♂].
　Diagnosis　20~25 mm. Body above dark brownish or dark reddish coppery; legs and palpi black. 1) Head weakly punctate; 2) male antennae short, barely extending middle of elytra, with weak hairless ventral areas from segment five to seven; 3) pronotum sparsely punctate in median portion, with three to four marginal setae, hind angles prominently protruded though rounded at tips; 4) elytra weakly convex above, with shoulders distinct, striae between intervals strongly notched; 5) inner margin of male protibia protruded at middle though not angulate; 6) apical part of penis thicker, with flat area on its outer margin; 7) digitulus with rounded apex, constricted at base; 8) outer plate of ligular apophysis gently narrowed posteriad, inner plate with shallow median groove and broad anterior rim.
　Distribution　Is. Awa-shima (Niigata Pref.).
　Etymology　Named after the type locality, Is. Awa-shima of Niigata Prefecture.

　タイプ産地　新潟県，粟島．

タイプ標本の所在　首都大学東京動物系統分類学研究室(八王子)[ホロタイプ, ♂].

亜種の特徴　20~25 mm. 背面の色彩は暗茶褐色ないし暗赤褐色で, 緑化型や黒化型は知られていない. 脚および口肢は黒色. 1)頭部の点刻は弱い; 2)♂触角は短く, 先端はかろうじて鞘翅中央に達し, 第5節から第7節の腹面に弱い無毛部を有する; 3)前胸背板は中央部で疎に点刻され, 3ないし4本の側縁剛毛を有し, 後角は顕著に後方へ突出するが先端は丸い; 4)鞘翅の膨隆は弱く, 肩部の張り出しは顕著で, 間室間線状は強い刻み目を有する; 5)♂前脛節内側縁は中央で膨隆するが角張らない; 6)陰茎先端は他亜種に比しやや太めで, 背面に平坦部を有する; 7)指状片の先端は鈍く丸まり, 側縁は基部におけるくびれが顕著; 8)♀交尾器膣底部節片外板は後方に向けて緩やかに狭まり, 内板は浅い中央溝を有し, 前縁は幅広く縁取られる.

分布　粟島(新潟県).

亜種小名の由来と通称　粟島に因む. アワシマクロオサムシという通称で呼ばれることもある.

20-9. *C. (O.) a. freyi* VAN EMDEN, 1932　クロオサムシ佐渡島亜種

Original description　*Carabus* (*Ohomopterus*) *Albrechti Freyi* VAN EMDEN, 1932, Neue Beitr. syst. Insektenkunde, 5 (4/5), p. 62, fig. 5 (tab. 1).

Type locality　Japan: Insel Sado.

Type depository　Unknown. (Staatliches Museum für Tierkunde, Dresden ?)

Diagnosis　19~25 mm. Body above variable in color, dark coppery to dark green; legs and palpi black. 1) Head hardly punctate; 2) male antennae slightly extending middle of elytra, with hairless ventral areas from segment five to seven; 3) pronotum with punctures in median portion fused with one another to form wrinkles, pronotal margins with three to four setae on each side, hind angles strongly produced posteriad; 4) elytra long and slender, weakly convex above, nearly parallel-sided in male, striae between intervals weakly notched; 5) inner margin of male protibia apparently angulate at middle; 6) apical part of penis with flat area on its outer margin; 7) digitulus long, feebly constricted at base; 8) outer plate of ligular apophysis broad and triangular, inner plate with anterior rim broad, smooth at bottom without median groove.

Distribution　Is. Sado-ga-shima (Niigata Pref.).

Etymology　Named after Georg FREY (1902~1976) of Munich.

タイプ産地　日本, 佐渡島(新潟県佐渡市).

タイプ標本の所在　不明. 著者 VAN EMDEN (ファン・エムデン)のコレクションは, 国立動物学博物館(ドレスデン)に保管されている可能性が高いが, 本亜種のタイプ標本の所在は確認できていない.

亜種の特徴　19~25 mm. 背面は多彩で, 暗銅色から暗緑色に至る変異が見られ, 脚および口肢は黒色. 1)頭部にはほぼ点刻を欠く; 2)♂触角先端は僅かに鞘翅中央を越え, 第5節から第7節の腹面に無毛部がある; 3)前胸背板中央部の点刻は互いに融合して皺をなし, 側縁剛毛は3~4本で, 後角は後方に向けて強く突出する 4)鞘翅は細長く, 膨隆は弱く, ♂では側縁が直線的で左右ほぼ平行になる. 間室間線状の刻み目は弱い; 5)♂前脛節内側縁は中央部で明らかに角張る; 6)陰茎先端外側縁には平坦部がある; 7)指状片は長く, 側縁は基部において僅かにくびれる; 8)♀交尾器膣底部節片外板は幅広い三角形. 内板の表面は平滑で中央溝を欠き, 前縁の縁取りは幅広い.

分布　佐渡島(新潟県).

亜種小名の由来と通称　亜種小名はミュンヘンの Georg FREY (ゲオルク・フレイ) (1902~1976)に因む. コガネムシ類の専門家である一方, 資産家で, 私設の博物館(G. フレイ博物館)を設立し, Entomologische Arbeiten という昆虫学の雑誌も刊行した. 15万種300万頭(うち約2万頭のタイプ標本を含む)を擁すると言われる同博物館のコレクションは世界的に有名で, 彼の死後, 若干の曲折を経て現在ではスイスのバーゼル自然史博物館に収められている. サドクロオサムシという通称が一般的だが, ホソアカガネオサムシ佐渡島亜種と同じサドオサムシという名が本亜種に対して用いられたこともあるため, 通称による呼称は紛らわしく, かつ意図する分類群を正しく示すことのできない危険性を孕んでいる.

20-10. *C. (O.) a. esakianus* (NAKANE, 1961)　クロオサムシ関東地方北西部亜種

Apotomopterus albrechti esakii: NAKANE, 1952a, Shin-Konchû, 5 (6), p. 13; range: "Kantô" [Nec CSIKI,1927].
Apotomopterus albrechti race *esakii*: NAKANE, 1952b, Shin-Konchû, 5 (11), p. 50, fig. 32; range: "Kantô Plain;Takao-san" [Nec CSIKI, 1927].
Apotomopterus albrechti esakii: NAKANE, 1953, Sci. Rep. Saikyo Univ., (Nat. Sci. & Liv. Sci.), 1 (2), 96, fig. 13, D-d, D-5 (Ibaragi (Hitachi)), D-e, D-6 (Masutomi, Kai); range: "Kwanto District including Tokyo" [Nec CSIKI,1927].
Apotomopterus albrechti esakianus NAKANE, 1961, Sci. Rep. Kyoto Pref. Univ. (Nat. Sci. & Liv. Sci.), 3 (2), p. A99 (= p. 27): "the Kwanto Plain".
Carabus (*Ohomopterus*) *albrechti esakianus* (NAKANE, 1961): TAKAMI & ISHIKAWA, 1997, TMU Bull. nat. Hist, Tokyo, (3), p. 71, (Lectotype (♂) designation; type locality: Takaosan).

Type locality　"Takaosan" (designated by TAKAMI & ISHIKAWA (1997)).

Type depository　Hokkaido University Museum, Sapporo (?) [Lectotype, ♂].

Diagnosis 19~24 mm. Body above polychromatic, bright-, dark- or greenish coppery, green or black; tibiae and tarsi blackish, sometimes a little reddish. 1) Male antennae long, extending middle of elytra, with distinct hairless ventral areas from segment five to seven; 2) head hardly punctate; 3) pronotum appearing round in outline, lateral margins with three to six setae, hind angles strongly produced and rounded at tips; 4) elytra relatively slender, less roundly arcuate in front, striae between intervals weakly notched; 5) inner margin of male protibia protruded at middle; 6) apical part of penis slender, weakly grooved on dorsal margin; 7) digitulus rather short; 8) outer margin of ligular apophysis triangular, inner plate with shallow median groove and broadly rimmed in front.

Distribution NW. Kanto Region (SW. Tochigi Pref., Gunma Pref., W. Saitama Pref., W. Tokyo Met., W. Kanagawa Pref., NE. Yamanashi Pref.) & CW. Bôsô Penins. of S. Chiba Pref., C. Honshu.

Etymology Named after Teizo ESAKI (1899~1957), one of the famous entomologists of Japan representing the first half of the Showa period.

タイプ産地　高尾山（東京都八王子市）．中根（1952）により初めて本亜種の解説がなされた時点では，「関東」という漠然とした地域名が指定されていたにすぎなかったが，TAKAMI & ISHIKAWA (1997) によって高尾山産の♂がレクトタイプに指定されている．

タイプ標本の所在　北海道大学総合博物館（札幌）（？）［レクトタイプ，♂］．TAKAMI & ISHIKAWA (1997) は「中根（1952b）において最初に指状片が図示された個体」を本亜種のレクトタイプに指定した．当該標本は恐らく，北海道大学総合博物館の中根コレクション中に保管されているものと思われるが，どの個体の指状片が図示されたかが判然としないため，レクトタイプを特定することができない．従って，その所在に関しては「？」を付けておく．

亜種の特徴　19~24 mm. 背面は多彩で，明銅色，暗銅色，緑銅色，緑色，黒色などの変化がある．脛節と附節はほぼ黒色で，時に多少赤味を帯びる．1)♂触角先端は鞘翅中央を越え，第5節から第7節の腹面に顕著な無毛部を持つ；2) 頭部は前頭から頸部にかけて殆ど点刻されない；3) 前胸背板側縁は丸く弧を描き，3~6本の側縁剛毛を持ち，後角は後方に強く突出し，先端は丸い；4) 鞘翅は比較的細長く，側縁前方の膨らみが弱く，間室間線条の刻み目は弱い；5)♂前脛節内側縁は中央で角張る；6) 陰茎先端は細長い指状で，背側縁には弱い溝がある；7) 指状片は比較的短い；8)♀交尾器膣底部節片外板は三角形．内板は浅い中央溝を有し，前縁の縁取りは広い．

分布　関東平野北西辺縁部の丘陵～山地帯（栃木県南西部～群馬県の大半～埼玉・東京・神奈川各都県の西部～山梨県北東端および千葉県房総半島中西部）．

亜種小名の由来と通称　亜種小名は，昭和時代前半を代表する昆虫学者の一人，江崎悌三（1899~1957）に因む．通称はエサキオサムシ．トウキョウオサムシという名が提唱されたこともあるが，殆ど使用されていない．

20-11. *C. (O.) a. tsukubanus* TAKAMI et ISHIKAWA, 1997　クロオサムシ関東地方北東部亜種

Original description *Carabus (Ohomopterus) albrechti tsukubanus* TAKAMI & ISHIKAWA, 1997, TMU Bull. nat. Hist, Tokyo, (3), p. 75, figs. 51~54 (p. 76) & 57, 58, 61, 62 (p. 77).

Type locality Mt. Tsukubasan, Ibaraki Prefecture.

Type depository Systematic Zoology Laboratory, Department of Biological Sciences, Tokyo Metropolitan University (Hachiôji) [Holotype, ♂].

Diagnosis 19~24 mm. Body above dark coppery or dark green and less lustrous; legs black, though tibiae are sometimes partly reddish. 1) Male antennae slightly extending middle of elytra, with hairless ventral areas from segment five to seven; 2) head almost impunctate; 3) pronotum a little more strongly convex above than in subsp. *esakianus*, with four to six marginal setae, hind angles strongly produced posteriad and rounded at tips; 4) elytra slender and less strongly convex above, with shoulders a little more effaced than in subsp. *esakianus*, striae between intervals weakly notched; 5) inner margin of male protibia protruded, though not strongly angulate; 6) apical part of penis a little slenderer than in subsp. *esakianus*, with dorsal margin flat or very weakly grooved; 7) digitulus relatively short; 8) outer plate of ligular apophysis triangular and broad, inner plate with shallow median groove and broad anterior rim.

Distribution N. Kanto Region to Fukushima Pref. incl. Tsukuba Mts., Yamizo Mts. & Abukuma Hills, C. Honshu.

Etymology Named after Mt. Tsukuba-san.

タイプ産地　茨城県，筑波山．

タイプ標本の所在　首都大学東京動物系統分類学研究室（八王子）［ホロタイプ，♂］．

亜種の特徴　19~24 mm. 背面の色彩は暗銅色ないし黒緑色で，光沢は比較的鈍い．分布域北部では黒緑色の個体は殆ど見られない．脚は黒色だが，時に脛節が部分的に赤みを帯びる個体が見られる．1)♂触角先端は鞘翅中央を僅かに越え，第5節から第7節の腹面に無毛部がある；2) 頭部には点刻をほぼ欠く；3) 前胸背板は関東地方北西部亜種のそれよりもやや強く膨隆し，4~6本の側縁剛毛を具え，後角は後方へ向けて強く突出し，先端は丸まる；4) 鞘翅は細長く，膨隆は強くなく，肩部の張り出しは関東地方北西部亜種のそれに比しやや弱く，間室間線状の刻み目は弱い；5)♂前脛節内側縁は膨らむが，あまり強く角張らない；6) 陰茎先端は関東地方北西部亜種のそれよりもやや細く，背側縁は平坦になるか弱く狭い溝を認める；7) 指状片は比較的短い；8)♀交尾器膣底部節片外板は幅広い三角形で，内板には浅い中央溝があり，前縁の縁取りは幅広い．

分布　関東地方北部山地から福島県のほぼ全域にかけての地域(筑波山, 八溝山, 阿武隈高地を含む).
亜種小名の由来と通称　筑波山に因む. ツクバクロオサムシという通称で呼ばれることもある.

20-12. *C. (O.) a. okumurai* (Ishikawa, 1966)　クロオサムシ山梨長野亜種

Original description　*Apotomopterus albrechti okumurai* Ishikawa, 1966, Bull. natn. Sci. Mus., Tokyo, 9 (4), p. 451, figs.1~5 (p. 452) & 1, 2 (pl. 2).
Type locality　Mt. Daibosatsu, Yamanashi Pref.
Type depository　National Museum of Nature and Science, Tokyo [Holotype, ♂].
Diagnosis　17~23 mm.　Body above variable in color, light-, dark-, greenish- or bluish coppery, or entirely black, and usually shiny; appendages reddish. 1) Male antennae long, apparently extending middle of elytra, with long and distinct hairless ventral areas from segment five to seven, each area almost as long as each antennal segment; 2) pronotum with three to six marginal setae on each side, hind angles strongly protruded with rounded tips; 3) elytra strongly convex above, with lateral margins evenly arcuate, apices remarkably re-entrant at middle because of rounded apex of each elytron, striae between intervals weakly notched; 4) inner margin of male protibia distinctly angulate; 5) apical part of penis robust, with marked groove outlined by ridge on dorsal margin; 6) digitulus slender, narrowed at base, with tip obtusely rounded; 7) outer plate of ligular apophysis broad triangular in shape, inner plate shallowly grooved in median portion and narrowly rimmed in front.
Distribution　Greater part of Nagano Pref. & Yamanashi Pref., C. Honshu.
Etymology　Named after Takashi Okumura (Yokohama), a member of the research group on carabid beetles in Keihin Insect Lovers' Club, Tokyo.

タイプ産地　山梨県, 大菩薩(嶺).
タイプ標本の所在　国立科学博物館[ホロタイプ, ♂].
亜種の特徴　17~23 mm. 背面は多彩で, 明銅色, 暗銅色, 緑銅色, 青銅色ないし黒色で, 一般に光沢が強い. 口肢, 脛節および附節は通常赤みを帯びる. 1)♂触角は長く, 先端は明らかに鞘翅中央を越え, 第5節から第7節の腹面に各節とほぼ等長の無毛部を認める; 2)前胸背板には3~6本の側縁剛毛を具え, 後角は強く突出し, 先端は丸い; 3)鞘翅の膨隆は強く, 側縁はほぼ均等に弧を描き, 先端が顕著に丸まるため, 左右の鞘翅端は解離する. 間室間線条の刻み目は弱い; 4)♂前脛節内側縁中央部は顕著に角張る; 5)陰茎先端部は太短く, 背側縁には陵によって縁取られた顕著な溝を有する; 6)指状片は細く, 側縁は基部で狭まり, 先端は鈍く丸まる; 7)♀交尾器膣底部節片外板は幅広い三角形で, 内板には浅い中央溝があり, 前縁の縁取りは狭い.
分布　長野・山梨両県の大部分. 本種の分布域の南西端を占める亜種で, 分布の北限である長野県北部(白馬村(はくばむら)~犀川~飯山市(いいやまし)一帯)で越後亜種に移行し, 同県の上田市では千曲川(ちくまがわ)の支流の神川(かんがわ)を挟んで関東地方北西部亜種と, また山梨県東部ではルイスオサムシと分布を接している.
亜種小名の由来と通称　亜種小名は京浜昆虫同好会オサムシグループの一員として活躍した奥村 尚に因む. 一般的な通称はマルバネオサムシだが, マルバネクロオサムシ, マルバネヒメオサムシ, オクムラクロオサムシ等の名で呼ばれることもある.

20. *Carabus* (*Ohomopterus*) *albrechti*

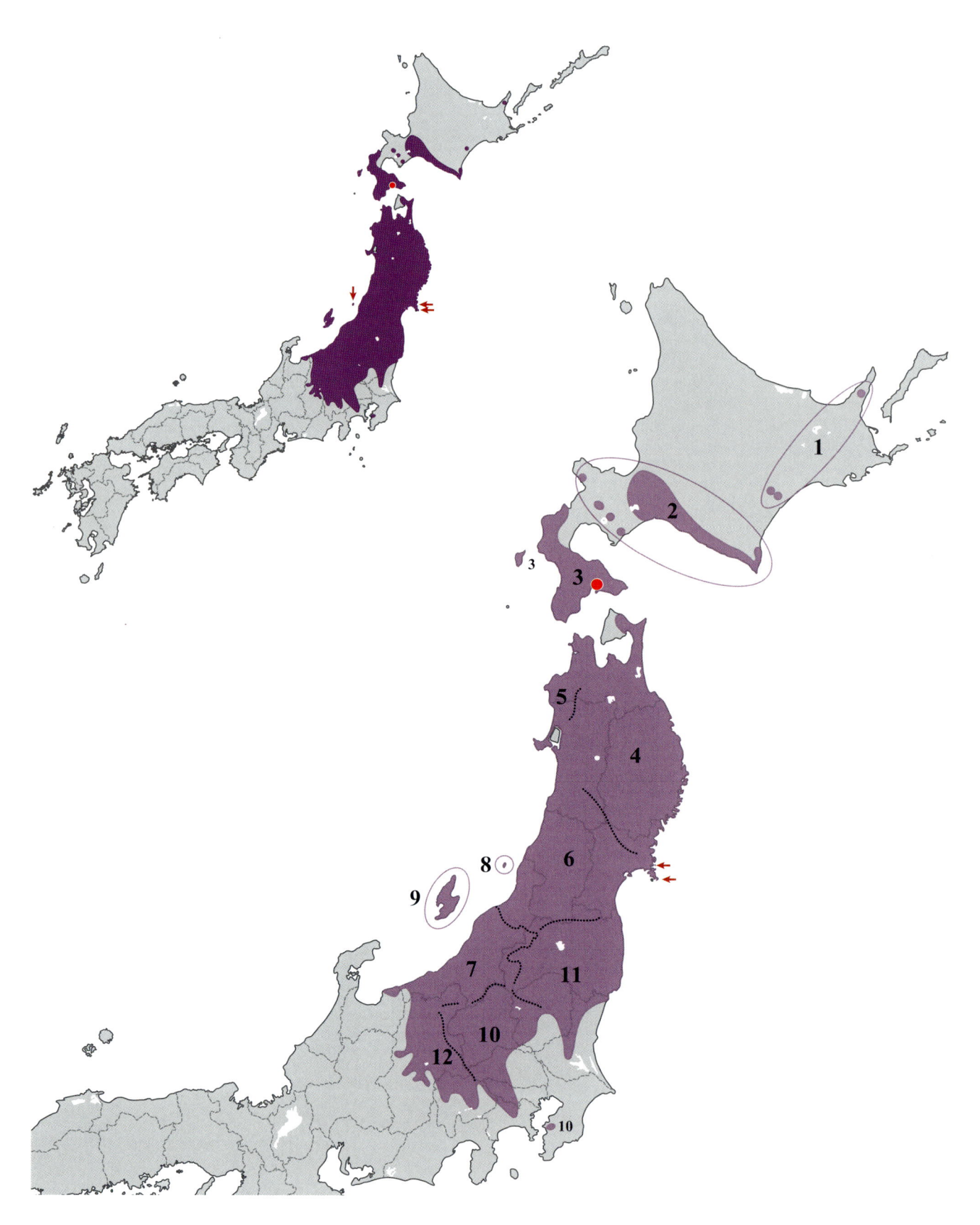

Fig. 42. Map showing the distributional range of ***Carabus*** (***Ohomopterus***) ***albrechti*** Morawitz, 1862. — 1, Subsp. ***itoi***; 2, subsp. ***hidakanus***; 3, subsp. ***albrechti***; 4, subsp. ***tohokuensis***; 5, subsp. ***yamauchii***; 6, subsp. ***hagai***; 7, subsp. ***echigo***; 8, subsp. ***awashimae***; 9, subsp. ***freyi***; 10, subsp. ***esakianus***; 11, subsp. *dentatohumus*; 12, subsp. *akaiwai*; ●, Type locality of the species (Hakodade = Hakodaté).

図 42. クロオサムシ分布図. — 1, 道東亜種; 2, 日高亜種; 3, 原亜種; 4, 東北地方東部亜種; 5, 古114山地亜種; 6, 木地帯, 11, 本地方中央部亜種; 7, 越後亜種; 8, 粟島亜種; 9, 佐渡島亜種; 10, 関東地方北西部亜種; 11, 関東地方北東部亜種; 12, 山梨長野亜種. ●・種のタイプ産地 (函館).

20. *Carabus* (*Ohomopterus*) *albrechti*

Fig. 43. Habitat of *Carabus* (*Ohomopterus*) *albrechti hidakanus*, *C.* (*Carabus*) *granulatus yezoensis*, *C.* (*Leptocarabus*) *arboreus pararboreus*, *C.* (*Asthenocarabus*) *opaculus opaculus* and *C.* (*Damaster*) *blaptoides rugipennis* (Shizunai-mauta, Shin-hidaka-chô, Hokkaido).

図43. クロオサムシ日高亜種, アカガネオサムシ北海道亜種, コクロナガオサムシ道央道東道北亜種, ヒメクロオサムシ基亜種, マイマイカブリ北海道亜種の生息地(北海道新ひだか町静内真歌).

Fig. 44. *Carabus* (*Ohomopterus*) *albrechti hidakanus* feeding on earthworm (Takaé, Niikappu-chô, Hokkaido).

図44. 側溝でミミズを捕食中のクロオサムシ日高亜種(北海道新冠町高江).

Fig. 45. *Carabus* (*Ohomopterus*) *albrechti esakianus* passing the winter in the soil (Mt. Karasawa-yama, Sano-shi, Tochigi Pref.).

図45. 土中で越冬しているクロオサムシ関東地方北西部亜種(栃木県佐野市唐沢山).

21. *Carabus* (*Ohomopterus*) *daisen* (Nakane, 1953)
ダイセンオサムシ

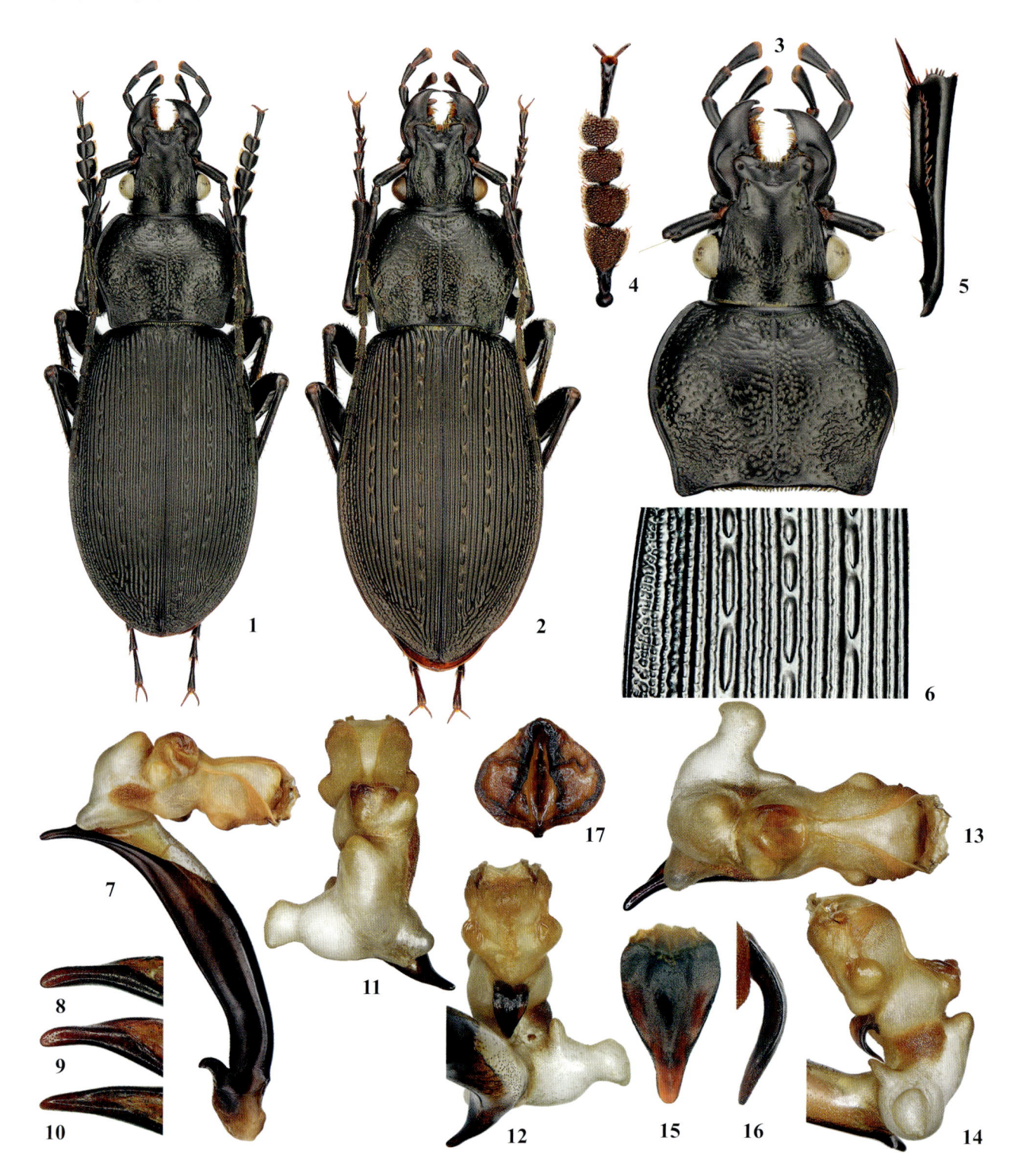

Figs. 1-17. *Carabus* (*Ohomopterus*) *daisen daisen* (Nakane, 1953) (Mt. Dai-sen, Daisen-chô, Tottori Pref.). — 1, ♂; 2, ♀; 3, male head & pronotum; 4, male right protarsus in ventral view; 5, male right protibia in subdorsal view; 6, male left elytron (median part); 7, penis & fully everted internal sac in right lateral view; 8, apical part of penis in right lateral view; 9, ditto in right subdorsal view; 10, ditto in dorsal view; 11, internal sac in caudal view; 12, ditto in frontal view; 13, ditto in dorsal view; 14, ditto in left sublateral view; 15, digitulus in frontal view; 16, ditto in right lateral view; 17, inner plate of ligular apophysis.

21. *Carabus* (*Ohomopterus*) *daisen*

Figs. 18-29. ***Carabus*** (***Ohomopterus***) ***daisen*** subspp. — 18-23, ***C***. (***O***.) ***d***. ***daisen*** (18-20, 22, ♂; 21, 23, ♀; 18, Kuto-yama, Shin-onsen-chô, Hyogo Pref.; 19, Yubara-onsen, Maniwa-shi, Okayama Pref.; 20, Hiruzen-shimo-tokuyama, Maniwa-shi, Okayama Pref.; 21, Mt. Makuragi-san, Matsué-shi, Shimane Pref.; 22-23, Mt. Sanbé-san, Ôda-shi, Shimane Pref.); 24-29, ***C***. (***O***.) ***d***. ***okianus*** (24, 26-29, ♂, 25, ♀; 24-25, Ikeda, Is. Dôgo-jima; 26, Fusé, Is. Dôgo-jima; 27, Nakazato, Is. Naka-no-shima; 28, Mita, Is. Nishi-no-shima; 29, Nibu, Is. Chiburi-jima; all from Iss. Oki). — a, Penis & fully everted internal sac in right lateral view; b, apical part of penis in the same view; c, ditto in right subdorsal view; d, ditto in dorsal view; e, internal sac in frontal view; f, ditto in dorsal view; g, digitulus in frontal view; h, ditto in right lateral view; i, inner plate of ligular apophysis.

図 18-29. ダイセンオサムシ各亜種. — 18-23, **基亜種**(18-20, 22, ♂, 21, 23, ♀; 18, 兵庫県新温泉町久斗山; 19, 岡山県真庭市湯原温泉; 20, 岡山県真庭市蒜山下徳山; 21, 島根県松江市枕木山; 22-23, 島根県大田市三瓶山); 24-29, **隠岐亜種**(24, 26-29, ♂, 25, ♀; 24-25, 島後島池田; 26, 同島布施; 27, 中ノ島中里; 28, 西ノ島三田; 29, 知夫里島仁夫; 以上すべて隠岐諸島産). — a, 陰茎と完全反転下の内嚢(右側面観); b, 陰茎先端(右側面観); c, 同(右斜背面観); d, 同(背面観); e, 内嚢(頭側面観); f, 同(背面観); g, 指状片(頭側面観); h, 同(右側面観); i, 膣底部節片内板.

21. *Carabus* (*Ohomopterus*) *daisen* (NAKANE, 1953) ダイセンオサムシ

Original description　*Apotomopterus japonicus* subsp.? *daisen* NAKANE, 1953, Scient. Rept. Saikyo Univ., Kyoto, (Nat. Sci. & Liv. Sci.), 1, p. 96, figs. 17, E-c, c' & E-3, 4 (p. 102).

History of research　This taxon was originally described by NAKANE (1953) as a subspecies of *Apotomopterus japonicus* (= *Carabus* (*Ohomopterus*) *japonicus* in the present sense) based on eight specimens without designation of the holotype. Of these, a male from Mt. Dai-sen was designated as the lectotype by IMURA (2005b). In current classification, this race is regarded as an independent species in view of its uniquely shaped penis. *Apotomopterus japonicus okianus* also described by NAKANE (1961) from the Oki Islands is regarded as its subspecies (*C.* (*O.*) *daisen okianus* in the present sense).

Type locality　Daisen (designated by IMURA (2005b)).

Type depository　Hokkaido University Museum (Sapporo) [Lectotype, ♂].

Morphology　20~25 mm. Body above black, black with greenish tinge or coppery. External features: Figs. 1~6. Genitalic features: Figs. 7~17. Allied to *C.* (*O.*) *albrechti*, but discriminated from that species in the following points: 1) body a little wider and robuster, above all in pronotum which is relatively larger and broader; 2) male antennae with wide and remarkable hairless ventral depressions from segment five to seven; 3) elytral sculpture with secondary and tertiary intervals more remarkably crenulate, striae between intervals more distinctly punctate; 4) metacoxa lacking inner seta; 5) apical lobe of penis a little robuster and a little more strongly bent ventrad; 6) digitulus with basal portion wider and more acutely narrowed apicad in frontal view, thinner and less strongly arcuate in lateral view; 7) female genitalia chestnut-shaped.

Molecular phylogeny　For an overview of the molecular phylogeny of *Ohomopterus*, see the same section of *C.* (*O.*) *yamato*. According to simultaneous analysis of six nuclear DNA sequences (4164 bp), this species forms a group together with six other species of the same subgenus, namely, *C.* (*O.*) *yaconinus*, *C.* (*O.*) *kawanoi*, *C.* (*O.*) *tosanus*, *C.* (*O.*) *japonicus*, *C.* (*O.*) *chugokuensis* and *C.* (*O.*) *dehaanii* (SOTA & NAGATA, 2008; classification of the species listed here is partly modified following that of this book). For details of the molecular genealogy of this species, also see SU *et al.* (2006) and OKAMOTO *et al.* (2007).

Distribution　N. Chugoku Region, W. Honshu / Oki Islands (Iss. Dôgo, Naka-no-shima, Nishi-no-shima & Chiburi-jima) (Shimane Pref.).

Habitat　Inhabiting hills or mountains consisting of the basement rock, and usually not found from an alluvial plain. Common in the edge of rather dry, thin forest composed of deciduous trees or pine trees, though also visible in the forest of planted trees such as cedar or cypress.

Bionomical notes　Considered to be a spring breeder, earthworm feeder.

Geographical variation　Classified into two subspecies.

Etymology　Named after Mt. Dai-sen, the type locality of this species, which means the Great Mountain in Japanese.

　研究史　本タクソンはNAKANE（1953）によりヒメオサムシの一亜種として記載された．原記載は8頭の標本に基づいて行われたが，ホロタイプの指定がなされていなかったため，IMURA（2005b）により後年，「大山」産の1♂がレクトタイプに指定されている．陰茎先端の形態が特徴的であるため，一般的には独立種として扱われることが多く，NAKANE（1961）により隠岐から同じくヒメオサムシの一亜種として記載された *okianus* という亜種小名によって代表されるタクソンは現在，本種の亜種とみなされている．

　タイプ産地　大山．NAKANE（1953）の原記載論文において示された基準産地は"Daisen"と"Tottori"の2箇所だが，IMURA（2005b）により大山産の♂がレクトタイプに指定されている．

　タイプ標本の所在　北海道大学総合博物館（札幌）［レクトタイプ，♂］．

　形態　20~25 mm. 背面は黒色，黒緑色，銅色，などの色彩型がある．外部形態：図1~6．交尾器形態：図7~17．クロオサムシに似るが，以下の諸点により識別される：1）体つきはやや幅広く，とりわけ前胸背板が大きく幅広い；2）♂触角第5節から第7節の腹面に広い無毛部がある；3）鞘翅第二次，第三次間室はより顕著に小鈍鋸歯状を呈し，間室間の条溝はより顕著に点刻される；4）後基節は内側剛毛を欠く；5）陰茎先端部がやや太く，腹側に向けての湾曲もやや強い；6）指状片は基部がより幅広く，先端に向けての狭まりがより急で，側面から見て基部がより薄く，湾曲が弱い；7）♀交尾器膣底部節片内板は栗の実形で，前縁中央は突出し，側縁の後半と内側の縦溝の縁は強く硬化する．

　分子系統　オホモプテルス亜属の分子系統の概要についてはヤマトオサムシの同項を参照．形態学的に本種は中国地方北部の基亜種と隠岐の隠岐亜種に分けられるが，核ITSとmtDNAを用いた両亜種およびヒメオサムシ，アキオサムシ等近縁種との系統関係については岡本ら（2007）に詳しい．6遺伝子4164塩基対の核遺伝子配列を用いた系統構築の結果からは，本種はヤコンオサムシ，アワオサムシ，トサオサムシ，ヒメオサムシ，アキオサムシ，オオオサムシら（種の分類は本書のそれに倣って一部改編）と共にダイセンオサムシ種群を形成し，群内では最も初期に分化したとされる（SOTA & NAGATA, 2008）．

　分布　本州（中国地方北部）／隠岐諸島（島後，中ノ島，西ノ島，知夫里島；以上，島根県）．

　生息環境　基本的には基板岩からなる山地～丘陵に生息し，疎林ないし林縁等の環境に多い．スギ・ヒノキ等の植林にも見られる．

　生態　春繁殖型で，幼虫は環形動物（フトミミズ等）食と思われる．

　地理的変異　中国地方北部の基亜種と隠岐の隠岐亜種に分けられる．

21. *Carabus* (*Ohomopterus*) *daisen*

種小名の由来　基準産地の大山に由来する.

21-1. *O. d. daisen* (Nakane, 1953)　ダイセンオサムシ基亜種

Original description, type locality & type depository　See explanation of the species.
Diagnosis　20~25 mm. Black, often with coppery, greenish or bluish tinge along body margins. Differs from subsp. *okianus* in the following points: 1) body above essentially black; 2) hairless ventral depression of male antennae more distinctly carved; 3) pronotum usually with three or more marginal setae; 4) punctures on striae between intervals more strongly impressed; 5) inner margin of male protibia more strongly angulated; 6) digitulus broader in basal portion and more acutely convergent toward apex; 7) left basal lateral lobe of internal sac a little longer, with tip more prominently protruded ventrad.
Distribution　N. Chugoku Region (W. Hyogo Pref., Tottori Pref., C. Shimane Pref. & N. Okayama Pref.), W. Honshu.

原記載・タイプ産地・タイプ標本の所在　種の項を参照.
亜種の特徴　20~25 mm. 黒色で, 体側縁はしばしば銅, 緑, 青銅色を帯びる. 隠岐亜種とは以下の諸点が異なる; 1) 背面の色彩は基本的に黒色; 2) ♂触角腹面無毛部のえぐれがより強い; 3) 前胸背板の側縁剛毛は3本以上; 4) 鞘翅間室間条溝の点刻はより顕著に刻まれ, 間室隆条の側部に及び, 時に横条刻となる; 5) 前脛節内側縁の角張りがより強い; 6) 指状片の基部がより幅広く, 先端に向けての狭まりが急; 7) 内囊の左基部側葉がやや長く, 先端は腹側に向けてより顕著に突出する.
分布　中国地方北部(兵庫県西部, 鳥取県, 島根県中部, 岡山県北部). 分布の東限は兵庫県北西部の香美町(かみちょう)付近, 西限は島根県中部の江の川(ごうのかわ)北東岸と思われる. 分布域の辺縁ではアキオサムシと分布を接しており, 一部の地域では同種との自然交雑に由来すると思われる個体が得られている.

21-2. *O. d. okianus* (Nakane, 1961)　ダイセンオサムシ隠岐亜種

Original description　*Apotomopterus japonicus okianus* Nakane, 1961, Fragm, coleopterol., Kyoto, (1), p. 1.
Type locality　Dogo, Oki Is., Shimane Pref., Honshu.
Type depository　Hokkaido University Museum (Sapporo) [Holotype, ♀].
Diagnosis　19~24 mm. For the differential diagnosis of this subspecies, see the same section of the nominotypical subspecies.
Distribution　Iss. Dôgo, Naka-no-shima, Nishi-no-shima & Chiburi-shima of the Oki Islands (Shiname Pref.).
Etymology　Named after the Oki Islands, the type area of this subspecies.

タイプ産地　本州, 島根県, 隠岐, 島後. 原記載には基準産地が島後であることしか記されていないが, ホロタイプのラベルによれば "Higashidani"(東谷)において採集されたもののようである.
タイプ標本の所在　北海道大学総合博物館(札幌)[ホロタイプ, ♀].
亜種の特徴　19~24 mm. 基亜種との識別については基亜種の項を参照.
分布　隠岐諸島(島後, 中ノ島, 西ノ島, 知夫里島; 以上, 島根県).
亜種小名の由来と通称　亜種小名は基準産地の隠岐に由来する. 通称はオキオサムシ.

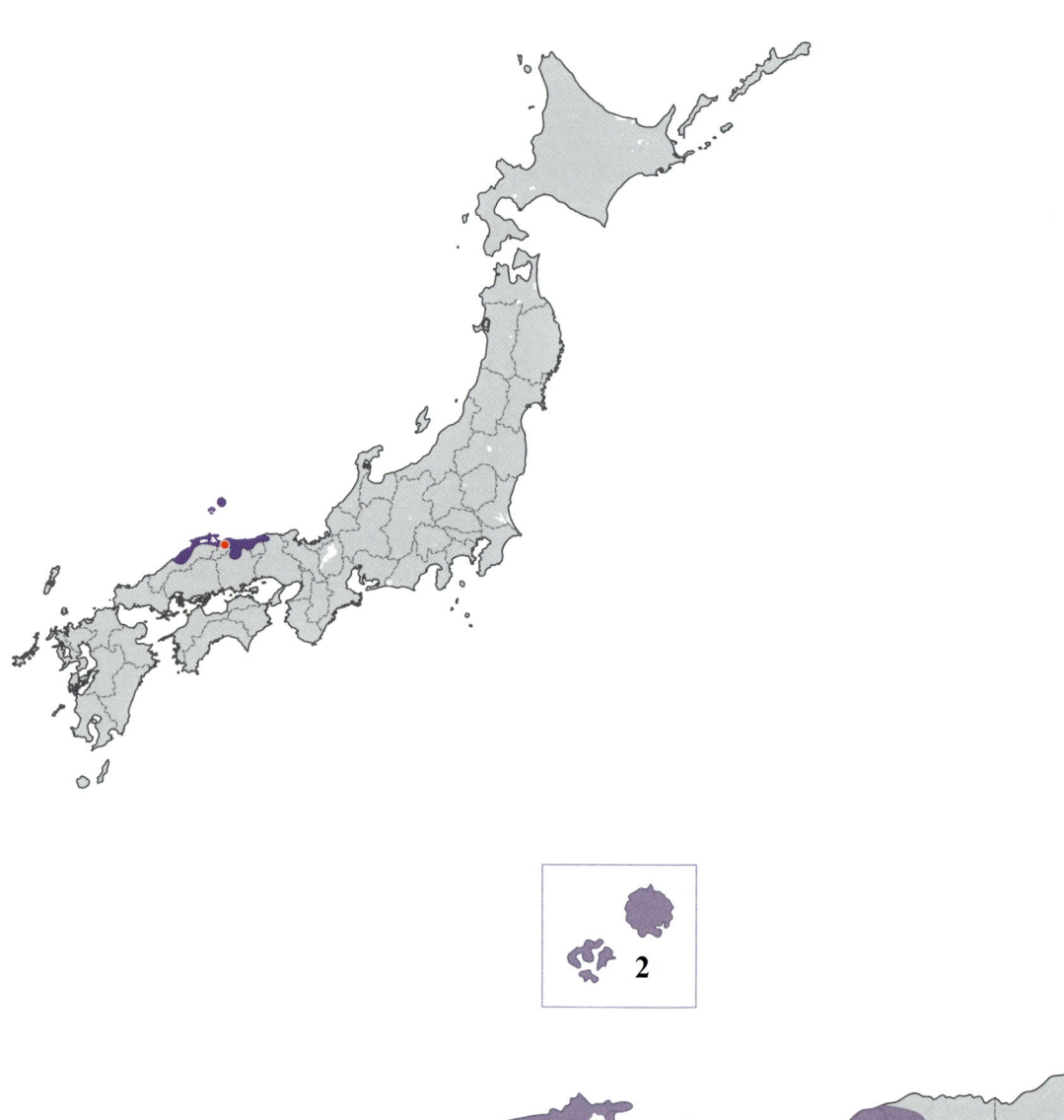

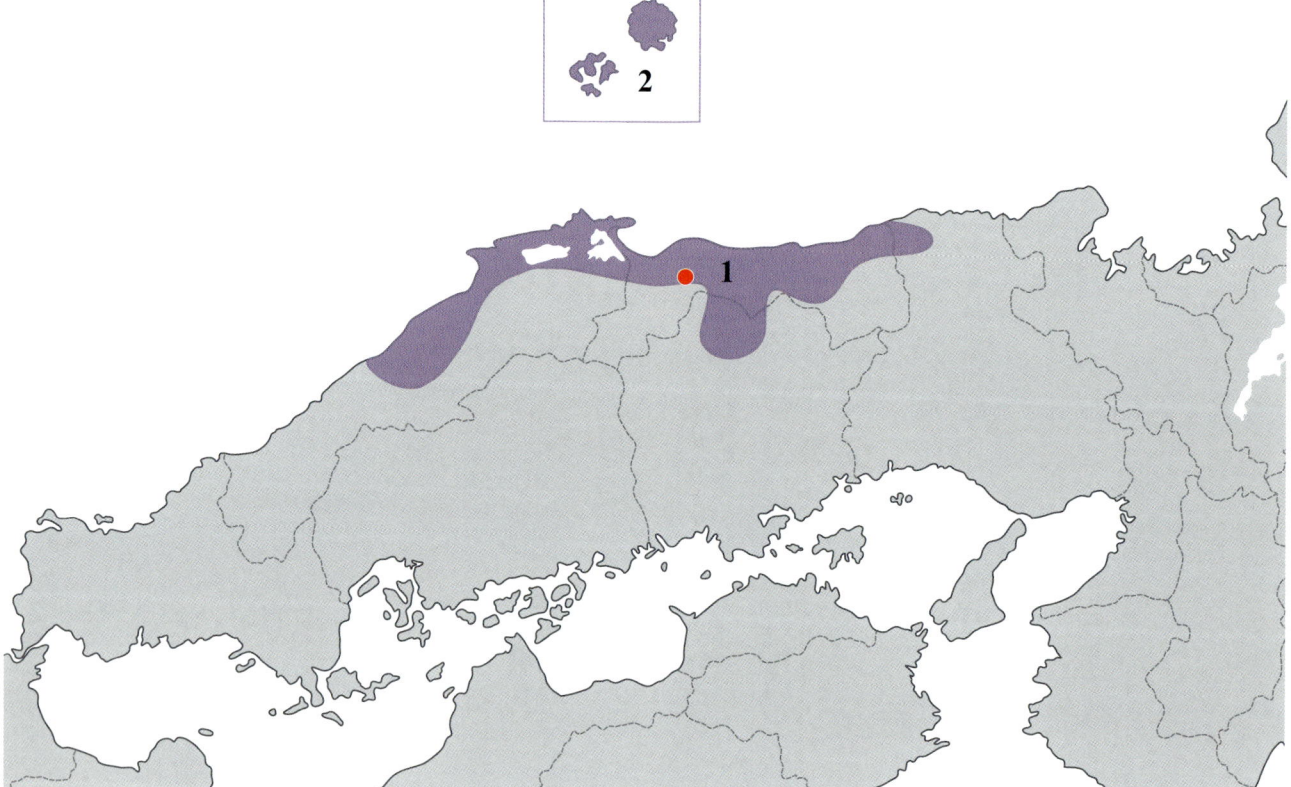

21. Carabus (Ohomopterus) daisen

Fig. 30. Map showing the distribution of *Carabus* (*Ohomopterus*) *daisen* (NAKANE, 1953) — 1, subsp. ***daisen***; 2, subsp. ***okianus***. ●: Type locality of the species (Daisen = Mt. Daisen).

図 30. ダイセンオサムシ分布図. — 1, 基亜種; 2, 隠岐亜種. ●: 種のタイプ産地(大山).

21. *Carabus* (*Ohomopterus*) *daisen*

Fig. 31. *Carabus* (*Ohomopterus*) *daisen daisen* (Akima, Akamatsu, Daisen-chô, Tottori Pref.).
図31. ダイセンオサムシ基亜種(鳥取県大山町赤松明間産).

Fig. 32. *Carabus* (*Ohomopterus*) *daisen daisen* (Akima, Akamatsu, Daisen-chô, Tottori Pref.).
図32. ダイセンオサムシ基亜種(鳥取県大山町赤松明間産).

22. *Carabus* (*Ohomopterus*) *chugokuensis* (Nakane, 1961)
アキオサムシ

Figs. 1-18. *Carabus* (*Ohomopterus*) *chugokuensis chugokuensis* (Nakane, 1961) (Mt. Takajô-san, Hamada-shi, Shimane Pref.). — 1, ♂; 2, ♀; 3, male head & pronotum; 4, male right protarsus in ventral view; 5, male right protibia in subdorsal view; 6, male left elytron (median part); 7, penis & fully everted internal sac in right lateral view; 8, apical part of penis in right lateral view; 9, ditto in right subdorsal view; 10, ditto in dorsal view; 11, internal sac in caudal view; 12, ditto in frontal view; 13, ditto in dorsal view; 14, ditto in left sublateral view; 15, digitulus in frontal view; 16, ditto in right lateral view; 17, terminal plate of internal sac in right lateral view; 18, inner plate of ligular apophysis.

22. *Carabus* (*Ohomopterus*) *chugokuensis*

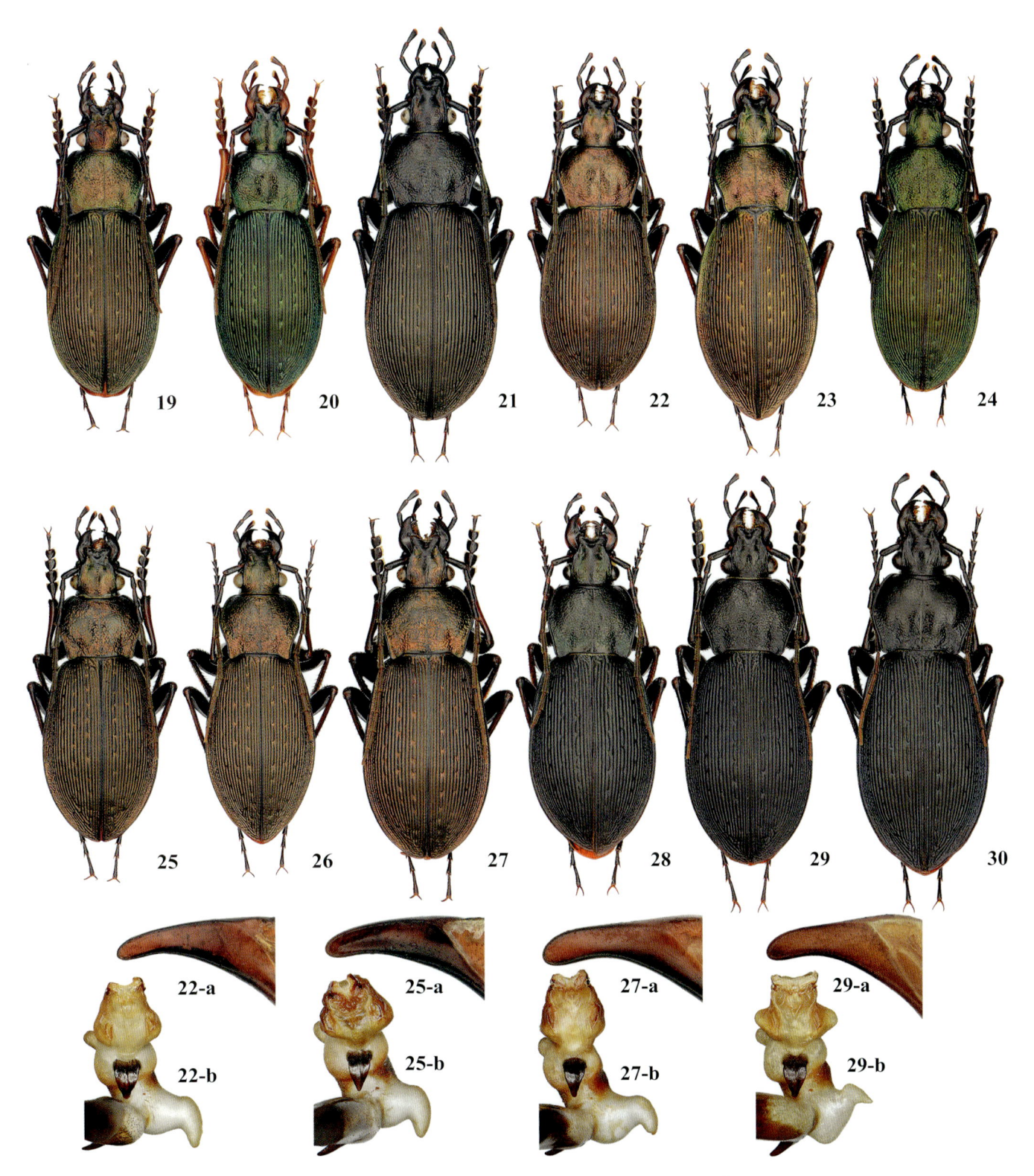

Figs. 19-30. ***Carabus*** (***Ohomopterus***) ***chugokuensis*** subspp. — 19-21, ***C.*** (***O.***) ***c. chugokuensis*** (19-21, ♂; 19, Mt. Aoba-san, Takahama-chô, Fukui Pref.; 20, Kaita, Mimasaka-shi, Okayama Pref.; 21, Is. Ômi-jima, Nagato-shi, Yamaguchi Pref.); 22-23, ***C.*** (***O.***) ***c. seizaburoi*** (22, ♂, HT, 23, ♀, PT; Kanka-kei Valley, Is. Shôdo-shima, Kagawa Pref.); 24-26, ***C.*** (***O.***) ***c. mikianus*** (24, ♂, natural hybrid between *C.* (*O.*) *c. mikianus* & *C.* (*O.*) *japonicus awajiensis* ?, 25, ♂, HT, 26, ♀, PT; 24, Pass Ôsaka-tôgé, Higashi-kagawa-shi, Kagawa Pref.; 25-26, Ôyamahata, Kami-ita-chô, Tokushima Pref.); 27-28, ***C.*** (***O.***) ***c. mochizukii*** (27, ♂, HT, 28, ♀, PT; Mt. Nenbutsu-yama, Is. Ô-shima, Imabari-shi, Ehime Pref.); 29-30, ***C.*** (***O.***) ***c. umekii*** (29, ♂, HT, 30, ♀, PT; Mt. Kanô-san, Is. Yashiro-jima, Yamaguchi Pref.). — a, Apical part of penis in right lateral view; b, internal sac in frontal view.

図 19-30. **アキオサムシ**各亜種. — 19-21, **基亜種**(19-21, ♂; 19, 福井県高浜町青葉山; 20, 岡山県美作市海田; 21, 山口県長門市青海島); 22-23, **小豆島亜種**(22, ♂, HT, 23, ♀, PT; 香川県小豆島寒霞渓); 24-26, **讃岐山脈亜種**(24, ♂, ヒメオサムシとの自然雑種?, 25, ♂, HT, 26, ♀, PT; 24, 香川県東香川市大坂峠; 25-26, 徳島県上板町大山畑); 27-28, **芸予諸島大島亜種**(27, ♂, HT, 28, ♀, PT; 愛媛県今治市大島念仏山); 29-30, **周防亜種**(29, ♂, HT, 30, ♀, PT; 山口県屋代島嘉納山). — a, 陰茎先端(右側面観); b, 内嚢(頭側面観).

185

22. Carabus (Ohomopterus) chugokuensis

Fig. 31. Map showing the distribution of *Carabus (Ohomopterus) chugokuensis* NAKANE, 1961 — 1, Subsp. ***chugokuensis***; 2, subsp. ***seizaburoi***; 3, subsp. *nakamurai*; 4, subsp. *moralii*; 5, subsp. *umekii*. ● : Type locality of the species (Mt. Takashiro = Mt. Takajō-San).

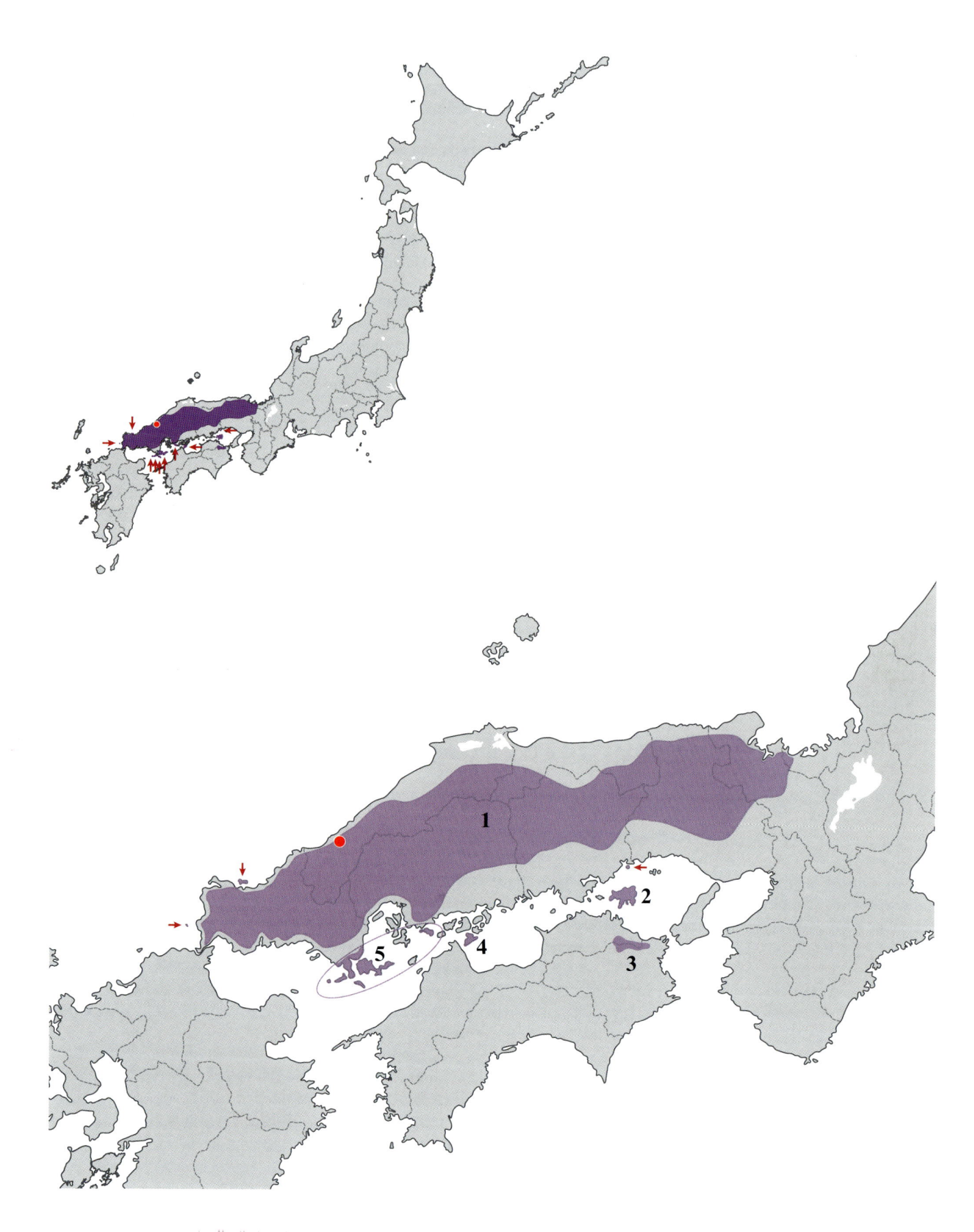

図 31. アキオサムシ分布図. — 1, 基亜種; 2, 小豆島亜種; 3, 讃岐山脈亜種; 4, 芸予諸島大島亜種; 5, 周防亜種. ● : 種のタイプ産地 (高城山).

22. Carabus (Ohomopterus) chugokuensis

Fig. 32. Habitat of *Carabus* (*Ohomopterus*) *chugokuensis seizaburoi*, *C.* (*O.*) *yaconinus* and *C.* (*Damaster*) *blaptoides blaptoides* (Is. Shôdo-shima, Kagawa Pref.).

図32. アキオサムシ小豆島亜種, ヤコンオサムシ, マイマイカブリ基亜種の生息地(香川県小豆島).

Fig. 33. *Carabus* (*Ohomopterus*) *chugokuensis seizaburoi* (Is. Shôdo-shima, Kagawa Pref.).

図33. アキオサムシ小豆島亜種(香川県小豆島).

Fig. 34. Habitat (type locality) of *Carabus* (*Ohomopterus*) *chugokuensis mikianus* (Ôyamahata, Kami-ita-chô, Tokushima Pref.).

図34. アキオサムシ讃岐山脈亜種の生息地(タイプ産地の徳島県上板町大山畑).

23. *Carabus (Ohomopterus) sue* IMURA, 2012
ホウオウオサムシ

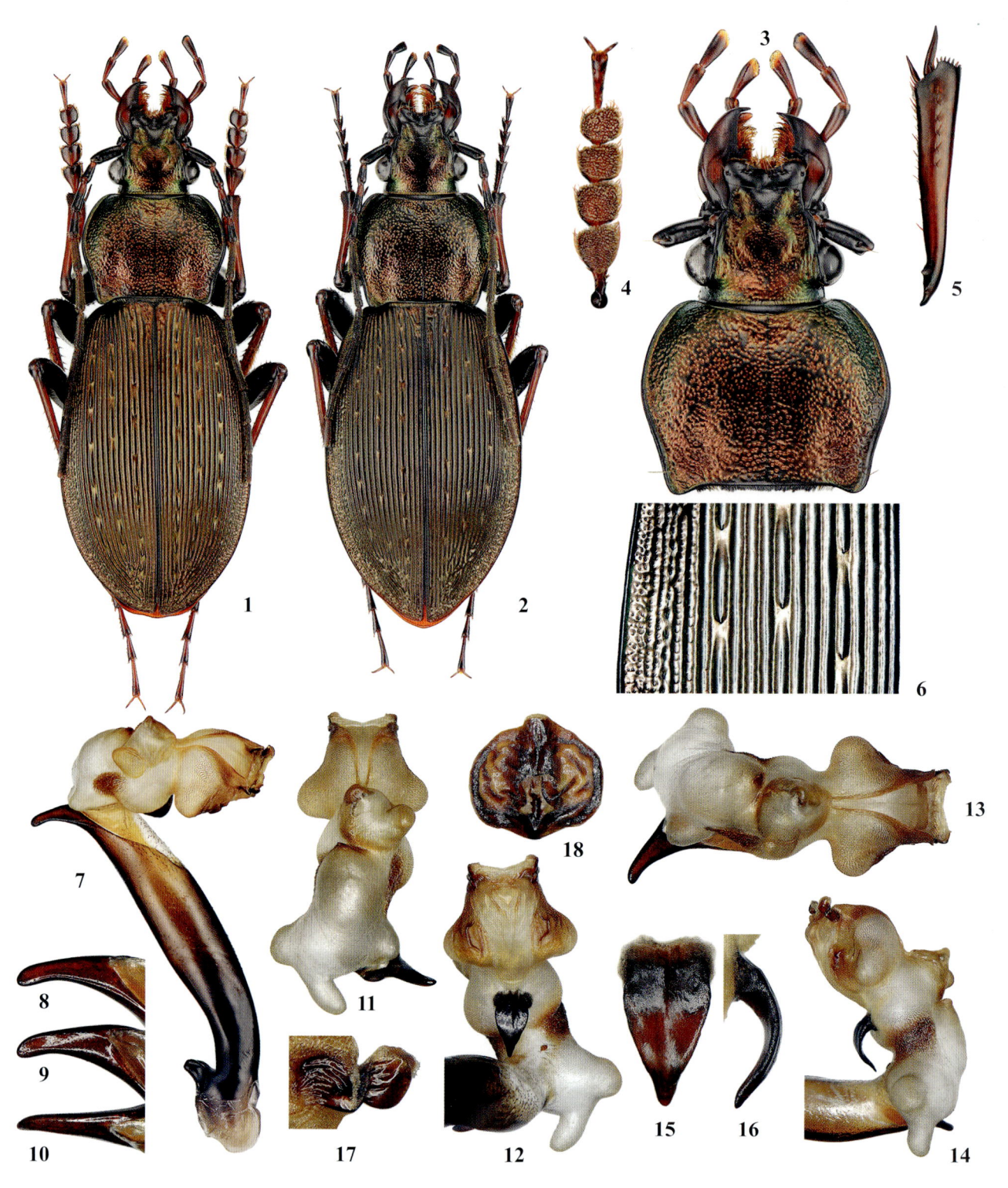

Figs. 1-18. *Carabus (Ohomopterus) sue* IMURA, 2012 (Kazasuki, N. of Mt. Ôgi-yama, Saijô-shi, Ehime Pref.). — 1, ♂, HT; 2, ♀, PT; 3, male head & pronotum; 4, male right protarsus in ventral view; 5, male right protibia in subdorsal view; 6, male left elytron (median part); 7, penis & fully everted internal sac in right lateral view; 8, apical part of penis in right lateral view; 9, ditto in right subdorsal view; 10, ditto in dorsal view; 11, internal sac in caudal view; 12, ditto in frontal view; 13, ditto in dorsal view; 14, ditto in left sublateral view; 15, digitulus in frontal view; 16, ditto in right lateral view; 17, terminal plate of internal sac in right lateral view; 18, inner plate of ligular apophysis.

23. *Carabus* (*Ohomopterus*) *sue*

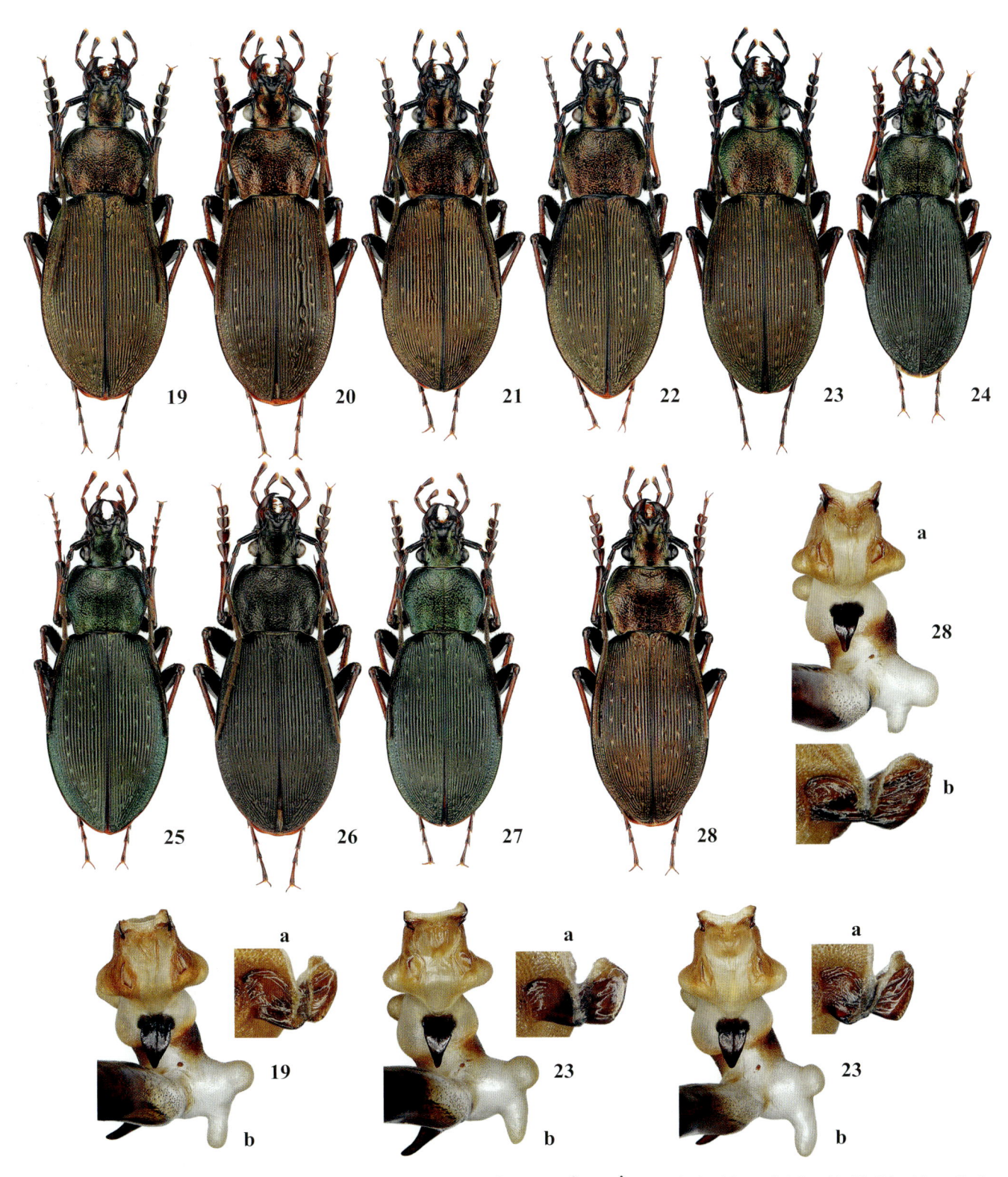

Figs. 19-27. ***Carabus*** (***Ohomopterus***) ***sue***. — 19-21, 23, 24, 26, 27, ♂, 22, 25, ♀; 19, Ôizumi, Ikeda-chô-umaji, Miyoshi-shi, Tokushisma Pref.; 20, SSW. of Pass Sakaimé-tôgé, Kawataki-chô-shimoyama, Shikoku-chûô-shi, Ehime Pref.; 21-22, Pass Horikiri-tôgé, Shikoku-chûô-shi, Ehime Pref.; 23, ENE. of Mt. Suiha-miné, Shikoku-chûô-shi, Ehime Pref.; 24, Mt. Akaboshi-yama, Shikoku-chûô-shi, Ehime Pref.; 25, Tônaru, Niihama-shi, Ehime Pref.; 26, Hosono, Nakaoku-yongô, Saijô-shi, Ehime Pref.; 27, Yokominé-ji Temple, Komatsu-chô-ishizuchi, Saijô-shi, Ehime Pref. — 28, Natural hybrid between *C.* (*O.*) *sue* & *C.* (*O.*) *japonicus awajiensis* (♂, Nishi-no-kawa-hei, Saijô-shi, Ehime Pref.). — a, Internal sac in frontal view; b, terminal plate of internal sac in right lateral view.

図 19-27. **ホウオウオサムシ**. — 19-21, 23, 24, 26, 27, ♂, 22, 25, ♀; 19, 徳島県三好市池田町馬路大泉; 20, 愛媛県四国中央市川滝町下山境目峠南南西; 21-22, 愛媛県四国中央市堀切峠; 23, 愛媛県四国中央市翠波峰東北東; 24, 愛媛県四国中央市赤星山; 25, 愛媛県新居浜市東平; 26, 愛媛県西条市中奥4号細野; 27, 愛媛県西条市小松町石鎚横峰寺. — 28, ホウオウオサムシとヒメオサムシ淡路島四国亜種の自然雑種 (♂, 愛媛県西条市西之川丙). — a, 内嚢(頭側面観); b, 内嚢頂板(右側面観).

193

23. *Carabus* (*Ohomopterus*) *sue* Imura, 2012 ホウオウオサムシ

Original description　*Carabus* (*Ohomopterus*) *sue* Imura, 2012, Coléoptères, Paris, 18 (7), p. 58, figs. 2 (p. 60) & 3 (p. 61).

History of research　The most recently discovered species of the Japanese Carabina (= genus *Carabus*), narrowly distributed in the Hô-ô Mountains and northern Ishizuchi Mountains of northern Shikoku. It had been overlooked by most Japanese carabidologists for many years simply because of similarity in external features to *C.* (*O.*) *japonicus*. Viewed from detailed structure of the internal sac of the male genital organ, however, it is apparently distinguishable from such allied species as *C.* (*O.*) *japonicus* and *C.* (*O.*) *chugokuensis*. Imura (2012) pointed this out first, and described the Hô-ô race as a new species. A new species of *Ohomopterus* was described for the first time in half a century after the discovery of *C.* (*O.*) *uenoi* in 1960.

Type locality　Southwestern Japan, northern Shikoku, above Kazasuki on the northern foot of Mt. Ôgi-yama in Saijô-shi of Ehime Prefecture, 680 m in altitude.

Type depository　Muséum National d'Histoire Naturelle, Paris [Holotype, ♂].

Morphology　19~24 mm.　Body above usually red brownish coppery often with faint greenish tinge, sometimes coppery greenish, greenish coppery, dark greenish black or dark brown; tibiae, tarsi and palpi dark rufous to rufous. External features: Figs. 1~6. Genitalic features: Figs. 7~18. Closely allied to *C.* (*O.*) *chugokuensis mikianus* or *C.* (*O.*) *japonicus awajiensis*, both distributed in northern Shikoku, but definitely differs from them in shape of the male genital organ. Above all, combination of the *japonicus*-like left basal lateral lobe and the *chugokuensis*-like short terminal plate is very unique and is the most outstanding characteristic of this species. Inner margin of male protibia a little less weakly protruded than in *C.* (*O.*) *japonicas*; apical part of penis slenderer, more strongly constricted near middle and more remarkably curved ventrad than in *C.* (*O.*) *japonicus*; left basal lateral lobe of internal sac similar to that of *C.* (*O.*) *japonicas*, but its lateral protrusion is almost spherical in shape, and ventral protrusion larger, longer and usually curved inward.

Distribution　N. Shikoku (Hô-ô Mts. & N. side of Ishizuchi Mts. in E. Ehime Pref., partly penetrating into NW. Tokushima Pref.). Some strange individuals most probably produced by crossbreeding with *C.* (*O.*) *japonicus* have been obtained in several stations such as N. slope of Mt. Ishizuchi-san and Pass Sakaimé-tôgé on the borders between Ehime and Tokushima Prefectures.

Habitat　Mesophilous forest species, inhabiting the woodland composed either of deciduous broadleaved trees or of planted cedar trees on the base rock mountain, recorded from the area between 180 m and 930 m in altitude.

Bionomical notes　In captivity, the adult is rather omnivorous, feeds on earthworm, raw meat, fruits, jelly, etc. Life cycle and feeding habit of the larva are unknown, though most probably similar to or almost the same as those of other species in the same subgenus, that means, spring breeder and earthworm feeder.

Geographical variation　Greenish individuals appear in a certain ratio in the western part of the distributional range, e.g., northern side of the Ishizuchi Mts., above all in high altitudinal area, but they are rarely or hardly found in the eastern part of the Hô-ô Mts.

Etymology　The specific name, *sue* ([su:] or [sû]), comes from the nickname of author's wife, Masumi Imura, who is a professor of the Japanese Red Cross College of Nursing in Tokyo.

　研究史　邦産オサムシの中では最も新しく発見, 記載された種で, 四国北部の法皇山脈から石鎚山塊北面にかけてのごく狭い範囲のみから知られている. 同地域からは, これまでにも少ないながらヒメオサムシ種群に属する小型のオサムシが記録されていたが, いずれも種としては「ヒメオサムシ」と同定され, その分類学的重要性が見過ごされてきた. しかし, 本集団の♂交尾器, とりわけ内嚢の左基部側葉と頂板の形態は近縁他種のものとは明らかに異なっており, 同じ四国北部に産し, 外見の酷似するヒメオサムシやアキオサムシなどから明確に識別できることがImura (2012) により始めて指摘され, 同年の春に新種として記載された. 邦産オサムシの種レベルにおける新タクソンの記載は, オオクロナガオサムシ (1974年) 以来38年ぶり, オホモプテルス亜属としてはドウキョウオサムシ (1960年) 以来52年ぶりのことであった.

　タイプ産地　西南日本, 北四国, 愛媛県西条市, 扇山北麓, 風透, 標高680 m.

　タイプ標本の所在　パリ自然史博物館 [ホロタイプ, ♂].

　形態　19~24 mm. 背面は銅色系主体だが, 黒褐色, 緑褐色, 黒緑色, 緑色等の型も出現する. 外部形態: 図1~6. 交尾器形態: 図7~18. 同じ北四国に分布するアキオサムシ讃岐山脈東部亜種やヒメオサムシ淡路島四国亜種に非常に近いが, 内嚢の形態は非常に特徴的で, ヒメオサムシ型の二又に分かれた左基部側葉とアキオサムシ型の短い頂板というユニークな組み合わせを持つ点で容易に識別できる. ただし, 本種の左基部側葉はヒメオサムシのそれとは似て非なるもので, 側方への突出部分はより球形に近く, 短く, 腹側突起はヒメオサムシのそれより通常大きく, より長く伸長し, 内側に向けて湾曲することが多い. また, 本種では内嚢先端部にある脚葉が一般にヒメオサムシのものより側方に向けての突出が弱い. 陰茎の形態もヒメオサムシとはやや異なり, 本種では先端部がより細く, 中央でより顕著にくびれ, 腹側への湾曲がより強いものが多い. ♀交尾器膣底部節片内板はヒメオサムシ, アキオサムシのものによく似ている. 外部形態の上では, 一般にヒメオサムシ淡路島四国亜種よりやや小型で, ♂前脛節内側縁の角張りがやや弱い.

　分布　四国北部 (愛媛県東部~徳島県北西端; 石鎚山塊北面~法皇山脈). 石鎚山塊の北面から法皇山脈のほぼ全域にかけての, 中央構造線に沿って東西に細長く伸びた特異な分布圏を持つ. これまでに確認されている分布の西限は石鎚山北西麓

23. *Carabus* (*Ohomopterus*) *sue*

の志河川(しこがわ)(中山川支流)南岸，東限は徳島県三好市池田町馬路(うまじ)付近．分布域の北縁は中央構造線にほぼ一致し，同構造線の北側にある，新居浜市街に近い瀬戸内海に面した低丘陵地帯からは発見されていない．分布域西部では広義の石鎚山脈(堂ヶ森〜石鎚山〜瓶ヶ森(かめがもり)〜笹ヶ峰〜東赤石山〜ハネヅル山)の高所を東西に結ぶ主稜線が南縁となっており，北側斜面に本種が，主稜線の高標高域から南側斜面にかけてヒメオサムシが分布する．この地域では銅山川(どうざんがわ)(別子(べっし)ダム〜法皇湖にかけての範囲)を北に越えてヒメオサムシが山脈の南斜面に侵入しており，赤星山南麓では中尾谷川(なかおだにがわ)を境界として，南岸にヒメオサムシが，北岸に本種が分布している．分布域東部の法皇山脈東部(赤星山〜翠波峰(すいはみね)〜徳島県三好市池田町馬路にかけての地域)では銅山川と馬路川〜金生川(きんせいがわ)に挟まれた山塊に分布し，愛媛・徳島県境の境目峠(さかいめとうげ)付近では，讃岐山脈西部の分布域から南進してきたヒメオサムシが本種と分布を接している．このように，瀬戸内海側(＝北側)を除く本種分布域の西・南・東の三方向をとり囲む形でヒメオサムシが側所的に分布しており，石鎚山主峰の北麓や境目峠南方など一部の地域では，両種の交雑により生じたと思われる個体(図28)が得られている．

生息環境 基盤山地の森林に生息し，標高180~930 mの範囲から記録されている．これまでに確認されている生息標高の上限(930 m)はヒメオサムシのそれ(約1,500 m)よりも低い．本来の生息環境は落葉広葉樹主体の自然林と思われるが，スギやヒノキの植林地にも広く見られ，標高500~700 m付近で最も個体数が多いようである．

生態 同亜属他種と同様，春繁殖型で環形動物(フトミミズ等)食と思われる．啓蟄(けいちつ)は4月下旬から5月上旬頃で，6~7月に個体数が多く，テネラルは一般に7月下旬以降に見られる．冬季には土中で越冬している成虫が得られているが，オサ掘りによる採集は比較的困難である．

地理的変異 緑色系の個体は分布域西部(石鎚山塊北面等)や標高の高い地域においてより高頻度に出現し，低標高の地域や分布域東部(法皇山脈東部)では大半の個体が銅色型となる．

種小名の由来 種小名の*sue*(スー)は，著者の妻，井村真澄(日本赤十字看護大学教授)の愛称に，また和名は本種の主たる分布域である法皇山脈に因む．

23. *Carabus* (*Ohomopterus*) *sue*

Fig. 29. Map showing the distributional range of ***Carabus*** (***Ohomopterus***) ***sue*** Imura, 2012. ●: Type localities of the species (Kazasuki).
図 29. **ホウオウオサムシ**分布図. ●: 種のタイプ産地（風透）.

23. *Carabus* (*Ohomopterus*) *sue*

Fig. 30. Habitat of *Carabus* (*Ohomopterus*) *sue*, *C.* (*O.*) *tosanus tosanus* and *C.* (*Damaster*) *blaptoides blaptoides* (Mt. Akaboshi-yama, Shikoku-chûô-shi, Ehime Pref.).

図30. ホウオウオサムシ, トサオオサムシ基亜種, マイマイカブリ基亜種の生息地(愛媛県四国中央市赤星山).

Fig. 31. *Carabus* (*Ohomopterus*) *sue* feeds on earthworm (N. slope of Mt. Ishizuchi-san, Saijô-shi, Ehime Pref.).

図31. ミミズを摂食中のホウオウオサムシ(愛媛県西条市石鎚山北麓).

Fig. 32. *Carabus* (*Ohomopterus*) *sue* passing the winter in the soil (Fuji-no-ishi, Saijô-shi, Ehime Pref.).

図32. 土中で越冬しているホウオウオサムシ(愛媛県西条市藤之石).

24. *Carabus* (*Ohomopterus*) *japonicus* Motschulsky, 1857
ヒメオサムシ

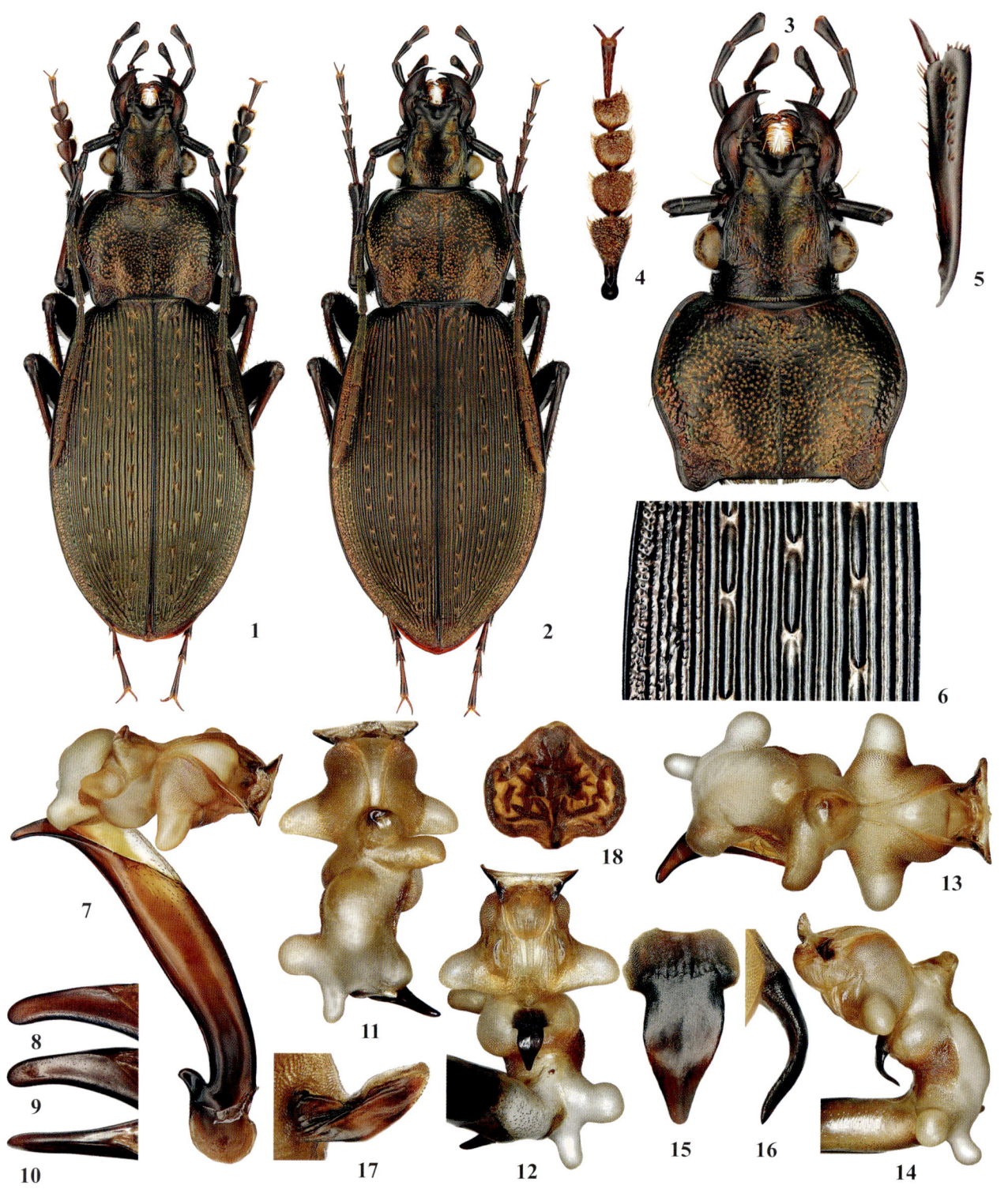

Figs. 1-18. *Carabus* (*Ohomopterus*) *japonicus japonicus* Motschulsky, 1857 (Unzen-onsen, Obama-chô, Nagasaki Pref.). — 1, ♂; 2, ♀; 3, male head & pronotum; 4, male right protarsus in ventral view; 5, male right protibia in subdorsal view; 6, male left elytron (median part); 7, penis & fully everted internal sac in right lateral view; 8, apical part of penis in right lateral view; 9, ditto in right subdorsal view; 10, ditto in dorsal view; 11, internal sac in caudal view; 12, ditto in frontal view; 13, ditto in dorsal view; 14, ditto in left sublateral view; 15, digitulus in frontal view; 16, ditto in right lateral view; 17, terminal plate of internal sac in right lateral view; 18, inner plate of ligular apophysis.

24. *Carabus* (*Ohomopterus*) *japonicus*

Figs. 19-30. ***Carabus*** (***Ohomopterus***) ***japonicus*** subspp. — 19-22, ***C***. (***O***.) ***j. awajiensis*** (19 (HT), 21, ♂, 20 (PT), 22, ♀; 19-20, Mt. Sen-zan, Is. Awaji-shima, Hyogo Pref.; 21-22, Mt. Jinga-yama, Kami-shi, Kochi Pref.); 23-24, ***C***. (***O***.) ***j. yoshiyukii*** (23, ♂, HT, 24, ♀, PT; Mt. Washio-yama, Kochi-shi, Kochi Pref.); 25-26, ***C***. (***O***.) ***j. okinoshimanus*** (25, ♂, HT, 26, ♀, PT; Moshima, Is. Oki-no-shima, Kochi Pref.); 27-28, ***C***. (***O***.) ***j. japonicus*** (27, ♂, Mt. Hiko-san, Nagasaki-shi, Nagasaki Pref.; 28, ♀, Mt. Kagami-yama, Karatsu-shi, Saga Pref.); 29-30, ***C***. (***O***.) ***j. tsushimae*** (29, ♂, 30, ♀; Mt. Tatera-san, Is. Tsushima, Nagasaki Pref.). — a, Apical part of penis in right lateral view; b, terminal plate of internal sac in right lateral view.

図 19-30. ヒメオサムシ各亜種. — 19-22, **淡路島四国亜種**(19 (HT), 21, ♂, 20 (PT), 22, ♀; 19-20, 兵庫県淡路島先山; 21-22, 高知県香美市神賀山); 23-24, **鷲尾山亜種**(23, ♂, HT, 24, ♀, PT; 高知県高知市鷲尾山); 25-26, **沖の島亜種**(25, ♂, HT, 26, ♀, PT; 高知県沖の島母島); 27-28, **基亜種**(27, ♂, 長崎県長崎市彦山; 28, ♀, 佐賀県唐津市鏡山); 29-30, **対馬亜種**(29, ♂, 30, ♀; 長崎県対馬龍良山). — a, 陰茎先端(右側面観); b, 内嚢頂板(右側面観).

24. *Carabus* (*Ohomopterus*) *japonicus*

Figs. 31-42. ***Carabus*** (***Ohomopterus***) ***japonicus*** subspp. — 31-32, ***C.*** (***O.***) ***j. ikiensis*** (31, ♂, 32, ♀; Mt. Také-no-tsuji, Is. Iki, Nagasaki Pref.); 33-34, ***C.*** (***O.***) ***j. hiradonis*** (33, ♂, HT, 34, ♀; Mt. Yasuman-daké, Is. Hirado-shima, Nagasaki Pref.); 35-36, ***C.*** (***O.***) ***j. chotaroi*** (35, ♂, HT, 36, ♀, PT; Is. Ikitsuki-shima, Nagasaki Pref.); 37-38, ***C.*** (***O.***) ***j. nozakicola*** (37, ♂, HT, 38, ♀; Is. Nozaki-jima, Nagasaki Pref.); 39-40, ***C.*** (***O.***) ***j. wakamatsuensis*** (39, ♂, HT, 40, ♀, PT; Wakamatsu-goé, Is. Wakamatsu-jima, Nagasaki Pref.); 41-42, ***C.*** (***O.***) ***j. onodai*** (41, ♂, HT, 42, ♀, PT; Mt. Kuchi-také, Is. Shimo-koshiki-jima, Kagoshima Pref.). — a, Apical part of penis in right lateral view; b, terminal plate of internal sac in right lateral view.

図 31-42. ヒメオサムシ各亜種. — 31-32, 壱岐亜種 (31, ♂, 32, ♀; 長崎県壱岐岳ノ辻); 33-34, 平戸島亜種 (33, ♂, HT, 34, ♀; 長崎県平戸島安満岳); 35-36, 生月島亜種 (35, ♂, HT, 36, ♀, PT; 長崎県生月島); 37-38, 野崎島亜種 (37, ♂, HT, 38, ♀; 長崎県野崎島); 39-40, 若松島亜種 (39, ♂, HT, 40, ♀, PT; 長崎県若松島若松郷); 41-42, 下甑島亜種 (41, ♂, HT, 42, ♀, PT; 鹿児島県下甑島口岳). — a, 陰茎右側面観; b, 内嚢頂板 (右側面観).

24. *Carabus* (*Ohomopterus*) *japonicus* MOTSCHULSKY, 1857 ヒメオサムシ

Original description   *Carabus japonicus* MOTSCHULSKY, 1857, Étud. ent., Helsingfors, 6 [1857], p. 111.

History of research   This species was originally described by the Russian entomologist V. DE MOTSCHULSKY (1857) from "Japon". NAKANE (1952a, b, 1953) regarded it, as well as *Carabus tsushimae* BREUNING, 1932, as a distinct species of the genus *Apotomopterus*, and later described *A. j. daisen* (NAKANE, 1953) and *A. j. chugokuensis* (NAKANE, 1961). NAKANE (1962) unified all the small-sized *Apotomopterus* species with triangular digitulus into a single species, *A. japonicus*, and classified it into 10 subspecies, namely, *japonicus*, *tsushimae*, *okianus*, *chugokuensis*, *daisen*, *yamato*, *lewisianus*, *esakianus*, *freyi* and *albrechti*. Later, subsp. *ikiensis* was additionally described also by NAKANE (1968). Of these, *albrechti*, *lewisianus* and *yamato* were raised to full species by ISHIKAWA (1969b), and a new name, the *japonicus* species-group, was proposed for the following five species, *A. albrechti*, *A. lewisianus*, *A. kimurai*, *A. yamato* and *A. japonicus*. In *A. japonicus*, ISHIKAWA (1969b) recognized four subspecies, *japonicus*, *tsushimae*, *chugokuensis* and *ikiensis*. ISHIKAWA (1973) reconsidered the higher classification of the subtribe Carabina, and all the Japanese species so far combined with *Apotomopterus* were transferred to *Ohomopterus* (ISHIKAWA regarded it as a subgenus of the genus *Carabus* in a narrow sense, though some authors regard it as a subgenus of the genus *Carabus* in a broad sense or as a distinct genus). IMURA et al. (1993) described six new subspecies of *C.* (*O.*) *japonicus* under the names of *chotaroi*, *hiradonis*, *onodai*, *okinoshimanus*, *awajiensis* and *seizaburoi* from several attached islands of Shikoku and Kyushu. IMURA & MIZUSAWA (1994) described three more subspecies of *C.* (*O.*) *japonicus* under the names of *nozakicola*, *wakamatsuensis* and *mochizukii* from the Gotô and Geiyo Islands of Kyushu and Shikoku. Finally in 1999, two more subspecies, *umekii* and *yoshiyukii* were described by IMURA & MIZUSAWA from Is. Yashiro-jima of Yamaguchi Prefecture and the Washio Hills of Kochi Prefecture, respectively. Of these, *seizaburoi*, *mochizukii* and *umekii* are now regarded as a subspecies of *C.* (*O.*) *chugokuensis* which was once described as a subspecies of *A. japonicus* but later raised its rank to a distinct species by IMURA (2003b). Thus, *C.* (*O.*) *japonicus* is currently classified into 11 subspecies.

Type locality   Japon (without designation of detailed locality). Judging from the historical background and morphological characters of the type specimen (female, 24 mm, with coppery brownish body color), it is highly plausible that the type specimen was collected from somewhere near Nagasaki of northwestern Kyushu, for example, Unzen in the Shimabara Peninsula which is a traditional hot spring resort.

Type depository   Zoological Museum of Moscow University [Holotype, ♀].

Morphology   20~32 mm. Body above coppery more or less with greenish tinge, or black with greenish or bluish tinge. External features: Figs.1~6. Genitalic features: Figs. 7~18. Closely allied to both *C.* (*O.*) *chugokuensis* and *C.* (*O.*) *sue*, but distinguished from them in the following points: 1) left basal lateral lobe of internal sac horizontally protruded and dome-like in shape, with an accessory protrusion at ventral side which is usually shorter than main lobe and not so strongly bent inward; 2) podian lobes strongly protruded laterad; 3) peripheral rim of gonopore strongly sclerotized bilaterad to form a pair of leaf-like terminal plates which are much longer than wide and most remarkably developed in all the three allied species.

Molecular phylogeny   For an overview of the molecular phylogeny of *Ohomopterus*, see the same section of *C.* (*O.*) *yamato*. For details of this species, above all of its relation with the *dehaanii* species-group, see SU et al. (2006) and OKAMOTO et al. (2007). According to simultaneous analysis of six nuclear DNA sequences (4164 bp), this species belongs to the same group as that consisting of *C.* (*O.*) *daisen*, *C.* (*O.*) *yaconinus*, *C.* (*O.*) *kawanoi*, *C.* (*O.*) *tosanus*, *C.* (*O.*) *chugokuensis* and *C.* (*O.*) *dehaanii*, showing the closest affinity to *C.* (*O.*) *dehaanii* (SOTA & NAGATA, 2008; classification of the species listed here is partly modified following that of this book).

Distribution   Shikoku, Kyushu, several islands in the Inland Sea of Setonaikai and southwestern tip of the Chugoku Region / Iss. Awaji-shima, Ié-shima, Tanga-shima (Hyogo Pref.), Hiko-shima of Shimonoseki-shi (Yamaguchi Pref.), Té-shima (Kagawa Pref.), Ao-shima of Ôzu-shi (Ehime Pref.), Oki-no-shima (Kochi Pref.), Ô-shima (= Chikuzen-ô-shima) of Munakata-shi, Genkai-jima, Shika-no-shima (Fukuoka Pref.), Kabé-shima, Ogawa-jima, Kakara-jima, Madara-shima, (Saga Pref.), Tsushima-kami-jima, Tshushima-shimo-jima, Iki, Taka-shima, Maki-shima, Hirado-shima, Ikitsuki-shima, Hario-jima, Nozaki-jima, Nakadôri-jima, Wakamatsu-jima, Naru-shima, Hisaka-jima, Fukué-jima, Saga-no-shima, (Nagasaki Pref.), Amakusa-kami-shima, Amakusa-shimo-shima, Gosho-no-ura-jima (or Gosho-ura-jima) (Kumamoto Pref.), Shoura-jima, Naga-shima, Kami-koshiki-jima, Naka-koshiki-jima & Shimo-koshiki-jima (Kagoshima Pref.).

Habitat   Rather widely distributed from the laurel forest in the low land to the deciduous forest about 1,500 m in altitude. In several attached islands of Shikoku and Kyushu lacking high mountains, this species is also found from the shrine or temple forest, fruit orchard, pine forest or the edge of farm field, etc.

Bionomical notes   Considered to be a spring breeder, earthworm feeder.

Geographical variation   This species shows marked local variation. Although classified into 11 subspecies at present, this classification is still provisional.

Etymology   The specific name means "of Japan" or "occurring in Japan".

研究史　本タクソンは江戸時代末期の1857年(安政4年)にロシアの昆虫研究家, V. de MOTSCHULSKY(モッチュルスキィ)により独立種として記載された．中根(1952a, b, 1953等)はこれを, BREUNING(ブロイニング)(1932)が対馬から記載した*Carabus tsushimae*と

24. *Carabus* (*Ohomopterus*) *japonicus*

共にアポトモプテルス属の独立種 Apotomopterus japonicus として扱い，その亜種として1953年に A. j. daisen, 1961年に A. j. chugokuensis を記載した．その後，中根(1962)は，邦産アポトモプテルス属のうち，体型が小型で小さい三角形の指状片を持つもののすべてを種 A. japonicus としてまとめ，同種を基亜種, tsushimae, okianus, chugokuensis, daisen, yamato, lewisianus, esakianus, freyi, albrechti の計10亜種に分類した. NAKANE (1968) はさらに壱岐から亜種 ikiensis を記載した. ISHIKAWA (1969b) は A. albrechti, A. lewisianus, A. kimurai, A. yamato A. japonicus の五つを独立種と認め，これらに対しヒメオサムシ種群という名を提唱し, A. japonicus には基亜種, tsushimae, chugokuensis, ikiensis の4亜種を認めた. ISHIKAWA (1973) は，♂交尾器内嚢の基本形態に基づいてオサムシ亜族の上位分類の再検討を行い，それまでアポトモプテルス属に置かれてきた日本産の全ての種を，狭義のカラブス属の中のオホモプテルス亜属に移した．以後，本種は基本的にオホモプテルス(著者により属とみなすかカラブス属(広義又は狭義)の亜属とみなすかの違いはあるが)の一員として扱われている．井村ら(1993)は本種の地理的変異について検討し，四国，九州の島嶼部から本種の6亜種(亜種小名はそれぞれ chotaroi, hiradonis, onodai, okinoshimanus, awajiensis, seizaburoi)を記載した．井村・水沢は翌1994年に3亜種(同 nozakicola, wakamatsuensis, mochizukii)を，1999年には2亜種(同 umekii, yoshiyukii)を追加記載した．それまで本種の亜種とみなされてきた，亜種小名 chugokuensis に代表される集団(アキオサムシ)は，IMURA (2003b) により独立種へと昇格され，ヒメオサムシの亜種として記載された小豆島の seizaburoi, 芸予諸島大島の mochizukii, 周防(すおう)地方の umekii の三者は現在，その亜種として扱われている．従って，現段階でヒメオサムシは基亜種を含め計11亜種に分類される．

タイプ産地 日本．原記載論文には「日本」と記されているだけで，具体的な産地名の記述はない．本種が記載された頃の時代背景(記載年は江戸時代末期の安政5年)から判断すると，タイプ標本は，当時外国人が比較的容易に上陸できた長崎市ないしその周辺地域から得られたものである可能性が最も高い．ただし，長崎市界隈に生息する本種は背面の色彩が青みを帯びた黒色で比較的大型のものが多い(長崎のものは MOTSCHULSKY (1865) により別種 Carabus corvinus として記載されている(一般には C. japonicus の異名とされる))のに対し，ホロタイプは銅褐色でやや小型(♀, 24 mm)であることを考慮すると，長崎市そのものよりも，そこから少し離れた，銅褐色で小型の個体の出現頻度がより高い地域，例えば，古くから保養地として有名で，幕末には欧米人にも紹介されていたという島原半島の雲仙温泉付近などをタイプ産地の候補として挙げておきたい．

タイプ標本の所在 モスクワ大学動物学博物館 [ホロタイプ, ♀].

形態 20~32 mm. 背面は銅色系で多少とも緑色光沢を帯びるものが多いが，黒色，黒緑色，黒青色等の型もしばしば出現し，地域によりその出現頻度が異なる．外部形態：図1~6. 交尾器形態：図7~18. アキオサムシ，ホウオウオサムシの2種に極めて近いが，以下に述べる♂交尾器形態の違いにより識別される：1)内嚢の左基部側葉は水平に突出した部分がドーム形で，腹側にはホウオウオサムシ同様，副突起があるが，ドーム形の左基部側葉本体よりは通常短く，内側に向けて殆ど湾曲しない；2)脚葉は側方へ強く張り出す；3)射精孔縁膜の左右両端部が硬化して形成される頂板は木の葉形で細長く，近縁3種の中では最大．

分子系統 オホモプテルス亜属の分子系統の概要についてはヤマトオサムシの同項を参照．本種を含むいわゆるヒメオサムシ種群の分化については，Su et al. (2006), 岡本ら(2007) 等を参照．これらの論文によれば，先ずヒメオサムシの祖先型に相当するものが西日本に拡がり，地域的に分かれて遺伝的な分岐(ミトコンドリア，核DNA共に)が起き，その後，オオオサムシ群が分化して西日本に分布を拡大し，ヒメオサムシ群と交雑しながらヒメオサムシ群のミトコンドリアを受け取ったというシナリオが強く示唆されるという．6遺伝子4164塩基対の核遺伝子配列を用いた系統構築の結果から，本種はオホモプテルス亜属の中ではヤコンオサムシ，アワオサムシ，トサオサムシ，アキオサムシ，オオオサムシら(種の分類は本書のそれに倣って一部改編)と共にダイセンオサムシ種群に入り，オオオサムシとの類縁が最も近いとされる(SOTA & NAGATA, 2008).

分布 四国，九州(一部，瀬戸内海および中国地方南西端の島嶼を含む)／淡路島，家島，男鹿島(以上，兵庫県)，彦島[下関市](山口県)，手島(香川県)，青島[大洲市](愛媛県)，沖の島(高知県)，大島(=筑前大島)[宗像市]，玄海島，志賀島(以上，福岡県)，加部島，小川島，加唐島，馬渡島(以上，佐賀県)，対馬(地峡の人工的開削により主島が分断された後の，現在で言う上島，下島双方に産する)，壱岐，鷹島，牧島，平戸島，生月島，針尾島，野崎島，中通島，若松島，奈留島，久賀島，福江島，嵯峨ノ島(以上，長崎県)，天草上島，天草下島，御所浦島，(以上，熊本県)，諸浦島，長島，上甑島，中甑島，下甑島(以上，鹿児島県).

生息環境 低地の照葉樹林から1,500 mを越す山地の落葉樹林まで，生息範囲の幅は広い．高い山を欠く島嶼部では，島状に残された社寺林や果樹園，松林，畑地の縁などにも生息している．

生態 春繁殖型で幼虫は環形動物(フトミミズ等)食と思われる．冬期には主として土中から越冬中の成虫が得られるが，朽木からの採集例もある．

地理的変異 西南日本一帯とその周辺島嶼に広く分布し，地理的変異が著しい．対馬や壱岐のものは古くから固有の亜種として知られてきたが，井村ら(1993)と井村・水沢(1994, 1999)により地理的変異に関する再検討が行われ，現在11亜種に分類されている．以前はアキオサムシとその亜種も本種に含めて扱われていたが，IMURA (2003b) によりヒメオサムシとアキオサムシは種のレベルで異なるものとして扱うべきであることが指摘されている．

種小名の由来 「日本の」を意味する．

24-1. *C. (O.) j. awajiensis* Imura, Dejima et Mizusawa, 1993 ヒメオサムシ淡路島四国亜種

Original description *Carabus* (*Ohomopterus*) *japonicus awajiensis* Imura, Dejima & Mizusawa, 1993, Gekkan-Mushi, Tokyo, (264), p. 16, figs. 19, 20 (pl. 1) & 32 (p. 12).
Type locality Mt. Sen-zan, Is. Awaji-shima, Sumoto-shi, Hyôgo Pref.
Type depository National Museum of Nature and Science, Tokyo [Holotype, ♂].
Diagnosis 22~26 mm. Body above coppery usually with greenish tinge along margins, sometimes entirely black or dark green; tibiae and tarsi dark rufous to black. 1) Antennae long, extending beyond middle of elytra in male and reaching middle of elytra in female; 2) pronotum strongly convex above, basal foveae shallow; 3) elytra luffa-like in profile, widest behind middle, more gradually narrowed toward bases than toward apices; 4) apical part of penis long and slender, faintly constricted at middle.
Distribution Greater part of Shikoku & several islands in the Inland Sea of Setonaikai (Iss. Awaji-shima, Ié-shima, Tanga-shima (Hyogo Pref.), Té-shima (Kagawa Pref.) & Ao-shima of Ôzu-shi (Ehime Pref.)). As to the population occurring in Iss. Awaji-shima, Ié-shima and Tanga-shima, there still remains suspicions that they might have been introduced artificially.
Etymology Named after the type locality, Is. Awaji-shima.

タイプ産地　兵庫県, 洲本市, 淡路島, 先山.
タイプ標本の所在　国立科学博物館 [ホロタイプ, ♂].
亜種の特徴　22~26 mm. 背面は銅色ないし暗銅褐色で, 前胸背板はしばしば弱い赤紫色の光沢を帯び, 鞘翅側縁は時に緑色光沢を帯びる. 黒色型, 緑色型も出現する. 脛節と附節は暗赤褐色ないし黒色. 1) 触角は長く, ♂では先端が鞘翅中央を越え, ♀でもほぼ鞘翅中央に達する; 2) 前胸背板は一様に強く膨隆し, 基部凹陥は浅い; 3) 鞘翅は最広部が中央よりやや後方にあるので, 下膨れの「へちま」形を呈するものが多い (♂においてより顕著); 4) 陰茎先端部は非常に細長く, 中央で軽くくびれる.
分布　四国 (アキオサムシの分布域である讃岐山脈東部とホウオウオサムシの分布域である石鎚山北面から法皇山脈にかけての地域を除く, 大半の地域) および瀬戸内海の一部の島嶼 (淡路島, 家島, 男鹿島 (以上, 兵庫県), 手島 (香川県), 青島 [大洲市] (愛媛県)). 本亜種は当初, 淡路島の特産亜種として記載されたが, 四国の大半に分布するヒメオサムシの集団も形態学的には本亜種と基本的な差がない (特に♂交尾器内嚢や頂板の形態) ことから, 本書では同じ亜種として扱った. 本亜種のタイプ産地である淡路島では, 同島中部の先山山地の千光寺 (せんこうじ) を中心とする狭い範囲のみに生息し, 南部の諭鶴羽 (ゆづるは) 山地からは記録がなく, その分布はやや不自然である. 家島と男鹿島の集団についても言えることであるが, 過去に人為的に移入されたものが定着している可能性も否定できないように思われる.
亜種小名の由来と通称　亜種小名はタイプ産地の淡路島に因む. 原記載論文中で提唱された通称はアワジヒメオサムシ.

24-2. *C. (O.) j. yoshiyukii* Imura et Mizusawa, 1999 ヒメオサムシ鷲尾山亜種

Original description *Carabus* (*Ohomopterus*) *japonicus yoshiyukii* Imura & Mizusawa, 1999, Gekkan-Mushi, Tokyo, (338), p. 4, figs. 3, 4 & 6.
Type locality Mt. Washio-yama (the Washio Hills, S. of Kôchi City in Kôchi Pref. of Shikoku, SW. Japan).
Type depository National Museum of Nature and Science, Tokyo [Holotype, ♂].
Diagnosis 21~29 mm. Body above brownish coppery, sometimes with faint greenish tinge; tibiae dark reddish black. Most closely allied to subsp. *awajiensis*, but differs from it as follows: 1) size larger; 2) pronotum broader, with disk more strongly convex above; 3) elytra robuster and more strongly convex above, with shoulders more distinct, lateral margins more acutely convergent toward apices; 4) male genitalia almost as in subsp. *awajiensis*.
Distribution Washio Hills, S. of Kôchi-shi, Kochi Pref., SC. Shikoku.
Etymology Named after Yoshiyuki Itô of Kôchi-shi who collected the most part of the type series.

タイプ産地　鷲尾山 (南西日本, 四国, 高知県, 高知市, 鷲尾丘陵).
タイプ標本の所在　国立科学博物館 [ホロタイプ, ♂].
亜種の特徴　21~29 mm. 背面は銅褐色で, 時に緑色光沢を帯びる. 黒~緑色型は知られていない (ただし, 全体的に黒味の強い個体は稀に出現する). 脛節は暗赤褐色を帯びた黒色. 淡路島四国亜種に近いが, 1) 大型で, ♂では25 mm前後, ♀では26 mm前後のものが多い; 2) 前胸背板は幅広く, 強く膨隆する; 3) 上翅も膨隆が顕著で, 肩, 最広部共に幅広く, 側縁は後方へ向けての狭まりが強い; 4) 陰茎先端部の形態は淡路島四国亜種のそれにほぼ等しく, 細長く, 中央で軽くくびれる. 内嚢, 指状片, 頂板の形態も淡路島四国亜種のそれに近い.
分布　鷲尾山を中心とする丘陵地帯 (高知県高知市南部). 鏡川 (かがみがわ), 高知市街地, 仁淀川 (によどがわ) によって囲まれた狭い地域に孤立分布する.
亜種小名の由来と通称　亜種小名は高知市のアマチュア昆虫研究家, 伊東善之に因む. 原記載論文中で提唱された通称はワシオヒメオサムシ.

24-3. *C. (O.) j. okinoshimanus* IMURA, DEJIMA et MIZUSAWA, 1993 ヒメオサムシ沖の島亜種

Original description   *Carabus* (*Ohomopterus*) *japonicus okinoshimanus* IMURA, DEJIMA & MIZUSAWA, 1993, Gekkan-Mushi, Tokyo, (264), p. 15, figs. 15, 16 (pl. 1) & 31 (p. 12).
Type locality   Moshima, Is. Oki-no-shima, Sukumo-shi, Kôchi Pref.
Type depository   National Museum of Nature and Science, Tokyo [Holotype, ♂].
Diagnosis   25~29 mm. Body above coppery, sometimes with faint greenish tinge on lateral margins; tibiae black. 1) Antennae not so long, barely reaching middle of elytra in male; 2) pronotum broad, about 1.35 times as wide as long, widest at middle, strongly narrowed toward base, with margins slightly sinuate at basal halves; hind angles short, with apices rounded; disk densely scattered with large punctures; basal foveae shallow; 3) elytra rather strongly convex above, widest at a little behind middle, with shoulders distinct; elytral margins almost parallel-sided above all in male; intervals wide and strongly raised, with primaries more frequently segmented than in other subspecies, primary foveoles deep and distinct; 4) apical part of penis almost as in subsp. *awajiensis*.
Distribution   Is. Oki-no-shima of Sukumo-shi (Kochi Pref.), SW. Shikoku.
Etymology   Named after the type locality, Is. Oki-no-shima.

タイプ産地　高知県，宿毛市，沖の島，母島．
タイプ標本の所在　国立科学博物館［ホロタイプ，♂］．
亜種の特徴　25~29 mm．背面は銅褐色で，体側縁は時に緑色光沢を帯びる．黒化型や緑化型は知られていない．脛節は黒色で赤みを欠く．1）触角は長くなく，先端は♂でも鞘翅中央を越えない；2）前胸背板は横長で，幅は長さの1.35倍前後．最広部はほぼ中央にあり，側縁は後方に向けて強く狭まり，中央より後方で弱く波曲する．後角は短く，先端は丸い．前胸背板の点刻はやや大きく，密で，基部凹陥は浅い；3）鞘翅は強く膨隆し，肩は強く張り出し，♂では側縁中央部が左右ほぼ平行で，最広部は中央よりやや後方にある．各間室は幅広く，強く隆起し，第一次原線は他亜種に比しより頻繁に分断され，第一次凹陥は深く明瞭に刻まれる．陰茎先端の形態は淡路島四国亜種のそれと大差ない．
分布　沖の島（高知県宿毛市）．
亜種小名の由来と通称　本亜種は香川県の出嶋利明によって最初に記録され，井村，水沢との共著で記載されたもので，亜種小名はタイプ産地の沖の島に因む．原記載論文中で提唱された通称はオキノシマヒメオサムシ．

24-4. *C. (O.) j. japonicus* MOTSCHULSKY, 1857 ヒメオサムシ基亜種

Original description, type locality & type depository   See explanation of the species.
Diagnosis   20~30 mm. As mentioned in the section of "History of research" of this species, it is most plausible that the type specimen of *Carabus japonicus* was collected from somewhere near Nagasaki of northwestern Kyushu. Although it is difficult to clearly define the distributional range of the nominotypical subspecies, here we tentatively regard the population distributed in the mainland Kyushu and several attached islands excluding those of already named populations as the nominotypical subspecies. The nominotypical *japonicus* is characterized mainly by well-developed podian lobes and terminal plates of internal sac of the male genital organ, but is variable greatly in the external appearance, suggesting that they might be classified into several more subspecies. For details on the geographical variation of this species in the mainland Kyushu and some attached islands, see IMURA & MIZUSAWA (1994).
Distribution   Kyushu & SW. tip of Chugoku Region / Iss. Hiko-shima of Simonoseki-shi (Yamaguchi Pref.), Ô-shima (= Chikuzen-ô-shima) of Munakata-shi, Genkai-jima, Shika-no-shima (Fukuoka Pref.), Kabé-shima, Ogawa-jima, Kakara-jima, Madara-shima, (Saga Pref.), Taka-shima, Maki-shima, Hario-jima, Nakadôri-jima, Naru-shima, Hisaka-jima, Fukué-jima, Saga-no-shima, (Nagasaki Pref.), Amakusa-kami-shima, Amakusa-shimo-jima, Gosho-no-ura-jima (or Gosho-ura-jima) (Kumamoto Pref.), Shoura-jima, Naga-shima, Kami-koshiki-jima & Naka-koshiki-jima (Kagoshima Pref.).

原記載・タイプ産地・タイプ標本の所在　種の項を参照．
亜種の特徴　20~30 mm．研究史の項で述べたように，ヒメオサムシのホロタイプは長崎近郊（例えば雲仙温泉等）から得られたものである可能性が高いが，基亜種の分布範囲を特定することはなかなか難しい．ここでは一応，九州本土とその付属島嶼の一部（ただし，すでに別亜種として記載されている集団の分布域を除く）に産するものを暫定的に基亜種とみなして解説するが，この範囲においても顕著な地理的変異が認められ，一亜種として括るには無理がある．背面は銅色を基本として，暗褐色，黒色，青黒色などのものが出現し，亜種としての最も顕著な特徴は，♂交尾器内嚢先端の脚葉の側方への突出が強く，頂板が木の葉形に細長く発達することにあるが，外部形態には地理的変異・個体変異が多い．佐賀県唐津のものはNAKANE（1961）により *Apotomopterus japonicus japonicus* f. *karatsuanus* と命名されていて，カラツオサムシなる通称まで与えられているが，この学名の中の *karatsuanus* は三語名に付加した第四名で，亜種よりも低位の実体に対して提唱された名なので，国際動物命名規約上，適格

ついては，井村・水沢（1994）に詳しい．
　　分布　九州および中国地方南西端／大島（＝筑前大島）［宗像市］，玄海島，志賀島（以上，福岡県），加部島，小川島，加唐島，馬渡島（以上，佐賀県），鷹島，牧島，針尾島，中通島，奈留島，久賀島，福江島，嵯峨ノ島（以上，長崎県），天草上島，天草下島，御所浦島（以上，熊本県），諸浦島，長島，上甑島，中甑島（以上，鹿児島県），彦島［下関市］（山口県）．山口県南西端の下関市にある彦島は，地方区分からは中国地方に属するが，この島に産する集団（暗青色を帯びた黒色のものが多く，かなり大型）はアキオサムシではなく，明らかにヒメオサムシ，それも基亜種群に近いものである．

24-5. *C. (O.) j. tsushimae* Breuning, 1932　ヒメオサムシ対馬亜種

Original description　*Carabus* (*Apotomopterus*) *tsuchimae* [sic] (misspelling of *tsushimae*) Breuning, 1932, Best.-Tab. eur. Coleopt., 104, p. 236.
Type locality　Insel Tsushima.
Type depository　Zoölogisch Museum Amsterdam [Holotype, ♂].
Diagnosis　21~26 mm. Body above coppery, head and pronotum partly with maroon reddish tinge, elytra bearing greenish tinge above all along margins; palpi, tibiae and tarsi dark rufous. 1) Antennae relatively long, extending middle of elytra in male; 2) pronotum cordate, strongly narrowed toward base, with hind angles subtriangularly produced; 3) terminal plate of internal sac apparently shorter than that of nominotypical *japonicus*, with ventral margin more strongly arcuate.
Distribution　Is. Tsushima (Kami-jima & Shimo-jima), Nagasaki Pref.
Etymology　Named after Iss. Tsushima, the type locality of this subspecies.

　　タイプ産地　対馬（長崎県）．
　　タイプ標本の所在　アムステルダム動物学博物館［ホロタイプ，♂］．
　　亜種の特徴　21~26 mm. 背面は銅色だが，頭部と前胸背板はえび茶色がかった赤褐色を帯び，鞘翅側縁は緑色光沢を帯びる．口肢，脛節，跗節は赤褐色．1) 触角は長く，♂では先端が鞘翅中央を越える；2) 前胸背板は心臓形で，後方に向けて強く狭まり，後角は三角形に突出する；3) ♂交尾器内嚢先端の頂板は，基亜種に比し明らかに短く，腹側縁はより強く弧を描く．
　　分布　対馬（地峡の人工的開削により主島が分断された後の，現在で言う上島，下島双方に産する）．
　　亜種小名の由来と通称　亜種小名はタイプ産地の対馬に因む．通称はツシマオサムシ．

24-6. *C. (O.) j. ikiensis* (Nakane, 1968)　ヒメオサムシ壱岐亜種

Original description　*Apotomopterus japonicus ikiensis* Nakane, 1968, Fragm. coleopterol., Kyoto, (21), p. 85.
Type locality　Iki Is., north of Kyushu, Japan.
Type depository　Hokkaido University Museum (Sapporo) [Holotype, ♂].
Diagnosis　24~31 mm. Body above coppery to reddish coppery bearing greenish tinge, or entirely black with bronze, dark green, dark blue or purplish tinge; tibiae dark brown. Characterized by the following points: 1) body broad and roundish in profile; 2) antennae short, not reaching middle of elytra even in male; 3) pronotum large and wide, with lateral margins strongly rounded, hardly emarginated behind, hind angles short and rounded at apices; 4) elytra long oval, widest apparently behind middle; 5) apical part of penis long and broad; terminal plate of internal sac apparently shorter and narrower than in nominotypical subspecies, with ventral margin gently rounded.
Distribution　Is. Iki (Nagasaki Pref.).
Etymology　Named after the type locality, Is. Iki.

　　タイプ産地　日本，九州の北，壱岐．
　　タイプ標本の所在　北海道大学総合博物館（札幌）［ホロタイプ，♂］．
　　亜種の特徴　24~31 mm. 背面は緑色光沢を帯びた銅～赤銅色，あるいは黒色で頭部と前胸背板が褐色又は暗緑色を，鞘翅側縁が暗青色又は紫褐色を帯びる．脛節は暗褐色．1) 体型は丸みを帯びて幅広い；2) 触角は短く，♂でも先端が鞘翅中央を越えない；3) 前胸背板は大きく，側縁は強く丸まり，後方で殆どえぐれず，後角の突出は弱く，先端は丸い；4) 鞘翅は長卵形で，最広部は明らかに中央より後方にある；5) 陰茎先端は太く長く，内嚢先端の頂板は基亜種より明らかに短く，やや幅が狭く，腹側縁は緩やかに弧を描く．
　　分布　壱岐（長崎県）．
　　亜種小名の由来と通称　亜種小名はタイプ産地の壱岐に因む．通称はイキオサムシ．

24-7. *C. (O.) j. hiradonis* Imura, Dejima et Mizusawa, 1993　ヒメオサムシ平戸島亜種

Original description　*Carabus* (*Ohomopterus*) *japonicus hiradonis* Imura, Dejima & Mizusawa, 1993, Gekkan-Mushi, Tokyo,

24. *Carabus* (*Ohomopterus*) *japonicus*

third (= outermost) primary intervals of elytra a little narrower than in other subspecies; 5) apical part of penis a little slenderer and more strongly tapered toward tip than in nominotypical subspecies of the mainland Kyushu, terminal plate of internal sac almost as in nominotypical subspecies.

Distribution　Is. Shimo-koshiki-jima (Kagoshima Pref.).

Etymology　Named after Shigeru ONODA who collected the type series of this subspecies.

タイプ産地　鹿児島県，薩摩郡，下甑島，口岳(現在は薩摩川内市(さつませんだいし)に属する).

タイプ標本の所在　国立科学博物館［ホロタイプ，♂］.

亜種の特徴　23~27 mm. 背面は金赤銅褐色で，一般に光沢が非常に強く，体側縁には顕著な緑色の金属光沢を帯びる．脛節は黒色で，赤みを欠く．1)前胸背板は中央の最広部から前方と後方に向けて強く狭まるため，やや縦長の六角形に近い輪郭を呈する；2)前胸背板の基部凹陥は小さく，あまり深く窪まない；3)鞘翅は，♂では左右の側縁中央部がほぼ平行に近くなるが，♀では紡錘形で中央で最も幅広くなり，翅端部はやや尖って見える；4)最外側の第一次間室は他亜種に比しやや細いものが多い；5)陰茎先端部は基亜種に比しやや細く，先端に向けてより顕著に細くなり，内嚢先端の頂板は基亜種のそれに近い．

分布　下甑島(鹿児島県).

亜種小名の由来と通称　亜種小名はタイプシリーズを採集した鹿児島の小野田 繁に因む．原記載論文中で提唱された通称はコシキヒメオサムシ．

24. *Carabus* (*Ohomopterus*) *japonicus*

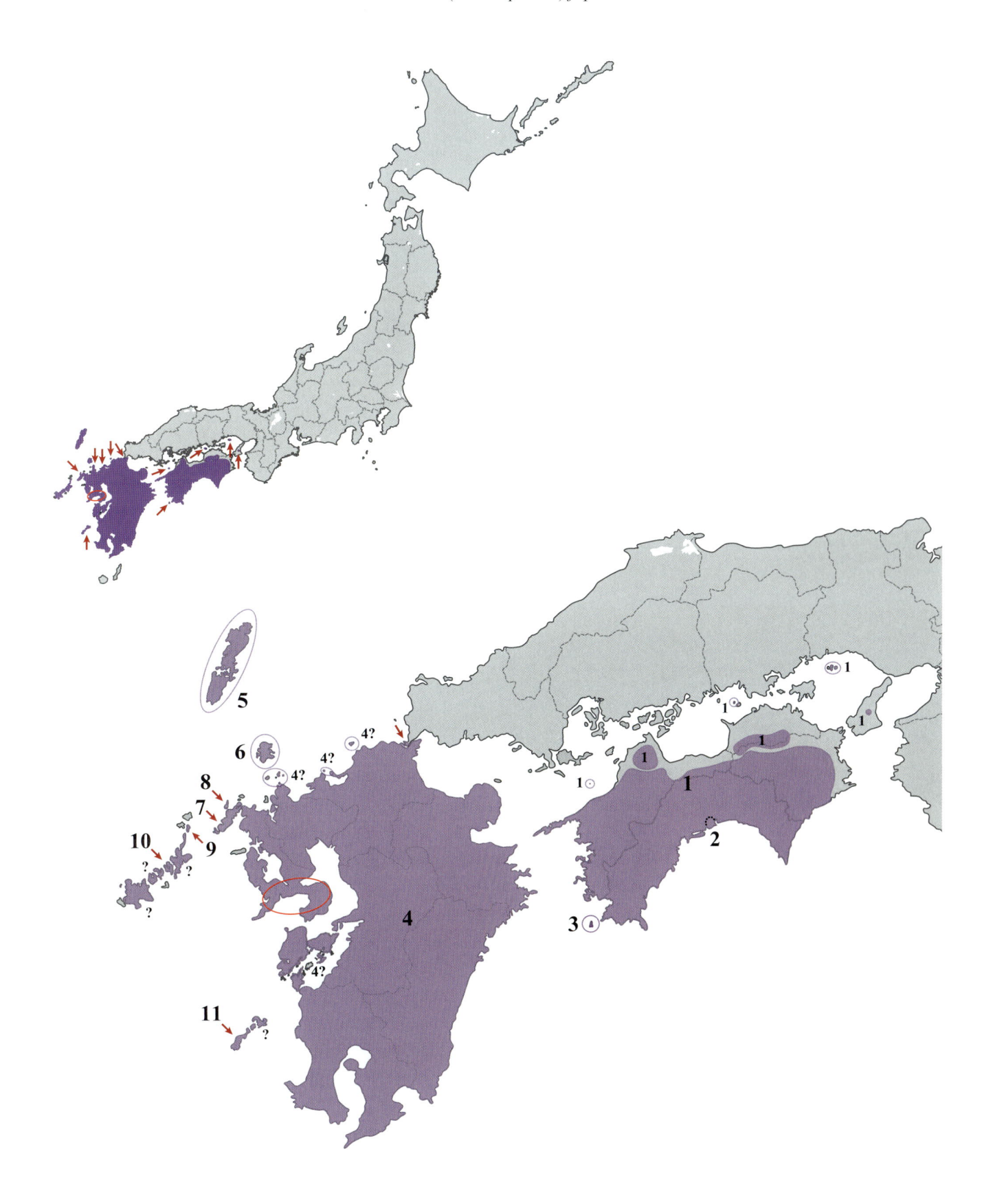

Fig. 43. Map showing the distributional range of ***Carabus*** (***Ohomopterus***) ***japonicus*** Motschulsky, 1858. — 1, Subsp. *awajiensis*; 2, subsp. *yoshiyukii*; 3, subsp. *okinoshimanus*; 4, subsp. *japonicus*; 5, subsp. *tsushimae*; 6, subsp. *ikiensis*; 7, subsp. *hiradonis*; 8, subsp. *chotaroi*; 9, subsp. *nozakicola*; 10, subsp. *wakamatsuensis*; 11, subsp. *onodai*. ◯: Possible type area of the species (near Nagasaki, e.g. Unzen).

図 43. ヒメオサムシ分布図. — 1, 淡路島四国亜種; 2, 鷲尾山亜種; 3, 沖の島亜種; 4, 基亜種; 5, 対馬亜種; 6, 壱岐亜種; 7, 平戸島亜種; 8, 生月島亜種; 9, 野崎島亜種; 10, 若松島亜種; 11, 下甑島亜種. ◯: 種のタイプ地域と思われる範囲（恐らく長崎からそう遠くない場所〜雲仙など〜）.

25. *Carabus* (*Ohomopterus*) *kawanoi* (Kamiyoshi et Mizoguchi, 1960)
アワオサムシ

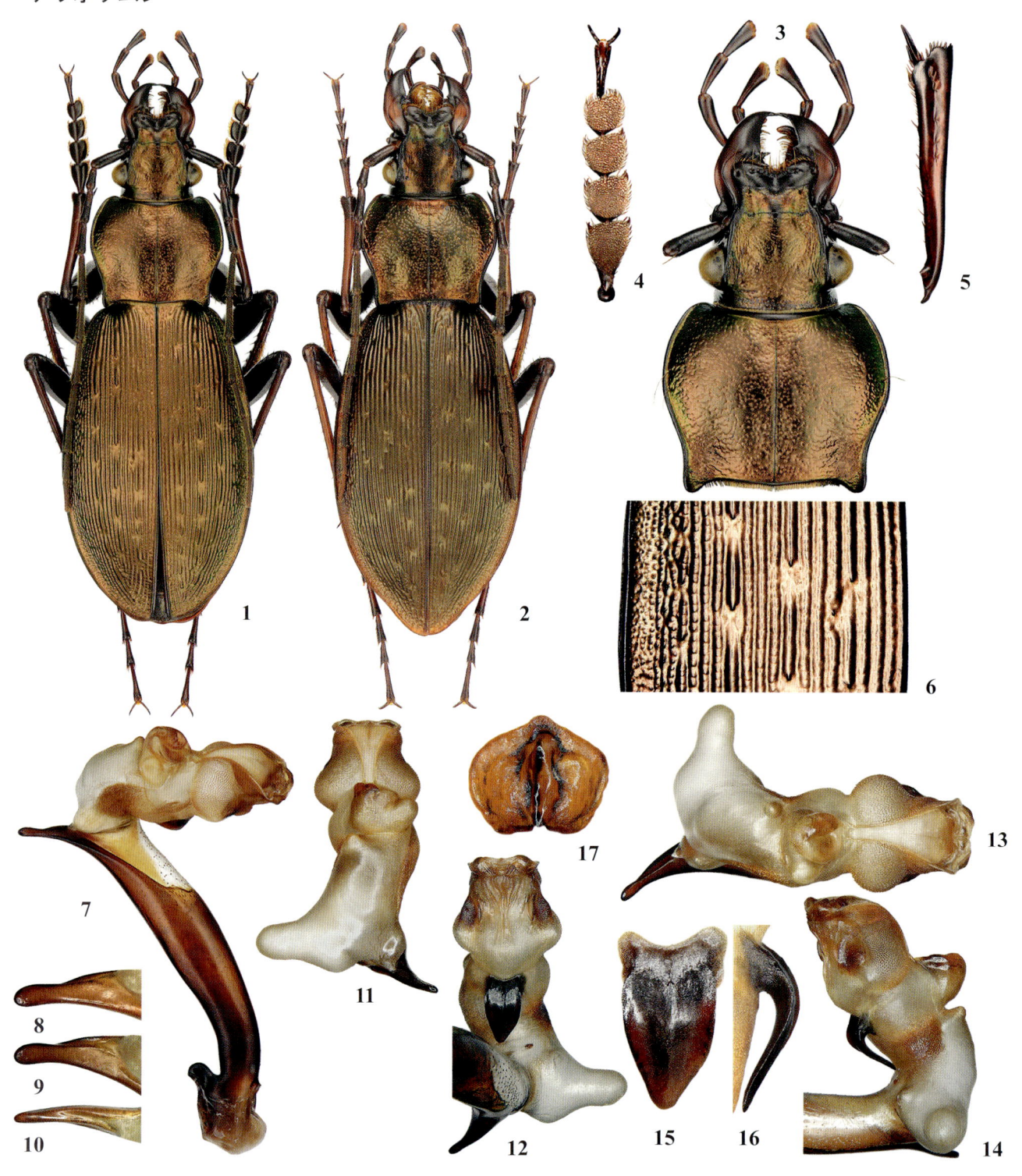

Figs. 1-17. *Carabus* (*Ohomopterus*) *kawanoi kawanoi* (Kamiyoshi et Mizoguchi, 1960) (Mt. Kôtsu-san, Yoshinogawa-shi, Tokushima Pref.). — 1, ♂; 2, ♀; 3, male head & pronotum; 4, male right protarsus in ventral view; 5, male right protibia in subdorsal view; 6, male left elytron (median part); 7, penis & fully everted internal sac in right lateral view; 8, apical part of penis in right lateral view; 9, ditto in right subdorsal view; 10, ditto in dorsal view; 11, internal sac in caudal view; 12, ditto in frontal view; 13, ditto in dorsal view; 14, ditto in left sublateral view; 15, digitulus in frontal view; 16, ditto in right lateral view; 17, inner plate of ligular apophysis.

25. *Carabus* (*Ohomopterus*) *kawanoi*

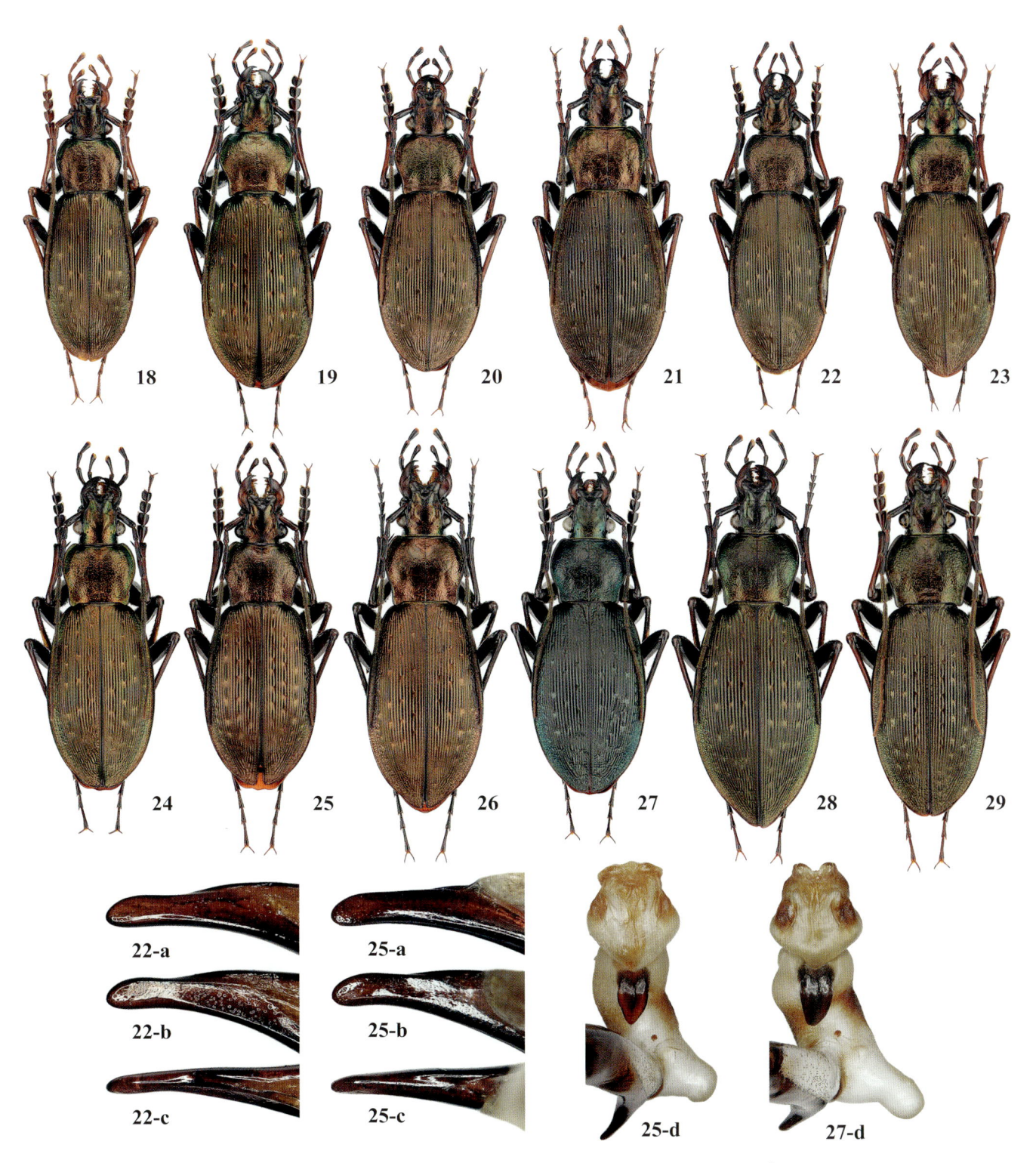

Figs. 18-26. *Carabus* (*Ohomopterus*) *kawanoi* subspp. — 18-24, *C*. (*O*.) *k*. *kawanoi* (18-20, 22, 24, ♂; 21, 23, ♀; 18, Mt. Tsurugi-san, Miyoshi-shi, Tokushima Pref.; 19, Higashi-iya-tsurui, Miyoshi-shi, Tokushima Pref.; 20, Mt. Shiraga-yama, Kami-shi, Kochi Pref.; 21, Tachibana, Sakihama-chô, Muroto-shi, Kochi Pref.; 22, Mt. Ôtaki-san, Takamatsu-shi, Kagawa Pref.; 23, Mt. Daisen-yama, Mannô-chô, Kagawa Pref.; 24, Mt. Jinga-yama, Kami-shi, Kochi Pref.); 25-26, *C*. (*O*.) *k*. *botchan* (25, ♂, HT, 26, ♀, PT; Mt. Inoko-yama, Matsuyama-shi, Ehime Pref.). — 27-28, Natural hybrid between *C*. (*O*.) *k*. *kawanoi* & *C*. (*O*.) *tosanus tosanus* (27, ♂, 28, ♀; Mt. Jinga-yama, Kami-shi, Kochi Pref.). — 29, Natural hybrid between *C*. (*O*.) *k*. *botchan* & *C*. (*O*.) *t*. *tosanus* (♂, S. of Mt. Higashi-sanpôga-mori, Tô'on-shi, Ehime Pref.). — a, Apical part of penis in right lateral view; b, ditto in right subdorsal view; c, ditto in dorsal view; d, fully everted internal sac in frontal view.

図 18-26. アワオサムシ各亜種. — 18-24, 基亜種(18-20, 22, 24, ♂; 21, 23, ♀; 18, 徳島県三好市剣山; 19, 徳島県三好市東祖谷釣井; 20, 高知県香美市白髪山; 21, 高知県室戸市佐喜浜町立花; 22, 香川県高松市大滝山; 23, 香川県まんのう町大川山; 24, 高知県香美市神賀山); 25-26, **高縄半島亜種**(25, ♂, HT, 26, ♀, PT; 愛媛県松山市伊之子山). — 27-28, アワオサムシ基亜種とトサオサムシ基亜種の自然雑種(27, ♂, 28, ♀; 高知県香美市神賀山). — 29, アワオサムシ高縄半島亜種とトサオサムシ基亜種の自然雑種(♂, 愛媛県東温市東三方ヶ森南方). — a, 陰茎先端(右側面観); b, 同(右斜背面観); c, 同(背面観); d, 内嚢(頭側面観).

25. *Carabus* (*Ohomopterus*) *kawanoi* (KAMIYOSHI et MIZOGUCHI, 1960)　アワオサムシ

Original description　*Apotomopterus kawanoi* KAMIYOSHI et O, [sic] MIZOGUCHI, 1960, Ins. Sci., p. 5, figs. A–C.

History of research　This taxon was originally described by KAMIYOSHI & MIZOGUCHI (1960) from Mt. Kôtsu-san of Tokushima Prefecture in eastern Shikoku as an independent species of the genus *Apotomopterus*. *Apotomopterus dehaanii hiraii* described by NAKANE (1961) from "Fukuwara near Mt. Tsurugi, Tokushima Pref., Shikoku" is its junior synonym. In current classification, this species is regarded as a member of the subgenus *Ohomopterus* of the genus *Carabus*. IMURA & MIZUSAWA (1995) described *C.* (*O.*) *kawanoi botchan* from the Takanawa Peninsula in Ehime Prefecture.

Type locality　Mt, [sic] Kotsu (= Kôtsu-san), Tokushima pref., Shikoku. [sic] Is, [sic] Japan.

Type depository　Osaka Museum of Natural History [Holotype, ♂].

Morphology　21~30 mm. Body above golden coppery to brownish coppery often with greenish tinge; palpi, mandibles, basal four segments of antennae, tibiae and tarsi are rufous, at least partly. External features: Figs. 1~6. Genitalic features: Figs. 7~17. Similar in external appearance to several species of the *japonicus* species-group, but a little larger and slenderer, and characterized by the following points: 1) antennae long, with apex extending middle of elytra even in female, palpi and legs also long; 2) male antennae without hairless area on ventral side; 3) pronotum cordiform, with lateral margins strongly emarginate behind middle, hind angles strongly protruded posteriad; 4) each elytron with four rows of primary intervals as chain striae, and umbilicate series is recognized at outer side of fourth primaries; 5) primary foveoles shallow but wide, partly invading adjacent tertiaries; 6) inner margin of male protibia faintly inflated though not angular; 7) apical part of penis hardly bent ventrad and gently dilated before apex in lateral view, with dorsum longitudinally grooved; digitulus long triangular, almost parallel-sided in basal half, with apex not sharply pointed; 8) inner plate of ligular apophysis peach-shaped or pentagonal, with apex obtusely rounded, inner wall rather smooth.

Molecular phylogeny　For details of the molecular phylogeny of *Ohomopterus*, see the same section of *C.* (*O.*) *yamato*. This species, together with *C.* (*O.*) *tosanus*, belongs to the *daisen* species-group, and forms the outgroup of the *japonicus-dehaanii* complex (SOTA & NAGATA, 2008).

Distribution　E. & NW. Shikoku (Tokushima Pref., S. Kagawa Pref., E. Kochi Pref. & N. Ehime Pref.).

Habitat　The range of vertical distributional of this species varies depending on locality; it mainly inhabits mountainous regions higher than 600 m in the Sanuki Hills, Takanawa Peninsula and Tsurugi-san Massif, but is found from much lower place in the southeastern part of its range, e.g., near the Cape Muroto-misaki of southeast Kochi Prefecture. Inhabitant of rather deep, matured forest, and prefers to pass the winter in the slope covered by the sand of weathered granite.

Bionomical notes　Spring breeder, earthworm feeder.

Geographical variation　Classified into two subspecies.

Etymology　Named after Ni-ichirô KAWANO (Tokushima).

研究史　本タクソンは神吉弘視（かみよしひろし）と溝口　修（みぞぐちおさむ）(1960)により徳島県中部の高越山からアポトモプテルス属の独立種として記載された．*Apotomopterus dehaanii hiraii* NAKANE, 1961（タイプ産地は徳島県の「剣山に近いふくわら」）は新参異名となるので使用できない．オオオサムシの亜種（NAKANE, 1961等），あるいは本書で言うトサオサムシの亜種（種和名としてはシコクオサムシ）（石川, 1985等）として扱われたこともあるが，本タクソンは一部の地域でオオオサムシ，トサオサムシそれぞれと混生していることから，独立種とみなすべきであろう．分布の中心は四国東部にあるが，愛媛県北部の高輪半島高所にもかけ離れた小分布圏を持つ．菅（かん）(1981)によれば，同所の集団は梅木　要らによって1964年，福見山（ふくみやま）から発見され，当時はフクミオサムシという名称で呼ばれていたと言う．小宮(1971)では*A. kawanoi*のn.(= natioナティオ)「高縄」として扱われたが，その後，高縄半島各地から生息地が発見され，井村・水沢(1995)により*botchan*という亜種小名を与えられて正式に記載された．

タイプ産地　日本，四国，徳島県，高越山（現徳島県吉野川市）．

タイプ標本の所在　大阪市立自然史博物館［ホロタイプ，♂］．原記載論文には本種のホロタイプの保管場所が「昆虫団体研究会オサムシグループ」と記されているが，著者らの意向によりその後，大阪市立自然史博物館に寄贈されたとのことである．

形態　21~30 mm. 背面は金銅色ないし銅褐色でしばしば緑色光沢を帯び，口肢，大顎，触角基部4節，および脛節と附節の少なくとも一部は赤褐色．外部形態：図1~6．交尾器形態：図7~17．ヒメオサムシ種群の各種に比し，より大型で細く，以下のような特徴を有する：1)触角は長く，♀でも先端は鞘翅中央を越え，口肢，四肢も長い；2)♂触角腹面に無毛部はない；3)前胸背板の両側縁は後方で強くえぐれて顕著な心臓形をなし，後角の突出も強い；4)鞘翅には各4条の第一次間室があり（この特徴は，以下に述べるトサオサムシとオオオサムシにも共有される，オオオサムシ種群に固有の形質），丘孔点列は第4第一次間室の外側にある；5)鞘翅第一次回陥は浅いが広く，隣接する第三次間室を侵す；6)♂前脛節内側縁中央は僅かに膨らむが，角張らない；7)陰茎先端部は側面から見て腹側へ殆ど湾曲せず，先端手前で軽く膨らみ，背側縁には溝状の窪みがある．指状片は長三角形で，基半部は左右ほぼ平行となり，先端は鈍く丸まる；8)♀交尾器腔底部節片内板は桃の実形ないし五角形で，先端部は鈍く丸まり，内側壁の表面は比較的滑らか．本種以下，オオオサムシまでの3種は，比較的大型で，各鞘翅には4条の第一次間室を持ち，指状片が小さい長三角形を呈するという形質を共有し，オオオサムシ種群としてまとめられる．

分子系統　オオモプテルス亜属の分子系統については*C.* (*O.*) *yamato*の同項を参照．核遺伝子配列を用いた系統情報の結果から，本種はトサオサムシと共にダイセンオサムシ種群に属し，ヒメオサムシ-オオオサムシ種群の外群を構成する(SOTA & NAGATA, 2008)．

分布 四国(徳島県, 香川県南部, 高知県東部, 愛媛県北部). 徳島県中西部の剣山地一帯から高知県東部, 及び讃岐山脈と高縄半島に分布する. 分布の南限は室戸岬北方の室戸市界隈(金剛頂寺(こんごうちょうじ)の南方あたりまで記録されているもよう). トサオサムシとはほぼ側所的に棲み分けており, 両者の分布境界は祖谷川(いやがわ)(吉野川中上流に南から流入する支流)と楮佐古川(かじさこがわ)(物部川(ものべがわ)中流に北から流入する支流)を結ぶ線にほぼ一致しており, 東方に本種, 西方にトサオサムシが分布する. 高知県香美市(かみし)中北部にある神賀山(じんがやま)では一部でトサオサムシと混生しており, 両種の交雑により生じたと思われる個体(図27, 28)が得られている. 愛媛県では高縄半島南東部の東三方が森(ひがしさんぼうがもり)と窓峠(まどとうげ)の間を境として北側に本種, 南側にトサオサムシが分布し, ここでも両者の交雑により生じたと思われる個体が得られている.

生息環境 分布域北部の讃岐山脈や高縄半島, 徳島県中部の剣山塊では主として標高600 m以上の高所に分布し, 垂直分布の上限は1,950 m近くに達するが, 分布域南東部では生息域の下限が低くなり, 高知県安芸郡(あきぐん)北川村(きたがわむら)の一部等では標高200 m足らずの低所からも得られている. 香川, 徳島, 高知の各県では, 最高点の標高にかかわらず, 生息する山塊の山頂付近に集中して生息している場合が多い. 高縄半島では主として標高600 m以上の成熟した森林に生息し, 暗く陰湿で下草の殆ど生えていないような環境を好む. 越冬中の成虫は主として土中から得られるが, 朽木中から発見されることもあり, また盛夏でもすでに土中で越冬態勢に入っている個体を掘り出すことができるという. 越冬場所としては, 花崗岩の風化した細かい砂に覆われた斜面を最も好む(菅, 1981).

生態 春繁殖・成虫越冬型で, 幼虫の主たる食餌は環形動物(フトミミズ等)であろうと思われる.

地理的変異 基亜種と高縄半島亜種の2亜種に分類されている.

種小名の由来 徳島県の昆虫愛好家, 河野仁一郎に因む. ツルギオサムシという名で呼ばれることもあり, また, 本タクソンをトサオサムシの亜種とみなし, トサオサムシ石鎚山脈亜種を合わせて *tosanus, ishizuchianus, kawanoi* の三者からなる分類群(種としては *C. tosanus*)に対してシコクオサムシという和名が用いられたこともある(石川, 1985等).

25-1. *C. (O.) k. kawanoi* (KAMIYOSHI et MIZOGUCHI, 1960) アワオサムシ基亜種

Original description, type locality & type depository　See explanation of the species.
Diagnosis　21~26 mm. Body above golden coppery or brownish coppery, often with greenish tinge. For diagnosis of this subspecies, see explanation of the species and that of subsp. *botchan*.
Distribution　E. Shikoku (Tokushima Pref., S. Kagawa Pref. (upper part of Sanuki Hills) & E. Kochi Pref.).

原記載, タイプ産地, タイプ標本の所在　種の項を参照.
亜種の特徴　21~26 mm. 背面は金銅色ないし銅褐色でしばしば緑色光沢を帯びる. 亜種としての特徴は, 種の解説ならびに高縄半島亜種の項を参照.
分布　四国東部(徳島県, 香川県南部(讃岐山脈高所), 高知県東部).

25-2. *C. (O.) k. botchan* IMURA et MIZUSAWA, 1995 アワオサムシ高縄半島亜種

Original description　*Carabus (Ohomopterus) kawanoi botchan* IMURA & MIZUSAWA, 1995, Gekkan-Mushi, Tokyo, (293), p. 13, figs. 7 & 14.
Type locality　Mt. Inoko-yama~Mt. Myôjin-ga-mori, 850m alt., Matsuyama-shi, Ehime Pref.
Type depository　National Museum of Nature and Science, Tokyo [Holotype, ♂].
Diagnosis　25~30mm. Discriminated from the nominotypical subspecies in the following points: 1) body above a little more strongly shiny; 2) size a little larger on an average, with wider pronotum and elytra, shoulders a little more distinct; 3) sutural part of elytra a little more strongly convex above; 4) apical part of penis less narrowly constricted at middle in lateral view, with median longitudinal groove on dorsum more shallowly impressed.
Distribution　NW. Shikoku (mountainous area of Takanawa Penins. in Ehime Pref.).
Etymology　"Botchan" means a son from a good family or so-called greenhorn in Japanese, and is the title of a well-known novel written by Soseki NATSUME, the scene of which is Matsuyama City lying near the distributional range of this subspecies.

タイプ産地　愛媛県, 松山市, 伊之子山～明神ヶ森, 標高850 m.
タイプ標本の所在　国立科学博物館[ホロタイプ, ♂].
亜種の特徴　25~30 mm. 基亜種からは以下の諸点により識別される: 1) 背面の金属光沢がより強い; 2) 平均してやや大型で, 前胸背板の幅がより広い. 鞘翅もやや幅が広く, 肩はより強く張り出す; 3) 鞘翅会合部はより強く膨隆する; 4) 陰茎先端はより太短く, 側方から見て中央部のくびれが弱く, 背面の溝はより浅い.
分布　高縄半島(東～南部). 高縄半島の特産で, 福見山, 白潰(しらつえ), 伊之子山, 東三方が森, 楢原山(ならばらさん), 今治市玉川町上木地(かみきじ), 同奥木地(おくきじ), 五葉が森(ごようがもり)などから記録があるが, 高縄山(たかなわさん)を中心とする山塊(大月山(おおつきやま), 北三方が森(きたさんぼうがもり)など)からは発見されていない.
亜種小名の由来と通称　本亜種の分布域に近い愛媛県松山を舞台とした夏目漱石の小説「坊っちゃん」に因む. フクミオサムシ, ボッチャンオサムシといった通称がある.

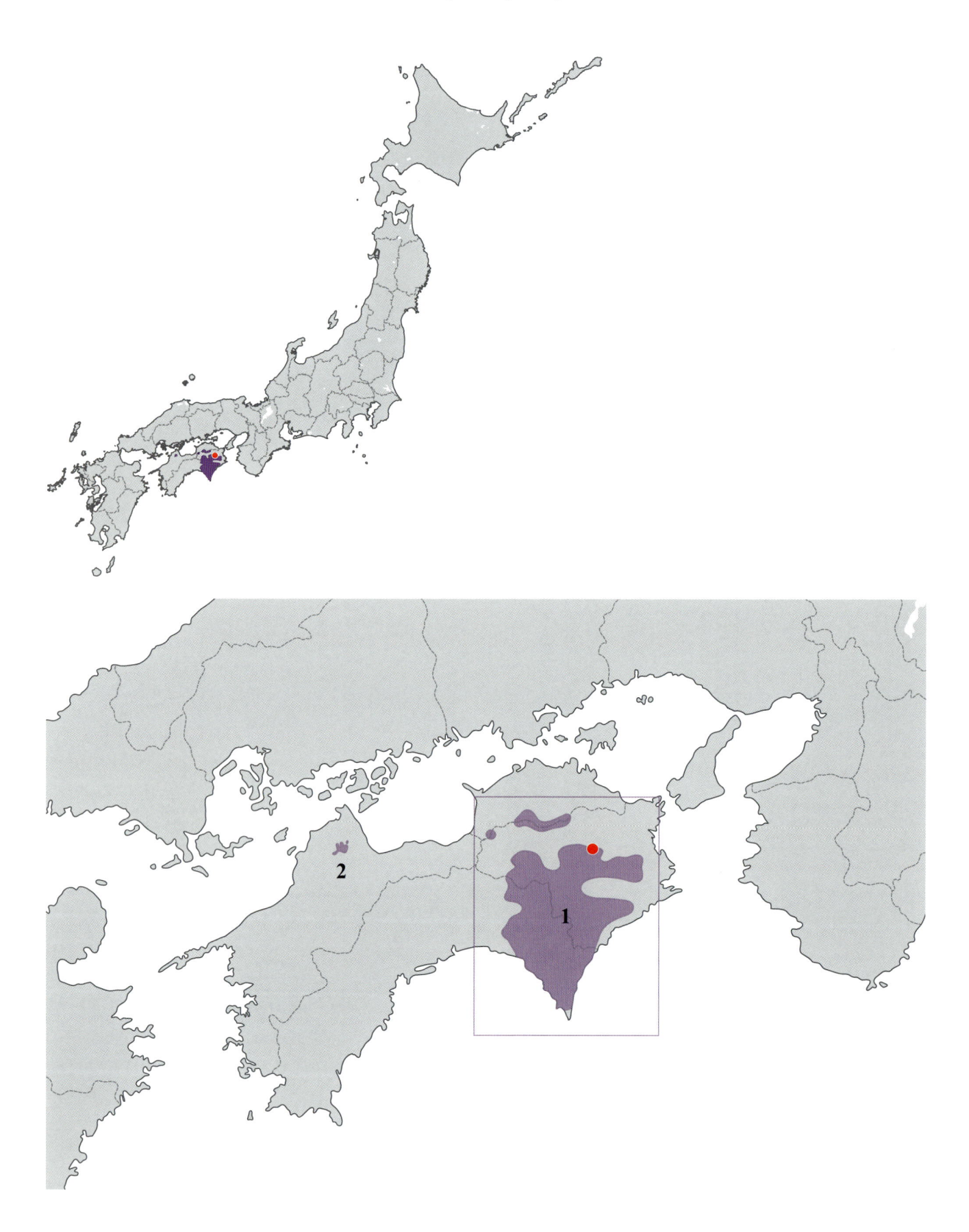

25. *Carabus* (*Ohomopterus*) *kawanoi*

Fig. 30. Map showing the distributional range of ***Carabus*** (***Ohomopterus***) ***kawanoi*** (Kamiyoshi et O. Mizoguchi, 1960). — 1, Subsp. ***kawanoi***; 2, subsp. *boichun*. ●: Type locality of the species (Mt. Kotsu = Mt. Kōtsu-san).
図 30. アワオサムシの分布図 — 1, 基亜種；2, 高縄半島亜種. ●, 種のタイプ産地（高越山）.

25. Carabus (Ohomopterus) kawanoi

Fig. 31. Habitat of *Carabus* (*Ohomopterus*) *kawanoi botchan*, *C.* (*O.*) *japonicus awajiensis*, *C.* (*O.*) *yaconinus yamaokai*, *C.* (*Leptocarabus*) *hiurai* and *C.* (*Damaster*) *blaptoides blaptoides* (Mt. Fukumi-yama, Tô'on-shi, Ehime Pref.).

図31. アワオサムシ高縄半島亜種, ヒメオサムシ淡路島四国亜種, ヤコンオサムシ四国和歌山亜種, シコククロナガオサムシ, マイマイカブリ基亜種の生息地(愛媛県東温市福見山).

Fig. 32. *Carabus* (*Ohomopterus*) *kawanoi botchan* (Mt. Fukumi-yama, Tô'on-shi, Ehime Pref.).

図32. アワオサムシ高縄半島亜種(愛媛県東温市福見山).

Fig. 33. *Carabus* (*Ohomopterus*) *kawanoi botchan* (Mt. Fukumi-yama, Tô'on-shi, Ehime Pref.).

図33. アワオサムシ高縄半島亜種(愛媛県東温市福見山).

26. *Carabus* (*Ohomopterus*) *tosanus* (Nakane, Iga et Uéno, 1953)
トサオサムシ

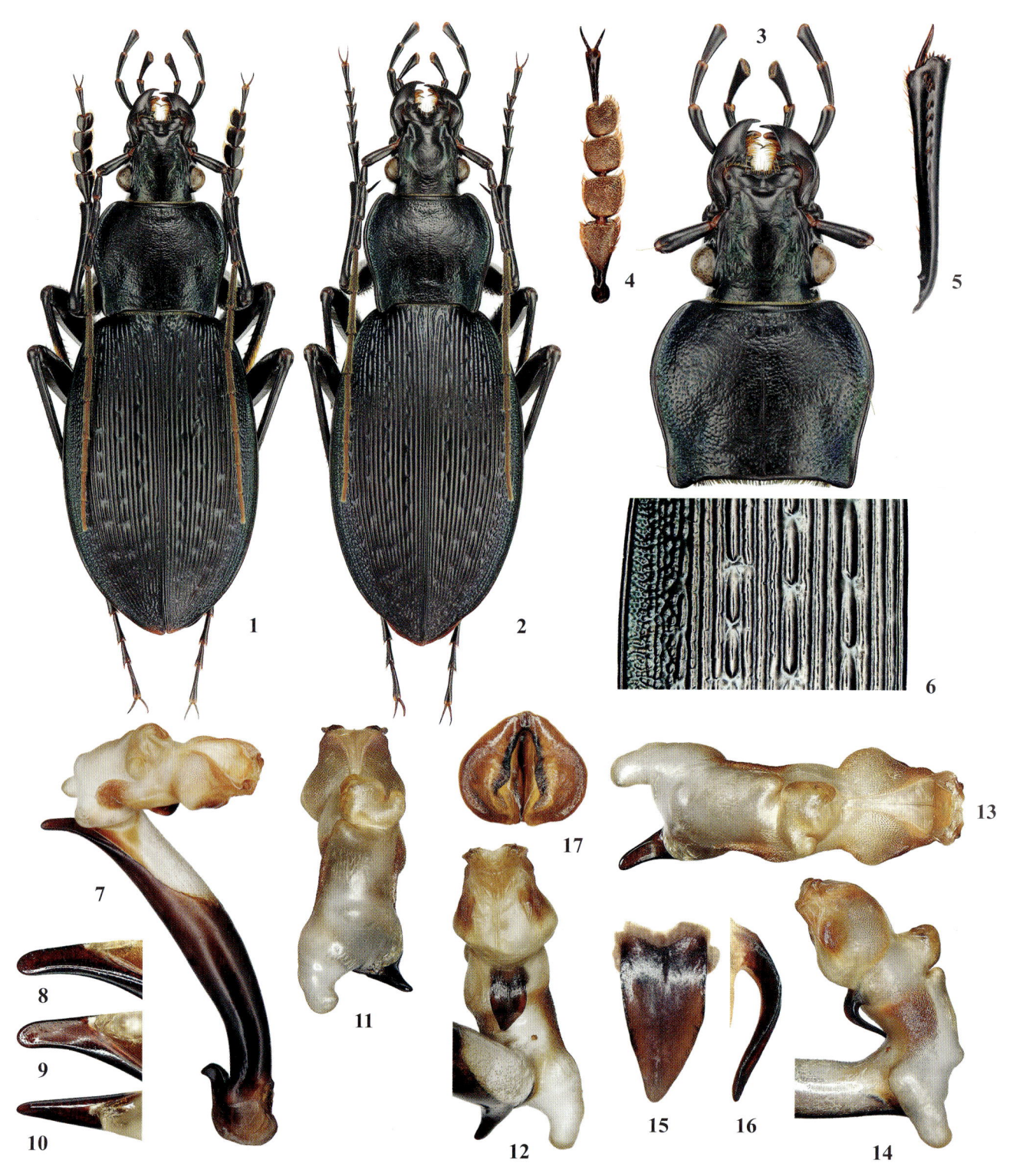

Figs. 1-17. *Carabus* (*Ohomopterus*) *tosanus tosanus* (Nakane, Iga et Uéno, 1953) (Mt. Kaji-ga-mori, Ôtoyo-chô, Kochi Pref.). — 1, ♂; 2, ♀; 3, male head & pronotum; 4, male right protarsus in ventral view; 5, male right protibia in subdorsal view; 6, male left elytron (median part); 7, penis & fully everted internal sac in right lateral view; 8, apical part of penis in right lateral view; 9, ditto in right subdorsal view; 10, ditto in dorsal view; 11, internal sac in caudal view; 12, ditto in frontal view; 13, ditto in dorsal view; 14, ditto in left sublateral view; 15, digitulus in frontal view; 16, ditto in right lateral view; 17, inner plate of ligular apophysis.

図 1-17. **トサオサムシ基亜種**(高知県大豊町梶ヶ森). — 1, ♂; 2, ♀; 3, ♂の頭部と前胸背板; 4, ♂右前跗節(腹面観); 5, ♂右前脛節(斜背面観); 6, ♂左鞘翅中央部; 7, 陰茎と完全反転下の内嚢(右側面観); 8, 陰茎先端(右側面観); 9, 同(右斜背面観); 10, 同(背面観); 11, 内嚢(尾側面観); 12, 同(頭側面観); 13, 同(背面観); 14, 同(左斜側面観); 15, 指状片(頭側面観); 16, 同(右側面観); 17, 腟底部節片内板.

26. *Carabus* (*Ohomopterus*) *tosanus*

Figs. 18-30. ***Carabus*** (***Ohomopterus***) ***tosanus*** subspp. — 18-25, ***C.*** (***O.***) ***t. tosanus*** (19-21, 24, ♂, 18, 22, 23, 25, ♀; 18, Chisei, Uchiko-chô, Ehime Pref.; 19, Is. Ô-shima, Sukumo-shi, Kochi Pref.; 20, Kawa-no-uchi, Yawatahama-shi, Ehime Pref.; 21, Pass Mado-tôgé, Saijô-shi, Ehime Pref.; 22, NW. slope of Mt. Ishizuchi-san, Saijô-shi, Ehime Pref.; 23, Mt. Akaboshi-yama, Shikoku-chûô-shi, Ehime Pref.; 24, Teragawa, Ino-chô, Kochi Pref.; 25, N. slope of Mt. Heiké-daira, Niihama-shi, Ehime Pref.); 26-30, ***C.*** (***O.***) ***t. ishizuchianus*** (26, 28-30, ♂, 27, ♀; 26-27, Mt. Dô-ga-mori, Saijô-shi, Ehime Pref.; 28, Mt. Ibuki-san, Saijô-shi, Ehime Pref.; 29, Mt. Kamé-ga-mori, Saijô-shi, Ehime Pref.; 30, Mt. Higashi-akaishi-yama, Niihama-shi, Ehime Pref.). — a, Apical part of penis in right subdorsal view; b, ditto in dorsal view; c, fully everted internal sac in frontal view.

図 18-30. トサオサムシ各亜種. — 18-25, 基亜種 (19-21, 24, ♂, 18, 22, 23, 25, ♀; 18, 愛媛県内子町知清; 19, 高知県宿毛市大島; 20, 愛媛県八幡浜市川之内; 21, 愛媛県西条市窓峠; 22, 愛媛県西条市石鎚山北西麓; 23, 愛媛県四国中央市赤星山; 24, 高知県いの町寺川; 25, 愛媛県新居浜市平家平北麓); 26-30, 石鎚山脈亜種 (26, 28-30, ♂, 27, ♀; 26-27, 愛媛県西条市堂ヶ森; 28, 愛媛県西条市伊吹山; 29, 愛媛県西条市瓶ヶ森; 30, 愛媛県新居浜市東赤石山). — a, 陰茎先端 (右斜背面観); b, 同 (背面観); c, 内嚢 (頭側面観).

26. *Carabus* (*Ohomopterus*) *tosanus* (Nakane, Iga et Uéno, 1953) トサオサムシ

Original description   *Apotomopterus dehaani* [sic] *tosanus* Nakane, Iga et Uéno, 1953, Sci. Rep. Saikyo Univ. (Nat. Sci. & Liv. Sci.), Kyoto, 1, p. 93, figs. 1, A-b, b' & 8 (p. 100).

History of research   This taxon was originally described by Nakane, Iga & Ueno (1953) as a subspecies of *Apotomopterus dehaani* [sic] (= *Carabus* (*Ohomopterus*) *dehaanii* in the present sense) from Mt. Kaji-ga-mori and Mt. Kuishi-yama of Kochi Prefecture (in Nakane, 1953), though the lectotype designation has not been made as yet. In the same paper, another subspecies of *A. dehaani* [sic] was described by Nakane under the name of *ishizuchianus* from the Ishizuchi Mountains of northwestern Shikoku. Komiya (1971) and Kan (1981) followed Nakane's view, and they regarded these two taxa as subspecies of *A. dehaanii*, and *A. kawanoi* as different species. Ishikawa (1985) treated *Carabus* (*Ohomopterus*) *tosanus* as a distinct species, and regarded *kawanoi* and *ishizuchianus* as its subspecies. In this book, *C.* (*O.*) *tosanus* and *C.* (*O.*) *kawanoi* are regarded as two different species, and the former is classified into two subspecies, the nominotypical *tosanus* and *ishizuchianus*.

Type locality   Kajigamori, Tosa; Mt. Kuishi, Tosa.

Type depository   Hokkaido University Museum (Sapporo) [Syntype (a male from "KAJIKAMORI [sic]")].

Morphology   21~34 mm. Coloration varies by subspecies or population; usually black with dark greenish or coppery tinge, occasionally coppery with greenish or reddish tinge in nominotypical subspecies, coppery brown more or less with greenish tinge in subsp. *ishizuchianus*; palpi and legs black in nominotypical subspecies, often dark rufous in subsp. *ishizuchianus*. External features: Figs. 1~6. Genitalic features: Figs. 7~17. Allied to *C.* (*O.*) *kawanoi* and both the species share a common characteristic of having four rows of primary intervals in each elytron, but distinguished from that species in the following points: 1) size larger on an average, above all in nominotypical subspecies; 2) color different as mentioned above excepting subsp. *ishizuchianus* whose coloration is very similar to that of *C.* (*O.*) *kawanoi*; 3) antennae extremely long, reaching apical third in male, extending middle of elytra even in female; palpi and legs also very long; 4) pronotum with shallower emargination of lateral margin behind middle, above all in nominotypical subspecies; disk flatter, with lateral grooves on both sides a little wider, basal foveae longitudinally elongate and convergent anteriad; 5) primary foveoles of elytra not invading adjacent tertiaries; 6) left basal lateral lobe of internal sac not protruding laterad as in *C.* (*O.*) *kawanoi*, but apparently bent ventrad and curved inward; 7) digitulus a little asymmetric, with tip not sharply pointed.

Molecular phylogeny   For details of the molecular phylogeny of *Ohomopterus*, see the same section of *C.* (*O.*) *yamato*. For the phylogenetic position of this species, see the same section of *C.* (*O.*) *kawanoi*.

Distribution   C.~W. Shikoku / Is. Ô-shima of Sukumo-shi (Kochi Pref.). This species is essentially parapatric with *C.* (*O.*) *kawanoi*, but their ranges partly overlap each other in a few localities such as Mt. Jinga-yama of Kami-shi, forming a narrow hybrid zone.

Habitat   The nominotypical subspecies is rather widely distributed from low hills to high mountains with the altitude of nearly 1,600 m. Subspecies *ishizuchianus* inhabits higher places, usually in the subalpine zone between 1,000 m and 1,900 m, though rarely recorded from lower place (under 600 m) in the northern slope of the Ishizuchi Mountains, e.g., northern slope of Mt. Kanpû-zan. Sylvicolous, inhabiting rather deep natural forest of evergreen- or deciduous broadleaved tree (low to middle altitudinal area) or mixed forest (high altitudinal area).

Bionomical notes   Spring breeder, earthworm feeder.

Etymology   Named after Tosa or Tosa-no-kuni, one of the ryôsei provinces of Japan used until the early Meiji period, which corresponds to the present Kochi Prefecture.

研究史　本タクソンは，Nakane (1953)による論文中で，Nakane, Iga & Uéno三名の共著によりApotomopterus dehaani [sic] tosanus，すなわちオオオサムシの亜種として記載された．原記載で示されたシンタイプの産地は高知県中東部の梶ヶ森と同北部の工石山であるが，レクトタイプの指定はまだなされていない．同じ論文の中でNakaneの単著により，やはりオオオサムシの亜種として愛媛県の石鎚山堂ヶ森からishizuchianusが記載された．小宮(1971)はNakane(1953)と同様，これらをオオオサムシの亜種とみなし，A. kawanoi (アワオサムシ)は別種とし，菅(かん)(1981)もこの扱いを踏襲したが，石川(1985)は独立種Carabus (Ohomopterus) tosanus (種和名はシコクオサムシ)を認め，kawanoiとishizuchianusをその亜種に位置付けた．本書ではtosanusとkawanoiの名によって代表される二つのタクソンが一部の地域で混生している事実を重視し，両者をそれぞれ独立種 (C. (O.) tosanusトサオサムシとC. (O.) kawanoiアワオサムシ)として扱い，ishizuchianusは前者の亜種とみなした．しかし，tosanusとishizuchianusの関係については，同種内の亜種同士とみなすべきか，あるいはそれぞれを別種とみなすべきか，判断に迷う面もあり，さらなる調査・研究が望まれる．

タイプ産地　土佐，梶ヶ森；土佐，工石山．

タイプ標本の所在　北海道大学総合博物館(札幌)[シンタイプ]．同館の中根コレクションには現在，梶ヶ森産(ラベルには"TOSA / KAJIKAMORI / (SHIKOKU) / VI. 21. 1938 / Coll M. IGA"と記されている)の本種1♂が収蔵されており，SYNTYPEのラベルが付されている．本種のレクトタイプとすべき標本であるが，まだその指定は行われていない．

形態　21~34 mm. 基亜種と石鎚山脈亜種で背面の色彩が大きく異なる．基亜種は基本的に暗緑色光沢を帯びた黒色のものが多いが，時には青紫や，銅褐色，もしくは紅色を帯びた細長で太めの色，紅色に近い独特の色彩を呈する個体も出現する．石鎚山脈亜種は基本的に銅褐色ないし赤銅褐色で，しばしば緑色光沢を帯び，ごく稀に全体が緑色を強く帯びた個体も出現する．口肢

と脚は基亜種では黒色だが，石鎚山脈亜種ではしばしば暗赤褐色となる．外部形態：図1~6．交尾器形態：図7~17．外部形態は基亜種ではオオオサムシに，石鎚山脈亜種ではアワオサムシに酷似し，交尾器形態は両亜種ともオオオサムシに近いが，以下の諸点で異なる：1) 基亜種はアワオサムシに比し大型；2) 色彩が異なる（ただし，石鎚山脈亜種はアワオサムシの色彩に近い）；3) 触角は非常に長く，その先端は♂では鞘翅先端1/3に達し，♀でも鞘翅中央を優に越す．また，口肢と脚も長い；4) 前胸背板側縁後方のえぐれがより浅く（特に基亜種），前胸背板表面は比較的平坦で，側溝はやや広く，基部凹陥は縦長で前方に収斂する；5) 鞘翅第一次凹陥は第三次間室を浸さない；6) ♂交尾器内嚢の左基部側葉は，アワオサムシのように側方に向けて水平に伸長せず，腹側に向けて伸長し，先端がやや内湾する．概形はオオオサムシのそれに近いが，本種の左基部側葉のほうがより太い；7) 指状片の両側縁は左右やや非対称で，先端は丸みを帯びる．これに対し，オオオサムシでは指状片が左右ほぼ対称で，先端はより鋭く尖る．

分子系統 オオオサムシ亜属の分子系統一般についてはヤマトオサムシの同項を，亜属内における本種の系統学的位置についてはアワオサムシの同項を参照．

分布 四国（中〜西部）／大島［宿毛市］（高知県）．分布の東限は徳島県三好市（みよし）西部から，高知県香美市（かみし）にかけての地域で，現在までに知られている東限は香美市神賀山（じんがやま）南麓．祖谷川（いやがわ）と楮佐古川（かじさこがわ）（物部川（ものべがわ）中流北側の支流）を結ぶ線が東縁となっているようで，それより東方からの記録はない．三好市の旧西祖谷山村（にしいややまそん）南西部から高知県大豊町東部にかけての地域では，本種とアワオサムシ基亜種が側所的に棲み分けているが，両者の分布が重なる一部の地域では交雑により生じたと思われる個体（図27, 28）が得られている（香美市神賀山など）．北限は愛媛県高縄半島（たかなわはんとう）基部の窓峠（まどとうげ）界隈で，同所ではアワオサムシ高縄半島亜種との自然雑種と思われる個体（図29）が得られている．北は石鎚山塊から法皇山脈（ほうおうさんみゃく）にかけて分布するが，瀬戸内海に面した北側斜面からの記録は少ない（全くないわけではないが，極めて散発的）．愛媛県東端の四国中央市旧新宮村（しんぐうむら）付近では，オオオサムシの分布域にかなり接近した地点からも記録されているが，両種の混生地はまだ発見されていない．分布の南西域では，高知県南西端の海岸近くまでほぼ普遍的に分布するようである．ただし，佐田岬（さたみさき）半島からの記録はなく，同半島基部付近における西限は八幡浜市（やわたはまし）田浪（たなみ）から古藪（ふるやぶ）にかけての地域と思われる．

生息環境 基亜種は平地に近い場所から，通常は1,000 m以下の中山帯にかけて分布しているが，標高1,600 m近い高所にも生息している．石鎚山脈亜種は通常，1,000 m以上の地域に産するが，石鎚山脈北側斜面では生息下限標高が低くなり，寒風山北麓では560~570 mの低所からも記録されている（冨永, 1984）．すなわち，これら2"亜種"は必ずしも標高差によって棲み分けているわけではなく，同一山塊において基亜種の方が石鎚山脈亜種よりも高所にまで進出して，両者の生息標高に逆転現象が見られる場合もある．両者の典型的な混棲地はまだ発見されていないが，水平距離にして僅か500 m程の至近距離をおいてこれら2"亜種"が側所的に分布し，一部の個体にかなり多彩な変異傾向の見られる場所もみつかっており，両者の関係を同一種内における亜種とみなすのが妥当なのか，或いはアワオサムシとトサオサムシの関係と同様，種のレベルで異なるものと捉えるべきかについては更なる検討を要する．森林性で，比較的湿潤な環境を好み，低地では照葉樹林や杉の植林地，高地では落葉広葉樹林や針広混交林，さらに亜高山帯針葉樹林にかけて生息する．

生態 春繁殖・成虫越冬型のオサムシで，幼虫は環形動物（フトミミズ等）食．石鎚山脈亜種では，越冬成虫の啓蟄は5月下旬頃で，この頃には日中，歩行中の個体が採集されることも多いという．交尾・産卵の後，新成虫は8月下旬から10月中旬にかけて出現し，9月にトラップ採集を行うと最も良好な成果が得られるという（菅, 1981）．基亜種の生態に関する報告は少ない．

地理的変異 基亜種と石鎚山脈亜種に分けられるが，生息環境の項でも述べたように，両者が同一種内の亜種関係にあるのか，あるいはそれぞれを独立種とみなすべきかについては検討が必要と思われる．

種小名の由来等 種小名は，かつての令制国（りょうせいこく）の一つで現在の高知県にほぼ相当する土佐国（とさのくに）に因む．本書では独立種として扱ったアワオサムシをトサオサムシの亜種とみなし，*tosanus*, *ishizuchianus*, *kawanoi* の三つからなる分類群（種としては *C. (O.) tosanus*）に対してシコクオサムシという和名が用いられることもある．

26-1. *C. (O.) t. tosanus* (NAKANE, IGA et UÉNO, 1953) トサオサムシ基亜種

Original description, type locality & type depository See explanation of the species.

Diagnosis 25~34 mm. Differs from subsp. *ishizuchianus* in the following points: 1) size larger on an average, with broader body; 2) body above usually black with dark greenish or coppery tinge, occasionally coppery with greenish or reddish tinge; palpi and legs black; 3) lateral margins of pronotum more shallowly emarginate behind middle; 4) apical part of penis shorter and robuster, with tip obtusely rounded; 5) left basal lateral lobe of internal sac more remarkably elongated ventrad.

Distribution C.~W. Shikoku except for high altitudinal area of Ishizuchi Mts. / Is. Ô-shima of Sukumo-shi (Kochi Pref.).

原記載・タイプ産地・タイプ標本の所在 種の項を参照．

亜種の特徴 25~34 mm．石鎚山脈亜種とは以下の諸点で異なる：1) より大型で体型は幅広い；2) 背面の色彩は通常，緑青色を帯びた黒色，ただし石鎚山系とその周辺地域では銅色や紅色に近い独特の色彩を有する個体も出現する．口肢と脚は基本的に黒色；3) 前胸背板側縁後方のえぐれがより浅い；3) 陰茎先端部は太短く，先端は鈍く丸まる；4) 内嚢の左基部側葉はより顕

27. *Carabus* (*Ohomopterus*) *dehaanii* Chaudoir, 1848
オオオサムシ

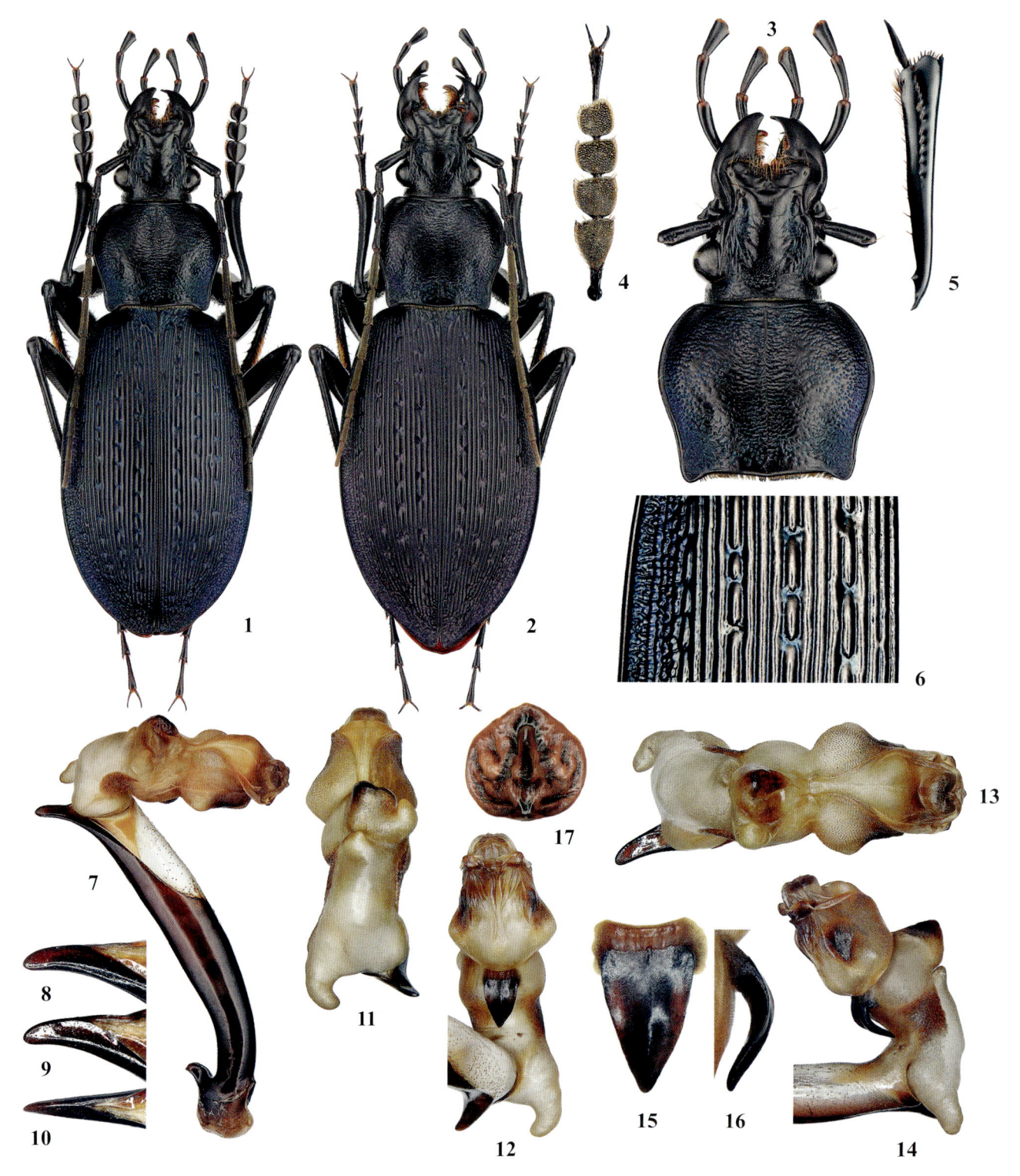

Figs. 1-17. *Carabus* (*Ohomopterus*) *dehaanii* Chaudoir, 1848 (♂, Konda, Habikino-shi, Osaka Pref.). — 1, ♂; 2, ♀; 3, male head & pronotum; 4, male right protarsus in ventral view; 5, male right protibia in subdorsal view; 6, male left elytron (median part); 7, penis & fully everted internal sac in right lateral view; 8, apical part of penis in right lateral view; 9, ditto in right subdorsal view; 10, ditto in dorsal view; 11, internal sac in caudal view; 12, ditto in frontal view; 13, ditto in dorsal view; 14, ditto in left sublateral view; 15, digitulus in frontal view; 16, ditto in right lateral view; 17, inner plate of ligular apophysis.

図 1-17. オオオサムシ (♂, 大阪府羽曳野市誉田) — 1, ♂; 2, ♀; 3, ♂の頭部と前胸背板; 4, ♂右前附節 (腹面観); 5, ♂右前脛節 (斜背面観); 6, ♂左鞘翅中央部; 7, 陰茎と完全反転下の内嚢 (右側面観); 8, 陰茎尖端 (右側面観); 9, 同 (右斜背面観); 10, 同 (背面観); 11, 内嚢 (尾側面観); 12, 同 (頭側面観); 13, 同 (背面観); 14, 同 (左斜側面観); 15, 指状片 (頭側面観); 16, 同 (右側面観); 17, 腟底部節片内板.

27. *Carabus* (*Ohomopterus*) *dehaanii*

Figs. 18-29. *Carabus* (*Ohomopterus*) *dehaanii* subspp. — 18-19, *C*. (*O*.) *d*. *punctatostriatus* (18, ♂, 19, ♀; 18, Mt. Ômi-yama, Suwa-shi, Nagano Pref.; 19, Hanaoka, Minato, Okaya-shi, Nagano Pref.); 20-21, *C*. (*O*.) *d*. *katsumai* (20, ♂, HT, 21, ♀, PT; Mt. Bi-zan, Tokushima-shi, Tokushima Pref.); 22-23, *C*. (*O*.) *d*. *strenuus* (22, ♂, 23, ♀; Kônosé, Kuma-mura, Kumamoto Pref.); 24-25, *C*. (*O*.) *d*. *kumaso* (24, ♂, HT, 25, ♀, PT; Ôkôbira, Ebino-shi, Miyazaki Pref.); 26-27, *C*. (*O*.) *d*. *ishidai* (26, ♂, HT, 27, ♀, PT; Pass Karasu-ga-tôgé, Is. Gosho-no-ura-jima, Kumamoto Pref.); 28-29, *C*. (*O*.) *d*. *koshikicola* (28, ♂, HT, 29, ♀, PT; Mt. Miné-no-yama, Is. Kami-koshiki-jima, Kagoshima Pref.). — a, Apical part of penis in right subdorsal view; b, ditto in dorsal view; c, digitulus in frontal view.

図 18-29. オオオサムシ各亜種. — 18-19, 本州中部亜種(18, ♂, 19, ♀; 18, 長野県諏訪市大見山; 19, 長野県岡谷市湊花岡); 20-21, 四国東部亜種(20, ♂, HT, 21, ♀, PT; 徳島県徳島市眉山); 22-23, 熊本県南西部亜種(22, ♂, 23, ♀; 熊本県球磨村神瀬); 24-25, 九州山地南部亜種(24, ♂, HT, 25, ♀, PT; 宮崎県えびの市大河平); 26-27, 御所浦島亜種(26, ♂, HT, 27, ♀, PT; 熊本県御所浦島烏峠); 28-29, 上甑島亜種(28, ♂, HT, 29, ♀, PT; 鹿児島県上甑島嶺の山). — a, 陰茎先端(右斜背面観); b, 同(背面観); c, 指状片(頭側面観).

27. *Carabus* (*Ohomopterus*) *dehaanii*

Figs. 30-41. *Carabus* (*Ohomopterus*) *dehaanii* subspp. — 30-31, **C. (O.) d. nakagomei** (30, ♂, 31, ♀; Hananoki, Nejimé-kawakita, Minami-ôsumi-chô, Kagoshima Pref.). — 32-41, *C.* (*O.*) *dehaanii* (the population whose subspecific account is still indecisive) (32, ♂, Takemori, Kôshû-shi, Yamanashi Pref.; 33, ♀, Tsuzura, Kawané-chô, Shimada-shi, Shizuoka Pref.; 34, ♂, Mt. Yuzuruha-san, Is. Awaji-shima, Hyogo Pref.; 35, ♂, 36, ♀, Mt. Dai-sen, Tottori Pref.; 37, ♂, Is. Tsuwaji-jima, Matsuyama-shi, Ehime Pref.; 38, ♂, Mt. Konpira-san, Nagasaki-shi, Nagasaki Pref.; 39, ♂, Mt. Shichirô-san, Is. Shishi-jima, Kagoshima Pref.; 40, ♀, Mt. Sasa-daké, Is. Fukué-jima, Nagasaki Pref.; 41, ♂, Mt. Ukizumi-yama, Is. Naka-koshiki-jima, Kagoshima Pref.). — a, Apical part of penis in right subdorsal view; b, ditto in dorsal view; c, digitulus in frontal view.

図 30-41. **オオオサムシ**各亜種. — 30-31, **大隅半島亜種**(30, ♂, 31, ♀; 鹿児島県南大隅町根占川北花ノ木). — 32-41, オオオサムシ(亜種分類が未だ確定的でないもの)(32, ♂, 山梨県甲州市竹森; 33, ♀, 静岡県島田市川根町葛篭; 34, ♂, 兵庫県淡路島論鶴羽山; 35, ♂, 36, ♀, 鳥取県大山; 37, ♂, 愛媛県松山市津和地島; 38, ♂, 長崎県長崎市金毘羅山; 39, ♂, 鹿児島県獅子島七郎山; 40, ♀, 長崎県福江島笹岳; 41, ♂, 鹿児島県中甑島河星山). — a, 陰茎先端(右側背面観), b, 同(背面観), c, 指状片(頭側面観).

27. *Carabus* (*Ohomopterus*) *dehaanii* Chaudoir, 1848 オオオサムシ

Original description   *Carabus De Haanii* Chaudoir, 1848, Bull. Soc. imp. Moscou, 21 (4), p. 452.

History of research   A large-sized well-known species widely distributed in the western half of Japan. However, its type locality cannot be identified, since the locality designated in the original description (Chaudoir, 1848) is "Java ou Japon" (Java is obviously wrong) without details on the collecting place. Bates (1873) described *Carabus De Haanii* var. *punctato-striatus* [sic] for such individuals as "the interstices of the elytra are distinctly crenated, the indentations increasing in strength towards the apex". Ishikawa (1985) regarded Bates' race as a subspecies of *C. dehaanii*, and applied it to the population distributed mainly in the Chubu Region of central Honshu. Breuning (1932) described *C.* (*Apotomopterus*) *dehaani* [sic] m. (= morpha) *strenua* (it must be spelled *strenuus* under the genus *Carabus* which is masculine) from Konosé of Kumamoto Prefecture in southwestern Kyushu. Nakane (1953) regarded Breuning's taxon as a subspecies of *Apotomopterus dehaanii* (= *Carabus* (*Ohomopterus*) *dehaanii* in the present sense) and described two new subspecies of the same species, namely, *tosanus* and *ishizuchianus*, though the latter two are currently regarded as an independent species (*C.* (*O.*) *tosanus*) and its subspecies, respectively. Nakane (1962) described *A. d. imafukui* from Mt. Kiso-koma-ga-také of Nagano Prefecture, which is currently regarded as a junior synonym of *C.* (*O.*) *d. punctatostriatus*. Imura & Mizusawa (1995) described four new subspecies of *C.* (*O.*) *dehaanii*, that is, *katsumai* from eastern Shikoku, *kumaso* from southern Kyushu Mountains, *ishidai* from Is. Gosho-no-ura-jima of the Amakusa Islands in western Kyushu, and *koshikicola* from Is. Kami-koshiki-jima, off the western coast of Kyushu. Kubota (2010) described another subspecies *nakagomei* from the Ôsumi Peninsula of southernmost Kyushu.

Type locality   Java ou Japon.

Type depository   Muséum Nationale d'Histoire Naturelle, Paris [Lectotype, ♂].

Morphology   23~41 mm. Body above bluish or purplish black sometimes with blue-greenish or coppery tinge, rarely dark brownish or entirely black. Tibiae usually black, though rarely red brownish. External features: Figs. 1~6. Genitalic features: 7~17. Closely allied to *C.* (*O.*) *tosanus tosanus*, but differs from that race in the following points: 1) dorsal coloration adifferent, usually not bearing greenish tinge as in *C.* (*O.*) *t. tosanus*, but mostly with bluish or purplish tinge; 2) antennae a little shorter, barely reaching middle of elytra in female; 3) basal foveae of pronotum a little shallower; 4) both pronotum and elytra less strongly convex above, and appear flatter; 5) primary foveoles of elytra smaller, hardly invading adjacent tertiaries; 6) left basal lateral lobe of internal sac slenderer and more strongly bent inward; 7) digitulus nearly symmetric, with apex more sharply pointed.

Molecular phylogeny   For the general overview of the molecular phylogeny of *Ohomopterus*, see the same section of *C.* (*O.*) *yamato*. On the molecular genealogical tree constructed by analyzing six nuclear DNA sequences, *C.* (*O.*) *dehaanii* is most closely related to *C.* (*O.*) *japonicus* and *C.* (*O.*) *chugokuensis* (Sota & Nagata, 2008). Analyses of mitochondrial ND5 gene and nuclear internal transcribed spacer I (ITS I) reveal that the population of *C.* (*O.*) *dehaanii* distributed in the area west of the Chugoku Region is considered to be authentic, while those in the Chubu and Kinki Regions are considered to be offspring of the hybrid between the male of *C.* (*O.*) *dehaanii* and the female of *C.* (*O.*) *maiyasanus*. For details, see Sota (2000c, 2003), Tominaga, Okamoto et al. (2005), Research group of molecular phylogeny on the *Ohomopterus* ground beetles (2005, 2006a), Su et al. (2006), etc.

Distribution   C.~W. Honshu, E. Shikoku & Kyushu; / Iss. Awaji-shima (Hyogo Pref.), Ôsaki-shimo-jima, Ôsaki-kami-jima, Kami-kamagari-jima, Itsuku-shima, Eta-jima+Nômi-shima (Hiroshima Pref.), Yashiro-jima, Naga-shima, Mukô-shima (Yamaguchi Pref.), Tsuwaji-shima (Ehime Pref.), Noko-no-shima (Fukuoka Pref.), Ô-nyû-jima (Oita Pref.), Nakadôri-jima, Wakamatsu-jima, Fukué-jima (Nagasaki Pref.), Gosho-no-ura-jima (or Goshoura-jima) (Kumamoto Pref.), Shishi-jima, Kami-koshiki-jima & Naka-koshiki-jima (Kagoshima Pref.).

Habitat   Silvicolous, preferring rather humid forest floor, generally restricted to the base rock mountain and not distributed on an alluvial plain, terrace and hills formed after the Pliocene.

Bionomical notes   Spring breeder, earthworm feeder.

Geographical variation   Morphologically classified into seven subspecies.

Etymology   This species is named after Wilhem De Haan, the Dutch Zoologist who is an expert of insects and crustaceans, and was the first keeper of invertebrates at the Rijksmuseum in Leiden, now Naturalis. He was forced to retire in 1846, when he was partially paralyzed by a spinal disease. He was responsible for the invertebrate volume of Siebold's *Fauna Japonica*, which was published in 1833, and introduced the western world for the first time to Japanese wildlife. He named a great many new taxa, and several taxa are named in his honor.

研究史　本種はChaudoir (ショドワー) (1848) により, ロシア産ハンミョウやトルコ産オサムシの記載論文の末尾において, いわば番外的な形で記載された. 原記載においてその産地は「ジャワもしくは日本」と記されており, パリ自然史博物館に保管されているレクトタイプにも採集場所に関するデータは一切付けられていないので, タイプ産地を特定することはできず, どの集団をもって基亜種とみなすべきか判断することも困難である. Bates (ベイツ) (1873) は, Lewis (ルイス) によってもたらされた本種の標本のうち, 「鞘翅の隙間 (線条) が顕著に鋸歯状を呈し, 刻み目が翅端に向けてより顕著になってゆく」2個体に対して *Carabus De Haanii* var. *punctato-striatus* と命名した. タイプ産地は記されていないが, おそらく諏訪湖周辺で得られた標本が記載に用いられたものと考えられる. 後年, 石川 (1985) は本州中部に産する集団にこの名を充て, オオオサムシの亜種 (本書で言うオオオサムシ本州中部亜

27. *Carabus* (*Ohomopterus*) *dehaanii*

種)とみなした．BREUNING(ブロイニング)(1932)は熊本県南西部の神瀬から*C.* (*Apotomopterus*) *dehaani* [sic] m.(モルファ) *strenua*を記載した．NAKANE(1953)はこれをオオオサムシの亜種*Apotomopterus d. strenuus*として扱い(所属する属の性が男性なので，亜種小名の語尾は-*us*とする必要がある)，同時に*A. d. tosanus*と*A. d. ishizuchianus*の2亜種を記載したが，後二者は今日では一般に独立種*C.* (*Ohomopterus*) *tosanus*及びその亜種として扱われている．中根(1962)は更に，長野県木曽駒ヶ岳からオオオサムシの亜種として*A. d. imafukui*を記載したが，現在ではオオオサムシ本州中部亜種の異名とみなされている．小宮(1971)は本種を*dehaanii*, *tosanus*, *ishizuchianus*の3亜種に分類し，BREUNING(1932)に倣い，n.(natioナティオ)やm.(morphaモルファ)といった階級を設けてその地理的変異に言及した．ISHIKAWA(1973)は♂交尾器内嚢の基本形態に基づいてオサムシ亜族の属・亜属の再定義を行い，それまでアポトモプテルス属に置かれていたすべての日本産種をカラブス属のオホモプテルス亜属へと移した．すなわち，オサムシ亜族を複数の属に分類する上位分類体系の下では*Apotomopterus dehaanii*から*Carabus* (*Ohomopterus*) *dehaanii*へ，また同亜族を一属*Carabus*とみなす，現在最も一般的な体系の下では*Carabus* (*Apotomopterus*) *dehaanii*から*C.* (*Ohomopterus*) *dehaanii*へと変更がなされ，現在ではこの形を踏襲する研究者が大半を占める．井村・水沢(1995)は各地から得られた多数の本種標本を検して，*katsumai*(四国東部)，*kumaso*(九州山地南部)，*ishidai*(御所浦島)，*koshikicola*(上甑島)の4亜種を新たに記載した．2000年代に入り，KUBOTA(2010)によって鹿児島県の大隅半島から亜種*nakagomei*が記載されている．

タイプ産地　ジャワもしくは日本．

タイプ標本の所在　パリ自然史博物館［レクトタイプ，♂］．

形態　23~41 mm．背面は青藍色ないし紫色を帯びた黒色を基調とするが，時に緑青色又は暗銅色を帯びる．極めて稀に，背面が紫褐色あるいは黒色の個体や，赤褐色の脛節を持つ個体も出現する．外部形態：図1~6．交尾器形態：図7~17．トサオサムシ基亜種に酷似するが，以下の諸点で異なる：1) 背面の色彩が異なり，通常は緑色を帯びることはなく，青藍色~紫色を帯びる場合が多い；2) 触角はやや短く，♀では先端が鞘翅中央に達する程度；3) 前胸背板の基部凹陥がより浅い；4) 前胸背板，鞘翅ともに膨らみがやや弱く，より平坦に見えるものが多い；5) 第一次凹陥はより小さく，隣接する第三次間室を殆ど侵さない；6) ♂交尾器内嚢の左基部側葉はより細く，より強く内湾する；7) 指状片は左右ほぼ対称で，先端はより鋭く尖る．

分子系統　核遺伝子の解析から，本種はヒメオサムシ(及びアキオサムシ)と姉妹種であることが推定されているが，ミトコンドリアDNAで見ると本種だけが独自に持つミトコンドリアの系統は存在せず，ミトコンドリアDNA，核遺伝子双方の解析結果から，中国地方以西の「オオオサムシ」はいわゆる純系とみなしうるものだが，中部・近畿地方の「オオオサムシ」は，東に分布を拡げる過程で他種(恐らくマヤサンオサムシ)との交雑(オオオサムシ♂×マヤサンオサムシ♀)によりミトコンドリアの系統が置換した集団であろうと考えられている(詳細は曽田，2000c, 2003；冨永・岡本ら，2005；オオオサムシ属の分子系統研究グループ，2005, 2006a；SU *et al*., 2006等を参照)．

分布　本州(中部以西)，四国，九州／淡路島(兵庫県)，大崎下島，大崎上島，上蒲刈島，厳島，江田島＋能美島(以上，広島県)，屋代島［＝周防大島］，長島，向島(以上，山口県)，津和地島(愛媛県)，能古島(福岡県)，大入島(大分県)，中通島，若松島，福江島(以上，長崎県)，御所浦島(熊本県)，獅子島，上甑島，中甑島(以上，鹿児島県)．中部地方以西の西南日本に広く分布するが，分布は必ずしも連続しておらず，空白域が多い．本州中部では長野県中北部から山梨県中北部，太平洋側では静岡県中部にかけて分布し，内陸部では，北は長野県北部の千曲市(ちくま)付近まで，東は長野県霧ヶ峰から山梨県甲府盆地北部にかけて記録されている．太平洋側では静岡市南西部の藁科川(わらしながわ)右岸(西岸)が東限となっており，同所から藤枝市北部にかけての数か所から記録されている．日本海側では石川県南部から美濃白鳥(みのしろとり)付近にかけての地域が北東限と思われる．近畿地方では中部から北東部にかけて分布するが，琵琶湖南東部や大阪平野を中心とした地域からは記録がない．同地域における南限はほぼ紀ノ川~櫛田川(くしだがわ)を結ぶ線で，それより南では生石山(おいしやま)から龍門山(りゅうもんざん)，九度山町(くどやまちょう)にかけてのごく狭い分布圏が知られているのみ．近畿地方北西部では琵琶湖北西部から敦賀湾南部，篠山盆地(ささやまぼんち)を経て瀬戸内海側の兵庫県中西部に至る帯状の空白域がある．これより西方の中国地方には広く分布するが，中国地方産集団の東限は京都府北部の由良川(ゆらがわ)西岸で，広島県東部~岡山県~兵庫県西部にかけての瀬戸内海側の地域からは記録がない．四国では北東部(香川県南部から徳島県中~南東部，および愛媛県東端のごく一部)のみに分布する．九州には比較的広く分布し，かつては分布の空白域と思われていた南端部の薩摩・大隅両半島にも先端部を除き局所的ながら比較的広く分布しているが，長崎県西半部および宮崎県南部から鹿児島県東部にかけての太平洋沿岸地域からは記録がない．

生息環境　近畿オサムシ研究グループ(1979)によれば，近畿地方では中~低山を中心に生息し，産地では個体数が多いにもかかわらず分布は基盤産地に限定され，沖積平野・段丘・丘陵など大阪層群とその相当層(鮮新~更新統)とそれより新期の第四紀層には生息しないという(ただし，鈴鹿川より北の鈴鹿山脈東麓では丘陵地帯にも見られる)．暗い林床や沢べりに個体数が多く，同所的に生息するヤコンオサムシやヤマトオサムシに比べ，より湿潤な環境を必要とする種であろうとされている．一方，菅(かん)(1981)によれば，愛媛県の津和地島(防予諸島(ぼうよしょとう))では平地のミカン畑に多産するという．

生態　春繁殖型．幼虫は環形動物(主としてフトミミズ類)食．

地理的変異　形態学的には8亜種に分類されるが，タイプ産地を特定できないため，どの地域集団を基亜種とみなすべきかという問題が本種の亜種分類を論じるうえでの最大の重石となっている．これまでに記載された各亜種は，少なくともレクトタイプからは明らかに識別できる形態学的特徴を有しているゆえ，いわば消去法的に記載されてきたものと言えるが，ほかにも島嶼部を中心にいくつか特徴的な形態を持つ集団が存在する．例えば，五島列島の福江島の集団は大型で背面が平坦なものが多く，屋代島

や津和地島など瀬戸内海西部に浮かぶ島嶼に産する集団は背面が緑色味を帯びたものが多い．本書では便宜上，計7亜種に分類したが，これはあくまで暫定的な措置である．今後は先ず，レクトタイプをより詳細に検討し，各地の集団の特徴とすり合わせたうえで，基亜種に相当する集団を特定する作業が必要であろう．そのうえで，地理的変異に関する総合的な再検討が望まれる．

　　種小名の由来　　種小名はオランダの動物学者，Wilhem De Haan（ウィレム・ドゥ・ハーン）（1801~1855）に因む．専門は昆虫類と甲殻類で，オランダ国立自然史博物館無脊椎動物部門の初代チーフを務め，1833年に出版されたシーボルトの「ファウナ・ヤポニカ」の無脊椎動物の巻の作成にも大きく貢献した．自ら多数の新種を記載したほか，本種をはじめ，彼に献名された分類群の数も多い．脊髄疾患による麻痺のため45歳の若さで現役引退を余儀なくされ，その10年後に亡くなったという．

27-1. *C. (O.) d. punctatostriatus* Bates, 1873　オオオサムシ本州中部亜種

　Original description　*Carabus De Haanii* var. *punctato-striatus* Bates, 1873, Trans. ent. Soc. Lond., 1873, p. 231.
　Type locality　Not described in the original description, but presumably the Lake Suwa-ko or its nearby regions of Nagano Prefecture in central Honshu.
　Type depository　Natural History Museum, London [Syntype].
　Diagnosis　27~39 mm. According to Bates' original description, this race is defined by the following characters: "the interstices of the elytra are distinctly crenated, the indentations increasing in strength towards the apex".
　Distribution　C. Honshu, though the distributional range of this subspecies is not yet clearly defined.
　Etymology　The subspecific name means "having punctate (elytral) striation".

　　研究史　本タクソンはBates（1873）により*Carabus De Haanii* var. *punctato-striatus*という名で記載された．この名は1961年よりも前に"変種"として公表されたものなので，原公表から亜種の階級であるとみなされる（国際動物命名規約第4版条10.2）．種小名*De Haanii*は単一語を形成しているとみなされるが分離した複数単語として公表された複合語の種階級群名なので，その要素語をハイフンなしでつなげる必要があり（条32.5.2.2），大文字の頭文字は小文字に置き換えなければならない（条32.5.2.5）ので，*dehaanii*となる．また，亜種小名中のハイフンは取り払って（条32.5.2.3），*punctatostriatus*と表記される．
　　タイプ産地　原記載に産地の記述はないが，Batesに標本を提供したLewisの日本における行程，並びに鞘翅彫刻の特徴などから考えて，長野県諏訪湖周辺の個体に基づいて記載された可能性が高い．
　　タイプ標本の所在　ロンドン自然史博物館［シンタイプ］．
　　亜種の特徴　27~39 mm．Batesの原記載によれば，本亜種は「鞘翅の隙間（線条）は顕著に鋸歯状を呈し，刻み目が翅端に向けてより顕著になってゆく」という特徴を有する．
　　分布　本州中部．長野県諏訪湖付近の集団に代表される本州中部産のものを便宜上，本亜種に充てておくが，どの範囲のものまでを本亜種に含めるべきかについては定見がない．
　　亜種小名の由来と通称　亜種小名は「点刻された（鞘翅）線条の」を意味する．石川（1985）によりチュウブオオオサムシという通称が提唱されている．

27-2. *C. (O.) d. dehaanii* Chaudoir, 1848　オオオサムシ基亜種

　Original description, type locality & type depository.　See explanation of the species.
　Distribution (the range shown here is just provisional)　W. Honshu (Kinki & Chugoku Regions) & Kyushu; / Iss. Awaji-shima (Hyogo Pref.), Ôsaki-shimo-jima, Ôsaki-kami-jima, Kami-kamagari-jima, Itsuku-shima, Eta-jima＋Nômi-shima (Hiroshima Pref.), Yashiro-jima, Naga-shima, Mukô-shima (Yamaguchi Pref.), Tsuwaji-shima (Ehime Pref.), Noko-no-shima (Fukuoka Pref.), Ô-nyû-jima (Oita Pref.), Nakadôri-jima, Wakamatsu-jima, Fukué-jima (Nagasaki Pref.) & Shishi-jima (Kagoshima Pref.).

　　原記載・タイプ産地・タイプ標本の所在　種の項を参照．
　　種の解説で述べたように，本種はタイプ産地を特定できないため，どの地域集団をもって基亜種とみなすべきかという問題が残されている．レクトタイプの特徴としては，前胸背板の皺が比較的弱く，主として中央部付近に認められ，前胸背板側縁は後方に向けてあまり強く狭まらず，波曲も弱く，鞘翅は比較的短く，肩部の張り出しが強く，各間室の幅はほぼ均等で，第一次凹陥は比較的顕著に刻まれ，第二次，第三次間室の刻み目は目立たない，といった点が挙げられる．本種が記載された当時の時代背景から考えて，記載に用いられた標本は長崎や神戸など，当時外国人の出入りが可能であった地域からもたらされたものである可能性が高いが，どこのものかを特定することは困難である．本書では，あくまで暫定的ながら，近畿，中国，九州の集団を基亜種として扱っておくが，これらの範囲においてもかなり顕著な地理的変異が認められ，単純に一亜種として括るには無理がある．色彩的には，屋代島や津和地島など，瀬戸内海西部の島嶼に産する集団は背面が緑色を帯びるものが多く，岐阜県南東部の中津川付近には，背面が紫茶色を帯びる独特の色彩を有する個体が出現する場所がある．全身がほぼ黒化した型も各地で得られている．
　　分布（あくまで暫定的）本州（近畿，中国地方），九州／淡路島（兵庫県），大崎下島，大崎上島，上蒲刈島，厳島，江田島＋能美島（以上，広島県），屋代島［＝周防大島］，長島，向島（以上，山口県），津和地島（愛媛県），能古島（福岡県），大入島（大分

県), 中通島, 若松島, 福江島 (以上, 長崎県), 獅子島 (鹿児島県).

27-3. *C. (O.) d. katsumai* IMURA et MIZUSAWA, 1995　オオオサムシ四国東部亜種

Original description　*Carabus (Ohomopterus) dehaanii katsumai* IMURA & MIZUSAWA, 1995, Gekkan-Mushi, Tokyo, (293), p. 12, figs. 3 & 9.
Type locality　Mt. Bi-zan, Tokushima-shi, Tokushima Pref.
Type depository　National Museum of Nature and Science, Tokyo [Holotype, ♂].
Diagnosis　29~39mm. Body above dark purple or dark bluish purple, more lightly so on lateral margins of pronotum and elytra. Identified by the following characters: 1) size considerably large on an average; 2) pronotum remarkably rugoso-punctate; 3) elytra wide and less strongly convex, above all in female; 4) primary intervals of elytra wider than secondaries and tertiaries, and the latter two obviously crenulate on surface; 5) apical portion of penis wide, not acutely narrowed toward apex in lateral view, and strongly compressed right laterad.
Distribution　E. Shikoku (Tokushima Pref.~S. Kagawa Pref.~easternmost Ehime Pref.).
Etymology　Named after Katsuma ÔTSUKA who is a grandfather of the senior author, Y. IMURA.

　タイプ産地　徳島県, 徳島市, 眉山.
　タイプ標本の所在　国立科学博物館 [ホロタイプ, ♂].
　亜種の特徴　29~39 mm. 背面の色彩は暗紫色～暗青紫色で, 前胸背板と鞘翅の側縁はより明色. 大型で, 前胸背板は皺と点刻が著しく, 鞘翅は幅広く平坦 (♀においてより顕著). 第一次間室は第二次, 第三次間室より幅広く, 後二者の表面は明らかに横の刻み目を有する. 陰茎先端は側面から見て幅広く, 右方向から押圧され, 先端に向けてあまり急に狭まらない.
　分布　四国東部. 東は讃岐山脈の東端～徳島市眉山～阿南市(あなん)南部(福井町の東方)にかけての地域, 南限はほぼ海部川(かいふがわ)に一致するが, 佐野(1993)によれば, 一か所のみながら海部川を南西に越えた地点(海陽町(かいようちょう); 記録された当時は海南町)若松の笹無谷)から記録されている. 北は讃岐山脈一帯に広く分布し, 北限はさぬき市中部の大川町(おおかわまち)界隈. 西限は徳島県三好市(みよしし)と愛媛県四国中央市の境界にある境目峠(さかいめとうげ)付近)で, 南西方向は徳島県旧西祖谷山村(にしいややまそん)(現在の三好市の一部)の祖谷川右岸(東岸)～剣山～海部川を北西から南東に結ぶ線にほぼ一致して斜めに分布が切れる.
　亜種小名の由来と通称　主著者井村の祖父, 大塚勝馬に因む. シコクオオオサムシという通称で呼ばれることもある.

27-4. *C. (O.) d. strenuus* BREUNING, 1932　オオオサムシ熊本県南西部亜種

Original description　*Carabus (Apotomopterus) dehaani* [sic] m. *strenua* BREUNING, 1932, Best.-Tab. eur. Coleopt., 104, p. 232.
Type locality　Konosé (= the name of a place in Kuma-mura of Kumamoto Pref., Kyushu).
Type depository　Zoölogisch Museum Amsterdam [Holotype, ♂].
Diagnosis　26~32mm. The characteristics of this taxon shown in the original description by BREUNING (1932) are as follows: "Kleiner und besonders schmäler als die Nominatform; Flügeldecken langoval; seitlich der series umbilicata der Raum unregelmäßig fein gekörnt".
Distribution　SW. Kumamoto Pref. Distributional range of this subspecies is not yet clearly defined.
Etymology　Subspecies name means "swift" or "active".

　研究史　本タクソンはBREUNING(1932)によりCarabus (Apotomopterus) dehaani m. strenuaという名で記載された(dehaaniはdehaaniiの後綴り). ルイスオサムシの研究史の項でも述べたように, このstrenuaという名はm.(= morphaモルファ)という階級で記載されたものの, その後の使用状況等を勘案し, 原公表から亜種の階級とみなすのが合理的だろう. ただし, 所属する属の性が男性なので, 性を一致させるため, 種小名の語尾は -usに換えてstrenuusとする必要がある. 基亜種のシノニムとみなされる場合もあるが, 本書では独立した亜種として扱った.
　タイプ産地　神瀬(＝熊本県球磨郡(くまぐん)球磨村(くまむら)にある地名).
　タイプ標本の所在　アムステルダム動物学博物館 [ホロタイプ, ♂].
　亜種の特徴　26~32mm. 原記載でBREUNINGにより示された本タクソンの特徴は「原型より小型で, また際立って細く, 鞘翅は長卵形で, 丘孔点列側方の顆粒が不規則で細かい」という点のみで, 「おそらくはdehaaniiの山地型とみなすべきもので, 基亜種との地理的な境界も明確ではないため, 真の地域型(亜種)ではない」と述べられている.
　分布　熊本県南西部(神瀬付近). ただし, 本亜種に該当する集団がどの範囲にまで分布しているのか, 分布域や変異の詳細, 次亜種との関係等については未調査の部分が多い.
　亜種小名の由来と通称　亜種小名は「敏捷な, 活動的な」を意味する. ツクシオサムシという通称で呼ばれることが多いが,

27-5. *C. (O.) d. kumaso* Imura et Mizusawa, 1995 オオオサムシ九州山地南部亜種

Original description *Carabus* (*Ohomopterus*) *dehaanii kumaso* Imura & Mizusawa, 1995, Gekkan-Mushi, Tokyo, (293), p. 13, figs. 4 & 10.
Type locality Ôkôbira, Ebino-shi, Miyazaki Pref.
Type depository National Museum of Nature and Science, Tokyo [Holotype, ♂].
Diagnosis 28~32mm. Body above black with greenish coppery tinge on lateral margins of pronotum and elytra. Identified by the following characters: 1) pronotal disc vaguely punctate, with central portion rather smooth; 2) elytra long and slender, spindle-shaped, with surface rather smooth; 3) primary foveoles of elytra very small, secondary and tertiary intervals hardly crenulate; 4) antennae and legs very long and slender; 5) apical portion of penis also slender, slightly constricted at middle in lateral view, and not obviously compressed right laterad.
Distribution S. Kyushu Mts. (W. Miyazaki Pref.~SE. Kumamoto Pref.~N. Kagoshima Pref.).
Etymology "Kumaso" is an old regional name indicating southern Kyushu, or the name of peoples who lived there.

タイプ産地　宮崎県, えびの市, 大河平.
タイプ標本の所在　国立科学博物館［ホロタイプ, ♂］.
亜種の特徴　28~32mm. 背面は黒色で, 前胸背板と鞘翅の辺縁は弱い銅緑色光沢を帯びる. 以下の特徴を有する：1) 前胸背板の点刻はごく弱く, とりわけ中央部が滑らかになる; 2) 鞘翅は細長く, 紡錘形に近く, 表面は凹凸が少なく滑らかである; 3) 鞘翅第一次凹陥は極めて小さく, 第二次, 第三次間室に横の刻み目は殆ど認められない; 4) 触角と脚は細長い; 5) 陰茎先端も細長く, 側面から見て中央部が軽くくびれ, 右側面は殆ど押圧されない.
分布　九州山地南部（宮崎西部〜熊本南東部〜鹿児島北部）.
亜種小名の由来と通称　亜種小名の *kumaso*（熊襲）は, 古代の南部九州の地域名, あるいはその地域の居住者の族名を意味する. クマソオオオサムシという通称がある.

27-6. *C. (O.) d. ishidai* Imura et Mizusawa, 1995 オオオサムシ御所浦島亜種

Original description *Carabus* (*Ohomopterus*) *dehaanii ishidai* Imura & Mizusawa, 1995, Gekkan-Mushi, Tokyo, (293), p. 13, figs. 5 & 11.
Type locality Pass Karasu-ga-tôgé, Is. Gosho-no-ura-jima, Amakusa-gun, Kumamoto Pref.
Type depository National Museum of Nature and Science, Tokyo [Holotype, ♂].
Diagnosis 32~38 mm. Body above dark bluish black, with violet bluish or coppery greenish tinge on lateral margins of pronotum and elytra. Identified by the following characters: 1) pronotal disc not remarkably rugose but strongly punctate, basal foveae small and shallow; 2) elytra with sutural part rather strongly convex above and is angulated in posterior view, shoulders distinct, lateral sides not roundly arcuate but rather straight in basal halves; 3) elytral intervals with primary foveoles small, secondary and tertiary intervals weakly crenulate near apical portions; 4) apical part of penis short and robust, not strongly compressed laterad, and rather acutely narrowed toward tip; 5) median longitudinal ridge of digitulus weakly convex and not remarkably outlined.
Distribution Is. Gosho-no-ura-jima (or Gosho-ura-jima) (Kumamoto Pref.)
Etymology Named after Toyoaki Ishida of Kumamoto Pref.

タイプ産地　熊本県, 天草郡, 御所浦島, 烏峠（からすがとうげ／からすとうげ）（現在は熊本県天草市に属する）.
タイプ標本の所在　国立科学博物館［ホロタイプ, ♂］.
亜種の特徴　32~38 mm. 背面は暗青色で, 側縁は青紫〜銅緑色の光沢を帯びる. 以下の特徴を有する：1) 前胸背板は強く点刻され, 皺はあまり顕著に刻まれず, 基部凹陥は小さく浅い; 2) 鞘翅は会合線付近で比較的強く隆起し, 尾側から見ると会合線が角張って見える. 肩の張り出しは顕著で, 鞘翅側縁の前半部分は直線的; 3) 鞘翅第一次凹陥は小さく, 第二次, 第三次間室は先端付近で弱い横の刻み目を有する; 4) 陰茎先端は側面から見て太短く, 側方からの押圧は強くなく, 先端に向けて比較的急に狭まる; 5) 指状片中央の稜は弱い.
分布　御所浦島（熊本県）. 近隣の獅子島にも本種を産するが, 詳しい形態学的検討はまだなされていない.
亜種小名の由来と通称　タイプシリーズを採集した熊本県の石田豊明に因む. ゴショノウラオオオサムシという通称がある.

27-7. *C. (O.) d. koshikicola* Imura et Mizusawa, 1995 オオオサムシ上甑島亜種

Original description *Carabus* (*Ohomopterus*) *dehaanii koshikicola* Imura & Mizusawa, 1995, Gekkan-Mushi, Tokyo, (293), p. 13, figs. 6 & 12.
Type locality Eastern slope of Mt. Miné-no-yama, Is. Kami-koshiki-jima, Satsuma-gun, Kagoshima Pref.
Type depository National Museum of Nature and Science, Tokyo [Holotype, ♂].

28. *Carabus* (*Ohomopterus*) *iwawakianus* (Nakane, 1953)
イワワキオサムシ

Figs. 1-17. *Carabus* (*Ohomopterus*) *iwawakianus iwawakianus* (Nakane, 1953) (Mt. Iwawaki-san, Kawachi-nagano-shi, Osaka Pref.). — 1, ♂; 2, ♀; 3, male head & pronotum; 4, male right protarsus in ventral view; 5, male right protibia in subdorsal view; 6, male left elytron (median part); 7, penis & fully everted internal sac in right lateral view; 8, apical part of penis in right lateral view; 9, ditto in subdorsal view; 10, ditto in dorsal view; 11, internal sac in caudal view; 12, ditto in frontal view; 13, ditto in dorsal view; 14, ditto in left sublateral view; 15, digitulus in frontal view; 16, ditto in right lateral view; 17, inner plate of ligular apophysis.

図 1-17. イワワキオサムシ基亜種（大阪府河内長野市岩湧山）— 1, ♂; 2, ♀; 3, ♂の頭部と前胸背板; 4, ♂右前跗節（腹面観）; 5, ♂右前脛節（斜背面観）; 6, ♂左鞘翅中央部; 7, 陰茎と完全反転下の内嚢（右側面観）; 8, 陰茎先端（右側面観）; 9, 同（斜背面観）; 10, 同（背面観）; 11, 内嚢（尾側面観）; 12, 同（頭側面観）; 13, 同（背面観）; 14, 同（左斜側面観）; 15, 指状片（頭側面観）; 16, 同（右側面観）; 17, 膣底部節片内板．

28. *Carabus* (*Ohomopterus*) *iwawakianus*

Figs. 18-26. *Carabus* (*Ohomopterus*) *iwawakianus* subspp. — 18-19, *C.* (*O.*) *i. narukawai* (18, ♂, 19, ♀; Azaka-jinja Shrine, Matsusaka-shi, Mie Pref.); 20-21, *C.* (*O.*) *i. shima* (20, ♂, 21, ♀; Mt. Hi-no-yama, Toba-shi, Mie Pref.); 22-23, *C.* (*O.*) *i. muro* (22, ♂, 23, ♀; Shôren-ji, Nabari-shi, Mie Pref.); 24-26, *C.* (*O.*) *i. kiiensis* (24, ♂, 25-26, ♀; 24-25, Sôgawa, Hidakagawa-chô, Wakayama Pref.; 26, Shimo-kitayama-mura, Nara Pref.). — 27-28, Natural hybrid between *C.* (*O.*) *i. narukawai* & *C.* (*O.*) *maiyasanus suzukanus* (27, 28, ♂; Tsubaki-ôkamiyashiro, Suzuka-shi, Mie Pref.). — 29, Natural hybrid between *C.* (*O.*) *i. shima* & *C.* (*O.*) *arrowianus murakii* (♂; Utsugiwara, Isé-shi, Mie Pref.). — a, Apical part of penis in right subdorsal view; b, ditto in dorsal view; c, digitulus in frontal view.

図 18-26. イワワキオサムシ各亜種. — 18-19, **布引山地亜種**(18, ♂, 19, ♀; 三重県松阪市阿射加神社); 20-21, **志摩半島亜種**(20, ♂, 21, ♀; 三重県鳥羽市樋ノ山); 22-23, **室生山地亜種**(22, ♂, 23, ♀; 三重県名張市青蓮寺); 24-26, **紀伊半島亜種**(24, ♂, 25-26, ♀; 24-25, 和歌山県日高川町寒川; 26, 奈良県下北山村). — 27-28, イワワキオサムシ布引山地亜種とマヤサンオサムシ鈴鹿山脈南東部亜種の自然雑種(27, 28, ♂; 三重県鈴鹿市椿大神社). — 29, イワワキオサムシ志摩半島亜種とミカワオサムシ志摩半島北部亜種の自然雑種(♂; 三重県伊勢市槍原). — a, 陰茎先端(右斜背面観); b, 同(背面観); c, 指状片(頭側面観).

28. *Carabus* (*Ohomopterus*) *iwawakianus* (NAKANE, 1953) イワワキオサムシ

Original description　*Apotomopterus yaconinus iwawakianus* NAKANE, 1953, Scient. Rept. Saikyo Univ. (Nat. Sci. & Liv. Sci.), Kyoto, 1, p. 59 [sic] (= misprint of 95), figs. 12, C-e, e',9, f,10, g & 11 (p. 101).

History of research　This taxon was originally described by NAKANE (1953) as a subspecies of *Apotomopterus yaconinus* (= *Carabus* (*Ohomopterus*) *yaconinus* in the present sense). Since NAKANE's taxon is radically different in genitalic features from *C. (O.) yaconinus*, and the former is partly sympatric with the latter, they are regarded as two different species in current classification. In the same paper (NAKANE, 1953), another subspecies of *A. yaconinus* was described by NAKANE & IGA under the name of *kiiensis*. In current classification, it is regarded as a subspecies of *C. (O.) iwawakianus* (HIURA, 1965, etc.). ISHIKAWA & KUBOTA (1995) morphologically revised *C. (O.) iwawakianus*, and classified it into five subspecies, namely, *iwawakianus*, *kiiensis*, *narukawai*, *shima* and *muro*. Of these, the latter three were newly described at that time. IMURA et al. (2005) revised this species molecular phylogenetically (refer to the "Molecular phylogeny" section of this species).

Type locality　"Iwawaki", "Kimi-toge", "Amami in Kawachi", "Kasuga, Nara" and "Yunoyama in Ise".

Type depository　Hokkaido University Museum (Sapporo) [Syntype].

Morphology　19~27mm. Body above coppery brown, greenish coppery or black with faint bluish or green-bluish tinge; tibiae and tarsi variable in color, varied from black to rufous, not rarely bright red. External features: Figs. 1~6. Genitalic features: Figs. 7~17. Externally similar to other medium-sized *Ohomopterus* species, but identified by the following characteristics: 1) male antennae with hairless areas on ventral side from segment five to seven; 2) each elytron with three rows of primary intervals; 3) inner margin of male protibia strongly angulated at middle; 4) digitulus broad and short, its apex sharply pointed like a short process to form peach- or acorn-shaped profile in frontal view; 5) inner plate of ligular apophysis of female genitalia with two-types, rhomboidal (type A) and subquadrate (type B) (KAMIYOSHI, 1963).

Molecular phylogeny　IMURA, TOMINAGA et al. (2005) revised this species molecular phylogenetically using mitochondrial ND5 gene and nuclear ITS I DNA, and suggested that the population represented by *C. (O.) iwawakianus kiiensis* and a part of *C. (O.) i. muro* should be regarded as an authentic species. On the other hand, *C. (O.) i. iwawakianus*, *C. (O.) i. narukawai* and *C. (O.) i. shima* are considered to be offspring of the hybrid between "*C. (O.) kiiensis*" and *C. (O.) maiyasanus*. These hybridized populations further made secondary hydridization between other *Ohomopterus* species such as *C. (O.) arrowianus* or *C. (O.) yaconinus*. In this manner, morphologically defined "*C. (O.) iwawakianus*" would have played an important role in the faunal formation of the subgenus *Ohomopterus* in the Kinki Region. According to a simultaneous analysis of six nuclear DNA sequences, this species does not belong to the *daisen* species-group to which *C. (O.) yaconinus* belongs, but belongs to the *iwawakianus-insulicola* species-group (SOTA & NAGATA, 2008).

Distribution　Honshu (Kinki Region).

Habitat　Distributed on hills and mountains consisting of the basement rock, and not found from the Quaternary strata. Inhabitant of woodland, common in the forest edge of rather old broadleaved tree or those of planted tree such as cedar or cypress, but generally not so common in the thin forest.

Bionomical notes　Spring breeder, earthworm feeder.

Geographical variation　Polytypical species. Morphologically classified into five subspecies.

Etymology　Named after one of the type localiies, Iwawaki or Mt. Iwawaki-san of Osaka Prefecture.

研究史　本タクソンはNAKANE(1953)によりヤコンオサムシの一亜種（記載時に与えられた学名はApotomopterus yaconinus iwawakianus）として記載されたが，ヤコンオサムシとは交尾器形態が異なるうえ，一部の地域で両者は混生しているため，一般に独立種イワワキオサムシとして扱われている．NAKANE(1953)による同じ論文中で，NAKANE & IGAにより同じくヤコンオサムシの一亜種として記載された紀伊半島産の集団（記載時に与えられた学名はA. y. kiiensis）は，イワワキオサムシの一亜種として扱われることが多い．本種の所属は当初，アポトモプテルス属に置かれていたが，ISHIKAWA(1973)以来，一般にカラブス属オホモプテルス亜属の一員とされている．ISHIKAWA & KUBOTA(1995)は本種の再検討を行い，上記2亜種の他にnarukawai, shima, muroの3亜種を追加して計5亜種に分類した．

タイプ産地　NAKANE(1953)によって指定されたシンタイプの産地は "Iwawaki"（岩湧；大阪府河内長野市岩湧山ないしその界隈），"Kimi-toge"（紀見峠；和歌山県橋本市），"Amami in Kawachi"（天見；大阪府河内長野市），"Kasuga, Nara"（春日；奈良県奈良市春日山ないしその界隈），"Yunoyama in Ise"（湯の山；三重県三重郡菰野町（こものちょう））の5箇所である．ISHIKAWA & KUBOTA(1995)は"Iwawaki"を含む地域（上記シンタイプの5産地のうち前4ヶ所）に産するものを基亜種とみなしているが，レクトタイプの指定が行われていないので，どの集団にどの亜種名を充てるかといった問題も含めて今後，命名規約上の処置に変更が生じる可能性も残されている．

タイプ標本の所在　北海道大学総合博物館（札幌）［シンタイプ］．

形態　体長19~27 mm. 背面は銅褐色ないし緑銅褐色，あるいは黒色で僅かに青ないし緑青色を帯びる．脛節と附節は黒から明赤色まで変化する．外部形態：図1~6．交尾器形態：図7~17．オホモプテルス亜属に属する他の中・小型種によく似ているが，以下のような特徴により識別される：1) ♂触角第5節から第7節の腹面には無毛部がある；2) 各鞘翅には3条の第一次間室があ

る；3）♂前脛節内側縁は中央で強く角ばる；4）指状片は幅広く，短く，先端は小突起様に鋭く尖り，全体として桃の実ないしドングリ形；5）♀交尾器腟底部節片内板には2型が認められ，幅広い菱形（A型），或いは前縁中央が突出し，その両側がややえぐれた四角形（B型）を呈する（神吉，1963）．内板後方の腟壁は辺縁の不整な三角形に硬化し，色素沈着を示す．本種と次のヤコノオサムシは，桃の実形ないし五角形の指状片を持つことから，形態学的にはヤコノオサムシ種群としてまとめられるが，複数の核遺伝子による分子系統樹では，イワワキオサムシはイワワキーアオオサムシ種群に，ヤコノオサムシはダイセンオサムシ種群に属するという結果も出ている（Sota & Nagata, 2008）．

分子系統 Imura, Tominaga et al. (2005)は本種の各亜種のミトコンドリアND5遺伝子と核ITS I DNAを解析し，これまで形態学的には多型を示す独立種とみなされてきた「イワワキオサムシ」の亜種のうち，紀伊半島亜種（いわゆるキイオサムシ）および室生山地亜種の少なくとも一部は純系の集団であるが，基亜種，布引山地亜種，志摩半島亜種の三者は，いわゆるキイオサムシの♂とマヤサンオサムシ♀との交雑に由来する集団である可能性が高いことを示唆した．この交雑由来集団は，さらにミカワオサムシやヤコノオサムシなど複数の他種との間に二次的な交雑を起こし，近畿地方におけるオホモプテルス相の形成過程において重要な役割を果たしたと考えられている．

分布 本州（近畿地方）．

生息環境 基盤山地に生息する種で，第四紀層には見られない．森林を主な生活の場とし，疎林では個体数が著しく少ない（近畿オサムシ研究グループ，1979）．

生態 同亜属他種と同様，春繁殖型・環形動物食であろう．

地理的変異 形態学的には多型を示す1種と見なされ，計5亜種に分類される．

種小名の由来 タイプ産地の一つである大阪府南部の岩湧（山）に因む．

28-1. *C. (O.) i. narukawai* Ishikawa et Kubota, 1995 イワワキオサムシ布引（ぬのびき）山地亜種

Original description *Carabus (Ohomopterus) iwawakianus narukawai* Ishikawa & Kubota, 1995, Bull. Biogeogr. Soc. Japan, Tokyo, 50 (1), p. 42, figs. 15~26 (p. 43).

Type locality Azaka-jinja, 48-63m, Koazaka-chô.

Type depository Systematic Zoology Laboratory, Department of Biological Sciences, Tokyo Metropolitan University (Hachiôji) [Holotype, ♂].

Diagnosis 22~27 mm. Body above black or coppery brown (the former color form is usually predominant over the latter); palpi black; tibiae and tarsi black to dark rufous. Allied to the nominotypical *iwawakianus*, but differs from it in the following points: 1) size a little larger on an average; 2) pronotum a little broader; 3) shoulders a little more distinct; 4) apical part of penis broader in lateral view, less sharply narrowed towards apex; 5) digitulus subquadrate in frontal view, with lateral sides almost parallel-sided.

Distribution S. Suzuka Mts.~Nunobiki Mts., and a part of Hakusan-chô and Matsusaka-shi.

Etymology Named after Nobuyuki Narukawa.

タイプ産地 小阿坂町，阿射加神社，標高48-63m（三重県松坂市（まつさかし）に属する）．

タイプ標本の所在 首都大学東京動物系統分類学研究室（八王子）［ホロタイプ，♂］．

亜種の特徴 22~27 mm. 背面は黒色または銅褐色で，黒色型が優勢となる地域が多い．口肢は黒色で，脛節と跗節は黒～暗赤色．基亜種に近いが，以下の諸点で異なる：1）平均してやや大型；2）前胸背板はより幅広い；3）鞘翅肩部の張り出しがやや強い；4）陰茎先端は側方から見てより太く，先端に向けてあまり顕著に細くならない；5）指状片の側縁は左右ほぼ平行で，先端より1/4付近に向けて僅かに広がり，先端は尖る．♀交尾器腟底部節片内板は横長で幅広く，前縁中央部は僅かに突出し，その両側でごく軽く陥入する．内板後方の腟壁は硬化し，色素沈着を示す．

分布 鈴鹿山脈南部～布引山地および津市白山町から松坂市にかけての丘陵地の一部．

亜種小名の由来と通称 亜種小名は生川展行に因む．通称はヌノビキオサムシ．

28-2. *C. (O.) i. shima* Ishikawa et Kubota, 1995 イワワキオサムシ志摩半島亜種

Original description *Carabus (Ohomopterus) iwawakianus shima* Ishikawa & Kubota, 1995, Bull. Biogeogr. Soc. Japan, Tokyo, 50 (1), p. 44, figs. 27~35 (p. 45).

Type locality Mt. Asamagatake, Toinoyama [sic] (misreading of Hi-no-yama), 100m, Toba-shi, Mie Pref.

Type depository Systematic Zoology Laboratory, Department of Biological Sciences, Tokyo Metropolitan University (Hachiôji) [Holotype, ♂].

Morphology 21~26 mm. Body above coppery brown, rarely bronzy with greenish tinge, or entirely black; tibiae and tarsi black to dark rufous. Identified by the following characters: 1) inner margin of male protibia more strongly convex and subangulated at middle than in nominotypical *iwawakianus*; 2) primary intervals of elytra higher than secondaries and tertiaries, with primary foveae larger; 3) apical part of penis a little wider than in nominotypical *iwawakianus* in lateral view, a little longer and slenderer than in

28. *Carabus* (*Ohomopterus*) *iwawakianus*

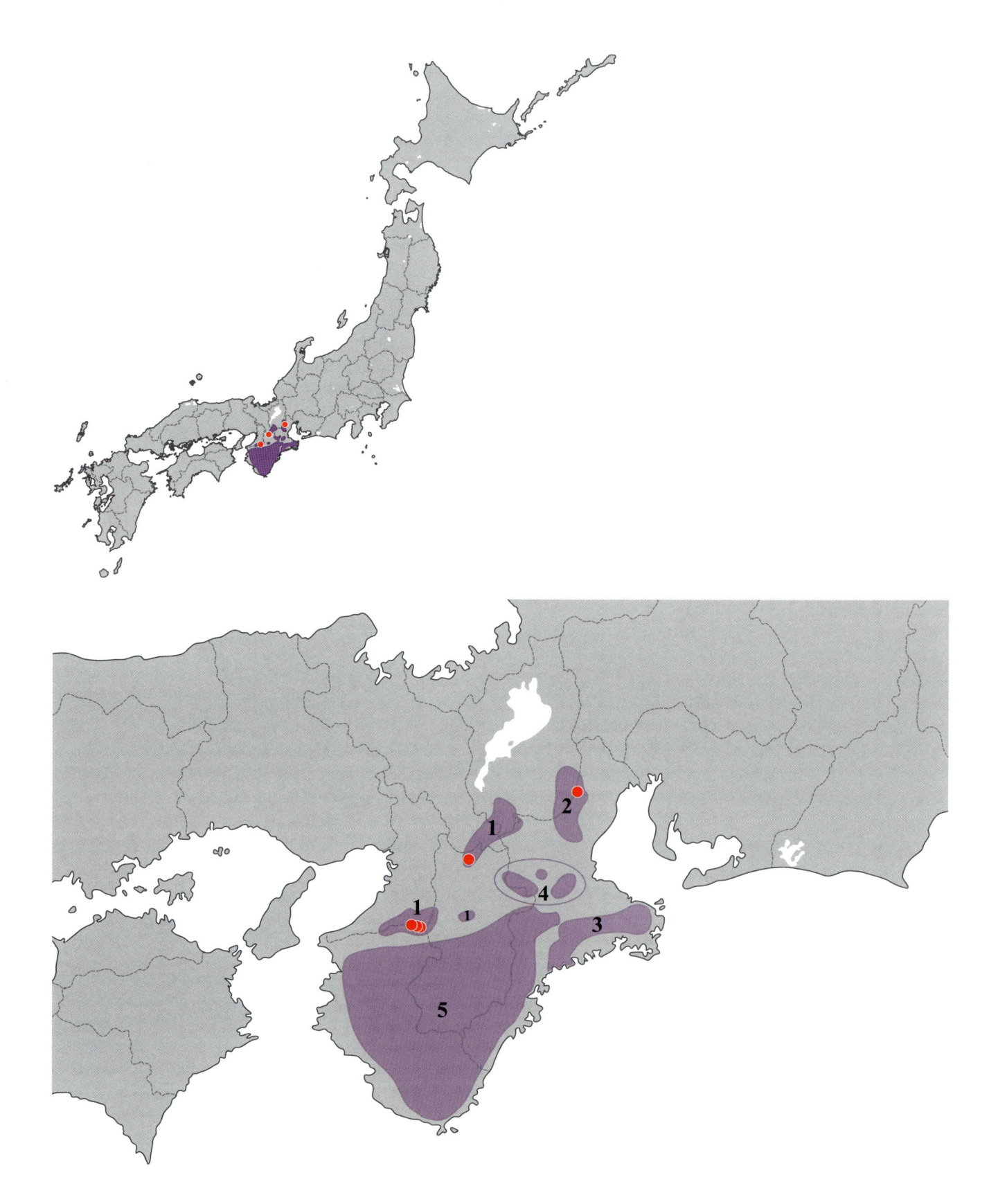

Fig. 30. Map showing the distributional range of ***Carabus*** (***Ohomopterus***) *iwawakianus* (Nakane, 1953). — 1, subsp. ***iwawakianus***; 2, subsp. ***narukawai***; 3, subsp. ***shima***; 4, subsp. ***muro***; 4, subsp. ***kiiensis***. ●: Type localities of the syntypes designated in the original description (Iwawaki, Kimi-toge, Amami in Kawachi, Kasuga, Nara & Yunoyama in Ise).

238

28. Carabus (Ohomopterus) iwawakianus

Fig. 31. Habitat of *Carabus* (*Ohomopterus*) *iwawakianus iwawakianus* (Mt. Iwawaki-san, Kawachi-nagano-shi, Osaka Pref., one of the type localities of the species). The adult usually pass the winter in the soil in such an environment.

図31. イワワキオサムシ基亜種の越冬環境（大阪府河内長野市岩湧山）.

Fig. 32. *Carabus* (*Ohomopterus*) *iwawakianus iwawakianus* (Mt. Iwawaki-san, Kawachi-nagano-shi, Osaka Pref.).

図32. イワワキオサムシ基亜種（大阪府河内長野市岩湧山）.

Fig. 33. *Carabus* (*Ohomopterus*) *iwawakianus iwawakianus* passing the winter in the soil (Mt. Iwawaki-san, Kawachi-nagano-shi, Osaka Pref.).

図33. 土中で越冬しているイワワキオサムシ（大阪府河内長野市岩湧山）.

29. *Carabus (Ohomopterus) yaconinus* BATES, 1873
ヤコンオサムシ

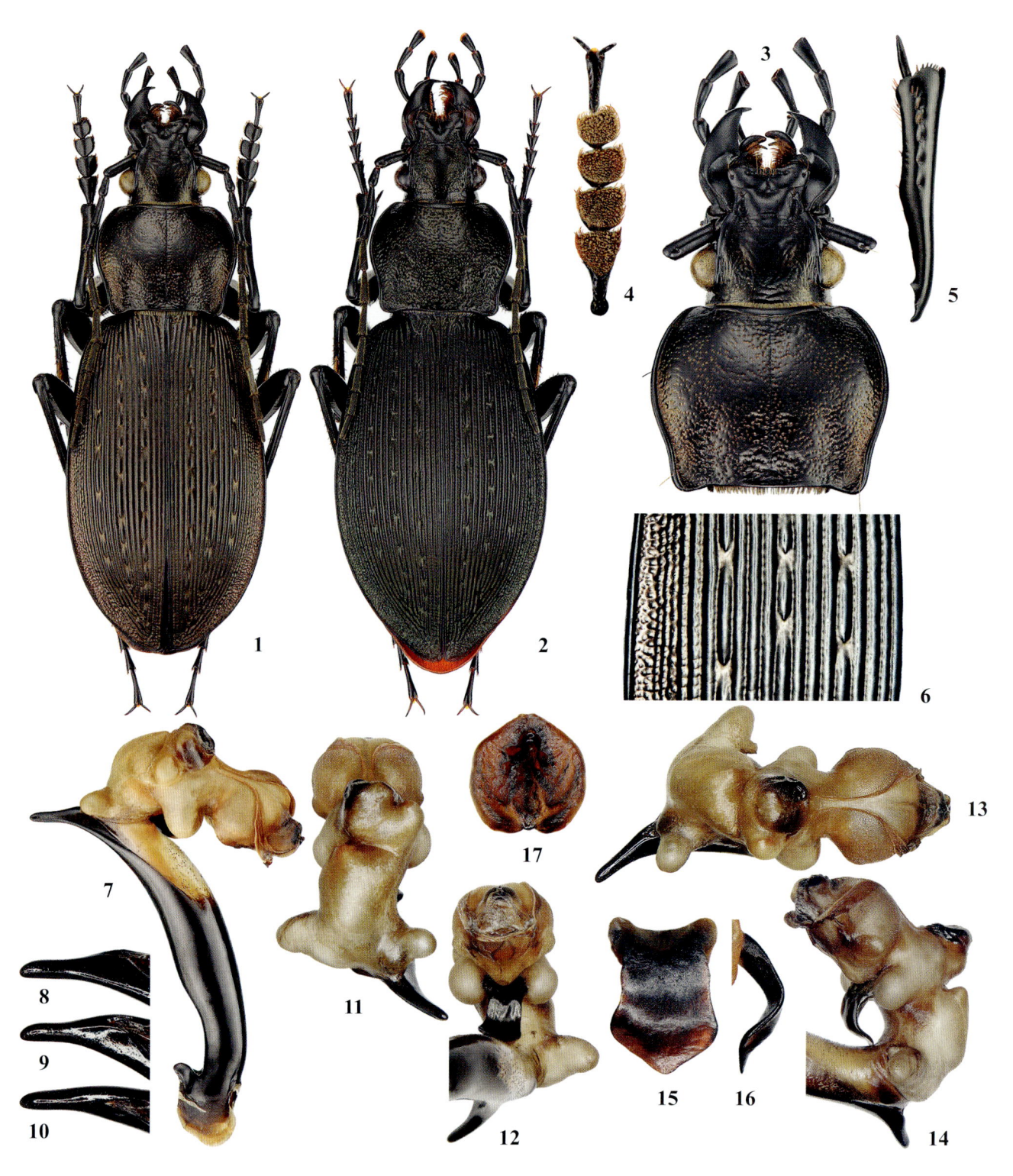

Figs. 1-17. *Carabus* (*Ohomopterus*) *yaconinus yaconinus* BATES, 1873 (Chûô-ku, Kôbé-shi, Hyogo Pref.). — 1, ♂; 2, ♀; 3, male head & pronotum; 4, male right protarsus in ventral view; 5, male right protibia in subdorsal view; 6, male left elytron (median part); 7, penis & fully everted internal sac in right lateral view; 8, apical part of penis in right lateral view; 9, ditto in right subdorsal view; 10, ditto in dorsal view; 11, internal sac in caudal view; 12, ditto in frontal view; 13, ditto in dorsal view; 14, ditto in left sublateral view; 15, digitulus in frontal view; 16, ditto in right lateral view; 17, inner plate of ligular apophysis.

図 1-17. ヤコンオサムシ基亜種（兵庫県神戸市中央区）．― 1, ♂; 2, ♀; 3, ♂の頭部と前胸背板; 4, ♂右前跗節（腹面観）; 5, ♂右前脛節（斜背面観）; 6, ♂左鞘翅中央部; 7, 陰茎と示全反転下の内嚢（右側面観）; 8, 陰茎先端（右側面観）; 9, 同（右斜背面観）; 10, 同（背面観）; 11, 内嚢（尾側面観）; 12, 同（頭側面観）; 13, 同（背面観）; 14, 同（左斜側面観）; 15, 指状片（頭側面観）; 16, 同（右側面観）; 17, 膣底部節片内板.

29. *Carabus* (*Ohomopterus*) *yaconinus*

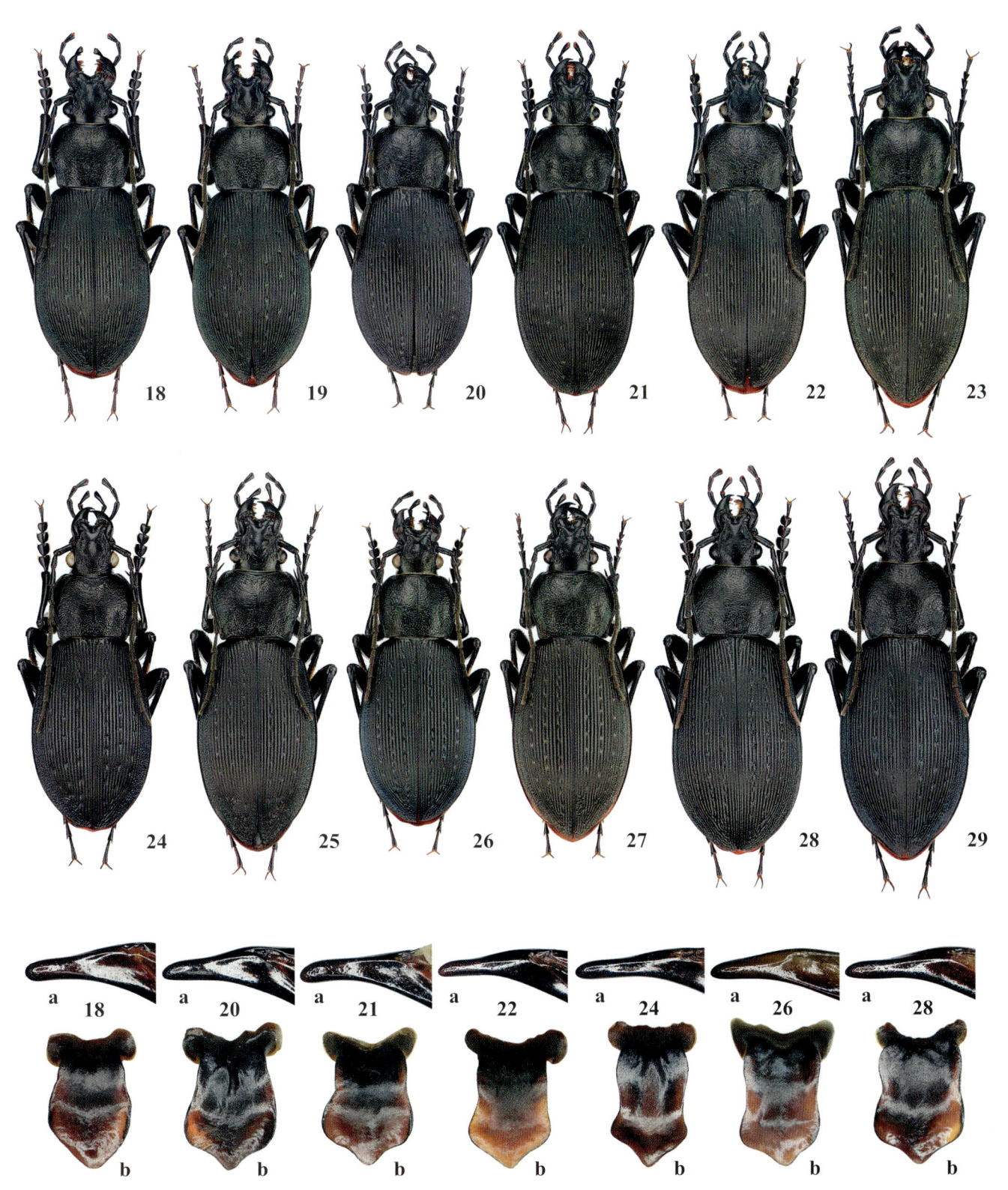

Figs. 18-29. ***Carabus*** (***Ohomopterus***) ***yaconinus*** subspp. — 18-19, **C**. (**O.**) ***y. blairi*** (18, ♂, 19, ♀; Fushiki-kofu, Takaoka-shi, Toyama Pref.); 20, **C**. (**O.**) ***y. sotai*** (♂; Kita-shirakawa, Kyoto-shi, Kyoto Pref.); 21, **C**. (**O.**) ***y. cupidicornis*** (♂; Shaden, Ureshino-kurono-chô, Matsusaka-shi, Mie Pref.); 22-23, **C**. (**O.**) ***y. oki*** (22, ♂, 23, ♀; Minami-dani, Fusé, Is. Dôgo, Iss. Oki, Shimane Pref.); 24-25, **C**. (**O.**) ***y. maetai*** (24, ♂, 25, ♀; Iké-no-tani, Aoya-chô, Tottori-shi, Tottori Pref.); 26-27, **C**. (**O.**) ***y. yamaokai*** (26, ♂, 27, ♀; Yoshifuji, Matsuyama-shi, Ehime Pref.); 28-29, **C**. (**O.**) ***y. seto*** (28, ♂, 29, ♀; Moto-nuwa, Is. Nuwa-jima, Matsuyama-shi, Ehime Pref.). — a, Apical part of penis in right subdorsal view; b, digitulus in frontal view.

図 18-29. ヤコンオサムシ各亜種. — 18-19, 北陸地方亜種(18, ♂, 19, ♀; 富山県高岡市伏木古府); 20, 近畿地方北部亜種(♂; 京都府京都市北白川); 21, 近畿地方中部亜種(♂; 三重県松阪市嬉野黒野町捨田); 22-23, 隠岐亜種(22, ♂, 23, ♀; 島根県隠岐島後布施南谷); 24-25, 山陰地方亜種(24, ♂, 25, ♀; 鳥取県鳥取市青谷町池ノ谷); 26-27, 四国・和歌山亜種(26, ♂, 27, ♀; 愛媛県松山市吉藤); 28-29, **忽那諸島西部亜種**(28, ♂, 29, ♀; 愛媛県松山市怒和島元怒和). — a, 陰茎先端(右斜背面観); b, 指状片(頭側面観).

29. *Carabus* (*Ohomopterus*) *yaconinus* BATES, 1873 ヤコンオサムシ

Original description   *Carabus Yaconinus* BATES, 1873, Trans. ent. Soc. Lond., 1873, p. 231.

History of research   This taxon was originally described by BATES (1873) as a distinct species of the genus *Carabus* (s. lat.) from "Nagasaki and Hiogo (= Hyôgo of Kôbe-shi in Hyôgo Prefecture)", though the former locality (Nagasaki) is most probably incorrect. BREUNING (1934) described *Apotomopterus Blairi* from "Fushiki" (Toyama Pref.). In his early works on the Japanese Carabina, NAKANE (1953) regarded these two taxa as belonging to a single species (*Apotomopterus yaconinus* in his sense), and divided it into four subspecies, namely, nominotypical *yaconinus*, *blairi*, *iwawakianus* and *kiiensis*. Of these, the latter two were described at that time and are currently regarded as another species. ISHIKAWA (1973) redefined the genera and subgenera of the subtribe Carabina, and all the Japanese species including BATES' species so far combined with *Apotomopterus* were transferred to *Ohomopterus*. ISHIKAWA & KUBOTA (1994a) made a tentative revision of *C.* (*O.*) *yaconinus* based on the morphology, and classified it into eight subspecies, six of which were newly described at that time.

Type locality   Nagasaki and Hiogo ("Nagasaki" is obviously erroneous).

Type depository   Museum Nationale d'Histoire Naturelle, Paris [Lectotype, ♂, body above coppery brown].

Morphology   24~32 mm.   Black with dark bluish or dark greenish tinge, or dark coppery. Rather large-sized *Ohomopterus* species with relatively short legs, bearing the following characteristics: 1) antennae relatively robust and short, with apices not reaching middle of elytra even in male; 2) male antennae with marked hairless ventral depression from segment four or five to seven; 3) each elytron with three rows of almost homogeneously raised primary intervals, with primary foveoles relatively large but not invading adjacent tertiaries, striae between intervals regularly notched; 4) inner margin of male protibia weakly angulated at middle; 5) penis large, about half as long as elytra; 6) digitulus pentagonal, with apex protruded though not sharply pointed; 7) inner plate of ligular apophysis circular or somewhat elongated longitudinally in dorsal view, with posterior margin thickly rimmed, inner wall irregularly bearing many pleats.

Molecular phylogeny   For an overview of the molecular phylogeny of *Ohomopterus*, refer to the same section of *C.* (*O.*) *yamato*. In the molecular genealogical trees, *C.* (*O.*) *yaconinus* splits into two clades. The western clade is monophyletic without contamination of any other species, and the other one, the Kinki clade, is intermingled with several other *Ohomopterus* species including *C.* (*O.*) *iwawakianus*. The ND5 gene of *C.* (*O.*) *yaconinus* in the Kinki Region is supposed to be replaced by that of *C.* (*O.*) *iwawakianus* when the western population invaded the Kinki Region. The nuclear ITS I of the Kinki population is clustered either with that of the western population of *C.* (*O.*) *yaconinus* or with that of *C.* (*O.*) *iwawakianus*. These results together with the distribution profiles of these two taxa suggest strongly that the western population of *C.* (*O.*) *yaconinus* would be the authentic strain within this species, while that in the Kinki Region is the offspring of hybrids between the authentic *C.* (*O.*) *yaconinus* and *C.* (*O.*) *iwawakianus*. *C.* (*O.*) *yanoninus* from the Noto Peninsula is considered to be the hybrid-derivative formed between the male of authentic *C.* (*O.*) *yaconinus* and the female of *C.* (*O.*) *arrowianus* (OKAMOTO et al., 2005).

Distribution   Honshu (Kinki & Chugoku Regions) & Shikoku / Iss. Noto-jima (Ishikawa Pref.), Oki-no-shima [Lake Biwa-ko] (Shiga Pref.), Kami-shima (Mie Pref.), Ié-shima, Awaji-shima, Nu-shima (Hyogo Pref.), Ao-shima [Lake Koyama-iké of Tottori-shi] (Tottori Pref.), Dôgo, Naka-no-shima (Shimane Pref.), Ôtabu-jima (Okayama Pref.), Sagi-shima, Ikuchi-jima, In-no-shima, Mukai-shima, Yoko-shima, Ta-shima (Hiroshima Pref.), Shôdo-shima, (Maé-jima (= Ômi Penins.)), Takami-shima, Awa-shima, Shishi-jima, (Ya-shima) (Kagawa Pref.), Shimada-jima, Ôgé-jima (Tokushima Pref.), Iwaki-jima, Ikina-jima, Yugé-shima, Uo-shima, Ômi-shima, Ô-shima in Niihama-shi, Nuwa-jima, Naka-jima, Futagami-jima, Nogutsuna-jima, Gogo-shima & Tsuru-shima (Ehime Pref.).

Habitat   Common species widely inhabiting various environment, and often predominant over other species not only in natural forest but also in open- to semi-open lands such as roadside, scrub, areas adjacent to a settlement or habitat, or even in such places as river beds or arable lands. Distributed mainly in the plain, hills and low mountains.

Bionomical notes   Spring breeder, earthworm feeder.

Geographical variation   Morphologically classified into eight subspecies.

Etymology   In the original description, BATES did not mention about the origin of the specific name "Yaconinus". If it is syllabicated as "Yacon-inus", it cannot be interpreted since there are no Japanese words corresponding to "Yacon" (or "Yakon"), so far as we know. However, if it is divided into such syllables as "Yaconin-us", it becomes possible to interpret this word reasonably, since "Yaconin" is considered to be a Latinized spelling of "Yakonin" (its correct spelling is "Yakunin") referring to officer, official or policeman in Japanese, and often used by western people who stayed in Japan from the late Edo Period to the early Meiji Period (the latter half of 1800s). The type specimens of *Carabus yaconinus* were collected by G. LEWIS and/or his Japanese assistants during his collecting trip made from the 1860s to the 1870s, and LEWIS must have known the word called "Yakonin", or possibly he was accompanied by "Yakonin" itself during his trip, though this story is nothing but a mere supposition.

研究史　本タクソンはBATES(ベイツ)(1873)により独立種*Carabus Yaconinus*として記載された．タイプ産地は「長崎と兵庫(現在の神戸市に相当する地名)」だが，前者は明らかに誤りであろう．BREUNING(ブロイニング)(1934)は「伏木」(富山県)から独立種*Apotomopterus blairi*を記載したが，NAKANE(1953)はこれをヤコンオサムシの亜種(*A. yaconinus blairi*)に降格し，さらに*A. y. iwawakianus*と*A. y. kiiensis*の二亜種を記載して，同種を計4亜種(基亜種を含む)に分類した．後二者は現在，ヤコンオサムシと

29. *Carabus* (*Ohomopterus*) *yaconinus*

は種のレベルで異なるもの(イワワキオサムシとその紀伊半島亜種)として扱われる場合が多い．所属に関しては，ISHIKAWA (1973)によるオサムシ亜族の上位分類群再検討以来，アポトモプテルス属からカラブス属オホモプテルス亜属(又はオホモプテルス属)に移されている．ISHIKAWA & KUBOTA (1994a)は外部形態と陰茎，指状片の形態に基づいて本種に再検討を加え，6亜種を新たに記載(両著者の連名による新亜種は2亜種のみで，残る4亜種はISHIKAWAの単著)し，計8亜種に分類した．

タイプ産地　長崎と兵庫(＝神戸市)．ただし，このうち長崎は明らかに誤りと思われる．

タイプ標本の所在　パリ自然史博物館[レクトタイプ，♂]．レクトタイプは銅褐色味を強く帯びた個体で，ラベルには"Nagasaki"と記されているが，実際には神戸付近から得られたものである可能性が高いとされる．

形態　24〜32 mm．黒色で体側縁に暗青色〜暗緑色を帯びるものを基本に，個体ないし地域によっては暗銅色となる．やや大型で脚の短いオホモプテルスで，以下の特徴を有する：1) 触角は太く，短く，♂でも先端が鞘翅中央に達しない；2) ♂触角第4節ないし第5節から第7節にかけての腹面には無毛凹陥部がある；3) 各鞘翅には3条の第一次間室があり，各間室は均一に隆起し，第一次凹陥は比較的大きいが隣接する第三次間室を侵すことはなく，間室間の線条には規則的な刻み目がある；4) ♂前脛節内側縁は中央で鈍く角ばる；5) 陰茎は長く，鞘翅長の約1/2；6) 指状片は五角形に近く，遠位端の中央は突出するが，あまり鋭く尖らず，鈍く丸まる場合が多い；7) ♀交尾器膣底部節片内板はやや縦に長い円形で，後縁の縁取りは厚く，内壁は襞状．

分子系統　オホモプテルス亜属の分子系統の概要については，ヤマトオサムシの同項を参照．ミトコンドリアND5遺伝子と核ITS I DNAの分析から，本種は，四国北部から中国地方(隠岐を含む)，さらに近畿北西部を経て北陸地方に至る純系の集団と，近畿地方を中心に分布し，恐らくイワワキオサムシとの交雑に由来する集団との2系統に大きく分けられ，両者の境界は琵琶湖－淀川線のやや西方から富山湾にかけての地域にあることが判明している．能登半島に分布する集団は純系のヤコンオサムシとミカワオサムシとの交雑に由来するものである可能性が高く，形態学的に北陸地方亜種と定義される集団は，分子系統学的観点から見ると，少なくとも二通りの異なる系統に由来する集団が混在したものであるらしい(OKAMOTO et al., 2005)．

分布　本州(近畿・中国地方)，四国／能登島(石川県)，沖島[琵琶湖](滋賀県)，神島(三重県)，家島，淡路島，沼島(以上，兵庫県)，青島[鳥取市湖山池]，島後，中ノ島(以上，島根県)，大多府島(岡山県)，佐木島，生口島，因島，向島，横島，田島(以上，広島県)，小豆島，(前島(＝大深半島))，高見島，粟島，志々島，(屋島)(以上，香川県)，島田島，大毛島(以上，徳島県)，岩城島，生名島，弓削島，魚島，大三島，大島[新居浜市]，怒和島，中島，二神島，野忽那島，興居島，釣島，(以上，愛媛県)．北東限は能登半島基部から富山盆地西端にかけての山地．近畿地方では琵琶湖北東部から鈴鹿山脈北部にかけての地域が東限となるが，南東部では雲出川(くもずがわ)沿いに伊勢平野に進出している．南限は櫛田川(くしだがわ)〜紀ノ川を結ぶ線になるが，紀伊半島西部の沿岸地域ではこれより南方にも分布する．西限は島根県西部から広島市付近にかけての地域．

生息環境　自然林は元より，人里に近い二次林から川原や耕地といった人為的影響の濃い場所に至る幅広い環境に適応した生活力の旺盛な種で，一般に個体数も多い．中国地方では山地にも生息しているが，近畿地方では東に行くほど平地性になる．

生態　春繁殖，成虫越冬型のオサムシで，幼虫は環形動物食(フトミミズ等)．

地理的変異　形態学的には目下，計8亜種に分類されている．

種小名の由来　原記載の中でBATESは*Yaconinus*という種小名の由来について何も触れていないが，頭文字のYが大文字で記されていることから，同論文中に現れるマヤサンオサムシ *C. Maiyasanus*(摩耶山に因む)やオオオサムシ *C. DeHaanii* (DE HAAN(ドゥ・ハーン)という人名に因む)等と同様，地名や人名，或いは何らかの「もの」を表す固有名詞に因む名と考えられる．今日，我々が使用している「ヤコンオサムシ」という和名は，*Yaconinus*という種小名の音節構成を*Yacon*という語幹の後に「類似」を示すラテン語の接尾辞 -inusがついた形，すなわち「ヤコンに似た，ヤコンのような」と解釈して作られたものであろう．では「ヤコン」とは一体何のことだろうか？ *Yacon*や*Yakon*ないしそれに準じた発音を持つ固有名詞は，当時を含め，少なくとも我が国には該当するものが見当たらないし，神話や伝説に登場する架空の「なにか」(神や怪獣，魔物等)を表す名称でもなさそうである．思い当たるのは南米原産の根菜，ヤーコンくらいだが，日本から発見された新種のオサムシにこのような名が唐突に持ち出されるはずもあるまい．これまであまりにも当然の如く「ヤコンオサムシ」と呼び慣わしてはきたものの，ヤコンという語に種小名の由来を求めることは困難であり，納得のいく説明も目にしたことがない．この件に関して筆者(井村)は，ロンドン自然史博物館やパリ自然史博物館のスタッフにも直接質問してみたが，彼らにも心当たりはなく，やはりなんらかの日本語に由来するものではないか，という答えであった．そこで，「ヤコン」という既成概念にとらわれず，音節の区切り方を変えて名詞*Yaconin*の語尾に-usを付けて形容詞化したもの(「ヤコニンの」)と考えてみよう．すると，別の解釈が可能になるのである．無論，「ヤコン」と同様，このままでは何を意味するかわからないが，140年も前に外国人が書いた論文であるから，現地民の発音を言わば「聞きなす」形でローマ字に転記された単語も多かったはずである．当時の欧米人が書き残した日本に関する文献，例えば1865年にロンドンで出版された "Araki the Daimio: A Japanese Story of the Olden Time"(大名アラキ：昔の日本の物語)や，1867年に出版された世界初の英文日本ガイドブックと言われる "The Treaty Ports of China and Japan"(中国と日本の条約港)等を紐解いてみると，現在よく用いられているヘボン式や日本式とは異なる，独特の綴りによってローマ字表記された単語が多く見受けられる．例えば，「侍」はSamourai，「将軍」はSho-goon，「お目付け」はOmetsky，といった具合である．そして，これらの語に混じって "Yakonin(s)" という単語がしばしば登場するのである．*Yakonin*とはすなわち「役人」のことであり(語尾にsを付けて複数形で示されている場合も多い)，当時の欧米人が日本の警官(Japanese police)あるいは二本差しの公務員を指す言葉として用いていた．「町中(まちなか)で多くの*Yakonins*を見かけた」といった記述が複数の文献に記されていることからも，当時の欧米人にとってそれほど珍しい言葉ではなかったように思われる．BATESに標

本をもたらしたのは，かの有名なLewis(ルイス)であるが，彼が日本を旅行したのは1860年代の終わりから1870年代にかけての5年間と，1880年から1881年にかけての2年間の計2回．このうち，ヤコンオサムシを得た旅行は初回の5年間になされたもので，時まさに明治維新の真っただ中．時代背景から考えて，旅行中多くのYakoninsを見かけたはずである．さらに，1883年に出版された論文の中で，BatesはLewisが「日本人助手たちの助けを借りて("with the aid of his Japanese assistants")日本各地で昆虫採集に精を出した」と明記しているので，Lewisの旅行に付き添い，採集の手助けをした日本人がいたことも確かである．この時代に外国人が民間人を自由に雇って長期間に亘り日本国内を採集して回れたとは到底思えないので，護衛と見張りを兼ねて，政府あるいは地方の行政府のお役人が彼と行動を共にしていた可能性は高い．つまり，Lewisは単にYakoninという言葉を知っていただけでなく，彼と行動を共にしていたアシスタントがまさにYakoninそのものであった可能性も高いのではないだろうか？Batesは我が国から新種を記載するにあたり，日本語に由来する名をいくつも用いている．地名に因んだ例が多いが，複数のゴミムシに対しては"Daimio"(大名)という学名を与えているので，新種のオサムシに対してYakonin(ラテン語化してYaconin)という興味深い異国語(Lewisからの伝聞によって知った可能性も高いだろう)を元に命名したとしても不思議ではない．ヤコニンオサムシ・・・あくまで推測の域を出ないとは言え，少なくとも，「ヤコン」という何を指すのか皆目見当もつかない語に種小名の由来を求めるよりは，よほど現実的な解釈ではないだろうか？(本項は，井村・水沢と親しい昆虫研究者からのアドバイスを元に作成した．脱稿後に出版された他の著書の中で，曽田(2013)も全く同様の見解を述べており，ヤコンオサムシ＝ヤコニン(ヤクニン)オサムシ説には一定の信頼度があるとみてよさそうに思われるが，このことを裏付けるデータはみつかっていない．本当のところは天国のBatesに聞くしかないのかもしれない．)

29-1. *C. (O.) y. blairi* (Breuning, 1934)　ヤコンオサムシ北陸地方亜種

Original description　*Apotomopterus Blairi* Breuning, 1934, Folia Zool. Hydrobiol., Riga, 6, p. 31.
Type locality　Fushiki (= it belongs to Takaoka-shi of Toyama Pref. at present).
Type depository　Zoölogisch Museum Amsterdam [Holotype, ♂].
Diagnosis　24~31 mm. Body above predominantly black with faint dark greenish, blue-greenish or brownish tinge, rarely entirely dark brown. Distinguished from other subspecies by the following points: 1) apical lobe of penis not strongly narrowed in median portion in lateral view; 2) digitulus with preapical widest part effaced, apical tip not sharply pointed.
Distribution　Toyama Pref. & Ishikawa Pref., NC. Honshu.
Etymology　Named after Kenneth Gloyne Blair (1882~1952) of London.

タイプ産地　伏木(現在の所属は富山県高岡市)．
タイプ標本の所在　アムステルダム動物学博物館［ホロタイプ，♂］．
亜種の特徴　24~31 mm．黒色で僅かに暗緑色，暗青緑色ないし暗銅色を帯びるものが大半を占めるが，全身が暗茶褐色の個体も出現する．陰茎先端中央部は側面から見てそれほど顕著にくびれず，指状片は先端やや基部よりにある最広部の張り出しが弱く，先端部全体はなだらかな三角形を呈し，中央はそれほど顕著に突出しないことにより他種から識別される．
分布　富山，石川両県．
亜種小名の由来と通称　亜種小名はゴミムシダマシ類の専門家であったロンドンのKenneth Gloyne Blair(ケネス・グロイン・ブレア)(1882~1952)に因む．最も広く用いられてきた通称はトヤマオサムシ．ホクリクオサムシという名が提唱されたこともあるが，マヤサンオサムシ北陸地方亜種にも同じ通称が用いられるので，注意が必要である．

29-2. *C. (O.) y. sotai* Ishikawa et Kubota, 1994　ヤコンオサムシ近畿地方北部亜種

Original description　*Carabus (Ohomopterus) yaconinus sotai* Ishikawa & Kubota, 1994, Proc. Japan. Syst. Zool., (50), p. 29, figs. 3 & 4 (p. 30).
Type locality　Kitashirakawa (the Botanical Garden of Kyoto University), Kyoto.
Type depository　Systematic Zoology Laboratory, Department of Biological Sciences, Tokyo Metropolitan University (Hachiôji) [Holotype, ♂].
Diagnosis　26~29 mm. Usually black with faint bluish tinge along elytral margins, rarely entirely dark brown. 1) Pronotum weakly convergent posteriad, with hind angles very short, basal foveae shallow; 2) primary foveoles of elytra small but deeply carved; 3) male antennae with hairless ventral areas from segment five to seven; 4) inner margin of male protibia weakly angulated; 5) apical part of digitulus not strongly protruded, with tip not sharply pointed, lateral sides not sinuately convergent from widest part to base.
Distribution　E. Kyoto Pref., Shiga Pref., N. Mie Pref. & SW. Gifu Pref.
Etymology　Named after Teiji Sota.

タイプ産地　京都，北白川(京都大学植物園)(京都府京都市)．
タイプ標本の所在　京都大学未来動物系統力観学研先室(八王子)［ホロタイプ，♂］．
亜種の特徴　26~29mm．通常黒色で，鞘翅辺縁は僅かに青色を帯びる．稀に全体が暗褐色の個体が見られる．1)前胸背板

側縁の後方への狭まりは弱く,後角の突出はごく弱く,基部凹陥は浅い;2)鞘翅第一次凹陥は小さいが深く明瞭に刻まれる;3)♂触角第5節から第7節の腹面に無毛部がある;4)♂前脛節内側縁中央の角ばりは弱い;5)指状片先端部は短い三角形を呈し,先端は鋭く尖らず,側縁は最広部から基部に向かってあまり波曲することなく狭まる.

分布 京都府東部〜滋賀県のほぼ全域〜三重北部〜岐阜県南西部.

亜種小名の由来と通称 亜種小名は曽田貞滋に因む.ソタヤコンオサムシと呼ばれることもある.

29-3. *C. (O.) y. cupidicornis* Ishikawa et Kubota, 1994 ヤコンオサムシ近畿地方中部亜種

Original description *Carabus* (*Ohomopterus*) *yaconinus cupidicornis* Ishikawa & Kubota, 1994, Proc. Japan. Syst. Zool., (50), p. 31, figs. 5 & 6 (p. 32).
Type locality Shaden, 15m, Ureshino-chô, Mie Pref. (Matsusaka-shi of Mie Pref. at present).
Type depository Systematic Zoology Laboratory, Department of Biological Sciences, Tokyo Metropolitan University (Hachiôji) [Holotype, ♂].
Diagnosis 24~32 mm. Usually black, rarely dark brownish. 1) Male antennae with hairless ventral areas from segment four to seven; 2) basal foveae of pronotum deeper and larger than in other races; 3) elytral shoulders more distinct; 4) inner margin of male protibia distinctly angulated; 5) apical lobe of penis slender, gradually narrowed toward tip in lateral view, feebly bent right laterad in dorsal view; 6) digitulus widest just before apex, where its lateral margins apparently convex laterad to form nipple-like appearance.
Distribution C. Mie Pref., N. Nara Pref., Osaka Pref. & N. Wakayama Pref.
Etymology Named after characteristically shaped apical tip of the digitulus.

タイプ産地 三重県,嬉野町,捨田,15 m(嬉野町は合併により消滅し,現在は松阪市の一部).
タイプ標本の所在 首都大学東京動物系統分類学研究室(八王子)[ホロタイプ,♂].
亜種の特徴 24~32mm.通常黒色で稀に暗褐色.1)♂触角第4節から第7節の腹面に無毛凹部がある;2)前胸背板の基部凹陥は通常大きく深い;3)鞘翅は肩部が張り出し,より箱形に近い輪郭を呈する;4)♂前脛節内側縁中央の角ばりは顕著;5)陰茎先端部は細長く,先端に向けて徐々に狭まり,背面から見て僅かに右側へ湾曲する;6)指状片は先端部両側縁が強く側方に突出し,先端部が乳頭状を呈する.
分布 三重県中部〜奈良県北部〜大阪府のほぼ全域〜和歌山県北部.
亜種小名の由来と通称 亜種小名は指状片先端の形態に因むものだろう.キューピーヤコンオサムシという奇妙な通称がある.

29-4. *C. (O.) y. yaconinus* Bates, 1873 ヤコンオサムシ基亜種

Original description, type locality & type depository See explanation of the species.
Diagnosis 26~31 mm. Dark brown or black with bluish margins. 1) Male antennae with marked hairless ventral areas from segment five to seven; 2) shape of pronotum variable according to individuals, though usually strongly narrowed toward base, hind angles short and rounded at tips; 3) pronotal disk coarsely and irregularly punctate, median longitudinal line narrow but clearly impressed, basal foveae shallow but large; 4) elytral shape also variable according to individuals, though usually with distinct shoulders, small but conspicuously curved primary foveoles; 5) inner margin of male protibiae weakly angulated; 6) apical lobe of penis slender and faintly bent right laterad in dorsal view; 7) digitulus widest near base, its lateral sides almost parallel-sided and weakly sinuate, widest again before apex, with tip strongly sinuate in lateral view.
Distribution W. Kyoto Pref., Hyogo Pref., S. Okayama Pref. & S. Hiroshima Pref. / Is. Awaji-shima (Hyogo Pref.).

原記載,タイプ産地,タイプ標本の所在 種の項を参照.
亜種の特徴 26~31mm.暗褐色または側縁部が微かに青味を帯びた黒色.1)♂触角第5節から第7節の腹面に顕著な無毛部がある;2)前胸背板は一般に後方に向けて強く狭まるが,個体差が著しく,側縁が波曲し心臓型を呈するものもある.後角は短く,先端は丸い;3)前胸背板表面の点刻は密ながらしばしば不規則で,正中線は細いが通常完全,基部凹陥は浅いが大きい;4)鞘翅の輪郭も個体差が大きいが,通常肩部の張り出しはやや強めで,第一次凹陥は小さいがよく目立つ;5)♂前脛節内側縁の角張りは弱い;6)陰茎先端は細く,背面から見て僅かに右側へ曲がる;7)指状片は基板近くで最も幅が狭く,中央部で両側縁は左右ほぼ平行になり,僅かに波曲し,先端手前で再度幅広くなり,同部で弧状に張り出し,側方から見て先端部は強く屈曲する.
分布 京都府西部〜兵庫県〜岡山・広島両県南部/淡路島(兵庫県).

29-5. *C. (O.) y. oki* Ishikawa, 1994 ヤコンオサムシ隠岐亜種

Original description *Carabus* (*Ohomopterus*) *yaconinus oki* Ishikawa, in Ishikawa & Kubota, 1994, Proc. Japan. Syst. Zool., (50), p. 33, fig. 12 (p. 35).
Type locality "Is. Oki Dôgo, Minamidani, 10m, Fuse-mura" (Okinoshima-chô of Shimane Pref. at present).

Type depository   Systematic Zoology Laboratory, Department of Biological Sciences, Tokyo Metropolitan University (Hachiôji) [Holotype, ♂].

Diagnosis   26~32 mm. Dark brown, mat, often faintly bluish along lateral margins of pronotum and elytra. 1) Male antennae with marked hairless ventral areas from segment five to seven; 2) pronotum cordiform, with hind angles triangularly protruded; disk evenly and weakly convex above, coarsely and regularly punctate as a whole and rather mat; median longitudinal line weak and partly unclear; basal foveae very shallow, sometimes unclear; 3) elytra elongated oval, intervals almost evenly convex, primary foveoles very small; 4) apical lobe of penis slender and thin, faintly bent right laterad in dorsal view, with left margin somewhat convex by apical part of membraneous area; 5) digitulus relatively slender, triangularly pointed at apex, with lateral sides rather strongly constricted before widest part and near base.

Distribution   Oki Islands (Iss. Dôgo & Naka-no-shima).

Etymology   Named after the Oki Islands.

タイプ産地　布施村, 隠岐, 島後, 南谷, 10 m (布施村は合併により消滅し, 現在は隠岐の島町 (おきのしまちょう)).

タイプ標本の所在　首都大学東京動物系統分類学研究室 (八王子) [ホロタイプ, ♂].

亜種の特徴　26~32 mm. 背面は暗褐色で, 光沢は鈍く, 前胸背板と鞘翅の側縁部は僅かに青味を帯びる. 1) ♂触角第5節から第7節の腹面には顕著な無毛部がある; 2) 前胸背板は顕著な心臓形を呈し, 後角は三角形に突出する. 背面は均等に軽度に膨隆し, 全面に亘り密な点刻を規則的に装い, 光沢は乏しい. 正中線は弱く, 部分的に不明瞭. 基部回陥は非常に浅く, 時に不鮮明; 3) 鞘翅は長卵形で, 各間室はほぼ均等に膨隆し, 第一次回陥は非常に小さい; 4) 陰茎先端部は細長く, 薄く, 背面から見て僅かに右へ曲がり, 基部左側は軽く膨らむ; 5) 指状片は細長く, 先端は三角形に尖り, 側縁は先端手前の最広部基部よりと基板近くの二箇所において比較的強くくびれる.

分布　隠岐諸島 (島後および中ノ島).

亜種小名の由来と通称　亜種小名は隠岐に因む. オキヤコンオサムシという通称で呼ばれることもある. Ishikawa & Kubota (1994a) の共著論文中で記載されたが, 本タクソンの著者は Ishikawa 単独となっているので注意が必要である.

29-6. *C.* (*O.*) *y. maetai* Ishikawa, 1994 ヤコンオサムシ山陰地方亜種

Original description   *Carabus* (*Ohomopterus*) *yaconinus maetai* Ishikawa, in Ishikawa & Kubota, 1994, Proc. Japan. Syst. Zool., (50), p. 36, figs. 13~15 (p. 35).

Type locality   Ikenotani, 80m, Aoya-chô, Tottori Pref.

Type depository   Systematic Zoology Laboratory, Department of Biological Sciences, Tokyo Metropolitan University (Hachiôji) [Holotype, ♂].

Diagnosis   26~31 mm. Black or brownish black, bearing bluish tinge along margins. 1) Male antennae with marked hairless ventral areas from segment five to seven; 2) pronotum with lateral sides strongly sinuate; hind angles triangularly protruded; median longitudinal line weak but almost completely recognized; disk irregularly and coarsely punctate, though impunctate and rugulose in median portion; basal foveae oblong, shallow but distinct; 3) elytra ovate, often narrower in male; intervals evenly raised; 4) inner margin of male protibiae weakly subangulated; 5) apical lobe of penis very slender in apical half in lateral view, feebly bent right laterad in dorsal view, with left margin gently arcuate; 6) digitulus long, narrowest of all the subspecies of *C.* (*O.*) *yaconinus*, with apex subtriangular and bluntly pointed, lateral margins strongly convex at preapical widest part, distinctly convergent therefrom, then more or less divergent toward base, and often divergent again behind basal plate.

Distribution   Tottori Pref., Shimane Pref. & N. Hiroshima Pref.

Etymology   Named after Yasuo Maeta of Shimane University.

タイプ産地　鳥取県, 青谷町, 池ノ谷, 80 m (現在は鳥取市青谷町).

タイプ標本の所在　首都大学東京動物系統分類学研究室 (八王子) [ホロタイプ, ♂].

亜種の特徴　26~31mm. 背面は黒色ないし僅かに褐色味を帯び, 前胸背板と鞘翅の側縁部は多少青味がかる. 1) ♂触角第5節から第7節の腹面には顕著な無毛部がある; 2) 前胸背板は側縁の波曲が顕著で, 後角は三角形に突出し, 正中線は弱いが通常完全. 表面には不規則な点刻を密に装うが, 中央部には点刻を欠き, 皺を有する. 基部回陥は縦長で, 浅いが明瞭; 3) 鞘翅は卵形で, ♂ではより細長く, 間室は均質; 4) ♂前脛節内側縁の角張りは弱い; 5) 陰茎先端部は側面から見て先端半分が非常に細長く, 背面から見て微かに右へ曲がり, 左側縁は緩やかに弧を描く; 6) 指状片は本種の各亜種中最も幅が狭く, 先端部は三角形だがそれほど鋭く尖らない. 側縁は先端手前の最広部で強く側方に突出し, 一旦狭まった後, 基部に向けて再度僅かに広がり, 基板近くで再び弱くくびれる.

分布　中国地方北半部 (鳥取・島根両県～広島県北部).

亜種小名の由来と通称　亜種小名は島根大学の前田泰生に因む. サンインヤコンオサムシと呼ばれることもある. 本亜種も Ishikawa & Kubota (1994a) の共著論文中で記載されたが, 本タクソンの著者は Ishikawa 単独となっている.

29-7. *C. (O.) y. yamaokai* Ishikawa, 1994 ヤコンオサムシ四国和歌山亜種

Original description *Carabus (Ohomopterus) yaconinus yamaokai* Ishikawa, in Ishikawa & Kubota, 1994, Proc. Japan. Syst. Zool., (50), p. 38, figs. 16~19 (p. 37).
Type locality Yoshifuji, Matsuyama-shi, Ehime Pref.
Type depository Systematic Zoology Laboratory, Department of Biological Sciences, Tokyo Metropolitan University (Hachiôji) [Holotype, ♂].
Diagnosis 25~31 mm. Dark brownish coppery, sometimes dark greenish. 1) Male antennal segment five to seven with hairless ventral areas; 2) pronotum broad, with lateral margins strongly arcuate toward apex, strongly narrowed toward base though hardly sinuate; hind angles bluntly angulate and barely produced; median longitudinal line often vaguely impressed; basal foveae oblong, usually clearly impressed; disk coarsely scattered with small punctures; 3) elytra with shoulders distinct so that lateral sides appear less arcuate above all in male, primary foveoles distinct; 4) apical lobe of penis broad in basal half and acutely narrowed therefrom toward tip in lateral view, strongly bent right laterad in dorsal view; 5) digitulus large, broad, with distance between preapical widest part and apex very short, lateral margins weakly convergent before widest part and near basal plate. The population distributed in NW. Wakayama Pref. is usually larger and more strongly polished.
Distribution N. Shikoku, NW. Wakayama Pref. / Kutsuna Islands (Iss. Naka-jima & Gogo-shima) (Ehime Pref.). This species is also distributed in several other small islands in the Inland Sea of Setonaikai near Shikoku, but a taxonomic account for these population has not been provided as yet.
Etymology Named after Yukio Yamaoka of Ehime Prefecture.

タイプ産地　愛媛県, 松山市, 吉藤.
タイプ標本の所在　首都大学東京動物系統分類学研究室（八王子）［ホロタイプ, ♂］.
亜種の特徴　25~31 mm. 背面は暗褐色ないし同褐色で, 時に強く暗緑色を帯びる. 1)♂触角第5節から第7節の腹面に無毛部がある；2) 前胸背板は幅広く, 側縁は前方に向けて強く弧を描き, 後方に向けて強く狭まるが殆ど波曲しない. 後角は鈍角で殆ど突出しない. 正中線は弱く, しばしば不鮮明. 基部凹陥は縦長で, 通常明瞭. 表面には小点刻を密に装う；3) 鞘翅は肩部が張り出すため, 両側縁の丸みが弱くなる傾向にある（♂においてより顕著）. 第一次凹陥は明瞭；4) 陰茎先端は側面から見て基半部の幅が広く, そこから比較的急激に先端に向けて狭まる. 背面から見ると右側への湾曲が顕著；5) 指状片は大きく, 幅広く, 最広部から先端までの距離が極めて短く, 側縁は最広部の手前と基板近くで軽くくびれる. 和歌山県北西沿岸部に産するもの（一応本亜種に分類されている）は一般により大型で, 背面の光沢が強い.
分布　四国北半部および和歌山県北西沿岸部／忽那諸島（中島・興居島）（愛媛県）. 本種は, 瀬戸内海に浮かぶいくつかの小島嶼からも記録されており（愛媛県の魚島, 大三島, 大島, 香川県の高見島, 粟島, 志々島, 小豆島, 前島（土庄町南西部にあり, 一見小豆島に接続した半島のように見えるため大深（おおみ）半島と言う名までついているが, 元々は海峡（世界最小の海峡として知られる土渕（どぶち）海峡）によって小豆島本体と隔てられた「島」である）, 兵庫県淡路島南方の沼島等）, 忽那諸島からも他に野忽那島と釣島から記録があるが（次に述べる忽那諸島西部亜種が分布する怒和島と二神島を除く）, これらの集団の形態学的特徴や亜種分類についてはまだ詳しい検討がなされていない. 瀬戸内海及び沿岸地域の本種の変異と亜種分類に関しては今後, 筆者ら（井村・水沢）が中心となって検討を進めてゆく予定である.
亜種小名の由来と通称　亜種小名は, 愛媛県の昆虫愛好家, 山岡幸雄に因む. シコクヤコンオサムシと呼ばれることもある. 本亜種も Ishikawa & Kubota (1994a) の共著論文中で記載されたが, タクソンの著者は Ishikawa 単独となっている.

29-8. *C. (O.) y. seto* Ishikawa, 1994 ヤコンオサムシ忽那（くつな）諸島西部亜種

Original description *Carabus (Ohomopterus) yaconinus seto* Ishikawa, in Ishikawa & Kubota, 1994, Proc. Japan. Syst. Zool., (50), p. 39, fig. 20 (p. 37).
Type locality Is. Nuwashima, Ehime Pref.
Type depository Systematic Zoology Laboratory, Department of Biological Sciences, Tokyo Metropolitan University (Hachiôji) [Holotype, ♂].
Diagnosis 30~32 mm. Black bearing a faint bluish tinge along elytral margins. Identified by the following characters: 1) size a little larger on an average than that of most closely allied subspecies, *yamaokai*; 2) pronotum broad, distinctly narrowed posteriad though at most slightly cordate; hind angles triangularly produced; median longitudinal line often vaguely impressed; basal foveae oblong and deeply concave; disk evenly and coarsely punctate; 3) elytra a little longer than other subspecies, oval in female, oblong and nearly subparallel-sided in male; 4) penis and digitulus almost as in subsp. *yamaokai*.
Distribution W. Kutsuna Islands (Iss. Nuwa-jima & Futagami-jima) (Ehime Pref.).
Etymology Named after the Inland Sea of Seto-naikai.

タイプ産地　愛媛県, 怒和島.

29. *Carabus* (*Ohomopterus*) *yaconinus*

タイプ標本の所在 首都大学東京動物系統分類学研究室(八王子)[ホロタイプ, ♂].

亜種の特徴 30~32 mm. 背面は黒色で, 鞘翅側縁は僅かに青味を帯びる. 1)平均してやや大型;2)前胸背板は幅広く, 側縁は後方に向けて強く狭まるが, 波曲は弱く, 後角は三角形に突出し, 正中線はしばしば不明瞭で, 基部回陥は縦に長く, 深く刻まれる. 背面はほぼ均等かつ密に点刻される;3)鞘翅は他亜種に比しやや長く, ♀では卵形, ♂では側縁が平行に近い箱形を呈する;4)陰茎および指状片の形態は四国和歌山亜種のそれと大差ない.

分布 忽那諸島西部(怒和島, 二神島)(愛媛県).

亜種小名の由来と通称 亜種小名は瀬戸内海の瀬戸に因む. セトヤコンオサムシと呼ばれることもある. 本亜種も ISHIKAWA & KUBOTA (1994a)の共著論文中で記載されたが, タクソンの著者は ISHIKAWA 単独となっている.

29. *Carabus* (*Ohomopterus*) *yaconinus*

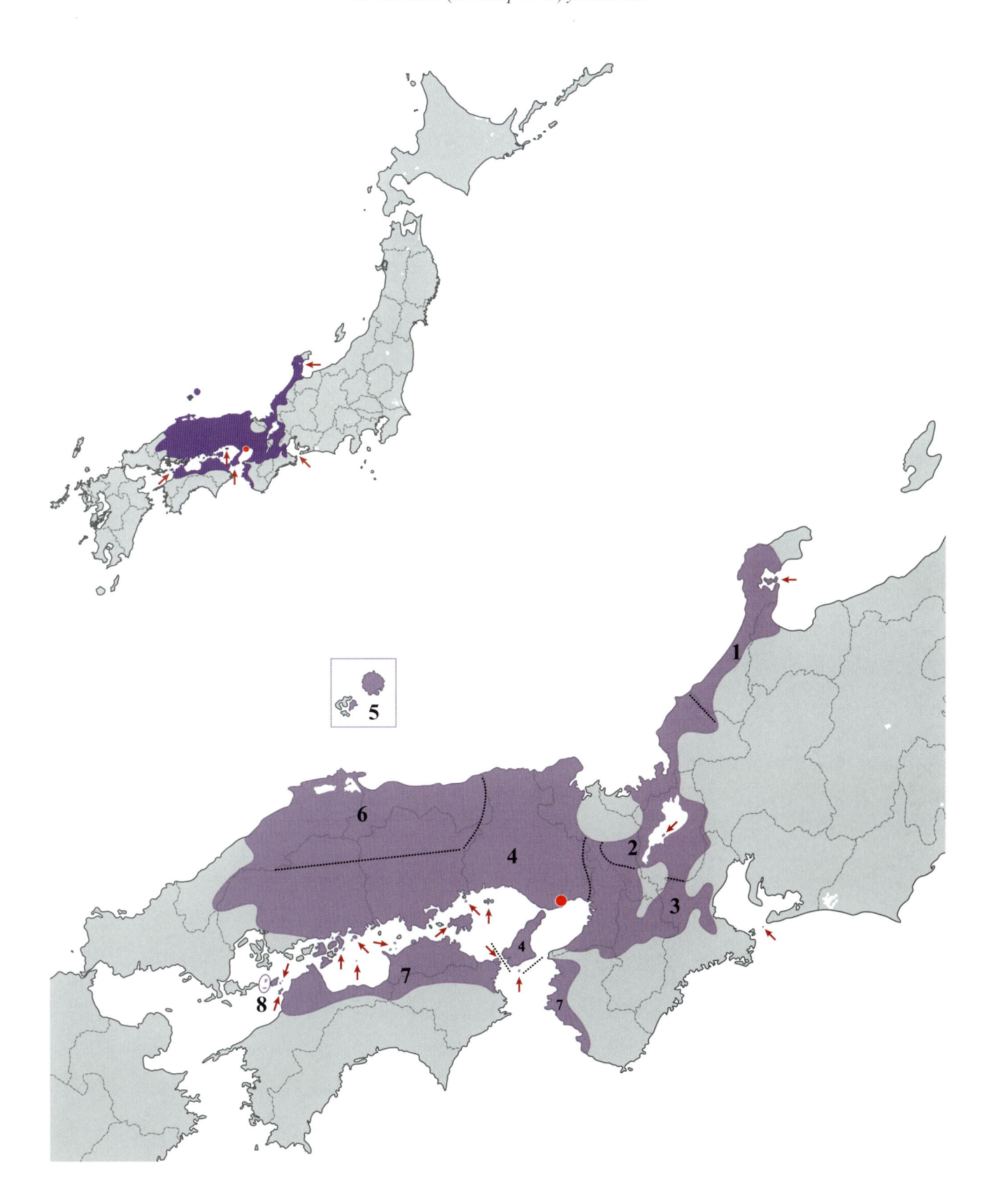

Fig. 30. Map showing the distributional range of ***Carabus*** (***Ohomopterus***) ***yaconinus*** Bates, 1873. — 1, Subsp. ***blairi***; 2, subsp. ***sotai***; 3, subsp. ***cupidicornis***; 4, subsp. ***yaconinus***; 5, subsp. ***oki***; 6, subsp. ***maetai***; 7, subsp. ***yamaokai***; 8, subsp. ***seto***. ●: Type locality of the species (Hiogo = Kôbé-shi).
図 30. ヤコンオサムシ分布図. — 1, 北陸地方亜種; 2, 近畿地方北部亜種; 3, 近畿地方南部亜種; 4, 基亜種; 5, 隠岐亜種; 6, 山陰地方亜種; 7, 四国和歌山亜種; 8, 忽那諸島西部亜種. ●: 種のタイプ産地(兵庫＝神戸市).

30. *Carabus* (*Ohomopterus*) *komiyai* (Ishikawa, 1966)
カケガワオサムシ

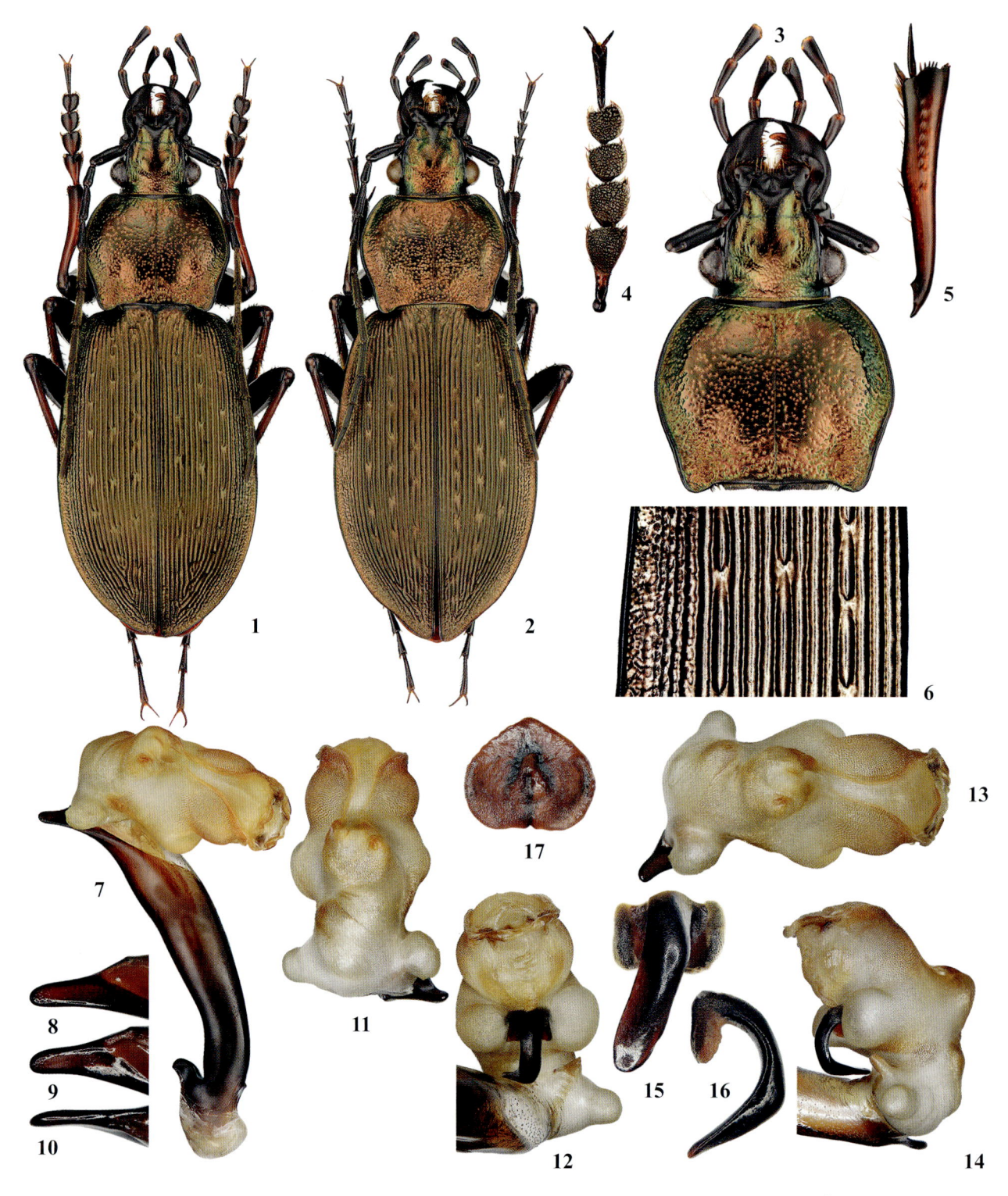

Figs. 1-17. *Carabus* (*Ohomopterus*) *komiyai komiyai* (Ishikawa, 1966) (Mt. Ogasa-yama, Fukuroi-shi, Shizuoka Pref.). — 1, ♂; 2, ♀; 3, male head & pronotum; 4, male right protarsus in ventral view; 5, male right protibia in subdorsal view; 6, male left elytron (median part); 7, penis & fully everted internal sac in right lateral view; 8, apical part of penis in right lateral view; 9, ditto in right subdorsal view; 10, ditto in dorsal view; 11, internal sac in caudal view; 12, ditto in frontal view; 13, ditto in dorsal view; 14, ditto in left sublateral view; 15, digitulus in frontal view; 16, ditto in right lateral view; 17, inner plate of ligular apophysis.

図 1-17. カケガワオサムシ基亜種(静岡県袋井市小笠山). — 1, ♂; 2, ♀; 3, ♂の頭部と前胸背板; 4, ♂右前跗節(腹面観); 5, ♂右前脛節(斜背面観); 6, ♂左上翅面(中央部); 7, 陰茎先端部(右側面観); 8, 陰茎先端部(右側面観); 9, 同(右亜背面観); 10, 同(背面観); 11, 内嚢(尾側面観); 12, 同(頭側面観); 13, 同(背面観); 14, 同(左斜側面観); 15, 括込片(頭側面観); 16, 同(右側面観); 17, 膜質鞭節片内板.

30. *Carabus* (*Ohomopterus*) *komiyai*

Figs. 18-26. ***Carabus*** (***Ohomopterus***) ***komiyai*** subspp. — 18-20, ***C***. (***O***.) ***k***. ***komiyai*** (18-19, ♂, 20, ♀; 18, Mt. Awa-ga-také, Kakegawa-shi; 19-20, Mt. Akiha-san, Tenryû-ku, Hamamatsu-shi); 21-23, ***C***. (***O***.) ***k***. ***matsunagai*** (21 (HT), 23, ♂, 22, ♀, PT; 21-22, Kamada, Iwata-shi; 23, Gôdaijima, Iwata-shi); 24-26, ***C***. (***O***.) ***k***. ***yamazumiensis*** (24, 26, ♂, PT, 25, ♀, PT; Pass Yamazumi-tôgé, Tenryû-ku, Hamamatsu-shi). — 27, Natural hybrid between *C*. (*O*.) *k*. *komiyai* & *C*. (*O*.) *esakii suruganus* (♂, Haruno-chô, Tenryû-ku, Hamamatsu-shi). — a, Digitulus in frontal view; b, ditto in right lateral view. All from Shizuoka Pref.

図 18-26. **カケガワオサムシ**各亜種. — 18-20, **基亜種**(18-19, ♂, 20, ♀; 18, 掛川市粟ヶ岳; 19-20, 浜松市天竜区秋葉山); 21-23, **磐田原亜種** (21 (HT), 23, ♂, 22, ♀, PT; 21-22, 磐田市鎌田; 23, 磐田市合代島); 24-26, **山住峠亜種**(24-26, ♂, PT, 24, ♀, PT; 浜松市天竜区山住峠). — 27, カケガワオサムシ基亜種とシズオカオサムシ富士川流域以西亜種の自然雑種(♂, 浜松市天竜区春野町). — a, 指状片(頭側面観); b, 同(右側面観). すべて静岡県産.

30. *Carabus* (*Ohomopterus*) *komiyai* (Ishikawa, 1966) カケガワオサムシ

Original description　*Apotomopterus insulicola komiyai* Ishikawa, 1966, Bull. natn. Sci. Mus., Tokyo, 9 (1), p. 12, fig. 2 on p. 12.

History of research　This taxon was originally described by Ishikawa (1966a) as one of the subspecies of *Apotomopterus insulicola* (= *Carabus* (*Ohomopterus*) *insulicola* in the present sense), and transferred later to a subspecies of *C.* (*O.*) *arrowianus* (Ishikawa, 1985). Imura & Matsunaga (2010) raised its rank to a full species, *Ohomopterus komiyai* (= *C.* (*O.*) *komiyai* in the present sense), based on radically different male genitalic features as well as on the distributional and reproductive aspects, and combined with it another lower taxon *matsunagai* which was described as a subspecies of *O. arrowianus* by Imura (2008). In the same paper (Imura & Matsunaga, 2010), a new subspecies of *O. komiyai* was described under the name of *yamazumiensis*. Following Imura & Matsunaga (2010), Ishikawa's taxon is regarded as a full species and is classified into three subspecies in this book.

Type locality　Mt. Ogasayama, Ogasa Co., Shizuoka Prefecture.

Type depository　National Museum of Nature and Science, Tokyo [Holotype, ♂].

Morphology　21~29 mm. Body above coppery brown or light coppery, often with greenish tinge on head and pronotum, occasionally black, very rarely entirely greenish; tibiae and tarsi usually dark rufous, at least partly. External features: Figs. 1~6. Genitalic features: Figs. 7~17. Allied to other medium-sized *Ohomopterus* species mentioned previously, but identified by the following characteristics: 1) male antennae with hairless areas on ventral side from segment five to seven; 2) primary foveoles of elytra not large, barely invading adjacent tertiaries; 3) inner margin of male protibia strongly convex though not remarkably angular; 4) digitulus short and narrow, nearly parallel-sided in frontal view, with apex basically bent toward right side of internal sac and rounded at tip; viewed laterally, it is L-shaped, thickest at flexed part, depressed dorso-ventrad in apical portion; 5) inner plate of ligular apophysis chestnut-like in profile in dorsal view, with inner wall hardly wrinkled and rather smooth.

Molecular phylogeny　For the molecular phylogeny of the subgenus *Ohomopterus*, see the same section of *C.* (*O.*) *yamato*.

Distribution　The area between the lower courses of the Ôi-gawa & Tenryû-gawa Rivers, W. Shizuoka Pref., SC. Honshu.

Habitat　Basically found from hilly to mountainous region consisting of the basement rock, but also found from the alluvial plain near the sea or river. The population in the latter area must have been reached the bank most probably on the occasion of flood, and colonized at each station. Commonly visible in the forest edge of broadleaved trees or those of planted trees such as cedar or cypress.

Bionomical notes　Presumed to be a spring breeder, earthworm feeder.

Geographical variation　Classified into three subspecies.

Etymology　Named after Jiro Komiya, a member of the research group on carabid beetles in Keihin Insect Lover's Club, Tokyo.

研究史　本タクソンはIshikawa（1966a）によりアオオサムシの一亜種*Apotomopterus insulicola komiyai*として記載されたが，Ishikawa（1985）以後，ミカワオサムシの一亜種として扱われることが多かった．しかし，本タクソンとミカワオサムシは♂交尾器の基本的形態が明らかに異なっている．すなわち，指状片の先端がミカワオサムシではマヤサンオサムシと同様，内嚢の左側に向かって伸長する（左巻き）のに対し，本タクソンではアオオサムシ，ドウキョウオサムシ，シズオカオサムシと同様，内嚢の右側に向かって伸長する（右巻き）ほか，指状片両側にある一対の内嚢膨隆部が，ミカワオサムシでは左右対称であるのに対し，本タクソンでは左右非対称（右＜左）である．また，天竜川左岸（東岸）の複数の地域では，ミカワオサムシ（主として河川流や沿岸流により漂着してコロニーを形成しているもの）と本タクソンが分布を接しており，いわば天然の交雑実験場の様相を呈しているが，これまでに両者の交雑により生じたと思われる個体は一頭も得られておらず，両者間の生殖的隔離がかなりの程度まで進んでいる可能性が示唆される．こうした点を鑑み，Imura & Matsunaga（2010）は本タクソンを独立種*Ohomopterus komiyai*に昇格し（本書での扱いは*Carabus* (*Ohomopterus*) *komiyai*），その一亜種として静岡県北西部の山住峠から*O. k. yamazumiensis*を記載した．また，同県南西部の磐田原台地からミカワオサムシの一亜種として記載された*O. arrowianus matsunagai*（Imura, 2008）の所属をカケガワオサムシの亜種へと移し，*O. k. matsunagai*とした．

タイプ産地　静岡県，小笠郡，小笠山（現在は掛川市と袋井市の境界付近に位置する）．

タイプ標本の所在　国立科学博物館［ホロタイプ，♂］．

形態　21~29 mm．背面は銅褐色ないし明銅色で，時に弱い緑色光沢を帯び，少ないながら黒色型も出現する．全身緑色の型も知られているが，極めて稀．脛節と跗節は少なくとも部分的に赤みを帯びる場合が多い．外部形態：図1~6．交尾器形態：図7~17．既述の中型オホモプテルス類によく似ているが，以下の諸形質により識別される：1）♂触角第5節から第7節の腹面には無毛部がある；2）鞘翅の第一次凹陥は大きくなく，隣接する第三次間室をごく僅かに侵す程度；3）♂前脛節内側縁は中央で強く膨らむが，顕著に角ばることはない；4）指状片は細く短く，背面から見て左右の側縁はほぼ平行で，先端は内嚢の右側に向かって伸長し（亜種によっては内嚢の縦軸に沿ってほぼまっすぐに伸びるものもあるが，少なくとも内嚢の左側に向かうことはない），先端は鈍く丸まる．側面から見ると「く」の字形に屈曲し，屈曲部で最も厚くなり，先端に向けて背腹方向に押圧される；5）♀交尾器膣底部節片は栗の実形で，前縁中央が尖り，壁面内側は滑らかであまり強く皺状にならない．本種以下，マヤサンオサムシまでの6種は，指状片がこれまでに述べた3種群（ヒメオサムシ種群，オオオサムシ種群，ヤコンオサムシ種群）のものより大きく伸長し，左右非対称で，翌細形，鋸形，の内嚢管を呈するという形質を共有し，形態学的にはアオオサムシ種群に近いと思われる．

分子系統　小型オホモプテルス亜属の分子系統の概要についてはヤマトオサムシの同項を参照．

分布　静岡県南西部(大井川右岸(西岸)から天竜川左岸(東岸))にかけての地域に分布し，北限は浜松市山住峠北方．

棲息環境　基本的には基盤岩によって形成された丘陵から山地にかけて生息するが，河川流及び海流に乗って漂着したと思われる集団が海岸や河川に近い平地に定着し，小コロニーを形成している場所もある．

生態　春繁殖型で，幼虫は環形動物(フトミミズ)食と思われる．

地理的変異　3亜種に分類される．

種小名の由来　京浜昆虫同好会オサムシグループの一員であった小宮次郎に因む．

29-1. *C. (O.) k. komiyai* (Ishikawa, 1966)　カケガワオサムシ基亜種

Original description, type locality & type depository　See explanation of the species.

Diagnosis　21~26 mm. Body above light reddish coppery to dark brownish coppery, often with weak greenish tinge on head and pronotum, rarely entirely blackish; tibiae and tarsi dark rufous. For the diagnosis of the nominotypical subspecies, see the "Morphology" section of the species and the diagnosis of the following two subspecies.

Distribution　The area between the lower courses of the Ôi-gawa & Tenryû-gawa Rivers except for the Iwatabara Plateau, W. Shizuoka Pref., SC. Honshu.

原記載・タイプ産地・タイプ標本の所在　種の項を参照．

亜種の特徴　21~26 mm. 背面は明赤銅色から(暗)銅褐色で，頭部と前胸背板にはしばしば弱い緑色光沢を帯び，稀に全身の黒化傾向の強い個体が見られるが，完全な緑化型は知られていない．脛節と跗節は赤みを帯びる．基亜種の形態学的特徴については，種の解説並びに以下の2亜種の特徴の項を参照．

分布　磐田原台地を除く，大井川下流右岸(西岸)から天竜川下流左岸(東岸)にかけての地域．

29-2. *C. (O.) k. matsunagai* (Imura, 2008)　カケガワオサムシ磐田原亜種

Original description　*Ohomopterus arrowianus matsunagai* Imura, 2008, Elytra, Tokyo, 22, p. 390, figs. 11~16 (p. 391).

Type locality　Kamada, ca. 10 m in altitude, on the Iwatabara (= Iwatahara or Iwata-ga-hara) Plateau, in Iwata-shi, of Southwest Shizuoka Prefecture.

Type depository　National Museum of Nature and Science, Tokyo [Holotype, ♂].

Diagnosis　24~29 mm. Body above light reddish coppery to dark coppery brown, often with greenish tinge on head and pronotum, very rarely entirely greenish; entirely blackish form has not been found as yet; tibiae and tarsi usually dark rufous. Closely allied to the nominotypical *komiyai*, but differs from that race in the following points: 1) size apparently larger on an average (mean body size is ca. 27 mm); 2) elytra usually more elongated, with lateral sides less acutely narrowed toward apices; 3) apical part of penis usually a little robuster; 4) digitulus much larger, with apical portion much more remarkably elongated, widely depressed and apparently dilated usually toward the left near apex which is obtusely rounded; its dorsal wall usully grooved longitudinally on both sides of central ridge.

Distribution　Iwatabara Plateau, SW. Shizuoka Pref.

Etymology　Named after Masamitsu Matsunaga (Shizuoka) who discovered this subspecies.

タイプ産地　静岡県南西部，磐田市，磐田原(いわたばら／いわたはら／いわたがはら)台地，鎌田，標高約10 m．

タイプ標本の所在　国立科学博物館 [ホロタイプ, ♂].

亜種の特徴　24~29 mm. 背面は明赤銅色から暗銅褐色で，頭部と前胸背板にはしばしば弱い緑色光沢を帯びる．筆者らの検しえた範囲では，過去に1例のみ全身が緑化した型が得られているが，黒色型は知られていない．脛節と跗節は通常，暗赤色．基亜種に近いが，以下の諸点で異なる：1) 明らかにより大型(平均体調は27 mm前後)；2) 鞘翅はより強く伸長し，側縁は先端に向けての狭まりがより緩やか；3) 陰茎先端は通常，より太短い；4) 指状片ははるかに大きく，先端部がより長く伸長し，幅広く押圧され，通常先端左方に向けて拡がり，先端は丸く，背面中央の稜の両側に溝を有するものが多い．

分布　磐田原台地(静岡県南西部)．東限は太田川とその支流の敷地川(しきじがわ)，西限は台地西縁の天竜川左岸(東岸)河岸段丘，南限はタイプ産地の鎌田付近．台地北部で基亜種へと移行する．

亜種小名の由来と通称　亜種小名は本亜種を発見した静岡市の松永正光に因む．通称はイワタオサムシ．

29-3. *C. (O.) k. yamazumiensis* (Imura et Matsunaga, 2010)　カケガワオサムシ山住峠亜種

Original description　*Ohomopterus komiyai yamazumiensis* Imura & Matsunaga, 2010, Coléoptères, Paris, 16 (14), p. 143, figs. 6-a~c (p. 145).

Type locality　Japan, central Honshu, Shizuoka Pref., Hamamatsu-shi, Tenryû-ku, Pass Yamazumi-tôgé, 1100 m.

Type depository　Muséum Nationale d'Histoire Naturelle, Paris [Holotype, ♂].

30. *Carabus* (*Ohomopterus*) *komiyai*

Diagnosis 21~25 mm. Most closely allied to the nominotypical *komiyai*, but discriminated from it in the following points: 1) size a little smaller on an average; 2) digitulus with apex less strongly turning rightward, often nearly straightly elongated along mid-line toward base of internal sac, though never turning leftward; it is nearly parallel-sided and faintly widened before apex in frontal view, median portion not rectangularly curved nor thickened as in nominotypical *komiyai* but roundly arcuate and almost parallel-sided in lateral view, dorsal margin not remarkably ridged nor longitudinally guttered. From subsp. *matsunagai*, this subspecies is readily distinguished by much smaller body size and much shorter digitulus.

Distribution High altitudinal area between the Misakubo-gawa and Keta-gawa Rivers, usually above 1,000 m in the altitude. Southernmost limit so far known is Mt. Idoguchi-yama. The northern margin may be defined by the upper course of the Misakubo-gawa River. In the periphery of its distributional range, this subspecies seems to be replaced to or hybridized with *C.* (*O.*) *esakii* in the north, *C.* (*O.*) *arrowianus* in the west and the nominotypical *C.* (*O.*) *komiyai* in the south.

Habitat Inhabitant of woodland consisting of the deciduous broadleaved tree. Also visible in the forest of planted cedar or cypress tree.

Etymology Named after the type locality, Pass Yamazumi-tôgé.

タイプ産地　日本，本州中部，静岡県，浜松市，天竜区，山住峠，1100 m.

タイプ標本の所在　パリ自然史博物館［ホロタイプ，♂］.

亜種の特徴　21~25 mm. 基亜種に最も近いが，以下の点で異なる：1）平均して小型；2）指状片先端が内嚢の右側に曲がる程度が弱く，正中線に沿ってまっすぐ内嚢基部に向かうことも多いが，左方に曲がることはない．頭側から見ると両側縁はほぼ平行で，先端手前で僅かに広がり，側方から見ると中央部は基亜種のように直角に曲がって肥厚することはなく，弧を描いて丸まり，外縁と内縁はほぼ平行で，背面に陵や溝を欠く．磐田原亜種とは，はるかに小型で指状片が短いことにより容易に識別される．

分布　静岡県浜松市天竜区の山住峠を中心とする，水窪川と気田川に挟まれた標高およそ1,000 m以上の地域に分布する．これまでに確認されている南限は井戸口山付近，北限は水窪川上流と思われる．分布域の北縁ではシズオカオサムシ，西縁ではミカワオサムシ，南縁ではカケガワオサムシ基亜種と分布を接しており，一部でこれら各タクソンとの自然交雑により生じたと思われる個体が得られている．

棲息環境　基盤山地上の落葉広葉樹林やスギ・ヒノキ等の植林地に生息する．

亜種小名の由来と通称　亜種小名はタイプ産地の山住峠に因む．ヤマズミオサムシという通称が提唱されている．

30. *Carabus* (*Ohomopterus*) *komiyai*

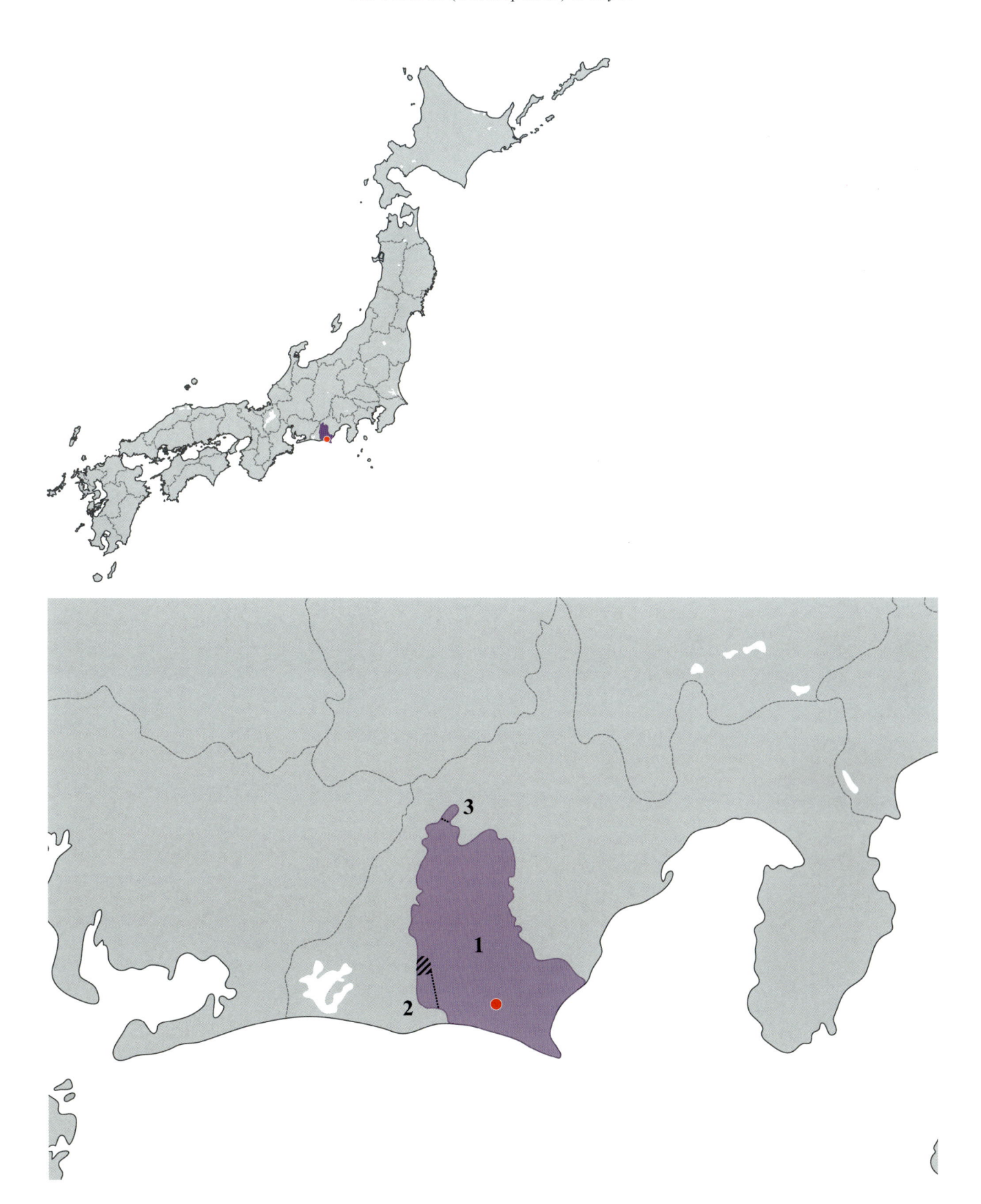

Fig. 28. Map showing the distributional range of ***Carabus*** (***Ohomopterus***) ***komiyai*** (Ishikawa, 1966). — 1, Subsp. ***komiyai***; 2, subsp. ***matsunagai***; 3, subsp. ***yamazumiensis***. Shaded area indicates the range of intergrading population between *komiyai* and *matsunagai*. ●: Type locality of the species (Mt. Ogasayama).
図 28. カケガワオサムシ分布図. — 1, 基亜種; 2, 磐田原亜種; 3, 山住峠亜種. 斜線部は基亜種と磐田原亜種の移行地帯. ●: 種のタイプ産地（小笠山）.

255

31. *Carabus* (*Ohomopterus*) *esakii* Csiki, 1927
シズオカオサムシ

Figs. 1-17. *Carabus* (*Ohomopterus*) *esakii esakii* Csiki, 1927 (Suyama, Susono-shi, Shizuoka Pref.). — 1, ♂; 2, ♀; 3, male head & pronotum; 4, male right protarsus in ventral view; 5, male right protibia in subdorsal view; 6, male left elytron (median part); 7, penis & fully everted internal sac in right lateral view; 8, apical part of penis in right lateral view; 9, ditto in right subdorsal view; 10, ditto in dorsal view; 11, internal sac in caudal view; 12, ditto in frontal view; 13, ditto in dorsal view; 14, ditto in left sublateral view; 15, digitulus in frontal view; 16, ditto in right lateral view; 17, inner plate of ligular apophysis.

図 1-17. シズオカオサムシ基亜種(静岡県裾野市須山) — 1, ♂; 2, ♀; 3, ♂の頭部と前胸背板; 4, ♂の右前跗節(腹面観); 5, ♂の右前脛節(斜背面観); 6, ♂の左上翅(中央部); 7, 陰茎と完全反転した内嚢(右側面観); 8, 陰茎先端(右側面観); 9, 同(右斜背面観); 10, 同(背面観); 11, 内嚢(尾側面観); 12, 同(頭側面観); 13, 同(背面観); 14, 同(左側面観); 15, 指状片(頭側面観); 16, 同(右側面観); 17, 膣底部節片内板.

31. *Carabus* (*Ohomopterus*) *esakii*

Figs. 18-23. ***Carabus* (*Ohomopterus*) *esakii*** subspp. — 18-19, *C.* (*O.*) *e. esakii* (18, ♂, 19, ♀; 18, Yugano, Kawazu-chô; 19, Sakuragi-chô, Atami-shi); 20-23, *C.* (*O.*) *e. suruganus* (20, 22, ♂, 21, 23, ♀; 20-21, Ichinosé, Fujieda-shi; 22, Pass Abé-tôgé, Shizuoka-shi; 23, Mt. Sotomori-yama, Sumata-kyô Valley, Kawané-honchô). — 24-27, Natural hybrid between *C.* (*O.*) *e. suruganus* & *C.* (*O.*) *komiyai yamazumiensis* (24-27, ♂; uppermost reaches of Riv. Misakubo-gawa, Tenryû-ku, Hamamatsu-shi). — a, Digitulus in frontal view; b, ditto in right lateral view. All from Shizuoka Pref.

図 18-23. シズオカオサムシ各亜種. — 18-19, **基亜種**(18, ♂, 19, ♀; 18, 河津町湯ヶ野; 19, 熱海市桜木町); 20-23, **富士川流域以西亜種**(20, 22, ♂, 21, 23, ♀; 20-21, 藤枝市市之瀬; 22, 静岡市安倍峠; 23, 川根本町寸又峡外森山). — 24-27, シズオカオサムシ富士川流域以西亜種とカケガワオサムシ山住峠亜種の自然雑種(24-27, ♂;浜松市天竜区水窪川最上流部). — a, 指状片(頭側面観); b, 同(右側面観).すべて静岡県産.

31. *Carabus* (*Ohomopterus*) *esakii* Csiki, 1927 シズオカオサムシ

Carabus Albrechti var.: Bates, 1883, Trans. ent. Soc. Lond., 1883, p. 229.
Carabus (*Morphocarabus*) *Albrechti* var. *Lewisi*: Géhin, 1885, Catalogue synonymique et systématique des Coléoptères de la tribu Carabides, avec des planches dessinées par Ch. Haury. p. 17 [Nec Rye, 1872].
Carabus (sect. *Ohomopterus*) *Albrechti* ab. *Esakii* Csiki, 1927, Col. Cat. Carabidae: Carabinae, 1 (Junk, Col. Cat., 91), p. 264.
Apotomopterus insulicola shizuokaensis (n. race MS.): Nakane, 1952a, Shin-Konchû, Tokyo, 5(6), p.12. — 1952b, idem, 5(11), p. 49.
Apotomopterus insulicola shizuokaensis: Nakane, 1953, Sci. Rep. Saikyo Univ., (Nat. Sci. & Liv. Sci.), 1(2), p. 95.
Apotomopterus insulicola esakii: Nakane, 1962, Ins. Jap., 2 (3), p. 26.
Apotomopterus esakii: Ishikawa, 1966a, Bull. natn. Sci. Mus., Tokyo, 9, p. 10.
Carabus (*Ohomopterus*) *esakii*: Ishikawa, 1985, Coleopt. Japan Col., Osaka, 2, p. 31.

History of research This species was first recorded by Bates (1883) as a variety of *Carabus Albrechti* based on the specimen collected by Lewis from Suyama on the southeastern foot of Mt. Fuji-san in Shizuoka Prefecture. Géhin (1885) named it *Lewisi* [sic] as a variety of *C.* (*Morphocarabus*) *Albrechti*. Since the name *lewisi* is a junior secondary homonym of *Damaster lewisi* (Rye, 1872) (= *C.* (*Damaster*) *blaptoides blaptoides* in the present sense), a new replacement name *Esakii* [sic] was given by Csiki (1927) as an aberrant form of *C.* (*Ohomopterus*) *Albrechti* [sic]. *Apotomopterus insulicola shizuokaensis* described by Nakae (1953) is a junior synonym of Csiki's taxon. Ishikawa (1966a) raised its rank to a full species, and this view is generally accepted in current classification, though adopted genus has been different according to the authors. In 2010, Ujiie et al. published a paper on the geographical variation of *C.* (*O.*) *esakii*, and described a new subspecies under the name of *suruganus*.

Type locality Suyama (= it belongs to Susono-shi in Shizuoka Pref. at present).
Type depository Unknown.
Morphology 20~29 mm. Body above yellowish to red-brownish coppery often with greenish tinge, or entirely black often with faint greenish tinge; tibiae and tarsi rufous, dark rufous or dark brown. External features: Figs. 1~6. Genitalic features: Figs. 7~17. Externally very similar to *C.* (*O.*) *komiyai*, but differs from that species in the following points: 1) inner margin of male protibia usually more strongly angular at middle; 2) apical part of penis a little slenderer in lateral view, its dorsal margin narrowly flat or faintly concave; 3) internal sac with right basal lateral lobe a little smaller and not constricted at base, left one not simply protruding laterad as in *C.* (*O.*) *komiyai*, but usually somewhat bent toward dorsal side, paraligula usually larger; 4) digitulus different in shape.

Molecular phylogeny For an overview of the molecular phylogeny of *Ohomopterus*, see the same section of *C.* (*O.*) *yamato*. Analysis of mitochondrial ND5 gene sequences reveals that the mitochondria of *C.* (*O.*) *insulicola* introgressed extensively into *C.* (*O.*) *esakii* through their distributional boundaries, so that most individuals of the latter species have the mitochondrial gene originated from the former (Sota et al., 2001).

Distribution E. Shizuoka Pref. and its nearby regions (SW. Kanagawa Pref. & W.~S. Yamanashi Pref.), SC. Honshu.
Habitat Rather widely distributed from low hills near the sea to the subalpine zone of mountainous region with a height of ca. 1,500 m, though usually not found on the alluvial plain. Inhabitant of woodland, and common in the edge of the forest of broadleaved trees or planted cedar or cypress trees.
Bionomical notes Spring breeder, earthworm feeder.
Geographical variation Classified into two subspecies, though the population distributed on the mountainous region near the westernmost area is not yet investigated satisfactorily.
Etymology Named after Teizo Esaki, the famous entomologist of Japan who was active in the first half of the Showa period.

研究史　本種は最初，Bates（ベイツ）(1883)により*Carabus Albrechti*（クロオサムシ）の一変種(variety)として記録された．Géhin（ジェアン）(1885)はこれに*Lewisi*という名を与えて記載したが，この名は*Damaster lewisi*（広義の*Carabus*属として扱われることも多い）の新参二次同名となるので，Csiki（チキ）(1927)により*Esakii*という置換名が与えられ，今日に至っている．Nakane(1953)によりアオオサムシの亜種として記載された*shizuokaensis*は*esakii*のシノニムとなる．このように，本種は当初，他種の一変種ないし亜種という位置付けであったが，Ishikawa(1966a)以来，一般に独立種として扱われている．長年に亘り単型種とみなされてきたが，2010年になってUjiie et al.により亜種*suruganus*が記載されている．

タイプ産地　須山(現在の所属は静岡県裾野市(すそのし))．

タイプ標本の所在　不明．

形態　20~29 mm. 背面は黄色～赤褐色を帯びた銅色で，頭部，前胸背板，鞘翅側縁はしばしば緑色光沢を帯び，時に全身が黒化した型や緑色の型も出現する．脛節と附節は赤褐色から黒褐色まで変化する．外部形態：図1~6．交尾器形態：図7~17．カケガワオサムシによく似ているが，以下の点で異なる：1)♂前脛節内側縁中央の角ばりがやや強い；2) 陰茎先端は側面から見てやや細く，背面には細長い平坦部ないしやや窪んだ部分がある；3) 内嚢の右基部側葉はやや小さく，カケガワオサムシのように基部でくびれず，左基部側葉はカケガワオサムシでは単純に側方へ突出するものが多いが，本種では先端が背側へと曲がるものが多く，側舌がやや大きい；4) 指状片の形態が異なる．

分子系統　オホモプテルス亜属の分子系統の概要についてはヤマトオサムシの同項を参照．Sota et al.(2001)によれば，本種とアオオサムシとの間に起きた浸透性交雑の結果，後者から前者へのmtDNAの浸透が見られるという．

分布　静岡県東部とその近隣地域（神奈川県南西部，山梨県西～南部）．東限は伊豆半島基部の相模湾沿岸部，西限はこれ

まで一般に大井川とされ，分布域南西部では確かに大井川下流東岸が本種の分布西限となっているが，上流では山地性となり，大井川をはるか西に越えて天竜川水系の水窪川（みさくぼがわ）上流域にまで分布を広げている（井村有希・松永正光，未発表）．北は山梨県西部韮崎市の釜無川（かまなしがわ）右岸地域に及ぶ．

　棲息環境　海岸に近い低所から標高1,500 mを越える針広混交林帯まで幅広く見られるが，基本的には基盤山地の低〜中標高域に生息する種と考えられる．自然林（落葉広葉樹，常緑照葉樹），人工林（スギ，ヒノキ，茶等）を問わず，林縁ないし周辺の明るい草地などによく見られる．

　生態　春繁殖，環形動物（フトミミズ等）食と考えられる．

　地理的変異　2亜種に分類されている．分布域東部では部分的にアオオサムシと混生し，時に両者の自然交雑により生じたと思われる個体（アオオサムシの図42）が見つかる．西部では一部の地域でカケガワオサムシ基亜種，同山住峠亜種と側所的に分布し，両者が分布を接する地域では自然交雑により生じたと思われる個体（図24-27およびカケガワオサムシの図27）が得られている（井村・松永，未発表）．

　種小名の由来　クロオサムシ関東地方北西部亜種のそれと同じく，江崎悌三に因む．

31-1. *C. (O.) e. esakii* CSIKI, 1927　シズオカオサムシ基亜種

Original description, type locality & type depository　See explanation of the species.
Diagnosis　22〜29 mm. Discriminated from subsp. *suruganus* by smaller primary foveoles of elytra and longer and thicker digitulus.
Distribution　E. Shizuoka〜SW. Kanagawa〜SE. Yamanashi (N. of Izu Penins.〜Hakone Mts., N.〜NE. of Mt. Fuji-san).

　原記載・タイプ産地・タイプ標本の所在　種の項を参照．
　亜種の特徴　22〜29 mm．黒化型は殆ど現れない．鞘翅の第一次凹陥がより小さく，指状片がより長く厚いことにより，富士川流域以西亜種から区別される．
　分布　静岡県東部〜神奈川県西部〜山梨県南東部（伊豆半島基部から箱根山地にかけての地域，および富士五湖から大月市にかけての地域）．富士五湖北方〜富士山西麓〜伊豆半島西・南部にかけての地域には次亜種への移行を示す集団が分布する．

31-2. *C. (O.) e. suruganus* UJIIE, ISHIKAWA et KUBOTA, 2010　シズオカオサムシ富士川流域以西亜種

Original description　*Carabus* (*Ohomopterus*) *esakii suruganus* UJIIE, ISHIKAWA, KUBOTA, 2010, Biogeography, 12, p. 56, figs. 3 (p. 57) & 20〜46 (p. 59).
Type locality　Ichinose, 400m, Fujieda-shi, Shizuoka Prefecture.
Type depository　Systematic Zoology Laboratory, Department of Biological Sciences, Tokyo Metropolitan University (Hachiôji) [Holotype, ♂].
Diagnosis　20〜28mm. Discriminated from the nominotypical *esakii* in the following points: 1) size a little smaller; 2) palpi and legs more remarkably reddish, above all in individuals occurring in high altitudinal area; 3) entirely blackish color form appears more frequently; 4) primary foveoles of elytra larger; 5) digitulus shorter and thinner.
Distribution　C. Shizuoka Pref. & SW.〜C. Yamanashi Pref., occupying the western part of the range of *C. (O.) esakii*, .
Etymology　Named after Suruga-no-kuni, one of the ryôsei provines of Japan used until the early Meiji period, which corresponds to the central part of Shizuoka Prefecture.

　タイプ産地　静岡県藤枝市市之瀬，400m．
　タイプ標本の所在　首都大学東京動物系統分類学研究室（八王子）[ホロタイプ，♂]．
　亜種の特徴　20〜28mm．基亜種に比し，一般により小型で，口肢と脚の赤みがより強い個体が多く（高標高地域に産するものにおいてより顕著），黒化型がより高頻度で出現する．鞘翅第一次凹陥はより大きく，指状片はより短く，薄い．大井川中流の寸又峡（すまたきょう）付近には，背面がしばしば強く緑色を帯び，口肢と脚の赤色味が強く，脚の長い独特の集団を産する．また，分布域西端の水窪川上流高所に生息する集団は，小型で光沢が非常に強く，口肢と脚が明赤色で，一部にカケガワオサムシとの自然雑種と思われる個体が混入する特異なものである（井村・松永，未発表）．シズオカオサムシの再検討を行ったUJIIE *et al.*(2010)は，この分布域西端部を占める集団についての知見がなかったと見え，論文中では全く触れられていない．本集団については目下，井村と松永により綿密な調査が行われており，近々その結果が発表される予定である．
　分布　静岡県中部〜山梨県南西部〜同県中部．富士川流域以西に分布し，分布域の南西部では大井川下流が西限となるが，北西部では山地性となり，大井川を西に大きく越えて天竜川水系の水窪川上流域域にまで分布を広げている．北限は山梨県西部韮崎市の釜無川右岸地域．
　亜種小名の由来と通称　亜種小名は駿河国（するがのくに）ないし駿河地方に因む．原記載論文中で提唱された通称はニシシズオカオサムシ．

31. *Carabus* (*Ohomopterus*) *esakii*

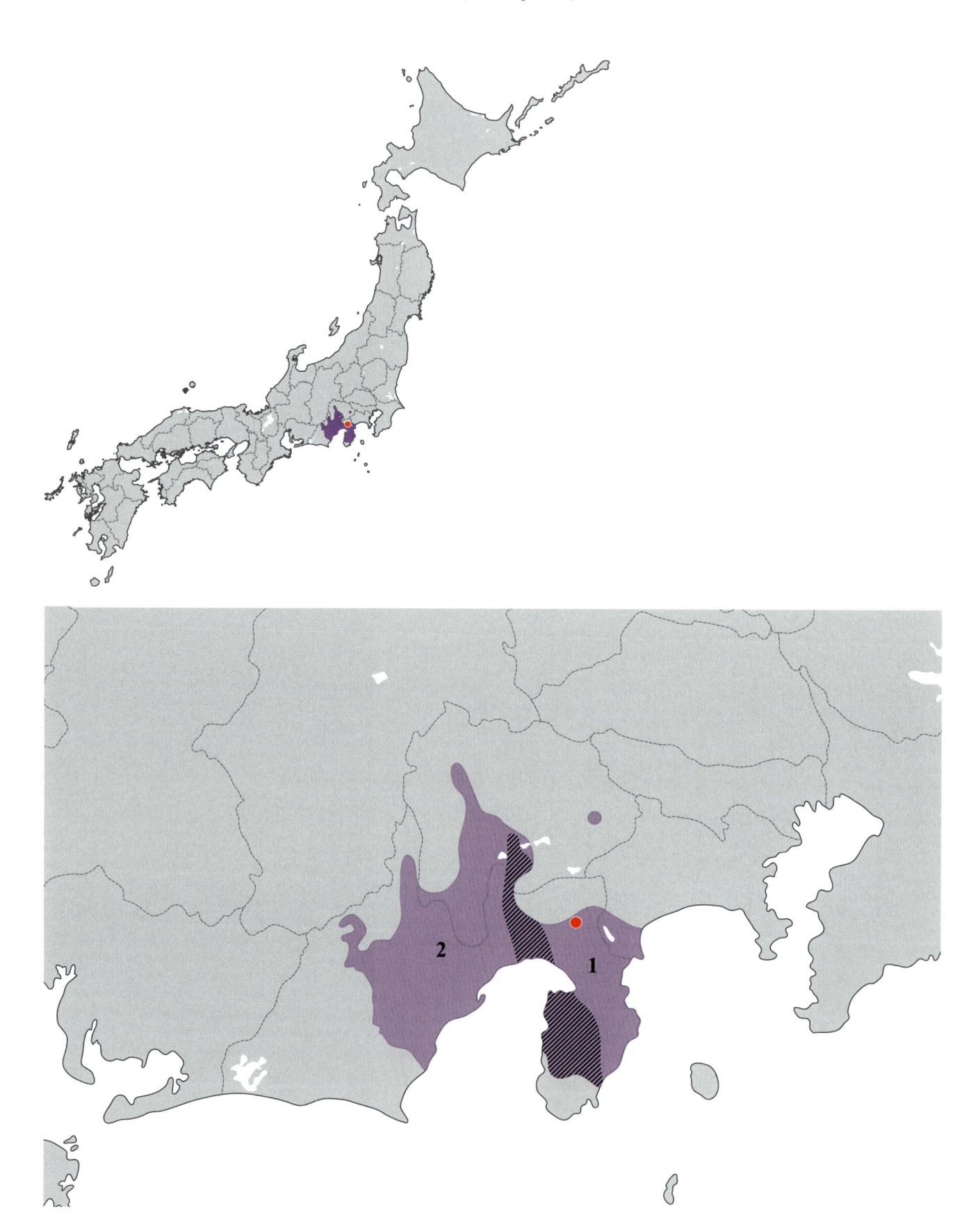

Fig. 28. Map showing the distributional range of *Carabus* (*Ohomopterus*) *esakii* Csiki, 1927. — 1, Subsp. *esakii*; 2, subsp. *suruganus*. Shaded area indicates the range of intergrading population between *esakii* and *suruganus*. ●, Type locality of the species (Suyama).

図 28. シズオカオサムシ分布図. — 1, 基亜種; 2, 富士川流域以西亜種. 斜線部は基亜種と磐山原亜種の移行地帯. ●, 種のタイプ産地(須山).

31. Carabus (Ohomopterus) esakii

Fig. 29. Habitat of *Carabus* (*Ohomopterus*) *esakii*, *C.* (*Pentacarabus*) *harmandi*, *C.* (*Leptocarabus*) *procerulus procerulus* and *C.* (*Damaster*) *blaptoides oxuroides* (uppermost reaches of Riv. Misakubo-gawa, Hamamatsu-shi, Shizuoka Pref.).

図29. シズオカオサムシ（分布域最西部の集団），ホソヒメクロオサムシ（分布域最南端の集団），クロナガオサムシ基亜種，マイマイカブリ関東・中部地方亜種の生息地（静岡県浜松市水窪川最上流部）.

Fig. 30. *Carabus* (*Ohomopterus*) *esakii suruganus* passing the winter in the soil (Mt. Ryûsô-zan, Shizuoka-shi, Shizuoka Pref.).

図30. 土中で越冬しているシズオカオサムシ富士川流域以西亜種（静岡県静岡市竜爪山）.

Fig. 31. *Carabus* (*Ohomopterus*) *esakii suruganus* passing the winter in the soil (Mt. Ryûsô-zan, Shizuoka-shi, Shizuoka Pref.).

図31. 土中で越冬しているシズオカオサムシ富士川流域以西亜種（静岡県静岡市竜爪山）.

32. *Carabus* (*Ohomopterus*) *insulicola* Chaudoir, 1869
アオオサムシ

Figs. 1-18. *Carabus* (*Ohomopterus*) *insulicola insulicola* Chaudoir, 1869 (Kôhoku-ku, Yokohama-shi, Kanagawa Pref.). — 1, ♂; 2, ♀; 3, male head & pronotum; 4, male right protarsus in ventral view; 5, male right protibia in subdorsal view; 6, male left elytron (median part); 7, penis & fully everted internal sac in right lateral view; 8, apical part of penis in right lateral view; 9, ditto in right subdorsal view; 10, ditto in dorsal view; 11, internal sac in caudal view; 12, ditto in frontal view; 13, ditto in dorsal view; 14, ditto in left sublateral view; 15, ditto in right lateral view; 16, digitulus in frontal view; 17, ditto in right lateral view; 18, inner plate of ligular apophysis.

図1-18 アオオサムシ基亜種(神奈川県横浜市港北区) 1, ♂; 2, ♀; 3, ♂の頭部と前胸背板; 4, ♂の右前跗節(腹面観); 5, ♂の右前脛節(斜背面観); 6, ♂の左鞘翅中央部; 7, 陰茎と完全反転下の内嚢(右側面観); 8, 陰茎先端(右側面観); 9, 同(右斜背面観); 10, 同(背面観); 11, 内嚢(尾側面観); 12, 同(頭側面観); 13, 同(背面観); 14, 同(左斜側面観); 15, 同(右側面観); 16, 指状片(頭側面観); 17, 同(右側面観); 18, 膣底部節片内板.

32. *Carabus* (*Ohomopterus*) *insulicola*

Figs. 19-30. ***Carabus* (*Ohomopterus*) *insulicola*** subspp. — 19-22, ***C.*** (***O.***) ***i. kita*** (19, 21, 22, ♂, 20, ♀; 19-20, Matsuzono, Morioka-shi, Iwate Pref.; 21, Jinkawa-chô, Hakodaté-shi, Hokkaido; 22, Is. Kinkazan, Ishinomaki-shi, Miyagi Pref.); 23-24, ***C.*** (***O.***) ***i. awashimaensis*** (23, ♂, 24, ♀; Uchiura, Is. Awa-shima, Niigata Pref.); 25-26, ***C.*** (***O.***) ***i. kantoensis*** (25, ♂, 26, ♀; Sayama-kyûryô, Tokorozawa-shi, Saitama Pref.); 27-28, ***C.*** (***O.***) ***i. nishikawai*** (27, ♂, 28, ♀; Mt. Tomi-san, Minami-bôsô-shi, Chiba Pref.); 29-30, ***C.*** (***O.***) ***i. okutonensis*** (29, ♂, 30, ♀; Ôana, Minakami-machi, Gunma Pref.). — a, Digitulus in frontal view; b, ditto in right lateral view.

図 19-30. **アオオサムシ各亜種**. — 19-22, **東北地方亜種**(19, 21, 22, ♂, 20, ♀; 19-20, 岩手県盛岡市松園; 21, 北海道函館市陣川町; 22, 宮城県石巻市金華山); 23-24, **粟島亜種**(23, ♂, 24, ♀; 新潟県粟島内浦); 25-26, **関東平野多摩川以北亜種**(25, ♂, 26, ♀; 埼玉県所沢市狭山丘陵); 27-28, **房総半島南部亜種**(27, ♂, 28, ♀; 千葉県南房総市富山); 29-30, **奥利根亜種**(29, ♂, 30, ♀; 群馬県みなかみ町大穴). — a, 指状片 (頭側面観); b, 同(右側面観).

32. *Carabus* (*Ohomopterus*) *insulicola*

Figs. 31-40. ***Carabus*** (***Ohomopterus***) ***insulicola*** subspp. — 31-32, ***C.*** (***O.***) ***i. sado*** (31, ♂, 32, ♀; Kanai-shinbo, Is. Sado-ga-shima, Niigata Pref.); 33-36, ***C.*** (***O.***) ***i. shinano*** (33, 35, 36, ♂, 34, ♀; 33-34, Asama-onsen, Matsumoto-shi, Nagano Pref.; 35, Kakeyu-onsen, Ueda-shi, Nagano Pref.; 36, Iwabuchi, Fujikawa-chô, Shizuoka Pref.); 37-38, ***C.*** (***O.***) ***i. kiso*** (37, ♂, 38, ♀; Hiyoshi-harano, Kiso-machi, Nagano Pref.); 39-40, ***C.*** (***O.***) ***i. komaganensis*** (39, ♂, 40, ♀; Akaho-fukuoka, Komagané-shi, Nagano Pref.). — 41, Natural hybrid between *C.* (*O.*) *i. shinano* & *C.* (*O.*) *arrowianus nakamurai* (= so called *C.* (*O.*) *insulicola pseudinsulicola*) (♂, Nosoko, Ina-shi, Nagano Pref.). — 42, Natural hybrid between *C.* (*O.*) *i. insulicola* & *C.* (*O.*) *esakii esakii* (♂, Tanna, Kannami-chô, Shizuoka Pref.). — a, Digitulus in frontal view; b, ditto in right lateral view.

図 31-40. **アオオサムシ**各亜種. — 31-32, **佐渡島亜種**(31, ♂, 32, ♀; 新潟県佐渡島金井新保); 33-36, **信濃亜種**(33, 35, 36, ♂, 34, ♀; 33-34, 長野県松本市浅間温泉; 35, 長野県上田市鹿教湯温泉; 36, 静岡県富士川町岩渕); 37-38, **木曽亜種**(37, ♂, 38, ♀; 長野県木曽町日義原野); 39-40, **駒ヶ根亜種**(39, ♂, 40, ♀; 長野県駒ヶ根市赤穂福岡). — 41, アオオサムシ信濃亜種とミカワオサムシ天竜川中流域亜種の自然雑種(いわゆるニセアオオサムシ)(♂, 長野県伊那市野底). — 42, アオオサムシ基亜種とシズオカオサムシ基亜種の自然雑種(♂, 静岡県函南町丹那). — a, 指状片(頭側面観); b, 同(右側面観).

32. *Carabus* (*Ohomopterus*) *insulicola* Chaudoir, 1869 アオオサムシ

Original description *Carabus insulicola* Chaudoir, 1869, Rev. Mag. Zool., 21, p. 26.

History of research This species was originally described by Chaudoir (1869) from "Japon" without designation of detailed type locality. Breuning (1932) placed it in the subgenus *Apotomopterus* of the genus *Carabus* (s. lat.). Nakane (1952b) regarded *Apotomopterus* as a distinct genus, and recognized in it five species from Japan, namely, *A. dehaani* [sic](subsequent spelling of *dehaanii*), *A. insulicola*, *A. yaconinus*, *A. albrechti* and *A. japonicus*. He recognized three geographical races in *A. insulicola*, namely, *maiyasanus*, *nagoyaensis* (MS: manuscript name) (= *Carabus arrowianus* (Breuning)) and *shizuokaensis* (MS) (= *C. esakii* Csiki). Of these, the former two were regarded by Hiura and his coworkers as distinct species, *A. maiyasanus* (Hiura, 1965, 1966; Hiura & Katsura, 1971; Hiura et al., 1971; Katsura et al., 1978) and *A. arrowianus* (Hiura et al., 1971; Katsura et al., 1978). Ishikawa (1966a) recognized *Apotomopterus esakii* as a distinct species because of sympatry with *A. insulicola* at some localities and difference in the genitalic features. In the same paper, Ishikawa described four new subspecies of *A. insulicola*, namely, *komiyai* (= *Carabus* (*Ohomopterus*) *komiyai komiyai* in the present sense), *nakamurai* (*C.* (*O.*) *arrowianus nakamurai*, ibid.), *pseudinsulicola* (= presumed to be a hybrid population between *C.* (*O.*) *insulicola shinano* and *C.* (*O.*) *arrowianus nakamurai*) and *nishikawai* (*C.* (*O.*) *i. nishikawai* in the present sense). Redefinition of the *Carabus* (*Ohomopterus* as its subgenus) and the genus *Apotomopterus* was made by Ishikawa (1973), and all the Japanese species so far combined with the latter were transferred to the former. Ishikawa (1985) defined *C.* (*O.*) *insulicola* by the characters of the digitulus (copulatory piece in his sense) and combined *komiyai* and *nakamurai* with *C. arrowianus* as its subspecies. Ishikawa & Ujiie (2000) revised *C.* (*O.*) *insulicola* morphologically, and classified it into nine subspecies, namely, *kita*, *awashimaensis*, *kantoensis*, *insulicola*, *okutonensis*, *sado*, *shinano*, *kiso* and *komaganensis*, eight of which were newly described at that time. The subspecies *nishikawai* was synonymized with the nominotypical subspecies in the same paper.

Type area "Japon" (most probably Yokohama or its nearby regions in central Honshu).

Type depository Museum Nationale d'Histoire Naturelle, Paris [Lectotype, ♂].

Morphology 22~34 mm. Body above basically greenish, though variable according to population or individual, e.g., golden green, yellowish green, dark coppery green, coppery with greenish tinge, dark reddish coppery, seldom mat chocolate-colored or black with faint bluish tinge; legs usually black, though tibiae are rarely dark rufous. External features: Figs. 1~6. Genitalic features: Figs. 7~18. Basically similar to other medium- to large-sized *Ohomopterus* species, but identified by the following characters: 1) head usually strongly punctate, vertex to neck remarkably wrinkled; 2) antennae relatively short, barely extending basal third of elytra in male; 3) ventral side of male antennae with hairless areas from segment five to seven; 4) pronotum with three to four marginal setae; 5) elytral intervals almost evenly raised and smooth, with primary foveoles invading adjacent tertiaries, above all in northern populations, striae between intervals shallowly but rather coarsely punctate; 6) inner margin of male protibia strongly protruded though not so remarkably angular; 7) penis large, about half the length of elyta, its apical part rather robust and somewhat spatulate; 8) internal sac as shown in Figs. 7, 11~15, digitulus large and hook-like, with apex turning to right side of internal sac as in *C.* (*O.*) *komiyai*, *C.* (*O.*) *esakii* and *C.* (*O.*) *uenoi*, apical portion compressed bilaterad to form vertical plate; 9) inner plate of ligular apophysis chestnut-shaped in dorsal view, with hind margin nearly linear though slightly bisinuate, inner wall irregularly wrinkled.

Molecular phylogeny For the general overview of the molecular phylogeny of *Ohomopterus*, see the same section of *C.* (*O.*) *yamato*. According to a simultaneous analysis of six nuclear DNA sequences, this species belongs to the *iwawakianus-insulicola* species-group (Sota & Nagata, 2008). For details of introgressive hybridization between *C.* (*O.*) *insulicola* and *C.* (*O.*) *arrowianus*, see Sota et al. (2000), Sota et al. (2001) and Ujiie et al. (2005). Also see Su et al. (2006) for the establishment of hybrid-derived offspring population in the *Ohomopterus* species through unidirectional hybridization.

Distribution C.~NE. Honshu, SW. Hokkaido (Hakodaté & its nearby regions; most probably introduced population) / Iss. Ô-shima of Kesennuma-shi, Kinkasan, Tashiro-jima, Aji-shima, Miyato-jima, Sabusawa-jima, Nono-shima, Katsura-shima (Miyagi Pref.), Awa-shima, Sado-ga-shima (Niigata Pref.), (Oki-no-shima), (Taka-no-shima) (Chiba Pref.), Izu-ô-shima (Tokyo Met.), É-no-shima & Jô-ga-shima (Kanagawa Pref.).

Habitat Common species inhabiting mainly plains, hills and low mountains preferably near human habitation and cultivated lands.

Bionomical notes Spring breeder, earthworm feeder.

Geographical variation Classified into 10 subspecies.

Etymology The specific name means "inhabitant of an island".

研究史　Chaudoir（ショドワー）(1869)により「日本」から記載された種で，原記載に詳しい場所の記載はないが，横浜界隈で得られた標本がタイプとして用いられたのではないかと推測されている．Breuning（ブロイニング）(1932)はその代表的名著「オサムシ属総説」の中で，本種を広義のオサムシ属の中のアポトモプテルス亜属に置いた．中根（1952b）はアポトモプテルスを属に昇格し，日本産のオサムシとしてはオオオサムシ，アオオサムシ，ヤコンオサムシ，クロオサムシ，ヒメオサムシの計5種をそこに含めた．そしてアオオサムシの「地理的品種」として*maiyasanus*, *nagoyaensis*（原稿名）（=*C. arrowianus*），*A. i. shizuokaensis*（同）（=*C. esakii*）の三つを認めた．これらはいずれも，現在では独立種マヤサンオサムシ，ミカワオサムシ，シズオカオサムシとして扱われている（前

roughly corresponds to the present Nagano Prefecture, the main distributional area of this subspecies.

タイプ産地　長野県, 松本市, 浅間温泉, 700 m.
タイプ標本の所在　首都大学東京動物系統分類学研究室（八王子）［ホロタイプ, ♂］.
亜種の特徴　22～32 mm. 体長は変異の幅が大きく, 本種中最も小型の個体が出現する. 背面は通常明緑色から緑褐色だが, 銅赤色の個体が優勢となる集団も見られ, 稀に全体黒色の個体も出現する. 脛節は時として部分的に赤みを帯びる. 前胸背板側縁後半部はほぼ直線状で, 後角は丸く, 後方へは殆ど突出しない. 前胸背板表面の点刻は変化の幅が大きいが, 一般に不規則で, 分布域の北端あるいは南端に近い地域から得られた個体では疎になる傾向が認められる. 鞘翅間室はよく隆起し, 第一次凹陥は小さく, 隣接する第三次凹陥を侵蝕しない. ♂交尾器指状片は側面から見て幅が狭く, 中央よりも先端よりで最も幅広くなるが, 個体差が著しい. 背面から見ると外側縁は幅が狭く, 右側縁は内湾し, 弧を描く. 基板は幅広く, 屈曲部に向けて明らかに狭まることが多い. 基板と指状片の間のスペースは広い.
分布　新潟（南部）, 長野, 山梨, 静岡各県.
亜種小名の由来と通称　亜種小名は, かつての令制国（りょうせいこく）の一つで現在の長野県にほぼ相当する信濃国（しなののくに）に因む. 通称はシナノアオオサムシ.

32-9. *C. (O.) i. kiso* Ishikawa et Ujiie, 2000　アオオサムシ木曽亜種

Original description　*Carabus* (*Ohomopterus*) *insulicola kiso* Ishikawa & Ujiie, 2000, Jpn. J. syst. Ent., 6 (2), p. 286, figs. 110 (p. 287) & 111~115-b (p. 287).
Type locality　Harano, 830-840 m, Hiyoshi-mura, Nagano Prefecture.
Type depository　Systematic Zoology Laboratory, Department of Biological Sciences, Tokyo Metropolitan University (Hachiôji) [Holotype, ♂].
Diagnosis　24~30 mm. Body above green, brassy green, brassy or rarely coppery with strong greenish tinge; median smooth part of pronotal disk largely black. Most closely allied to subsp. *shinano*, but distinguishable from that race by less sparsely punctate disk and deeper basal foveae of pronotum and more strongly convex elytra.
Distribution　Upper course of Riv. Kiso-gawa (Kiso-machi & Agematsu-machi, Nagano Pref.).
Etymology　Named after its distributional range which belongs to the Kiso area of Nagano Pref.

タイプ産地　長野県, 日義村, 原野, 830-840 m（日義村は合併により消滅し, 原野は現在, 木曽町（きそまち）に属する）.
タイプ標本の所在　首都大学東京動物系統分類学研究室（八王子）［ホロタイプ, ♂］.
亜種の特徴　24～30 mm. 背面は緑色ないし緑褐色で, 前胸背板中央の黒色無点刻部が広い. 前胸背板はほぼ中央で最も幅広く, 側縁後半部は僅かに内湾する程度. 後角は丸く, 後方への突出は弱い. 前胸背板表面には不規則で弱く疎な点刻を装う. 基部凹陥は大きく深い. 鞘翅の膨隆は強く, 間室は強く隆起し, 第一次凹陥は小さく, 隣接する第三次凹陥を侵蝕することは殆どない. ♂交尾器指状片の形態は個体差が著しく, 多くは奇形的な変形を伴い, 表面の皺が顕著. 基板は幅広い長方形. 指状片と基板との間隙は広い. 信濃亜種に近いが, 前胸背板の点刻が弱く, 同基部凹陥が深く, 鞘翅の膨隆が顕著なことなどにより識別される.
分布　木曽川上流地域（木曽町と上松町）. 木曽川上流のごく狭い範囲に分布する集団で, 北東部の鳥居峠（とりいとうげ）が信濃亜種との分布境界とされている.
亜種小名の由来と通称　亜種小名は本亜種の分布域である木曽地方に因む. 通称はキソアオオサムシ.

32-10. *C. (O.) i. komaganensis* Ishikawa et Ujiie, 2000　アオオサムシ駒ヶ根亜種

Original description　*Carabus* (*Ohomopterus*) *insulicola komaganensis* Ishikawa & Ujiie, 2000, Jpn. J. syst. Ent., 6 (2), p. 288, figs. 116 (p. 289) & 117~125 (p. 289).
Type locality　Fukuoka, 700 m, Komagane-shi, Nagano Prefecture.
Type depository　Systematic Zoology Laboratory, Department of Biological Sciences, Tokyo Metropolitan University (Hachiôji) [Holotype, ♂].
Diagnosis　24~29 mm. Body above uniformly bright to dark reddish coppery, or brassy with weak greenish tinge; legs black. Closely allied to subsp. *shinano* and barely distinguishable superficially from that race of adjacent areas by smaller body size, but male antennae uniquely bear hairless ventral depression on segment 8 in addition to those on segments 5 to 7. Digitulus smallest in size and simplest in shape in all races of this species.
Distribution　Western bank of Riv. Tenryû-gawa between Riv. Oguro-gawa & Riv. Nakatagiri-gawa.
Etymology　Named after its distributional range, Komagane of Nagano Pref.

タイプ産地　長野県, 駒ヶ根市, 福岡, 700 m.

32. *Carabus (Ohomopterus) insulicola*

タイプ標本の所在　首都大学東京動物系統分類学研究室(八王子)［ホロタイプ, ♂］.

亜種の特徴　24~29 mm. 背面は赤銅色ないし緑色味を帯びた褐色で, 脚は黒色. 近隣地域に産する信濃亜種に近いが, やや小型で, ♂触角腹面の無毛部は本種で通常見られる第5節から第7節に加えて第8節にも見られ, ♂交尾器指状片が全亜種中最も小型で単純な形態を有することなどにより識別される. 前胸背板の形態は信濃亜種のそれと大差ないが, 表面の点刻は密で, せいぜい前方にごく狭い無点刻部を認める程度. 鞘翅間室の隆起は顕著で, 第一次間室は第二次, 第三次間室と幅・高さ共にほぼ等しく, 第一次凹陥は小さく, 隣接する第三次間室を侵蝕することはなく, 間室間の線条にはごく弱い横の刻み目を有する. ♂前脛節内側縁中央は僅かに隆起する. ♂交尾器指状片は短く, 外側縁の幅は狭い. 基板も小さく短い.

分布　小黒川～中田切川間の天竜川西岸. 天竜川上流の西岸に分布し, 同河川の支流である小黒川と中田切川に挟まれた地域に分布する.

亜種小名の由来と通称　亜種小名は本亜種の分布域である駒ヶ根に因む. 通称はコマガネアオオサムシ.

32. *Carabus* (*Ohomopterus*) *insulicola*

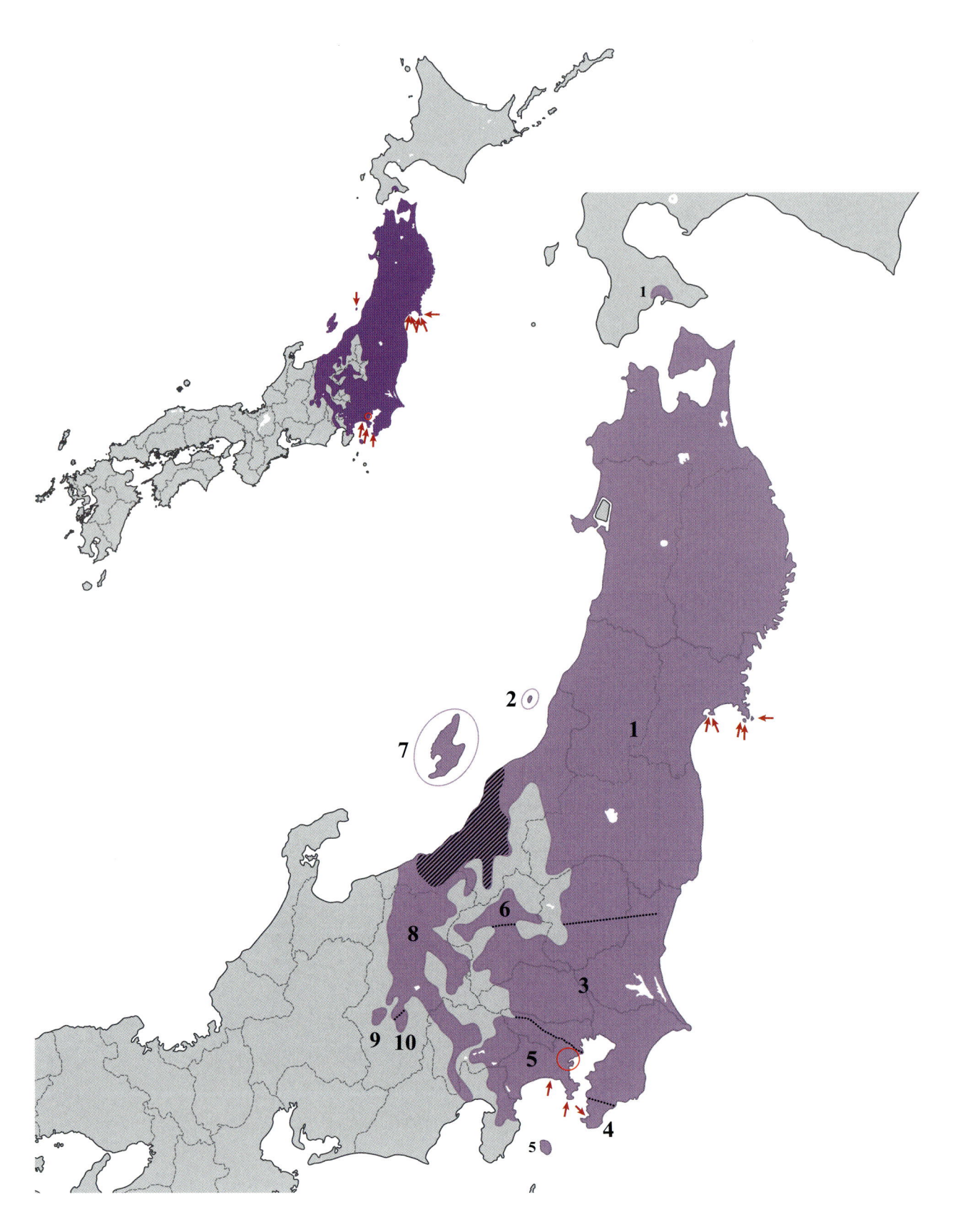

Fig. 43. Map showing the distributional range of ***Carabus*** (***Ohomopterus***) ***insulicola*** Chaudoir, 1869. — 1, Subsp. ***kita***; 2, subsp. ***awashimaensis***; 3, subsp. ***kantoensis***; 4, subsp. ***nishikawai***; 5, subsp. ***insulicola***; 6, subsp. ***okutonensis***; 7, subsp. ***sado***; 8, subsp. ***shinano***; 9, subsp. ***kiso***; 10, subsp. ***komaganensis***. Shaded area indicates the range of intergrading population between the two subspp. (?). The area of the species (most probably vicinity of its type locality).
図 43. アトリウムシの分布図. — 1, 東北地方亜種; 2, 粟島亜種; 3, 関東平野および利根川以北亜種; 4, 房総半島市部亜種; 5, 基亜種; 6, 奥利根亜種; 7, 佐渡島亜種; 8, 信濃亜種; 9, 木曽亜種; 10, 駒ヶ根亜種. 斜線部は東北地方亜種と信濃亜種の移行地帯. ◯: 種のタイプ地域（おそらく横浜界隈）.

32. *Carabus* (*Ohomopterus*) *insulicola*

Fig. 44. *Carabus* (*Ohomopterus*) *insulicola komaganensis* (Nishiharuchika, Ina-shi, Nagano Pref.).

図44. アオオサムシ駒ヶ根亜種(長野県伊那市西春近).

Fig. 45. *Carabus* (*Ohomopterus*) *insulicola kiso* (Hiyoshi, Kiso-machi, Nagano Pref.).

図45. アオオサムシ木曽亜種(長野県木曽町日義).

Fig. 46. *Carabus* (*Ohomopterus*) *insulicola kantoensis* passing the winter in the soil (Konuma, Sakado-shi, Saitama Pref.).

図46. 土中で越冬しているアオオサムシ関東平野多摩川以北亜種(埼玉県坂戸市小沼).

273

33. *Carabus* (*Ohomopterus*) *uenoi* (Ishikawa, 1960)
ドウキョウオサムシ

Figs. 1-15. *Carabus* (*Ohomopterus*) *uenoi* (Ishikawa, 1960) (Mt. Kongô-zan (or Kongô-san), Gosé-shi, Nara Pref.). — 1, ♂; 2, ♀; 3, male head & pronotum; 4, male right protarsus in ventral view; 5, male right protibia in subdorsal view; 6, male left elytron (median part); 7, penis & fully everted internal sac in right lateral view; 8, apical part of penis in right lateral view; 9, ditto in right subdorsal view; 10, ditto in dorsal view; 11, internal sac in caudal view; 12, ditto in frontal view; 13, ditto in dorsal view; 14, ditto in left sublateral view; 15, inner plate of ligular apophysis.

33. *Carabus* (*Ohomopterus*) *uenoi* (Ishikawa, 1960) ドウキョウオサムシ

Original description　*Apotomopterus uenoi* Ishikawa, 1960, Kontyû, 28, p. 168, figs. 1~4 (p. 169).

History of research　This species is endemic to the upper part of the Kongô Mountain Range in the Kinki Region, and is unique in having monstrously enlarged penis and digitulus. It had been confused with *C. (O.) yaconinus* simply because of similarity in the external features until Ishikawa (1960) described it as a distinct new species.

Type locality　Mt. Kongô-zan (or Kongô-san), Yamato, in central Honshu (on the borders between Nara & Osaka Prefectures).

Type depository　According to the original description, the holotype (♂) is preserved in R. Ishikawa's collection (now preserved in the Systematic Zoology Laboratory, Department of Biological Sciences, Tokyo Metropolitan University (Hachiôji) ?).

Morphology　23~28 mm. Usually black with weak blue-greenish tinge along margins, or sometimes brownish coppery; legs black, though tibiae rarely dark rufous. External features: Figs. 1~6. Genitalic features: Figs. 7~15. Allied to other medium-sized *Ohomopterus* species, above all to *C. (O.) arrowianus* or *C. (O.) yaconinus*, but identified by the following characteristics: 1) male antennae reaching middle of elytra, with hairless areas from segment five to seven on ventral side; 2) pronotum widest a little before middle, sides rather strongly convergent toward both apex and base, hind angles short, lateral margin usually with three setae; 3) elytra widest at about middle, each interval nearly equally raised and smooth on surface, though weakly notched around sutural part near bases and apices, primary foveoles small but slightly invading adjacent tertiaries, striae between intervals weakly punctate; 4) sides of sternites at most vaguely punctate; 5) metacoxa bisetose, lacking anterior seta, which is exceptional for a member of *Ohomopterus*; 6) inner margin of male protibia strongly convex and weakly angulated near middle; 7) penis monstrously large, about three-fifths as long as elytra, its details are shown in Figs. 7~10; 8) digitulus also monstrously elongated, about seven-eighths as long as penis, turning toward right side of internal sac, with marked deformity as shown in Figs. 7 and 11~14; 9) internal sac tube-like and very long due to extraordinarily elongated digitulus, lacking bilateral basal lateral lobes, bearing well-developed paraligula; peripheral rim of gonopore strongly sclerotized and pigmented, forming a pair of semicircular terminal plates; 10) inner plate of ligular apophysis top- or funnel-shaped, not deeply concave, similar to that of *C. (O.) iwawakianus*.

Molecular phylogeny　For the phylogenetic position of this species in the subgenus *Ohomopterus*, see Sota & Vogler (2003), Sota & Nagata (2008) and Sota (2011), etc. According to the analyses of the mitochondrial ND5 gene and the nuclear ITS I sequences (Tominaga, Imura *et al.*, 2005), this species is considered to be a descendant from a hybrid between the female of *C. (O.) arrowianus* and the male of *C. (O.) iwawakianus* (above all its "subspecies" *kiiensis*).

Distribution　Upper part of Mt. Kongô-zan, Mt. Yamato-katsuragi-san & Mt. Iwahashi-yama in Kinki Region, W. Honshu.

Habitat　Inhabitant of matured forest, both natural and planted, in the upper part of the Kongô Mountains.

Bionomical notes　Spring breeder and earthworm feeder.

Geographical variation　No marked geographical variation is visible in the distributional range, and considered to be a monotypic species.

Conservation　The distribution of this species is confined to the upper part of the Kongô Mountain Range. It was abundant until the 1960s, but gradually became rarer, probably due to excessive hunting by collectors. This species is designated as threatened species (VU; vulnerable species) in the 4th edition of Red List (Threatened Wildlife of Japan) of Ministry of the Environment (2012).

Etymology　The species name is dedicated to Shun-Ichi Uéno of National Museum of Nature and Science, Tokyo, who is well-known as an authority on carabidology (trechine beetles) and cave biology.

研究史　♂交尾器が著しく巨大化することでよく知られた種で，大阪府と奈良県の境にまたがる金剛山地高所の特産だが，Ishikawaによって1960年に新種として記載されるまでは，外観のよく似たヤコンオサムシと混同されていたようである．

タイプ産地　本州中部，大和，金剛山（奈良県御所市（ごせし）／大阪府南河内郡（みなみかわちぐん）千早赤阪村（ちはやあかさかむら））．

タイプ標本の所在　原記載論文によれば，本種のタイプ標本は計5♂5♀で，九州大学昆虫学研究室に保管される1ペアを除き，他のすべての標本（すなわちホロタイプ（♂）も）は石川良輔コレクション保管とされている（現在は首都大学東京動物系統分類学研究室（八王子）に保管？）．

形態　23~28 mm．黒色で側縁部に弱い青緑色の光沢を帯びるものが多いが，時に褐色型も見られる．脚は黒色で，稀に脛節が暗赤色を帯びるものがある．外部形態：図1~6．交尾器形態：図7~15．同亜属他種，とりわけミカワオサムシ黒色型やヤコンオサムシによく似ているが，以下のような特徴を持つ：1)♂触角は先端が鞘翅中央に達し，第5節から第7節の腹面に無毛部がある；2)前胸背板の最広部は中央よりやや前方にあり，側縁は前後に比較的強く狭まり，後角は短く，側縁には通常3本の剛毛がある；3)鞘翅の最広部は中央付近にあり，間室は均質に隆起し，平滑だが，会合線付近の前方と後方では弱い横条溝が認められ，第一次回陥は小さいが，隣接する第三次間室を僅かに侵し，間室間条溝の刻み目は弱い；4)腹部腹板の側面はせいぜいごく弱く点刻される程度で平滑(この形質は，ヤコンオサムシやマヤサンオサムシ等，外見的によく似た中型黒色の同属他種からの重要な鑑別点となる)；5)後基節はオホモプテルス亜属としては例外的に前方剛毛を欠く；6)♂前脛節内側縁はほぼ中央で強く膨らみ，鈍く角ばる；7)陰茎は巨大で，長さは鞘翅長の約3/5に達し，基部は太く強壮で，左近位端から右遠位端方向に向けて強くよじれ，中央部は側方から見ると緩やかに弧を描きつつ内湾し，背面から見ると波曲する．先端部は太短く，あまり鋭く尖らない；8)指状片も著しく長大で，その長さは陰茎全長の約7/8に達する．基部は縦方向にほぼ180°，横方向にはほぼ直角に屈曲し，先端

35. *Carabus (Ohomopterus) maiyasanus* Bates, 1873
マヤサンオサムシ

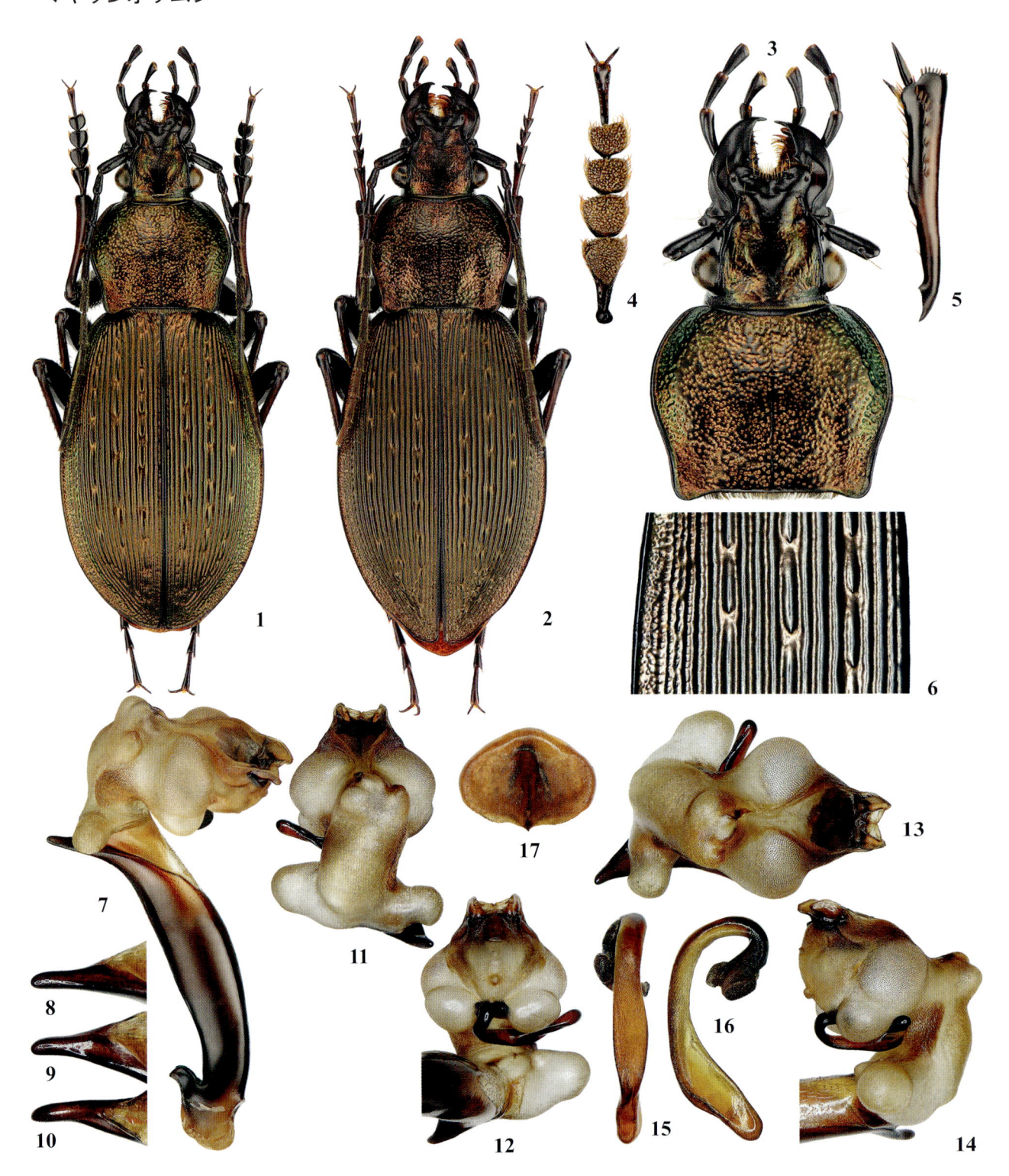

Figs. 1-17. *Carabus (Ohomopterus) maiyasanus maiyasanus* Bates, 1873 (Mt. Maya-san, Kôbé-shi, Hyogo Pref.). — 1, ♂; 2, ♀; 3, male head & pronotum; 4, male right protarsus in ventral view; 5, male right protibia in subdorsal view; 6, male left elytron (median part); 7, penis & fully everted internal sac in right lateral view; 8, apical part of penis in right lateral view; 9, ditto in right subdorsal view; 10, ditto in dorsal view; 11, internal sac in caudal view; 12, ditto in frontal view; 13, ditto in dorsal view; 14, ditto in left sublateral view; 15, digitulus in frontal view; 16, ditto in left lateral view; 17, inner plate of ligular apophysis.

図 1-17. マヤサンオサムシ基亜種(兵庫県神戸市摩耶山). — 1, ♂; 2, ♀; 3, ♂の頭部と前胸背板; 4, ♂右前跗節(腹面観); 5, ♂右前脛節(斜背面観); 6, ♂の左鞘翅中央部; 7, 陰茎と完全反転下の内嚢(右側面観); 8, 陰茎先端(右側面観); 9, 同(右斜背面観); 10, 同(背面観); 11, 内嚢(尾側面観); 12, 同(頭側面観); 13, 同(背面観); 14, 同(左斜側面観); 15, 指状片(頭側面観); 16, 同(左側面観); 17, 膣底部節片内板.

33. *Carabus* (*Ohomopterus*) *uenoi* (Ishikawa, 1960) ドウキョウオサムシ

Original description   *Apotomopterus uenoi* Ishikawa, 1960, Kontyû, 28, p. 168, figs. 1~4 (p. 169).

History of research   This species is endemic to the upper part of the Kongô Mountain Range in the Kinki Region, and is unique in having monstrously enlarged penis and digitulus. It had been confused with *C.* (*O.*) *yaconinus* simply because of similarity in the external features until Ishikawa (1960) described it as a distinct new species.

Type locality   Mt. Kongô-zan (or Kongô-san), Yamato, in central Honshu (on the borders between Nara & Osaka Prefectures).

Type depository   According to the original description, the holotype (♂) is preserved in R. Ishikawa's collection (now preserved in the Systematic Zoology Laboratory, Department of Biological Sciences, Tokyo Metropolitan University (Hachiôji) ?).

Morphology   23~28 mm. Usually black with weak blue-greenish tinge along margins, or sometimes brownish coppery; legs black, though tibiae rarely dark rufous. External features: Figs. 1~6. Genitalic features: Figs. 7~15. Allied to other medium-sized *Ohomopterus* species, above all to *C.* (*O.*) *arrowianus* or *C.* (*O.*) *yaconinus*, but identified by the following characteristics: 1) male antennae reaching middle of elytra, with hairless areas from segment five to seven on ventral side; 2) pronotum widest a little before middle, sides rather strongly convergent toward both apex and base, hind angles short, lateral margin usually with three setae; 3) elytra widest at about middle, each interval nearly equally raised and smooth on surface, though weakly notched around sutural part near bases and apices, primary foveoles small but slightly invading adjacent tertiaries, striae between intervals weakly punctate; 4) sides of sternites at most vaguely punctate; 5) metacoxa bisetose, lacking anterior seta, which is exceptional for a member of *Ohomopterus*; 6) inner margin of male protibia strongly convex and weakly angulated near middle; 7) penis monstrously large, about three-fifths as long as elytra, its details are shown in Figs. 7~10; 8) digitulus also monstrously elongated, about seven-eighths as long as penis, turning toward right side of internal sac, with marked deformity as shown in Figs. 7 and 11~14; 9) internal sac tube-like and very long due to extraordinarily elongated digitulus, lacking bilateral basal lateral lobes, bearing well-developed paraligula; peripheral rim of gonopore strongly sclerotized and pigmented, forming a pair of semicircular terminal plates; 10) inner plate of ligular apophysis top- or funnel-shaped, not deeply concave, similar to that of *C.* (*O.*) *iwawakianus*.

Molecular phylogeny   For the phylogenetic position of this species in the subgenus *Ohomopterus*, see Sota & Vogler (2003), Sota & Nagata (2008) and Sota (2011), etc. According to the analyses of the mitochondrial ND5 gene and the nuclear ITS I sequences (Tominaga, Imura et. al., 2005), this species is considered to be a descendant from a hybrid between the female of *C.* (*O.*) *arrowianus* and the male of *C.* (*O.*) *iwawakianus* (above all its "subspecies" *kiiensis*).

Distribution   Upper part of Mt. Kongô-zan, Mt. Yamato-katsuragi-san & Mt. Iwahashi-yama in Kinki Region, W. Honshu.

Habitat   Inhabitant of matured forest, both natural and planted, in the upper part of the Kongô Mountains.

Bionomical notes   Spring breeder and earthworm feeder.

Geographical variation   No marked geographical variation is visible in the distributional range, and considered to be a monotypic species.

Conservation   The distribution of this species is confined to the upper part of the Kongô Mountain Range. It was abundant until the 1960s, but gradually became rarer, probably due to excessive hunting by collectors. This species is designated as threatened species (VU; vulnerable species) in the 4th edition of Red List (Threatened Wildlife of Japan) of Ministry of the Environment (2012).

Etymology   The species name is dedicated to Shun-Ichi Ueno of National Museum of Nature and Science, Tokyo, who is well-known as an authority on carabidology (trechine beetles) and cave biology.

研究史　♂交尾器が著しく巨大化することでよく知られた種で，大阪府と奈良県の境にまたがる金剛山地高所の特産だが，Ishikawaによって1960年に新種として記載されるまでは，外観のよく似たヤコンオサムシと混同されていたようである．

　タイプ産地　本州中部，大和，金剛山（奈良県御所市(ごせし)／大阪府南河内郡(みなみかわちぐん)千早赤阪村(ちはやあかさかむら))．

　タイプ標本の所在　原記載論文によれば，本種のタイプ標本は計5♂5♀で，九州大学昆虫学研究室に保管される1ペアを除き，他のすべての標本（すなわちホロタイプ(♂)も）は石川良輔コレクション保管とされている（現在は首都大学東京動物系統分類学研究室（八王子）に保管？）．

　形態　23~28 mm. 黒色で側縁部に弱い青緑色の光沢を帯びるものが多いが，時に褐色型も見られる．脚は黒色で，稀に脛節が暗赤色を帯びるものがある．外部形態：図1~6．交尾器形態：図7~15．同亜属他種，とりわけミカワオサムシ黒色型やヤコンオサムシによく似ているが，以下のような特徴を持つ：1) ♂触角は先端が鞘翅中央に達し，第5節から第7節の腹面に無毛部がある；2) 前胸背板の最広部は中央よりやや前方にあり，側縁は前後に比較的強く狭まり，後角は短く，側縁には通常3本の剛毛がある；3) 鞘翅の最広部は中央付近にあり，間室は均質に隆起し，平滑だが，会合線付近の前方と後方では弱い横条溝が認められ，第一次凹陥は小さいが，隣接する第三次間室を僅かに侵し，間室間条溝の刻み目は弱い；4) 腹部腹板の側面はせいぜいごく弱く点刻される程度で平滑（この形質は，ヤコンオサムシやマヤサンオサムシ等，外見的によく似た中型黒色の同属他種からの重要な鑑別点となる）；5) 後基節はオホモプテルス亜属としては例外的に前方剛毛を欠く；6) ♂前脛節内側縁はほぼ中央で強く膨らみ，鈍く角ばる；7) 陰茎は巨大で，長さは鞘翅長の約3/5に達し，基部は太く強壮で，左近位端から右遠位端方向に向けて強くよじれ，中央部は側方から見ると緩やかに弧を描きつつ内湾し，背面から見ると波曲する．先端部は太短く，あまり鋭く尖らない；8) 指状片も著しく長大で，その長さは陰茎全長の約7/8に達する．基部は縦方向にほぼ180°，横方向にはほぼ直角に屈曲し，先端

は内嚢の右側方を向く．中央部は細長い板状で，外側縁は僅かに弧を描き，先端手前でやや突出して結節状になり，先端に向けて徐々に細まり，鈍く丸まって終わる；10）内嚢は指状片の巨大化に伴う変形が著しく，全体として非常に細長く，筒状で，オホモプテルス亜属において通常認められる左右の基部側葉を全く欠く一方，側舌はよく発達する．射精孔縁膜は硬化と色素沈着が著しく，半円形の頂板を形成する；11）♀交尾器膣底部節片内板は独楽あるいは漏斗を側面から見たような，いわゆるイワワキオサムシ型で，中央部はあまり強く窪まない．

分子系統 分子系統解析による本種の系統学的位置については Sota & Vogler (2003) を参照．それによれば，本種はオホモプテルス亜属の中でイワワキーアオオサムシ種群に属し，イワワキオサムシ，本種，マヤサンオサムシの共通祖先から先ずイワワキが分化，次いで本種とマヤサンが分化したと想定されているが，その一方で本種群においては種の分岐関係を推定することが困難であるとも述べられている (Sota & Nagata, 2008; 曽田, 2011等)．一方，Tominaga, Imura et. al. (2005) によれば，本種のミトコンドリア ND5 遺伝子はミカワオサムシ型，核 ITS I DNA はイワワキオサムシ紀伊半島亜種型であることが判明しており，恐らくは古い時代に起きたミカワオサムシ系の♀とイワワキオサムシ系の♂との交雑に由来するものであろうと考えられている．

分布 金剛山および大和葛城山から岩橋山にかけての高所．大阪府と奈良県の境にまたがる金剛山地の特産で，金剛山と大和葛城山の高所のみに分布すると考えられていたが，ごく最近になって更に北方に位置する岩橋山 (標高 659 m) にも生息していることが確認された (森岡・東, 2013)．

生息環境 金剛山地高所の成熟した森林を好むが，自然林のみならず，成熟した林であれば人工林からも得られる．

生態 春繁殖・成虫越冬型で，幼虫は環形動物 (ミミズ) 食．

地理的変異 亜種分類はなされておらず，単型種として扱われている．金剛山の集団と大和葛城山から岩橋山にかけての集団の間にもさしたる違いは見られないようである．

保全 金剛山地の上部のみから知られ，生息範囲が限られているうえ，同亜属他種に類を見ない巨大な♂交尾器を持つという珍奇性から，マニア，愛好家の間ではその希少価値が高く評価されている．1960年代前半頃までは比較的多くの個体が見られたようで，現在でも本種の生息環境自体は比較的良好に保たれているが，ロープウェイの設置や登山者，観光客の増加に加え，愛好家による採集圧 (冬季のオサ掘りと夏季のトラップ採集) により近年，徐々にその数を減じており，環境省第4次レッドリスト (2012) では，カテゴリー VU (絶滅危惧 II 類) に指定されている．

種小名の由来 チビゴミムシ類と洞窟生物学の権威として知られる国立科学博物館の上野俊一に献名された種．コンゴウオサムシという別称もあるが，殆ど使用されていない．

33. *Carabus* (*Ohomopterus*) *uenoi*

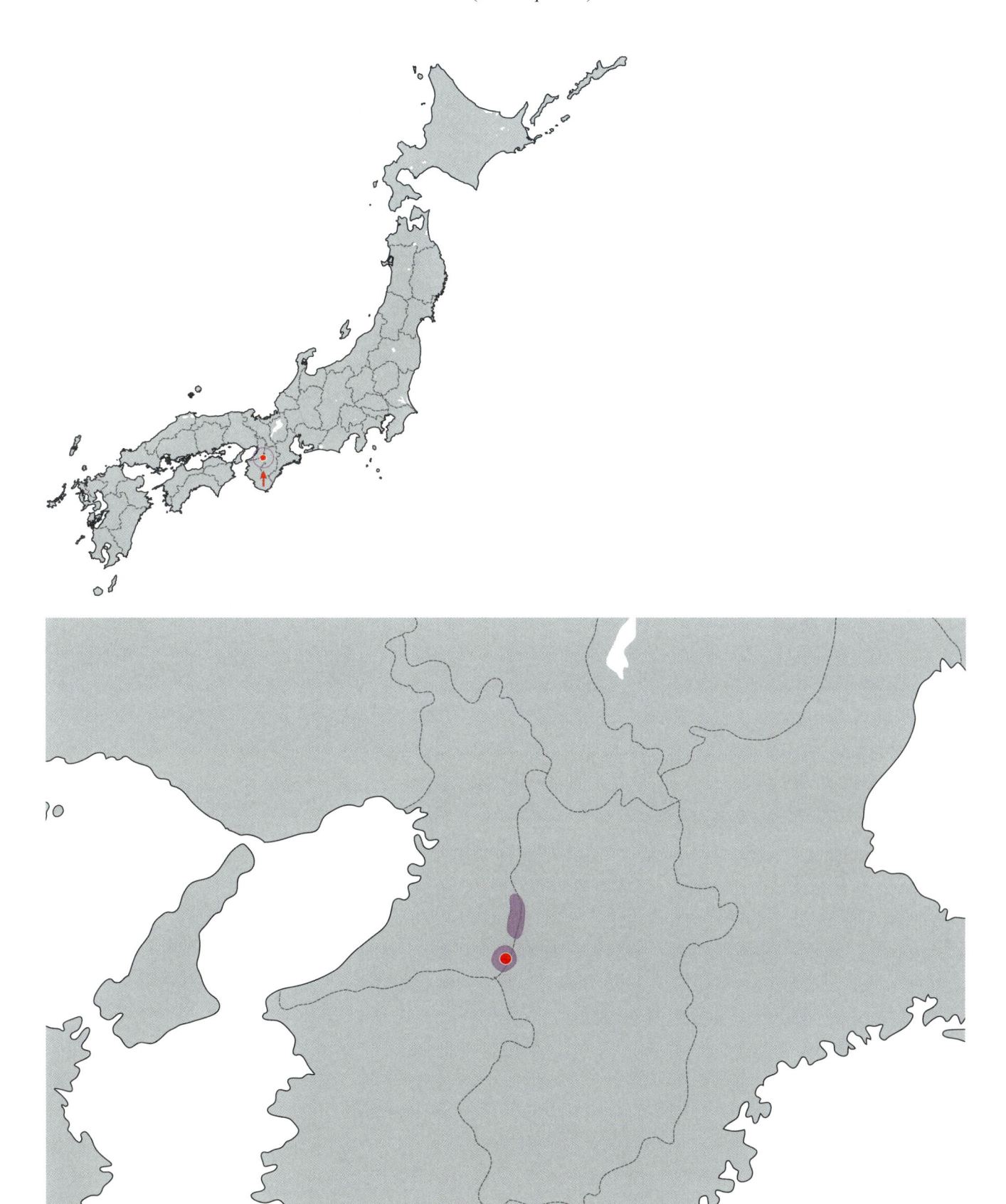

Fig. 16. Map showing the distributional range of ***Carabus*** (***Ohomopterus***) ***uenoi*** (Ishikawa, 1960). ●: Type locality of the species (Mt. Kongô-zan).
図 16. **ドウキョウオサムシ**分布図. ●: 種のタイプ産地（金剛山）.

34. *Carabus* (*Ohomopterus*) *arrowianus* (Breuning, 1934)
ミカワオサムシ

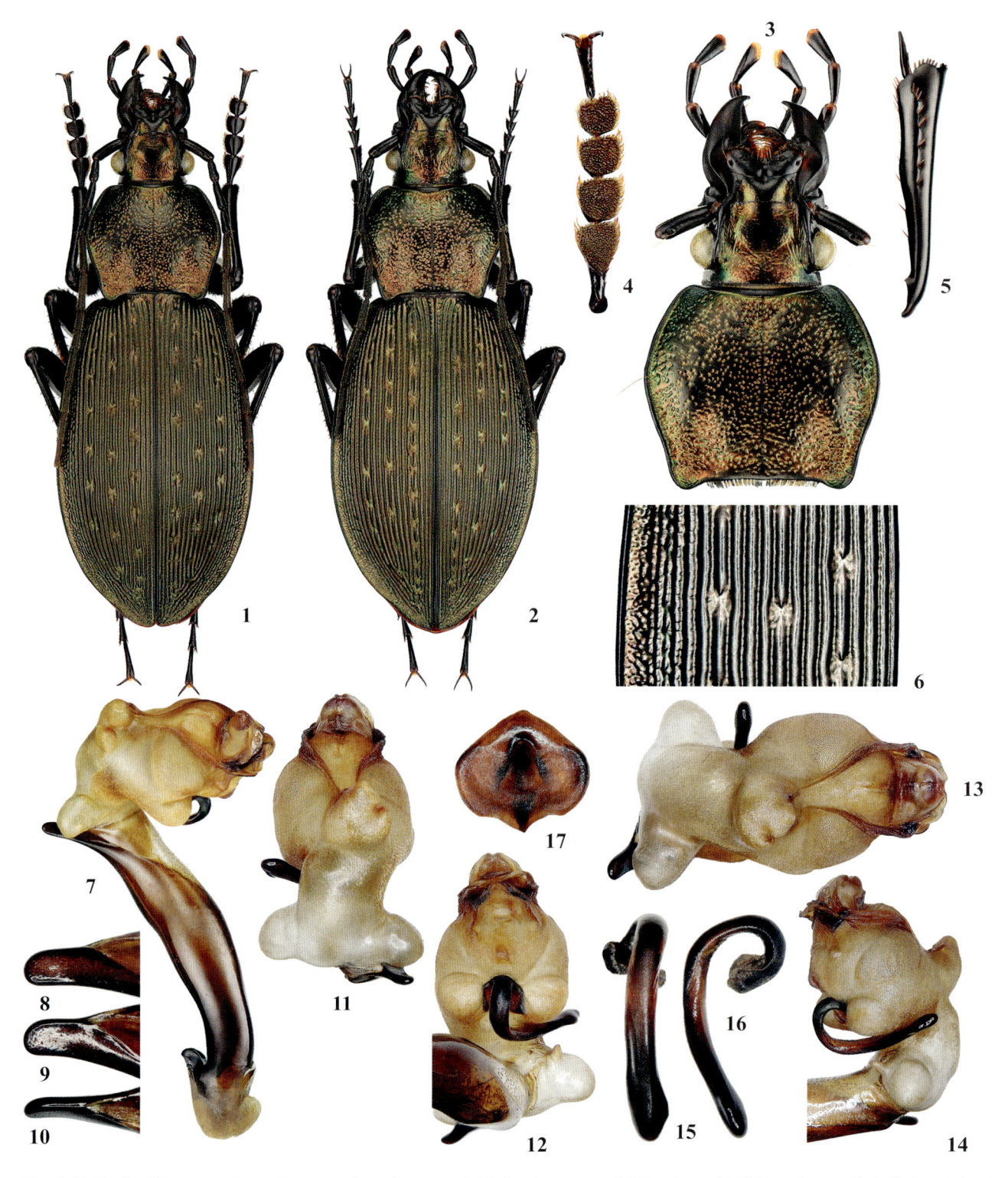

Figs. 1-17. *Carabus* (*Ohomopterus*) *arrowianus arrowianus* (Breuning, 1934) (Kami-nagayama-chô, Toyokawa-shi, Aichi Pref.). — 1, ♂; 2, ♀; 3, male head & pronotum; 4, male right protarsus in ventral view; 5, male right protibia in subdorsal view; 6, male left elytron (median part); 7, penis & fully everted internal sac in right lateral view; 8, apical part of penis in right lateral view; 9, ditto in right subdorsal view; 10, ditto in dorsal view; 11, internal sac in caudal view; 12, ditto in frontal view; 13, ditto in dorsal view; 14, ditto in left sublateral view; 15, digitulus in frontal view; 16, ditto in left lateral view; 17, inner plate of ligular apophysis.

図 1-17. ミカワオサムシ基亜種(愛知県豊川市上長山町). — 1, ♂; 2, ♀; 3, ♂の頭部と前胸背板; 4, ♂右前跗節(腹面観); 5, ♂右前脛節(斜背面観); 6, ♂左鞘翅中央部; 7, 陰茎と完全反転下の内嚢(右側面観); 8, 陰茎先端(右側面観); 9, 同(右斜背面観); 10, 同(背面観); 11, 内嚢(尾側面観); 12, 同(頭側面観); 13, 同(背面観); 14, 同(左斜側面観); 15, 指状片(頭側面観); 16, 同(左側面観); 17, 膣底部節片内板.

34. *Carabus* (*Ohomopterus*) *arrowianus*

Figs. 18-29. ***Carabus*** (***Ohomopterus***) ***arrowianus*** subspp. — 18-19, ***C***. (***O***.) ***a***. ***nakamurai*** (18, ♂, 19, ♀; Mt. Jinbagata-yama, Nakagawa-mura, Nagano Pref.); 20-21, ***C***. (***O***.) ***a***. ***shichinoi*** (20, ♂, PT, 21, ♀, PT; Miné, Tenryû-ku, Hamamatsu-shi, Shizuoka Pref.); 22-23, ***C***. (***O***.) ***a***. ***hidaosa*** (22, ♂, 23, ♀; Ôbora-machi, Takayama-shi, Gifu Pref.); 24-25, ***C***. (***O***.) ***a***. ***minoensis*** (24, ♂, 25, ♀; Mt. Kinka-zan, Gifu-shi, Gifu Pref.); 26-27, ***C***. (***O***.) ***a***. ***murakii*** (26, ♂, 27, ♀; Mishiodono-jinja Shrine, Futami-chô-shô, Isé-shi, Mie Pref.); 28-29, ***C***. (***O***.) ***a***. ***kirimurai*** (28, ♂, 29, ♀; Kami-ichigi, Mihama-chô, Mie Pref.). — a, Digitulus in frontal view; b, ditto in left lateral view.

図 18-29. **ミカワオサムシ**各亜種. — 18-19, **天竜川中流域亜種**(18, ♂, 19, ♀; 長野県中川村陣馬形山); 20-21, **佐久間亜種**(20, ♂, 21, ♀; 静岡県浜松市天竜区峰); 22-23, **岐阜県中北部亜種**(22, ♂, 23, ♀; 岐阜県高山市大洞町); 24-25, **岐阜県南部亜種**(24, ♂, 25, ♀; 岐阜県岐阜市金華山); 26-27, **志摩半島北部亜種**(26, ♂, 27, ♀; 三重県伊勢市二見町荘御塩殿神社); 28-29, **御浜町亜種**(28, ♂, 29, ♀; 三重県御浜町上市木). — a, 指状片(頭側面観); b, 同(左側面観).

34. *Carabus* (*Ohomopterus*) *arrowianus* (Breuning, 1934) ミカワオサムシ

Original description　*Apotomopterus arrowianus* Breuning, 1934, Folia zool. hydrobiol., Riga, 6, p. 31.

History of research　This taxon was first recorded as *Carabus* (*Apotomopterus*) *yaconinus* (= *C.* (*Ohomopterus*) *yaconinus* in the present sense) by Breuning (1932, p. 233 [Partim]), then described also by Breuning (1934) as a new species, though he did not designate the holotype. Later, Ishikawa & Kubota (1984) designated a male with the label "v. Bodemeyer Kobe Japan" as the lectotype of this species. Nakane (1952a) described a geographical race of *Apotomopterus insulicola* (= *C.* (*O.*) *insulicola* in the present sense) from Nagoya in Aichi Prefecture under the manuscript name of *nagoyaensis*. Anyway, it is synonymized with *C.* (*O.*) *arrowianus arrowianus* in current classification. Ishikawa (1966a) described *A. i. nakamurai* and *A. i. komiyai* from Nagano and Shizuoka Prefectures, respectively, but they were later transferred to subspecies of *C.* (*O.*) *arrowianus* by the same author. Of these, *C.* (*O.*) *a. komiyai* was raised its rank to a full species *Ohomopterus komiyai* (= *C.* (*O.*) *komiyai* in the present sense) based on the male genitalic morphology as well as on the distributional and reproductive aspects (Imura & Matsunaga, 2010). Ishikawa & Kubota (1984) described a new subspecies of *C.* (*O.*) *arrowianus* from the northern part of the Shima Peninsula in Mie Prefecture under the name of *murakii*. Ishikawa & Kubota (1994b) revised *C.* (*O.*) *arrowianus* morphologically, and additionally described two new subspecies under the names of *hidaosa* and *minoensis*. In 2003, an isolated population of *C.* (*O.*) *arrowianus* was unexpectedly found from Mihama-chô of southwestern Mie Prefecture, and it was named subsp. *kirimurai* (Kubota & Yahiro, 2003). Another isolated small population of the same species was found from Sakuma on the eastern bank of the Tenryû-gawa River in Shizuoka Prefecture, and it was described as subsp. *shichinoi* (Imura & Matsunaga, 2010).

Type locality　So-called Mikawa Region (Eastern part of Aichi Pref. in SC. Honshu). Judging from the characters of the digitulus of the type specimen (labeled "v. Bodemeyer Kobe Japan"), the type locality (area) of *C.* (*O.*) *arrowianus* is presumed to be the Mikawa Region which roughly corresponds to eastern Aichi Prefecture. The locality name "Kobe" written on the label is most probably erroneous (Ishikawa & Kubota, 1984).

Type depository　Zoölogisch Museum Amsterdam [Lectotype, ♂].

Morphology　21~31 mm. Body above reddish coppery or dark brown and less strongly greenish, sometimes entirely black with greenish or bluish tinge, or dark greenish; tibiae and tarsi black, though they are dark rufous in some subspecies or individuals. External features: Figs. 1~6. Genitalic features: Figs. 7~17. Similar to other medium- to large-sized *Ohomopterus* species, but identified by the following characters: 1) male antennae reaching or slightly extending middle of elytra; 2) lateral margins of pronotum less strongly emarginate before hind angles; 3) elytra long oval, widest behind middle, with shoulders rather distinct; 4) primary foveoles of elytra relatively large, more or less invading adjacent tertiaries; 5) apical part of penis short and robust as in *C.* (*O.*) *insulicola*, though much less strongly bent ventrad; 6) digitulus hook-like and narrowly elongated, with apex turning to left side of internal sac as in *C.* (*O.*) *maiyasanus*, and either depressed dorso-ventrad or cylindrical in apical portion, sometimes sinuate, swollen or partly protruded left laterad; 7) inner plate of ligular apophysis chestnut-shaped in dorsal view, with hind margin slightly emarginate on both sides, inner wall smooth.

Molecular phylogeny　An overview of the molecular phylogeny of *Ohomopterus* is given in the same section of *C.* (*O.*) *yamato*. According to a simultaneous analysis of six nuclear DNA sequences, this species belongs to the *iwawakianus-insulicola* species-group (Sota & Nagata, 2008). For details of the molecular phylogenetical and phylogeographical studies on and related to this species, see Sota *et al.* (2000), Sota (2000c, 2002b, 2003), Sota *et al.* (2001), Imura, Akita *et al.* (2005), Tominaga, Okamoto *et al.* (2005), Ujiie *et al.* (2005), Research group of molecular phylogeny on the *Ohomopterus* ground beetles (2005, 2006a), Su *et al.* (2006) and Nagata *et al.* (2009) etc.

Distribution　Honshu (Chubu Region & E. Kinki Region) / Iss. Také-shima, Saku-shima, Himaka-jima, Shino-jima (Aichi Pref.) & Tôshi-jima (Mie Pref.).

Habitat　The main habitat of this species is hills or mountains consisting of the basement rock formed after the Quaternary period, but also visible in such environment as alluvial plain, riverbed and fluvial terrace. Common in the edge of secondary forest, and also found from such dry environment as pine forest or tea plantation. It is also found from the coastal area near the mouths of large rivers flowing through the distributional range of this species. They must have been reached the bank most probably on the occasion of flood, and colonized at each station.

Bionomical notes　Spring breeder, earthworm feeder.

Geographical variation　Morphologically classified into seven subspecies (including the nominotypical one).

Etymology　Named after Gilbert John Arrow (1873~1938) of the Natural History Museum, London.

研究史　本タクソンはBreuning(ブロイニング)(1932)によってヤコンオサムシとして記録されたのち，同著者(1934)により新種として記載された．原記載ではホロタイプの指定が行われていないが，中根(1962, p. 26)によれば，ブロイニングから中根宛てに送られてきた私信の中に，タイプ標本はすべて「ミカワ」産である由の記述があったという．後年，Ishikawa & Kubota(1984)により，アムステルダム動物分類学研究所に保管されている"v. Bodemeyer Kobe Japan"というラベルのついた♂がレクトタイプに指定されているが，同標本の♂交尾器指状片の特徴も愛知県東部から静岡県西部にかけて産する集団のそれにほぼ一致しているので，"Kobe"(神戸)という産地名は恐らく誤りで，レクトタイプの真の産地はやはり三河地方(詳しい場所の特定は困難)であろうと考えら

れている．中根（1952a）は「愛知，岐阜」からアオオサムシの "geographical race" として*nagoyaensis*（原稿名）を記載したが，これはミカワオサムシ基亜種のシノニムとみなされている．ISHIKAWA（1966a）は長野県と静岡県からアオオサムシの亜種として*Apotomopterus insulicola nakamurai*と*A. i. komiyai*を記載したが，これらは同著者により後年，ミカワオサムシの亜種へとその所属を移された（石川，1985）．このうち後者は，♂交尾器の基本的形態がミカワオサムシとは明らかに異なっている点，また野外においてミカワオサムシとの間に生殖的隔離が進んでいる可能性が高い点等に基づき，独立種*Ohomopterus komiyai*（本書での扱いは*C. (O.) komiyai*）へと昇格されている（IMURA & MATSUNAGA, 2010）．ISHIKAWA & KUBOTA（1984）は志摩半島北部からミカワオサムシの亜種*murakii*を記載し，更に主として外部形態と指状片の形態に基づく本種の再検討を行い，*hidaosa*と*minoensis*の2亜種を記載した（ISHIKAWA & KUBOTA, 1994b）．21世紀に入り，三重県南部の御浜町から本種の孤立した小集団が発見され，KUBOTA & YAHIRO（2003）により*C. (O.) a. kirimurai*と命名された．さらにごく最近，天竜川下流東岸の佐久間から，体長，指状片とも著しく小型化した特異な小集団が発見され，*shichinoi*という亜種小名の下に記載されている（IMURA & MATSUNAGA, 2010）．

タイプ産地 三河地方（愛知県東部）．

タイプ標本の所在 アムステルダム動物学博物館［レクトタイプ，♂］．原記載論文中ではホロタイプが指定されておらず，ISHIKAWA & KUBOTA（1984）により "Typus // *arrowianus* BREUG. t. BREUNING c. // v. BODEMEYER Kobe Japan" というラベルを持つ♂がレクトタイプに指定されている．

形態 21〜31 mm．背面の色彩は一般に赤銅色から暗銅色で緑色光沢の乏しいものが多いが，黒色系や緑色系，更に青味を帯びた個体も一定比率で出現する．脛節と跗節は通常黒色だが，亜種によっては暗赤褐色となる．外部形態：図1〜6．交尾器形態：図7〜17．既述の中〜大型オホモプテルス属各種に似るが，以下の特徴を有する：1）♂触角の先端は鞘翅中央に達するか，同所を僅かに越す；2）前胸背板側縁後方のえぐれは弱く，直線状に近いものが多い；3）鞘翅は長卵形で，最広部は中央より後方にあり，肩部の張り出しが比較的強い；4）鞘翅第一次凹陥は比較的大きく，隣接する第三次間室を侵す；5）陰茎先端はアオオサムシのそれに近く太短いが，腹側への湾曲がはるかに弱い；6）指状片は細長い鉤状ないし棍棒状で，先端は内嚢の左側に向けて伸長し，亜種によっては背腹方向に押圧される．先端部の形状は亜種ないし個体によって波曲，肥大，左側の一部突出といった変化が見られる；7）♀交尾器膣底部節片内板は栗の実形で，後縁は中央の両側でややぐれ，内側壁は平滑．

分子系統 オホモプテルス亜属の分子系統の概要についてはヤマトオサムシの同項を参照．6種の核遺伝子配列を用いて構築された系統樹上で，本種はイワワキーアオオサムシ種群に属する（SOTA & NAGATA, 2008）．本種及び近縁種の分子系統，系統地理学的研究の詳細に関してはSOTA *et al.* (2000)，曽田（2000c, 2003），SOTA（2002b），SOTA *et al.* (2001)，IMURA, AKITA *et al.* (2005)，冨永ら（2005），UJIIE *et al.* (2005)，オオオサムシ属の分子系統研究グループ（2005, 2006a），SU *et al.* (2006) 等を参照されたい．

分布 本州（中部〜近畿東部）／竹島，佐久島，日間賀島，篠島（以上，愛知県），答志島（三重県）．

生息環境 海岸近くの沖積層や河川敷，河岸段丘から，ある程度の標高を有する基盤山地にまで広く生息しているが，近畿オサムシ研究グループ（1979）によれば生息地の主体は第四期層であるという．天竜川を始めとする大河川の下流域や河口近くの海岸線に沿った地域等では，河川流及び海流に乗って漂着したと思われる集団が定着している例も多く見られる．

生態 他のオホモプテルス各種と同様，春繁殖型，環形動物（フトミミズ）食と思われる．

地理的変異 形態学的には計7亜種（基亜種を含む）に分類される．当初アオオサムシの亜種として記載された*komiyai*（カケガワオサムシ）は本種の一亜種とみなされることもあるが，本書ではIMURA & MATSUNAGA（2010）に従い，独立種として扱った．本種の色彩パターンとその出現頻度については，長谷川ら（2001）による愛知県，岐阜県南部，静岡県西部，長野県南部における詳細な調査がある．それによると，本種の最も基本的な色彩型は赤銅色型で，ほぼすべての産地において出現し，濃尾平野，豊橋平野，伊那谷の平坦部ではほぼこの型のみが見られるという．黒色型の出現には明らかに地域的な偏りが見られ，浜名湖の西から豊橋市にかけての太平洋沿岸を中心とした地域と，西三河地方の瀬戸市東部から豊田市北部，旧額田町（ぬかたちょう；現在の岡崎市の一部），下山村（しもやまむら；現在の岡崎市・豊田市の一部），足助町（あすけちょう），旭町（共に現在の豊田市の一部）等の，矢作川（やはぎがわ））中流域東岸にかけての二地域において高頻度に出現し，さらに知多半島の一部と奥三河地方にも断続的に出現する地域があるという．また，天竜川東岸の山地帯（ミカワオサムシ天竜川中流域亜種の分布域）においても黒色型が高頻度で出現する地域がある．緑色型は，豊橋市南部の台地などで比較的高頻度に出現するが，同型が出現するのは基本的に黒色型が優先的に出現する地域であり，黒色型の現れない地域において単独で出現することはないという．

種小名の由来 種小名はロンドン自然史博物館のGilbert John ARROW（ギルバート・ジョン・アロウ）（1873〜1938）に献名されたもの．ナゴヤオサムシという名が種和名として用いられたこともある．

34-1. *C. (O.) a. nakamurai* (ISHIKAWA, 1966) ミカワオサムシ天竜川中流域亜種

Original description *Apotomopterus insulicola nakamurai* ISHIKAWA, 1966, Bull. natn. Sci. Mus., Tokyo, 9 (1), p. 13, fig. 3.

Type locality Mt. Jinbagatayama, on borders between Komagane City and Kamiina Co., Nagano Prefecture.

Type depository National Museum of Nature and Science, Tokyo [Holotype, ♂].

Diagnosis 22〜27 mm. Body above reddish coppery often with greenish tinge, sometimes entirely blackish or dark greenish; legs black though tibiae are often dark rufous. 1) Pronotum wide, widest before middle, narrowed toward base, with lateral margins not strongly emarginate behind middle, disk strongly punctate; 2) elytra oval, widest near middle, with shoulders not distinct, striae between intervals remarkably notched, primary foveoles large, often invading adjacent tertiaries; 3) inner margin of male protibia weakly protruded and not sharply angular; 4) digitulus short, almost parallel-sided though slightly dilated toward apex in frontal view, strongly curved and depressed near base, with lateral margins strongly ridged.

Distribution Both banks of middle reaches of Riv. Tenryû-gawa (S. Nagano Pref. & NW. Shizuoka Pref.). Northern limit on the eastern bank is defined by the former Takatô-machi of Ina-shi where this race is hybridized with *C. (O.) insulicola shinano* (this hybridized population was once named *Apotomopterus insulicola pseudinsulicola* by Ishikawa (1966a)). Northern limit on the western bank is Riv. Nakatagiri-gawa, the northern side of which is a distributional range of *C. (O.) i. komaganensis*. In the southern part of the distributional range, it is intergraded to the nominotypical *arrowianus*.

Etymology Named after Toshihiko Nakamura (Tokyo) who collected the holotype and a part of the paratypes.

タイプ産地　長野県，駒ヶ根市と上伊那郡の境界上，陣馬形山．

タイプ標本の所在　国立科学博物館［ホロタイプ，♂］．

亜種の特徴　22~27 mm．背面は赤銅褐色で，緑色を帯びることが多く，光沢はやや強い．黒色型や暗緑色型も出現する．脚は黒色だが，脛節はしばしば赤みを帯びる．1）前胸背板は幅広く，最広部は中央よりやや前方にあり，後方に向けて強く狭まるが側縁後方の内湾は弱い．前胸背板表面の点刻は顕著；2）鞘翅は卵型で，ほぼ中央で最も幅広く，肩部の張り出しは弱い．間室間線条の刻み目は顕著で，第一次回陥は大きく，しばしば隣接する第三次間室を侵蝕する；3）♂前脛節内側縁中央の張り出しは弱く，角張らない；4）♂交尾器指状片は本種としては短く，基部で強く湾曲し，同部で背腹方向に押圧され，側縁は顕著に稜をなす．両側縁は背面から見てほぼ平行だが，先端に向け軽く広がる．

分布　天竜川中流域の東西両岸（長野県伊那市旧高遠町・駒ヶ根市〜静岡県北西部）．東岸における北限は伊那市の旧高遠町で，同所においてアオオサムシと交雑しており，アオオサムシの亜種として伊那市の野底から記載された*Apotomopterus insulicola pseudinsulicola* Ishikawa, 1966は，これら両者の雑種集団に対して与えられた名称とされている．西岸における北限は駒ヶ根市で，中田切川を境として南に本亜種，北にアオオサムシ駒ヶ根亜種が分布する．分布域の南西部では基亜種へと移行する．同南東部では佐久間亜種へと移行するが，一部でカケガワオサムシ山住峠亜種とも分布を接している．

亜種小名の由来と通称　亜種小名は，タイプ標本を採集した東京の昆虫愛好家（専門は主としてカミキリムシ類），中村俊彦に因む．通称はテンリュウオサムシ．

34-2. *C. (O.) a. shichinoi* (Imura et Matsunaga, 2010) ミカワオサムシ佐久間亜種

Original description *Ohomopterus arrowianus shichinoi* Imura & Matsunaga, 2010, Coléoptères, Paris, 16 (14), p. 143, figs. 6-a~c (p. 145).

Type locality Japan, central Honshu, Shizuoka Pref., Hamamatsu-shi, Tenryû-ku, Miné, 500 m, on the western slope of Mt. Atago-yama.

Type depository Muséum Nationale d'Histoire Naturelle, Paris [Holotype, ♂].

Diagnosis 21~25 mm. Dorsal surface coppery to dark reddish coppery, sometimes with faint greenish tinge on head and/or pronotum; entirely dark greenish or blackish individuals are seldom found; tibiae and tarsi rufous except for apical part of each segment which is dark brown. Most closely allied to subsp. *nakamurai*, but readily distinguishable from that race by the following points: 1) size much smaller; 2) digitulus also much smaller and different in shape, flatter, spatulate and hardly twisted in apical portion, with dorsal margin not remarkably ridged nor longitudinally grooved but gently swollen, basal plate relatively large.

Distribution Miné and its nearby regions on Mt. Atago-yama, in the western part of Shizuoka Pref., SC. Honshu. Narrowly restricted to the high altitudinal area of the Atago-yama Massif lying between the Kôchi-gawa and Misakubo-gawa Rivers. The southern margin is defined by the left (= eastern) bank of the main course of the Tenryû-gawa, northern limit so far known is Minami-noda of Okuryôké.

Etymology Named after Yoshihiko Shichino, a member of the Japan Association for the Research of Carabid Beetles, who first collected this subspecies from the type locality.

タイプ産地　日本，本州中部，静岡県，浜松市，天竜区，愛宕山西麓，峰，500 m．

タイプ標本の所在　パリ自然史博物館［ホロタイプ，♂］．

亜種の特徴　21~25 mm．背面は銅褐色ないし赤銅色で頭部と前胸背板側縁は時に緑色光沢を帯びる．全身濃緑色ないし黒色の個体も稀に見られる．脛節と跗節は赤褐色だが，各節の先端部は黒褐色．天竜川中流域亜種に近いが，はるかに小型で，指状片はより短く，その伸長部分はより扁平で先端部でねじれず，外側は緩やかに膨隆して稜や縦溝を形成せず，基板が相対的に大きい，等の諸点により識別は容易である．

分布　愛宕山山塊の高所，峰集落とその周辺（静岡県浜松市天竜区佐久間町佐久間）．河内川と水窪川（いずれも天竜川の支流）に挟まれた地域に分布し，南限は天竜川本流左岸，北は奥領家の南野田付近まで分布することが確認されている．さらにその北方の大津峠付近にかけての地域では，天竜川中流域亜種への移行を示す個体が得られている．

亜種小名の由来と通称　亜種小名は，タイプ産地の峰から初めて本亜種を採集した日本オサムシ研究会の七野芳彦に因む．ヒメミカワオサムシという通称が提唱されている．

34-3. *C. (O.) a. hidaosa* Ishikawa et Kubota, 1994　ミカワオサムシ岐阜県中北部亜種

Original description　*Carabus (Ohomopterus) arrowianus hidaosa* Ishikawa & Kubota, 1994, Bull. Biogeogr. Soc. Japan, Tokyo, 49 (2), p. 124, figs. 85~99 (p. 125).
Type locality　Ohbora-chô (more precisely, Ôbora-machi), 660m. alt., Takayama-shi.
Type depository　Systematic Zoology Laboratory, Department of Biological Sciences, Tokyo Metropolitan University (Hachiôji) [Holotype, ♂].
Diagnosis　22~26 mm. Body above brownish coppery, often bearing greenish tinge; entirely black form is rarely found in the northern part of the distributional range; tibiae and tarsi usually black, though rarely dark rufous. 1) Pronotum cordiform, widest at middle, narrowed therefrom toward base, with hind angles triangularly produced, disk coarsely and often very irregularly punctate; 2) elytra long oval, widest around middle, striae between intervals very weakly notched and primary foveoles small, so that elytral surface appears rather smooth; 3) inner margin of male protibia weakly protruded and not angulated; 4) digitulus relatively short, more or less dilated near tip in lateral view.
Distribution　C.~N. Gifu Pref. Known from the area between Nagara-gawa and Hida-gawa Rivers in the south, from the area drained by the Miya-gawa River (= upper courses of the Jintsû-gawa River) in the north.
Etymology　The subspecific name *hidaosa* is an abbreviated Japanese name of this race, meaning "the carabid beetle of the Hida region".

タイプ産地　高山市，大洞町（おおぼらまち），標高660m.
タイプ標本の所在　首都大学東京動物系統分類学研究室（八王子）[ホロタイプ，♂].
亜種の特徴　22~26 mm. 背面は銅褐色で，頭胸部と鞘翅側縁は時に緑色光沢を帯びる．分布域北部では稀に辺縁に青色を帯びた黒色型も出現する．脛節と跗節は赤味を帯びる．1) 前胸背板は幅広く，最広部はほぼ中央にあり，辺縁は後方に向けて狭まり，心臓形の輪郭を呈する．後角は三角形に突出する．表面の点刻は密で，しばしば非常に不規則；2) 鞘翅は長卵形で，ほぼ中央で最も幅広く，間室間線条の刻み目はごく弱く，間室表面は滑らかで，第一次凹陥は小さい；3) ♂前脛節内側縁の突出は弱く，中央部で角張らない；4) ♂交尾器指状片はやや短めで，側面から見て先端付近で多少とも拡がる．
分布　岐阜県中北部．岐阜県南部亜種の分布域北方に当たる岐阜県中～北部を占める亜種で，分布域南部では長良川と飛騨川に挟まれた地域に，また北部では宮川（＝神通川上流）流域一帯から知られている．
亜種小名の由来と通称　通称はヒダオサムシで，亜種小名はその略称に由来する．

34-4. *C. (O.) a. arrowianus* (Breuning, 1934)　ミカワオサムシ基亜種

Original description, type locality & type depository　See explanation of the species.
Diagnosis　24~31 mm. Body above reddish coppery or dark brown, sometimes entirely black or greenish; legs black though tibiae and tarsi often partly dark rufous. Pronotum wide, widest near middle, with lateral sides sinuately narrowed toward base; hind angles wide and robust, weakly protruded posteriad. Elytra long oval, widest near middle, with lateral sides nearly parallel-sided in median portion above all in male; disk rather smooth, with striae between intervals weakly notched; primary foveoles variable in size, sometimes invading adjacent tertiaries. Inner margin of male protibia weakly protruded. Digitulus longer than that of subsp. *nakamurai*, above all in apical portion of digitate part, with basal portion less strongly curved in lateral view, lateral margins not carinate, apical portion often with a small accessory projection on left side.
Distribution　Nôbi-heiya Plan and its nearby regions (SE.Gifu Pref.~SW.Nagano Pref.~Aichi Pref.~W.Shizuoka Pref.), SC. Honshu / Iss. Také-shima, Saku-shima, Himaka-jima, Shino-jima (Aichi Pref.).

原記載・タイプ産地・タイプ標本の所在　種の項を参照.
亜種の特徴　24~31 mm. 背面は銅色ないし暗銅色で，緑色光沢はあってもごく僅か．黒化型や緑化型，青味を帯びた型も出現し，地域により各色彩型の出現頻度が異なる．脚は黒色だが，脛節と跗節は部分的に暗赤色を帯びる．前胸背板は広く，最広部は中央にあり，側縁後方はほぼ直線状ないし僅かに内湾し，後方に向けて波曲する．後角は幅広く，後方への突出は弱い．鞘翅は長卵形で，ほぼ中央で最も幅広く，側縁は中央部で左右ほぼ平行．表面の彫刻は比較的滑らかで，間室間線条の刻み目は弱い．第一次凹陥の大きさには変化があり，大きい場合には隣接する第三次間室を侵蝕する．♂前脛節内側縁は中央でごく弱く張り出す．♂交尾器指状片は前2亜種より長く，とりわけ遠位端における伸張が天竜川中流域亜種のそれに比し顕著．側面から見ると基部の湾曲がより緩やかで，側縁は陵をなさない．先端部は背面から見て左側方にやや張り出す．
分布　愛知県，岐阜県南東部，長野県南西部，静岡県西部／竹島，佐久島，日間賀島，篠島（以上，愛知県）．濃尾平野を中心とする一帯に分布し，西～北西限は木曽川で，同河川が岐阜南部亜種との境界となる．南は知多・渥美両半島部に達し，分布

34-5. *C. (O.) a. minoensis* Ishikawa et Kubota, 1994 ミカワオサムシ岐阜県南部亜種

Original description *Carabus* (*Ohomopterus*) *arrowianus minoensis* Ishikawa & Kubota, 1994, Bull. Biogeogr. Soc. Japan, Tokyo, 49 (2), p. 123, figs. 74~84 (p. 124).
Type locality Mt. Kinkazan, 140 m. alt., Gifu-shi.
Type depository Systematic Zoology Laboratory, Department of Biological Sciences, Tokyo Metropolitan University (Hachiôji) [Holotype, ♂].
Diagnosis 24~30 mm. Body above light or dark coppery brown often with greenish tinge; legs black, though rarely dark rufous. Most closely allied to subsp. *arrowianus*, but discriminated from it in differently shaped digitulus whose apical part is not so remarkably dilated laterad as in *arrowianus*, and usually lacking small accessory projection on left side near apex.
Distribution S. Gifu Pref., C. Honshu (areas between the lower courses of Kiso-gawa and Nagara-gawa Rivers).
Etymology Named after its main distributional range, the Mino area in the southern part of Gifu Prefecture.

タイプ産地　岐阜市, 金華山, 標高 140 m.
タイプ標本の所在　首都大学東京動物系統分類学研究室（八王子）［ホロタイプ, ♂］.
亜種の特徴　24~30 mm. 背面は明銅色ないし銅褐色で, しばしば緑色光沢を帯びる. 脚は黒色だが, 稀に暗赤色. ミカワオサムシ基亜種に最も近いが, 主として指状片先端の側方への広がりが弱く, 先端左側が突出しないことにより識別される.
分布　岐阜県南部（木曽川と長良川の下流域に挟まれた地域）.
亜種小名の由来と通称　亜種小名は, 本亜種が岐阜県南部の美濃地方に分布することに因む. 通称はミノオサムシ.

34-6. *C. (O.) a. murakii* Ishikawa et Kubota, 1984 ミカワオサムシ志摩半島北部亜種

Original description *Carabus* (*Ohomopterus*) *arrowianus murakii* Ishikawa & Kubota, 1984, Akitu, Kyoto, New Series, (65), p. 2, figs. 1~10 (p. 3).
Type locality Shô, 3 m. alt., Futami-chô, Watarai-gun, Mie Prefecture (= Futami-chô-shô of Isé-shi at present).
Type depository According to the original description, the holotype (♂) of this taxon is preserved in the collection of R. Ishikawa. Now it may be preserved in the Systematic Zoology Laboratory, Department of Biological Sciences, Tokyo Metropolitan University (Hachiôji).
Diagnosis 25~30 mm. Body above dull brownish coppery often with greenish tinge, or entirely black or dark green; legs black. Pronotum wide, with sides roundly arcuate and less strongly sinuate toward base, hind angles very short. Elyra elongate oval, widest apparently behind middle. Characterized by uniquely shaped digitulus which is the longest and thinnest in this species, with apical portion nearly cylindrical.
Distribution Northern part of Shima Penins. (flatland around the river mouth of Kushida-gawa & Miya-gawa) in Mie Pref., SW. Honshu / Is. Tôshi-jima (Mie Pref.).
Etymology Named after Takenori Muraki.

タイプ産地　三重県, 度会郡, 二見町, 荘, 標高 3 m（現在は伊勢市二見町荘）.
タイプ標本の所在　原記載ではホロタイプ（♂）は石川良輔コレクション保管となっている. 現在は首都大学東京動物系統分類学研究室（八王子）に保管されている可能性が高いが, 確認できなかった.
亜種の特徴　25~30 mm. 背面は鈍い暗銅褐色で, 辺縁はしばしば弱い緑色の光沢を帯びる. 黒色あるいは暗黒緑色の型も一定比率で出現する. 脚は黒色. 前胸背板は幅広く, 辺縁の丸みが強く, 側縁後方の波曲は弱く, 後角は短い. 鞘翅は長卵形で, 最広部は明らかに中央より後方にある. ♂交尾器指状片は本種中最も細長く, シンプルな棒状で, 側面から見て先端付近で僅かに波曲する場合もあるが, 板状に押圧されたり突出する部分は殆ど見られない.
分布　志摩半島北部（櫛田川および宮川の河口付近の平野部）／答志島（三重県）.
亜種小名の由来と通称　亜種小名は村木武則に因む. 通称はイセオサムシ.

34-7. *C. (O.) a. kirimurai* Kubota et Yahiro, 2003 ミカワオサムシ御浜町亜種

Original description *Carabus* (*Ohomopterus*) *arrowianus kirimurai* Kubota & Yahiro, 2003, Biogeography, (5), p. 9, fig. 3 (p. 11).
Type locality Kamiichigi, 20 m alt., Mihama-cho, Minamimuro-gun, Mie Prefecture.
Type depository Lake Biwa Museum (Shiga Pref.) [Holotype, ♂].

34. *Carabus* (*Ohomopterus*) *arrowianus*

Diagnosis 23~27 mm. Body above brownish coppery; tibiae and tarsi usually reddish. Pronotum broad, widest before middle, sides roundly arcuate near widest part and slightly narrowed behind, hind angles subtriangularly protruded; disk densely or rather irregularly punctate. Elytra long oval, widest behind middle, with surface of intervals smooth, striae between intervals weakly notched, primary foveoles large, often invading adjacent tertiaries. Inner margin of male protibia weakly angulated. Digitulus shorter than that of nominotypical *arrowianus*, subsp. *minoensis* and subsp. *murakii*, median portion of digitate part a little broader in dorsal view, more straight in lateral view, a little depressed dorso-ventrally, with left margin a little protruded before apex, and irregularly wrinkled on surface. Inner plate of ligular apophysis of female genitalia with anterior margin hardly protruding, looks more like that of *C.* (*O.*) *iwawakianus* than that of nominotypical *arrowianus*.

Distribution　Mihama-chô (SW. Mie Pref.).

Conservation　This subspecies is designated as the threatened species (CR; critically endangered species) in the 4th edition of Red List (Threatened Wildlife of Japan) of Ministry of the Environment (2012).

Etymology　Named after Nobuyuki Kirimura who discovered this subspecies.

タイプ産地　三重県, 南牟婁郡, 御浜町, 上市木, 標高20 m.

タイプ標本の所在　琵琶湖博物館［ホロタイプ, ♂］.

亜種の特徴　23~27 mm. 背面は銅褐色で, 脛節と跗節は赤味を帯びる. 前胸背板は幅広く, 中央より前方で最も幅広く, 辺縁は最広部において弧を描きつつ強く張り出し, 後方に向けて僅かに狭まる. 後角は亜三角形に突出する. 表面の点刻は密ないしやや不規則. 鞘翅は長卵形で, 中央よりも後方において最も幅広く, 間室間線条の刻み目は弱く, 間室表面は滑らか. 第一次凹陥は大きく, しばしば隣接する第三次間室を侵す. ♂前脛節内側縁は弱く角張る. ♂交尾器指状片は基亜種, 岐阜県南部亜種, 志摩半島北部亜種などに比しやや短く, 指状の伸張部分は背面から見てやや幅が広く, 側面から見てより直線的で, 背腹方向にやや押圧され, 先端よりやや手前で左側縁が僅かに張り出し, 表面には不規則な皺を装う. ♀交尾器膣底部節片内板は栗の実形にはならず, 前縁がより直線的で, イワワキオサムシのそれに近い.

分布　御浜町(三重県南西部). 三重県の南西端に近い御浜町の, 海岸に近い丘陵地のごく狭い範囲のみから知られており, 個体数も多くない. 原記載論文の中で本亜種の著者らは, 一見不自然に見える孤立した分布状況から, 海流に乗って岐阜・愛知方面から漂着したものがごく狭い範囲内に定着している可能性を述べているが, 紀伊半島東部における沿岸流は, 南西から北東へ, すなわち御浜町沖からミカワオサムシの母集団が分布する伊勢湾方面に向かっているため, 流れとしては逆方向で, この推論には無理がある. むしろ, 植物の根土や土壌などについて移入されたものが定着しているといった人為分布の可能性を考慮に入れる必要があるかもしれない.

保全　本亜種は, 環境省第4次レッドリスト(2012)で絶滅危惧IA類(カテゴリー CR; ごく近い将来における野生での絶滅の危険性が極めて高いもの)に指定されている.

亜種小名の由来と通称　亜種小名は本亜種を発見した桐村信行に因む. 通称はミハマオサムシ.

34. *Carabus* (*Ohomopterus*) *arrowianus*

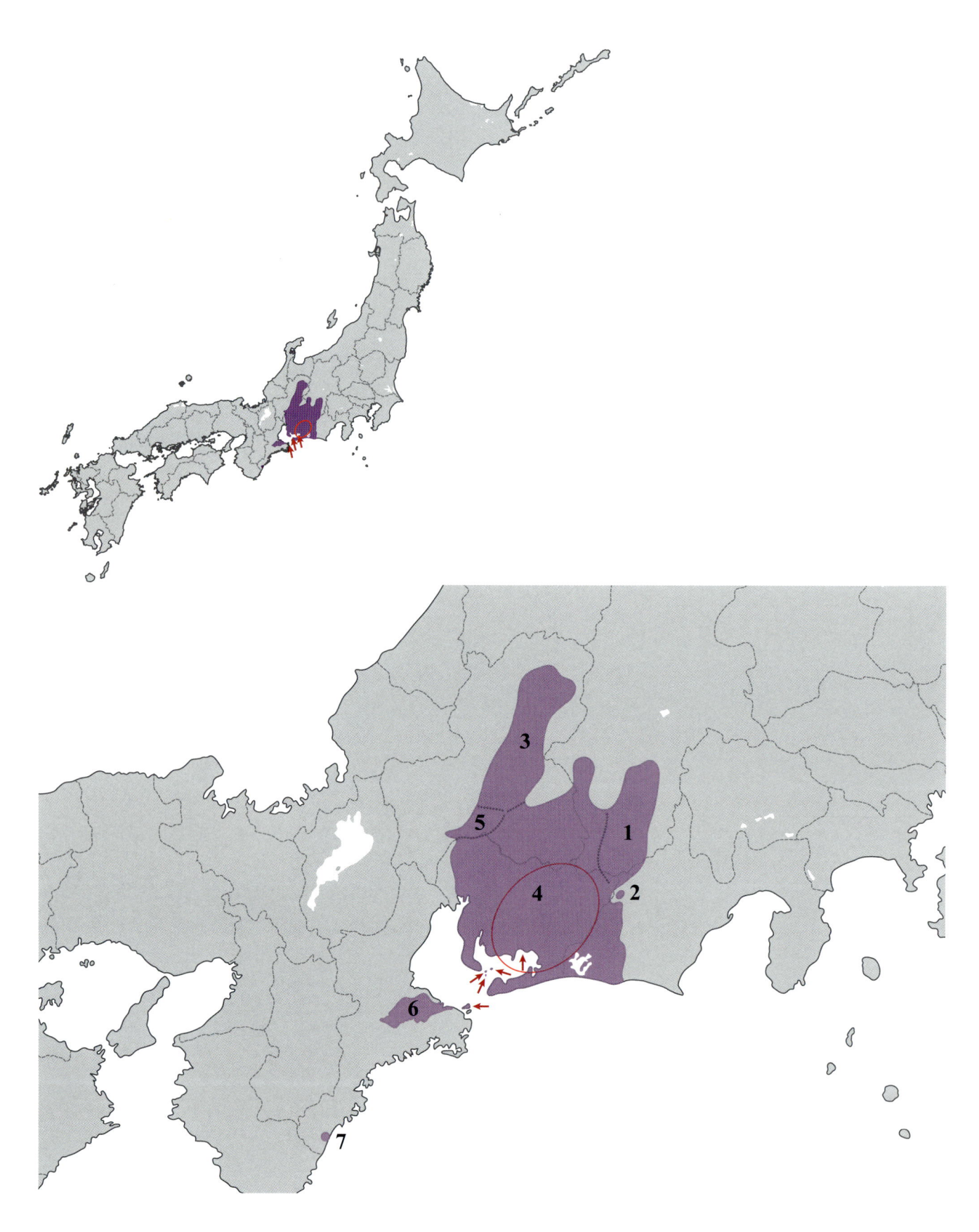

Fig. 30. Map showing the distributional range of ***Carabus*** (***Ohomopterus***) ***arrowianus*** (Breuning, 1934). — 1, Subsp. ***nakamurai***; 2, subsp. ***shichinoi***; 3, subsp. ***hidaosa***; 4, subsp. ***arrowianus***; 5, subsp. ***minoensis***; 6, subsp. ***murakii***; 7, subsp. ***kirimurai***. ◯: Type area of the species (Mikawa Region).

図 30. ミカワオサムシ分布図. — 1, 天竜川中流域亜種; 2, 佐久間亜種; 3, 岐阜県中北部亜種; 4, 基亜種; 5, 岐阜県南部亜種; 6, 志摩半島北部亜種; 7, 御浜町亜種. ◯: 種のタイプ地域（三河地方）.

34. *Carabus* (*Ohomopterus*) *arrowianus*

Fig. 31. Type locality of *Carabus* (*Ohomopterus*) *arrowianus shichinoi* (Miné, Tenryû-ku, Hamamatsu-shi, Shizuoka Pref.).

図31. ミカワオサムシ佐久間亜種のタイプ産地(静岡県浜松市天竜区峰).

Fig. 32. *Carabus* (*Ohomopterus*) *arrowianus shichinoi* passing the winter in the soil (Miné, Tenryû-ku, Hamamatsu-shi, Shizuoka Pref.).

図32. 土中で越冬しているミカワオサムシ佐久間亜種(静岡県浜松市天竜区峰).

Fig. 33. Type locality of *Carabus* (*Ohomopterus*) *arrowianus murakii* (Mishiodono-jinja Shrine, Futami-chô-shô, Isé-shi, Mie Pref., taken in 1987).

図33. ミカワオサムシ志摩半島北部亜種のタイプ産地(三重県伊勢市二見町荘御塩殿神社, 1987年撮影).

35. *Carabus (Ohomopterus) maiyasanus* Bates, 1873
マヤサンオサムシ

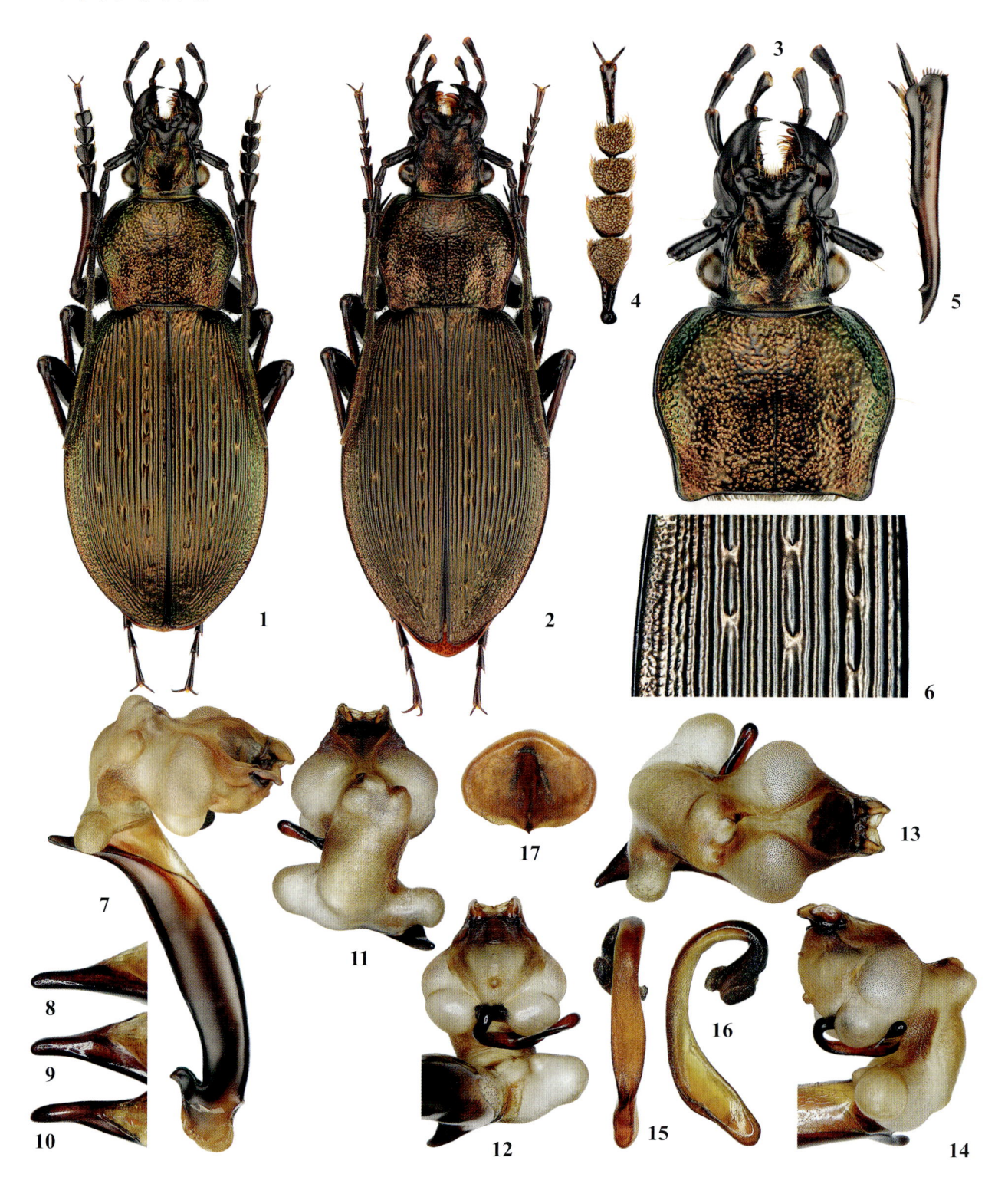

Figs. 1-17. *Carabus* (*Ohomopterus*) *maiyasanus maiyasanus* Bates, 1873 (Mt. Maya-san, Kôbé-shi, Hyogo Pref.). — 1, ♂; 2, ♀; 3, male head & pronotum; 4, male right protarsus in ventral view; 5, male right protibia in subdorsal view; 6, male left elytron (median part); 7, penis & fully everted internal sac in right lateral view; 8, apical part of penis in right lateral view; 9, ditto in right subdorsal view; 10, ditto in dorsal view; 11, internal sac in caudal view; 12, ditto in frontal view; 13, ditto in dorsal view; 14, ditto in left sublateral view; 15, digitulus in frontal view; 16, ditto in left lateral view; 17, inner plate of ligular apophysis.

図 1-17. **マヤサンオサムシ基亜種**(兵庫県神戸市摩耶山). — 1, ♂; 2, ♀; 3, ♂の頭部と前胸背板; 4, ♂右前跗節(腹面観); 5, ♂右前脛節(斜背面観); 6, ♂左鞘翅中央部; 7, 陰茎と完全反転下の内嚢(右側面観); 8, 陰茎先端(右側面観); 9, 同(右斜背面観); 10, 同(背面観); 11, 内嚢(尾側面観); 12, 同(頭側面観); 13, 同(背面観); 14, 同(左斜側面観); 15, 指状片(頭側面観); 16, 同(左側面観); 17, 膣底部節片内板.

35. *Carabus* (*Ohomopterus*) *maiyasanus*

Figs. 18-29. *Carabus* (*Ohomopterus*) *maiyasanus* subspp. — 18-19, *C*. (*O*.) *m. hokurikuensis* (18, ♂, 19, ♀; Mt. Shiro-yama, Asahi-machi, Toyama Pref.); 20-21, *C*. (*O*.) *m. yoroensis* (20, ♂, 21, ♀; Mt. Tado-yama, Kuwana-shi, Mie Pref.); 22-23, *C*. (*O*.) *m. suzukanus* (22, ♂, 23, ♀; Nagasawa-shinden, Suzuka-shi, Mie Pref.); 24-25, *C*. (*O*.) *m. shigaraki* (24, ♂, 25, ♀; Mt. Aboshi-yama, Rittô-shi, Shiga Pref.); 26-27, *C*. (*O*.) *m. takiharensis* (26, ♂, 27, ♀; Shimo-suga, Ôdai-chô, Mie Pref.); 28-29, *C*. (*O*.) *m. ohkawai* (28, ♂, 29, ♀; Ago-chô-ugata, Shima-shi, Mie Pref.). — a, Digitulus in frontal view; b, ditto in left lateral view.

図 18-29. マヤサンオサムシ各亜種． — 18-19, 北陸地方亜種(18, ♂, 19, ♀; 富山県朝日町城山); 20-21, 養老山地亜種(20, ♂, 21, ♀; 三重県桑名市多度山); 22-23, 鈴鹿山脈南東部亜種(22, ♂, 23, ♀; 三重県鈴鹿市長沢新田); 24-25, 信楽亜種(24, ♂, 25, ♀; 滋賀県栗東市阿星山); 26-27, 滝原亜種(26, ♂, 27, ♀; 三重県大台町下菅); 28-29, 鵜方亜種(28, ♂, 29, ♀; 三重県志摩市阿児町鵜方). — a, 指状片（頭側面観）; b, 同（左側面観）.

35. *Carabus* (*Ohomopterus*) *maiyasanus* BATES, 1873 マヤサンオサムシ

Original description *Carabus Maiyasanus* BATES, 1873, Trans. ent. Soc. Lond., 1873, p. 232.

History of research This species was described by the British coleopterologist H. W. BATES (1873) based on the specimens collected by G. LEWIS. Most of the medium-sized *Ohomopterus* species in Japan had been temporarily identified as BATES' race until NAKANE revised all the Japanese Carabina based on the male genitalic characters from the 1950s to the 1960s. NAKANE (1968) described *Apotomopterus insulicola ohkawai* from Ugata of Mie Prefecture, which is now regarded as a subspecies of *C.* (*O.*) *maiyasanus*. HIURA & KATSURA (1971) described *A. maiyasanus shigaraki* from the Shigaraki Hills of Shiga Prefecture. KATSURA & TOMINAGA (1978) described *Ohomopterus maiyasanus takiharensis* from the Takihara area of Mie Prefecture. ISHIKAWA & KUBOTA (1994b) revised *C.* (*O.*) *maiyasanus*, and additionally described three more subspecies, namely, *hokurikuensis*, *yoroensis* and *suzukanus*.

Type locality Moon-temple (Maiyasan), Kôbé; alt. 2000 feet. ("Maiyasan" means Mt. Maya-san of Kôbé-shi, Hyogo Pref.)

Type depository Museum Nationale d'Histoire Naturelle, Paris [Lectotype, ♂].

Morphology 21~32 mm. Body above variable in color, coppery, green or black; tibiae and tarsi black or dark rufous. External features: Figs. 1~6. Genitalic features: Figs. 7~17. Similar to other medium- to large-sized *Ohomopterus* species, above all to *C.* (*O.*) *arrowianus* or *C.* (*O.*) *iwawakianus*, but discriminated from them in the following characteristics: 1) antennae relatively short, with apices usually not reaching middle of elytra even in male; 2) ventral side of male antennae with hairless areas from segment five to seven; 3) pronotum broad, usually with two pairs of marginal setae; 4) elytral intervals smooth, with primary foveoles small and barely invading adjacent tertiaries, striae between intervals rather remarkably notched; 5) inner margin of male protibia protruded in various degree; 6) apical part of penis slender, most closely allied to that of *C.* (*O.*) *esakii*; 7) digitulus as in *C.* (*O.*) *insulicola*, hook-like and long, with apical part compressed bilaterad to form vertical plate, though its apex turns toward left side of internal sac, ventral side weakly sclerotized and light-colored; 8) internal sac short and robust; 9) inner plate of ligular apophysis transverse chestnut-shaped, with apex not strongly protruded and obtusely rounded, inner wall smooth.

Molecular phylogeny For the general overview of the molecular phylogeny of *Ohomopterus*, see the same section of *C.* (*O.*) *yamato*. For details on this species, see Research group of molecular phylogeny on the *Ohomopterus* ground beetles (2006b).

Distribution WC. Honshu / Is. Noto-jima (Ishikawa Pref.).

Habitat Distributed on hills and mountains consisting of the basement rock, and usually not found from an alluvium, though some subspecies such as *ohkawai* inhabit the terrace surface. Common in the edge of rather humid forest, though also found from dry, rather thin forest.

Bionomical notes Spring breeder, earthworm feeder.

Geographical variation Morphologically classified into seven subspecies.

Etymology Named after the type locality, Mt. Maya-san of Kôbé in Hyogo Prefecture, which was wrongly spelled as "Maiyasan" by BATES.

研究史　英国人採集家LEWIS（ルイス）の採集品に基づいて，同国のBATES（ベイツ）(1873)により神戸の摩耶山から記載された種で，1950~1960年代に中根により♂交尾器を用いた分類が取り入れられるまで，我が国に産する中型のオホモプテルス類（当時の認識としてはアポトモプテルス類）の多くが本種と同定されていたという．NAKANE(1968)は志摩半島の鵜方からアオオサムシの亜種として*Apotomopterus insulicola ohkawai*を記載したが，今日ではマヤサンオサムシの一亜種として扱われている．日浦・桂(1971)は信楽山地からマヤサンオサムシの亜種として*A. maiyasanus shigaraki*を記載し，桂・冨永(1978；論文は桂・冨永・日浦・土井・春沢・谷の6名による共著)は三重県の滝原地方から同じくマヤサンオサムシの亜種として*Ohomopterus maiyasanus takiharensis*を記載した．ISHIKAWA & KUBOTA(1994b)は外部形態並びに♂交尾器指状片の形態に基づいて本種（彼らの扱いは*Carabus* (*Ohomopterus*) *maiyasanus*）に再検討を加え，*hokurikuensis*, *yoroensis*, *suzukanus*の3亜種を記載した．

タイプ産地　ムーンテンプル（まいやさん），神戸，標高2000フィート（=現在の兵庫県神戸市摩耶山）．

タイプ標本の所在　パリ自然史博物館［レクトタイプ，♂］．

形態　21~32 mm. 背面の色彩は比較的多彩で，銅色系，緑色系，黒色系のものが出現する．脛節と附節は黒色ないし暗赤褐色．外部形態：図1~6. 交尾器形態：図7~17. 既述の中~大型オホモプテルス亜属各種，とりわけミカワオサムシやイワワキオサムシ等に似るが，以下の特徴により識別される：1)触角は短く，♂でも先端が鞘翅中央に達しない場合が多い；2)♂触角第5節から第7節の腹面に無毛部がある；3)前胸背板は幅広く，通常2対の側縁剛毛を有する；4)鞘翅間室は平滑で，第一次凹陥は小さく，隣接する第三次間室を殆ど侵さず，間室間線条の刻み目は比較的顕著；5)♂前脛節内側縁の突出と角張りの程度は亜種により変化する；6)陰茎先端は前3種のそれよりはるかに細長く，シズオカオサムシのものに最も近い；7)指状片は細長い鉤状で，先端部は側圧され，垂直板を形成し，概形はアオオサムシのそれに似るが，先端が内嚢の左側に向かって伸長する点で大きく異なる．また，指状片の内側縁に沿った部分と垂直板を形成する先端部分は硬化が弱く，淡色で，乾燥標本では変形が生じることがある；8)内嚢は太短く，先端部の横幅が広い；9)♀交尾器膣底部節片内板は横に長い栗の実形で，前方中央の突出は弱く，先端は鈍く丸まり，内側壁は比較的平滑．

分子系統 オホモプテルス亜属の分子系統の概要についてはヤマトオサムシの同項を参照．ミトコンドリア遺伝子及び核DNAの解析から，本種は同所的，側所的に生息するオホモプテルス亜属の他種（オオオサムシ，ヤコンオサムシ，イワワキオサムシ等）との間に交雑を起し，近畿地方の各地で交雑由来集団の形成に大きな役割を果たしたと考えられている．詳細はオオオサムシ属の分子系統研究グループ（2006b）等を参照．

分布 本州中西部／能登島（石川県）．

生息環境 山間の小盆地を除き，基本的には基盤山地に生息しており，沖積層には見られないが，高位段丘面上に生息している場合もある（鵜方亜種等）．一般に湿潤な森林を好むが，比較的乾いた疎林やスギ・ヒノキ等の植林にも見られる．

生態 春繁殖，成虫越冬型のオサムシで，幼虫は基本的に環形動物（主としてフトミミズ類）食であろうと思われる．

地理的変異 形態学的には7亜種に分類される多型種とみなされているが，分子系統解析の結果，これらマヤサンオサムシの各"亜種"は，純系種の地理的隔離によって生じたものと言うよりも，その多くが多種との交雑により生じた雑種由来の集団であるらしいことが判明している．

種小名の由来 タイプ産地の名に由来する．

35-1. *C. (O.) m. hokurikuensis* Ishikawa et Kubota, 1994 マヤサンオサムシ北陸地方亜種

Original description *Carabus (Ohomopterus) maiyasanus hokurikuensis* Ishikawa & Kubota, 1994, Bull. Biogeogr. Soc. Japan, Tokyo, 49 (2), p. 108, figs. 15~28 (p. 109).

Type locality Mt. Shiroyama, 100-120m. alt., near Miyazaki, Asahi-machi, Toyama Pref.

Type depository Systematic Zoology Laboratory, Department of Biological Sciences, Tokyo Metropolitan University (Hachiôji) [Holotype, ♂].

Diagnosis 21~29 mm. Color as in nominotypical *maiyasanus*. Pronotum robust, widest before middle, with lateral sides strongly and sinuately narrowed posteriad, hind angles very short, only weakly produced posteriad. Elytra oval, widest at middle; intervals smooth, with striae regularly and weakly notched; primary foveae large, invading adjacent tertiaries. Inner margin of male protibia not angulated but weakly convex at middle. Digitulus of male genitalia simpler than that of nominotypical *maiyasanus*, elongated rod-like in general view, with apical portion weakly bent inward, weakly dilated vertically where it is weakly compressed laterally and rather distinctly concave.

Distribution This subspecies occupies the northernmost part of the distributional rage of *C. (O.) maiyasanus*, extended down south in Gifu Prefecture to Gifu-shi, where it is defined in the east by Riv. Nagara-gawa from the range of *C. (O.) arrowianus*, and is extended by the Sea of Japan so far north to Itoigawa-shi in Niigata Prefecture.

Etymology Named after the main distributional range of this subspecies, the Hokuriku Region.

タイプ産地 富山県，朝日町，宮崎近郊，城山，標高100-120m．

タイプ標本の所在 首都大学東京動物系統分類学研究室（八王子）［ホロタイプ，♂］．

亜種の特徴 21~29 mm．背面の色彩は基亜種のそれに準じる．前胸背板は太短い心臓形で，最広部は中央より前方にあり，側縁は波曲しつつ後方に向けて強く狭まる．後角は短く，ごく僅かに後方へ突出する程度．鞘翅は卵形で，最広部はほぼ中央にあり，間室は平滑で間室間の線条は規則的に弱い刻み目を有する．第一次凹陥は大きく，隣接する第三次間室を侵触する．♂前脛節内側縁中央は軽く膨らむが角張らない．♂交尾器指状片は基亜種のそれより単純な形をしており，棒状に近く，先端部の湾曲とヘラ状の広がりは共に弱いが，側面の押圧は比較的顕著．

分布 基亜種の分布域東～北東方（南東方は岐阜市付近まで分布し，東限は長良川で，北東方は日本海沿いに新潟県糸魚川市付近まで分布する）．

亜種小名の由来と通称 亜種小名は主な分布域である北陸地方に因む．ホクリクオサムシという通称があるが，ヤコンオサムシ北陸地方亜種にも同じ名が提唱されたことがあるため，紛らわしい．

35-2. *C. (O.) m. maiyasanus* Bates, 1873 マヤサンオサムシ基亜種

Original description, type locality & type depository See explanation of the species.

Diagnosis 21~29 mm. Body above usually brownish coppery, often with greenish tinge on pronotum and elytral margins; some individuals are black with weak bluish or greenish tinge, or entirely greenish; tibiae rufous or dark rufous. Pronotum relatively small, a little longer than wide, widest at or before middle, with lateral sides rather evenly arcuate in front, sinuately narrowed posteriad behind middle, with hind angles very weakly protruded. Elytra elongated oval, widest near middle, with intervals smooth, primary foveae small, striae between intervals regularly and weakly notched. Inner margin of male protibia weakly angulated at middle. Digitulus of male genitalia with basal portion short, median portion weakly depressed, strongly curved inward near base, and parallel to basal piece in lateral view, apical portion curved at base in various degree, strongly compressed laterad to form vertical plate which is apparently concave left laterad.

Distribution NE. Chugoku Region, CN. Kinki Region & SW. Hokuriku Region.

原記載・タイプ産地・タイプ標本の所在　種の項を参照.
亜種の特徴　21〜29 mm. 背面は通常，銅褐色で，前胸背板や鞘翅の側縁部はしばしば緑色光沢を帯びる．黒色で弱い青色もしくは緑色光沢を帯びるものや緑化型も出現する．脛節と跗節は赤ないし暗赤色．前胸背板は小さく，僅かに縦長で，最広部は中央ないしそのやや前方にあり，側縁は前方では均等に弧を描いて丸く突出し，後方に向けて波曲しつつ狭まり，後角の突出はごく弱い．鞘翅は長卵形で，ほぼ中央で最も幅広く，側縁の基部に向けての狭まりは弱い．各間室は平滑で，第一次凹陥は小さく，間室間線条には規則的に弱い横の刻み目を有する．♂前脛節内側縁中央部の角張りは弱い．♂交尾器指状片は基部が短く，中央葉は基部近くで軽く押圧され，同部で強く湾曲し，側面から見て中部で基片とほぼ平行になり，先端部は基部において様々な程度に湾曲し，側方から強く圧され，左側面は明らかに窪み，垂直方向に明らかに広がる．
分布　中国地方北東部〜近畿地方中・北部〜北陸地方南西部．日本海側では鳥取県（西限は鳥取市南西部の佐治町(さじちょう)から岡山県境の恩原高原(おんばらこうげん)にかけての一帯と思われる）から福井県にかけて産し，福井県から南方に向けては琵琶湖東岸を経て三重県北部に達し，養老山地亜種および鈴鹿山脈南東部亜種に変化する．鈴鹿山脈南東部亜種との分布境界は移行地帯があるためはっきりとしないが，養老山地亜種とは牧田川(まきたがわ)（北部）および員弁川(いなべがわ)（南部）を境に明瞭に棲み分ける（ただし，これら河川近くの集団には，交雑によると思われる移行的あるいは中間的個体が含まれる）．

35-3. *C. (O.) m. yoroensis* Ishikawa et Kubota, 1994　マヤサンオサムシ養老山地亜種

Original description　*Carabus* (*Ohomopterus*) *maiyasanus yoroensis* Ishikawa & Kubota, 1994, Bull. Biogeogr. Soc. Japan, Tokyo, 49 (2), p. 110, figs. 29〜34 (p. 111).
Type locality　Mt. Tadoyama, 150m. alt., Tado-chô, Kuwana-gun, Mie Pref. (it belongs to Kuwana-shi at present).
Type depository　Systematic Zoology Laboratory, Department of Biological Sciences, Tokyo Metropolitan University (Hachiôji) [Holotype, ♂].
Diagnosis　23〜28 mm. Body above not strongly shiny, brownish coppery sometimes bearing greenish tinge on head, pronotum and elytral margins, or rarely entirely blackish or greenish; tibiae and tarsi partly dark rufous. Pronotum usually rather wide, with sides strongly and cordately sinuate, hind angles weakly produced posteriad. Elytra broad oval, widest behind middle, with shoulders not distinct, disk rather smooth; striae between intervals weakly and regularly notched; primary foveoles large, slightly invading adjacent tertiaries. Inner margin of male protibia bluntly angulated or only weakly convex. Digitulus of male genitalia with basal plate shorter, basal portion of digitate part thick, somewhat depressed, strongly arcuate, nowhere parallel with basal plate in lateral view, distal portion long, moderately bent inward, dilated vertically and weakly compressed left laterad.
Distribution　Yôrô Mts. (Gifu Pref. / Mie Pref.). Distributed in the area between the Ibi-gawa and Inabé-gawa Rivers.
Etymology　Named after the type locality (Yôrô Mts.).

タイプ産地　三重県，桑名郡，多度町，多度山，標高150m（現在は桑名市に属する）．
タイプ標本の所在　首都大学東京動物系統分類学研究室（八王子）[ホロタイプ，♂]．
亜種の特徴　23〜28 mm. 背面は通常銅褐色で，頭部，前胸背板，鞘翅側縁は時に緑色光沢を帯び，光沢は比較的鈍い．黒化型（体側縁に僅かに青色味を帯びる），或いはごく稀に緑化型も出現する．脛節と跗節は暗赤色を帯びる．前胸背板は比較的幅広く，側縁は強く波曲し，後角の後方への突出は弱い．鞘翅はやや幅の広い卵形で，肩の張り出しは弱く，最広部は中央より後方にあり，間室は滑らかで，間室間の線条は規則的に弱い刻み目を有する．第一次凹陥はやや大きく，隣接する第三次間室を僅かに侵蝕する．♂前脛節内側縁中央は鈍く角張るか，軽く膨らむ程度．♂交尾器指状片は基板が短く，指状部の基部は厚く，背腹方向に僅かに圧され，強く湾曲し，側面から見て基板と平行な部分はなく，先端部は長く，内湾はそれほど強くなく，垂直方向にやや広がり，左側方から弱く押圧される．
分布　養老山地（岐阜県〜三重県）．揖斐川と員弁川に挟まれた地域に産する．
亜種小名の由来と通称　養老山地に因む．ヨウロウオサムシという通称で呼ばれることがある．

35-4. *C. (O.) m. suzukanus* Ishikawa et Kubota, 1994　マヤサンオサムシ鈴鹿山脈南東部亜種

Original description　*Carabus* (*Ohomopterus*) *maiyasanus suzukanus* Ishikawa & Kubota, 1994, Bull. Biogeogr. Soc. Japan, Tokyo, 49 (2), p. 111, figs. 36〜43 (p. 113).
Type locality　Nagasawashinden, 90m. alt., Suzuka-shi, Mie Prefecture.
Type depository　Systematic Zoology Laboratory, Department of Biological Sciences, Tokyo Metropolitan University (Hachiôji) [Holotype, ♂].
Diagnosis　22〜32 mm. Body above brownish coppery, entirely green or black bearing greenish or bluish tinge along margins. Legs black or dark rufous. Pronotum relatively small, with lateral margins weakly sinuate, hind angles triangularly protruded with obtusely rounded tips. Elytra widest behind middle, with intervals smooth, primary foveoles small, and striae between intervals weakly notched. Inner margin of male protibia bluntly but distinctly angulated. Digitulus with basal plate long, digitate part strongly

depressed dorso-ventrally, strongly arcuate in lateral view, median portion rather short, apical portion very long, membraneous and pale-colored on ventral side, apparently hooked at apical third, strongly compressed laterad, and tapered toward rounded apex.

Distribution　SE. Suzuka Mts. (N. Mie Pref.).
Etymology　Named after the type locality (Suzuka Mts.).

タイプ産地　三重県, 鈴鹿市, 長沢新田, 標高90 m.
タイプ標本の所在　首都大学東京動物系統分類学研究室(八王子)[ホロタイプ, ♂].
亜種の特徴　22~32 mm. 背面は銅褐色で, 稀に微かな緑色光沢を帯びる. 分布域の北部では黒化型(側縁に弱い青色ないし緑色光沢を帯びる)の出現頻度が高い. 時に全身緑色の個体も出現する. 脚は黒色ないし暗赤色. 前胸背板は比較的小さく, 側縁の波曲は弱い. 後角は三角形に突出するが先端は鈍く丸まる. 鞘翅は中央より後方で最も幅広く, 間室は平滑で, 間室間の線条は弱い刻み目を有し, 第一次凹陥は小さい. ♂前脛節内側縁は比較的明瞭に角張る. ♂交尾器指状片は基板が長く, 指状部の基部は背腹方向から強く押圧され, 湾曲は強く, 基板に平行となる中央部の距離は短く, 先端部は非常に長く, 薄い膜状で, 腹面は明色になり, 先端1/3付近で屈曲し, 側方から強く圧され, 垂直方向にやや広がり, 丸い先端に向けて僅かに狭まる.
分布　鈴鹿山脈南東麓(菰野町(こものちょう)～四日市市～鈴鹿市). 鈴鹿山脈の南東麓(鈴鹿川支流の御幣川(おんべがわ)以北)に分布し, 湯の山温泉から朝明渓谷(あさけいこく)にかけての地域で基亜種に移行する. なお, 多賀(2002)により, やや南方に飛び離れた鈴鹿川南方の亀山市和賀町(わがちょう)と野村町にまたがる河岸段丘上部から, 本亜種に近い隔離集団が報告されているが, 人為的に移入された個体群である可能性が高いとされている.
亜種小名の由来と通称　亜種小名は鈴鹿山脈に因む. スズカオサムシという通称で呼ばれることがある.

35-5. *C. (O.) m. shigaraki* (Hiura et Katsura, 1971) マヤサンオサムシ信楽亜種

Original description　*Apotomopterus maiyasanus shigaraki* Hiura et Katsura, 1971, Bull. Osaka Mus. nat. Hist., (24), p. 15, figs. V, W & X (pp. 18~19).
Type locality　Western slope of Mt. Aboshiyama, Kurita-gun, Shiga Prefecture (= Rittô-shi of Shiga Pref. at present).
Type depository　Osaka Museum of Natural History [Holotype, ♂].
Diagnosis　22~27 mm. Body above brownish coppery sometimes with greenish tinge, black with weak bluish tinge, or entirely greenish. Black-colored individuals are more frequently found than in other subspecies. Tibiae usually rufous with darker tips, occasionally entirely black. Pronotum widest before middle, with sides weakly sinuate, hind angles short. Elytra rather smooth, with sides less roundly arcuate than in other subspecies; striae between intervals faintly notched; primary foveoles more or less invading adjacent tertiaries, inner margin of male protibia strongly angulated. Digitulus of male genitalia with basal plate longer, digitate part strongly curved near base, distinctly depressed from basal to median portion, nearly straight in median portion in lateral view, then remarkably hooked inward near apical third to form vertically dilated apical portion which is thin, largely pale colored and weakly compressed left laterad.
Distribution　Shigaraki mountain range (S. Shiga Pref.).
Etymology　Named after the Shigaraki mountain range, the main distribution range of this subspecies.

タイプ産地　滋賀県, 栗太郡, 阿星山西麓(現滋賀県栗東市).
タイプ標本の所在　大阪市立自然史博物館[ホロタイプ, ♂].
亜種の特徴　22~27 mm. 背面は銅褐色で, 時に緑色光沢を帯びる. 他の亜種に比べ黒化型(側縁部には通常, 弱い青色光沢を帯びる)の出現頻度が高く, 時に全身が緑銅色の個体も出現する. 脛節は通常, 赤みを帯びる. 前胸背板の最広部は中央より前方にあり, 側縁の波曲は弱く, 後角の突出も弱い. 鞘翅は側縁がより直線的で, 間室は平滑, 間室間の線条はごく弱い刻み目を有するにすぎない. 第一次凹陥は隣接する第三次間室を多少とも侵触する. ♂前脛節内側縁は強く角張る. ♂交尾器指状片は基板が長く, 基部で強く内湾し, 基部から中部にかけて著しく押圧され, 先端1/3近くで内側に屈曲し, 先端部は薄く, 膜様部分を多く含み, 腹面は明色で, 側方からの押圧は比較的弱い.
分布　信楽山地(滋賀県南部).
亜種小名の由来と通称　亜種小名は信楽山地(ないし信楽地方)に因む. 通称はシガラキオサムシ.

35-6. *C. (O.) m. takiharensis* (Katsura et Tominaga, 1978) マヤサンオサムシ滝原亜種

Original description　*Ohomopterus maiyasanus takiharensis* Katsura et Tominaga, 1978, Bull. Osaka Mus. nat. Hist., (31), p. 47, figs. 49~71 (p. 59).
Type locality　Shimosuga, Ohdai-cho, Taki-gun, Mie Prefecture.
Type depository　Osaka Museum of Natural History [Holotype, ♂].
Diagnosis　21~26 mm. Body above coppery and shiny, often with remarkable greenish tinge on head, pronotum and marginal area of elytra; tibiae and tarsi dark reddish. Pronotum relatively small, widest around or a little before middle, strongly narrowed

toward base, with hind angles weakly protruded posteriad. Elytra long oval, widest behind middle, with shoulders effaced. Striae between elytral intervals strongly notched, primary foveoles remarkably carved, invading adjacent tertiaries. Inner margin of male protibia strongly protruded though not sharply angulated. Digitulus similar to that of subsp. *ohkawai*, but less strongly hooked and dilated at apical portion.

Distribution　So-called Takihara area (the area between the Miya-gawa River & its tributary, the Ôuchiyama-gawa River in SC. Mie Pref.).

Etymology　Named after its main distributional range, the Takihara area.

タイプ産地　三重県, 多気郡, 大台町, 下菅.

タイプ標本の所在　大阪市立自然史博物館［ホロタイプ, ♂］.

亜種の特徴　21〜26 mm. 背面は光沢の強い銅褐色ないし明銅色で, 頭胸部と鞘翅側縁にはしばしば顕著な緑色光沢を帯びる. 脛節と跗節は暗赤色. 前胸背板は比較的小型で, 最広部は中央ないしやや前方にあり, 側縁は後方へ向けて強く狭まる. 後角の後方への突出は弱い. 鞘翅は長卵形で, 中央より後方で最も幅広く, 肩の張り出しは弱い. 鞘翅は間室間線条の刻み目が強いため, やや凹凸が顕著で, 第一次凹陥はよく目立ち, 隣接する第三次間室を侵蝕する. ♂前脛節内側縁は鈍いながら顕著に張り出す. ♂交尾器指状片は鵜方亜種のそれに近いが, 先端部における屈曲と広がりがやや弱い.

分布　三重県中南部, 宮川中流南岸〜大内山川下流西岸（大台町〜大紀町(たいきちょう)）. 志摩半島の中北部を東西に流れる宮川とその一支流である大内山川に挟まれたごく狭い範囲（滝原地方）のみに分布する.

亜種小名の由来と通称　亜種小名は本亜種が分布する滝原地方に因む. 通称はタキハラオサムシ.

35-7. *C. (O.) m. ohkawai* (Nakane, 1968)　マヤサンオサムシ鵜方亜種

Original description　*Apotomopterus insulicola ohkawai* Nakane, 1968, Fragm, coleopterol., Kyoto, (21), p. 85.

Type locality　Ukata [sic] (misreading of Ugata), Shima, Mie Pref., Honshu, Japan (= Ugata of Ago-chô in Shima-shi at present).

Type depository　Hokkaido University Museum (Sapporo) [Holotype, ♂].

Diagnosis　24〜30 mm. Body above coppery with faint greenish tinge; tibiae black though rarely dark reddish brown. Size larger on an average, with body rather flat. Pronotum relatively large and wide, widest nearly at middle, with lateral margins almost evenly arcuate, their posterior halves almost straight or faintly emarginated, hind angles short, robust and rounded at tips; pronotal disk coarsely punctate and transversely wrinkled. Elytra not strongly convex above, with prominent shoulders, striae between intervals strongly notched, primary foveoles small. Inner margin of male protibiae bluntly but distinctly protruded. Digitulus with basal plate long, basal portion of digitate part strongly depressed, median portion almost straight and almost parallel to basal plate, apical portion thin and elongated, slightly dilated in lateral view and twisted; apical portion of digitulus is somewhat membraneous, pale-colored and easily deformed in dried condition.

Distribution　Narrowly restricted to Ugata & its nearby regions (Ugata in Ago-chô and a part of Isobé-chô, SE. Shima Penins., Mie Pref.).

Habitat　Evergreen broadleaf forest in and around the town area of Ugata which is built on the marine terrace near the tip of the Shima Peninsula.

Conservation　The distributional range of this subspecies is narrowly limited to Ugata of Mie Prefecture. It was abundant until the first half of the 1990s, but has become much rarer, particularly over the past few years, probably due to the decrease of habitable environment and excessive hunting by collectors. This subspecies is designated as threatened species (VU; vulnerable species) in the 4th edition of Red List (Threatened Wildlife of Japan) of Ministry of the Environment (2012).

Etymology　Named after Chikao Ohkawa, a medical doctor and an amateur entomologist who first collected the specimen of this subspecies in the premises of Mie Prefectural Shima Hospital.

タイプ産地　日本, 本州, 三重県, 志摩, ウカタ（鵜方(ウガタ)の誤記）（現三重県志摩市阿児町鵜方）. ホロタイプは, 三重県の医師, 大川親雄により志摩病院において採集されたもので, ホロタイプのラベルにも採集地は「志摩病院」（正式名称は三重県立志摩病院）と明記されている. 原記載では「鵜方」の読み方が「Ukataうかた」と誤記されているが,「うがた」と濁るのが正しい.

タイプ標本の所在　北海道大学総合博物館（札幌）［ホロタイプ, ♂］.

亜種の特徴　24〜30 mm. 背面は銅褐色で, 光沢は強くなく, 斜めから見ると僅かに緑色の光沢を帯びる. 脛節は通常黒色だが, 稀に暗赤色. 体つきは幅広く, 平たい. 前胸背板は大きく幅広く, 最広部はほぼ中央にあり, 側縁はほぼ均等に弧を描いて張り出し, 後半部はほぼ直線状ないし僅かに内湾する. 後角は鈍角で先端は丸まり, 後方への突出は弱い. 前胸背板表面は密な点刻と横皺を装う. 鞘翅は幅広く, 膨隆は弱く, 肩部はよく張り出し, ♂ではほぼ長方形に近い輪郭を呈する. 間室には横の刻み目を有し, 間室間の線条にも強い刻み目を有する. 第一次凹陥は小さい. ♂前脛節の内側縁は鈍いながら顕著に張り出す. ♂交尾器指状片は基板が長く, 指状部分の基部は背腹方向に顕著に押圧され, 中央部は直線状で基板とほぼ平行で, 先端部は薄く長く, 側方から見て僅かに広がり, 軽く屈曲する. 同部は大部分が膜状で, 腹面から見ると明色を呈する. 指状片先端の腹面が膜状であるため, 乾燥標本にすると同部が萎縮し, 指状片の変形が著しくなる.

35. *Carabus* (*Ohomopterus*) *maiyasanus*

分布　志摩半島南東部（鵜方付近）．志摩半島の先端部に近い阿児町鵜方から磯部町にかけてのごく限られた範囲のみに分布する．

生息環境　鵜方の市街地に点在する社寺林や照葉樹林の林床に生息する．個体数の多かった1990年代前半頃までは，民家の庭先にある狭小な草付きなどでも比較的簡単に見られた．

保全　前述の如く，本亜種の分布域は鵜方の市街地にほぼ一致した狭い範囲に限られているため，市街地の拡大と都市基盤の整備に伴い，生息地は年々狭められている．1990年代前半頃までは比較的多くの個体が見られたが，生息環境の狭小化，悪化に加え，愛好家の採集圧により近年，著しくその数を減じており，環境省第4次レッドリスト（2012）では絶滅危惧II類（カテゴリーVU）に指定されている．

亜種小名の由来と通称　亜種小名は本亜種の発見者である三重県の医師，大川親雄に因む．通称はウガタオサムシ．

35. *Carabus* (*Ohomopterus*) *maiyasanus*

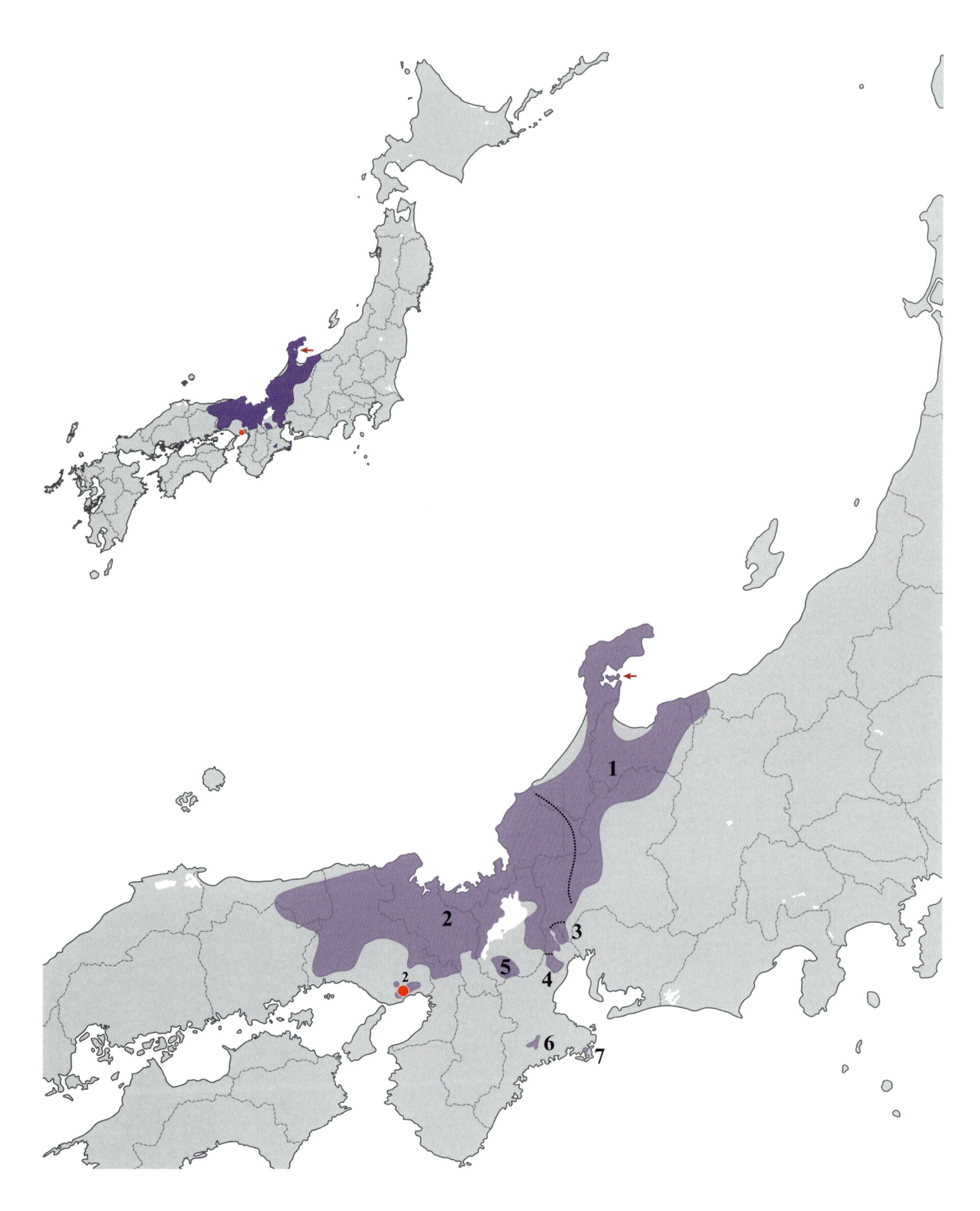

Fig. 30. Map showing the distributional range of ***Carabus*** (***Ohomopterus***) ***maiyasanus*** Bates, 1873. — 1, Subsp. ***hokurikuensis***; 2, subsp. ***maiyasanus***; 3, subsp. ***yoroensis***; 4, subsp. ***suzukanus***; 5, subsp. ***shigaraki***; 6, subsp. ***takiharensis***; 7, subsp. ***ohkawai***. ● : Type locality of the species (Maiyasan = Mt. Maya-san).

図 30. **マヤサンオサムシ**分布図. — 1, 北陸地方亜種; 2, 基亜種; 3, 養老山地亜種; 4, 鈴鹿山脈南東部亜種; 5, 信楽亜種; 6, 滝原亜種; 7, 鵜方亜種. ● : 種のタイプ産地 (摩耶山).

35. *Carabus* (*Ohomopterus*) *maiyasanus*

Fig. 31. Type locality of *Carabus* (*Ohomopterus*) *maiyasanus* (the top of Mt. Maya-san, Nada-ku, Kôbé-shi, Hyogo Pref.).

図31. マヤサンオサムシのタイプ産地(兵庫県神戸市灘区摩耶山).

Fig. 32. *Carabus* (*Ohomopterus*) *maiyasanus maiyasanus* passing the winter in the soil (Mt. Shakunagé-yama, Kita-ku, Kôbé-shi, Hyogo Pref.).

図32. 土中で越冬しているマヤサンオサムシ基亜種(兵庫県神戸市北区石楠花山).

Fig. 33. *Carabus* (*Ohomopterus*) *maiyasanus maiyasanus* (Mt. Maya-san, Nada-ku, Kôbé-shi, Hyogo Pref.).

図33. マヤサンオサムシ基亜種(兵庫県神戸市灘区摩耶山).

36. *Carabus* (*Megodontus*) *kolbei*

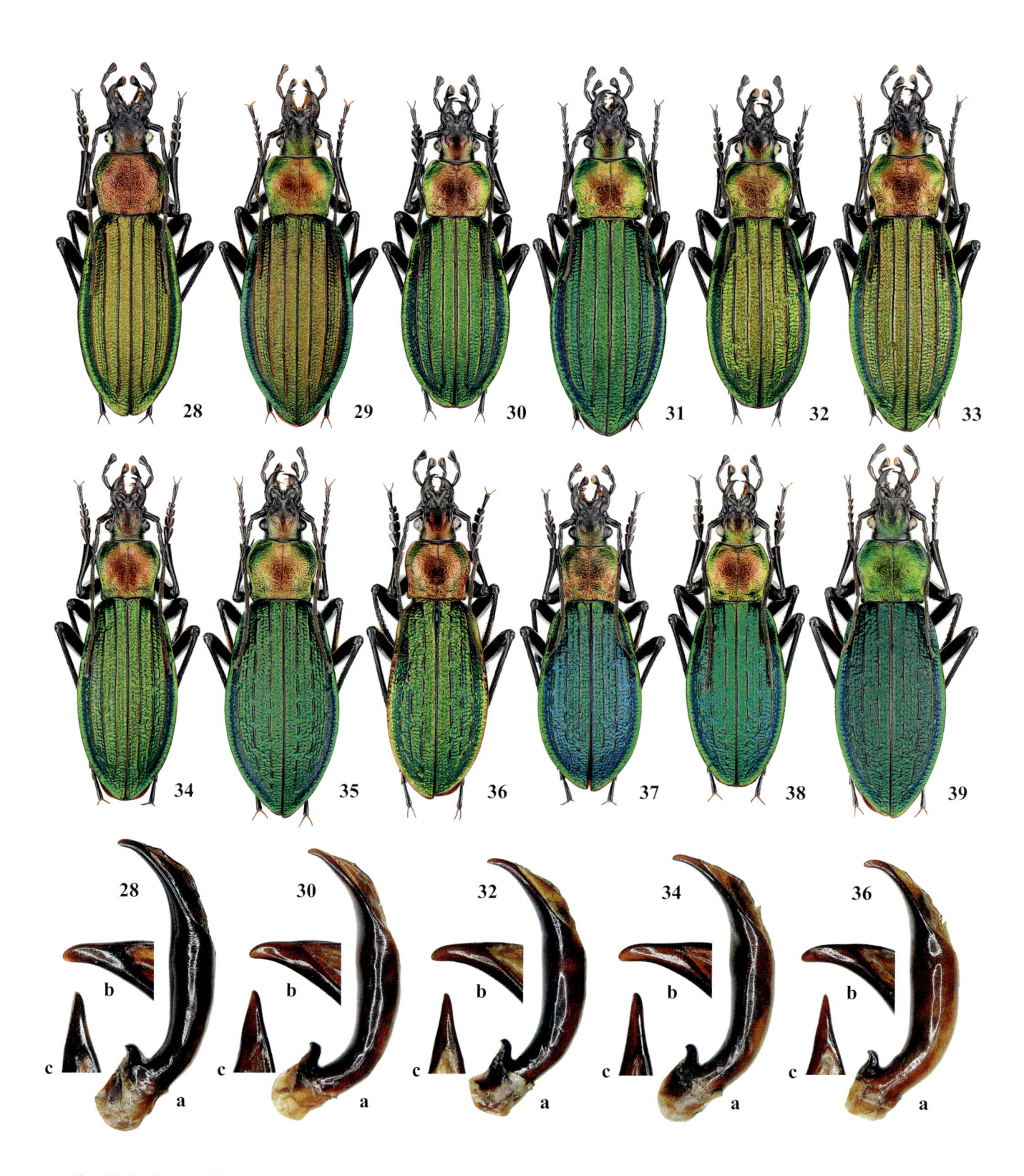

Figs. 28-39. *Carabus* (*Megodontus*) *kolbei* subspp. — 28-29, *C*. (*M*.) *k. mitsumasai* (28, ♂, HT, 29, ♀, PT; Hattari, Atsuta-ku-atsuta, Ishikari-shi); 30-31, *C*. (*M*.) *k. nitidipunctatus* (30, ♂, 31, ♀; Kami-tomamu, Shimukappu-mura); 32-33, *C*. (*M*.) *k. yubariensis* (32, ♂, 33, ♀; Mt. Yûbari-daké, Yûbari-shi); 34-35, *C*. (*M*.) *k. taikinus* (34, ♂, 35, ♀; Taiki-chô); 36-39, *C*. (*M*.) *k. hidakamontanus* (36-38, ♂, 39, ♀; 36, Riv. Tottabetsu-gawa, Obihiro-shi; 37, Mt. Petegari-daké, Shin-hidaka-chô; 38-39, Mt. Apoi-daké, Samani-chô). — a, Penis in right lateral view; b, apical part of penis in right subdorsal view; c, ditto in dorsal view. All from Hokkaido.

図 28-39. アイヌキンオサムシ各亜種. — 28-29, 増毛山地南西部亜種 (28, ♂, HT, 29, ♀, PT; 石狩市厚田区厚田発足); 30-31, トマム亜種 (30, ♂, 31, ♀; 占冠村上トマム); 32-33, 夕張山地亜種 (32, ♂, 33, ♀; 夕張市夕張岳); 34-35, 大樹町亜種 (34, ♂, 35, ♀; 大樹町); 36-39, 日高山脈亜種 (36-38, ♂, 39, ♀; 36, 帯広市戸蔦別川; 37, 新ひだか町ペテガリ岳; 38-39, 様似町アポイ岳). — a, 陰茎 (右側面観); b, 陰茎先端部 (右斜背面観); c, 同 (背面観). すべて北海道産.

36. *Carabus* (*Megodontus*) *kolbei*

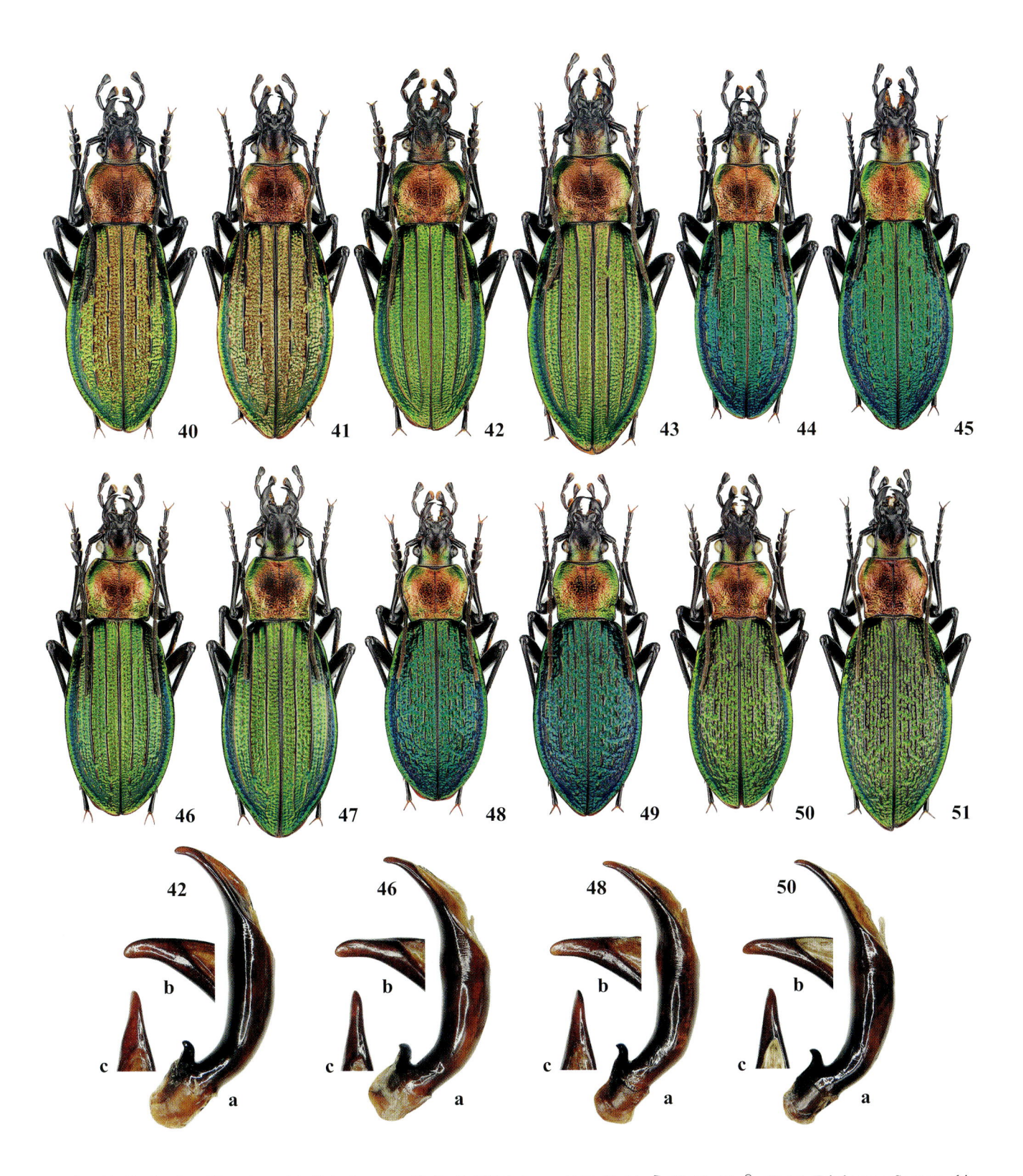

Figs. 40-51. *Carabus* (*Megodontus*) *kolbei* subspp. — 40-45, *C*. (*M*.) *k. kosugei* (40, 42, 44, ♂, 41, 43, 45, ♀; 40, Mt. Teiné-yama, Sapporo-shi; 41, Mt. Yoichi-daké, Akaigawa-mura; 42-43, Nagahashi, Otaru-shi; 44-45, Mt. Shakotan-daké, Shakotan-chô); 46-47, *C*. (*M*.) *k. kuniakii* (46, ♂, 47, ♀; Mt. Iwao-nupuri, Kucchan-chô); 48-49, *C*. (*M*.) *k. yasudai* (48, ♂, 49, ♀; Mt. Kariba-yama, Shimamaki-mura); 50-51, *C*. (*M*.) *k. munakataorum* (50, ♂, 51, ♀; Mt. Daisengen-daké, Matsumae-chô / Kami-no-kuni-chô). — a, Penis in right lateral view; b, apical part of penis in right subdorsal view; c, ditto in dorsal view. All from Hokkaido.

図 40-51. アイヌキンオサムシ各亜種. — 40-45, 積丹半島亜種(40, 42, 44, ♂, 41, 43, 45, ♀; 40, 札幌市手稲山; 41, 赤井川村余市岳; 42-43, 小樽市長橋; 44-45, 積丹町積丹岳); 46-47, ニセコ亜種(46, ♂, 47, ♀; 倶知安町イワオヌプリ); 48-49, 狩場山亜種(48, ♂, 49, ♀; 島牧村狩場山); 50-51, 大千軒岳亜種(50, ♂, 51, ♀; 松前町/上ノ国町大千軒岳). — a, 陰茎(右側面観); b, 陰茎先端部(右斜背面観); c, 同(背面観). すべて北海道産.

36. *Carabus (Megodontus) kolbei* Roeschke, 1897 アイヌキンオサムシ

Original description *Carabus kolbei* Roeschke, 1897, Ent. Nachr., Berlin, p. 116.

History of research This is one of the most beautiful species representing Hokkaido, as well as *Carabus (Acoptolabrus) gehinii*. It was described as *Carabus aino* by Rost (1908), and this name had been adopted by most authors for a long time. Nakane (1955) regarded it as belonging to the subgenus *Megodontus* of the genus *Procrustes*, and described a new subspecies under the name of *P. (M.) aino kosugei* from Otaru. Another subspecies, *chishimanus*, was described also by Nakane (1960) from Is. Etorofu (= Iturup) of South Kurils. Ishikawa followed Nakane's view as to the generic and subgeneric names, and described seven more subspecies, *nitidipunctatus*, *yubariensis*, *hidakamontanus* and *futabae* in 1966, *munakataorum* in 1969, and *kuniakii* and *yasudai* in 1971. Mandl (1975) examined the holotype of *Carabus kolbei* described by Roeschke (1897) from "China", and realized that so-called *C. aino* is identical with Roeschke's species. Since *C. kolbei* precedes *C. aino* in the year of description, the former must be used as the specific name of this taxon. Based on this view, Ishikawa (1985) adopted *Procrustes (Pachycranion) kolbei* as the scientific name of this taxon, and described a new subspecies under the name of *taikinus*. Two more subspecies were described by Imura, that is, *Carabus (Megodontus) kolbei hanatanii* from Is. Rishiri-tô in 1991 and *C. (M.) k. mitsumasai* from the Mashiké Mountains in 1994.

Type locality China. This locality is most probably erroneous. The holotype is assumed to be collected from the Pacific coast area of the Hidaka Region in south-central Hokkaido.

Type depository Museum für Naturkunde der Humboldt-Universität zu Berlin [Holotype, ♀].

Morphology 19~29 mm. Body above reddish coppery, yellowish coppery, green, blue or bluish purple with strong metallic luster. External features: Figs. 1~6. Description on details: penultimate segment of labial palpus multisetose, with four to seven setae; median tooth of mentum strongly protruded antero-ventrad; submentum asetose; male antennae without hairless area on ventral surface; pronotal margins basically bisetose; elytra long and narrow, strongly convex above and somewhat cylindrical; preapical emargination of elytra sometimes faintly recognized in female; elytral sculpture triploid heterodyname; primaries indicated by strongly raised costae either contiguous or irregularly segmented by primary foveoles of various sizes; both secondaries and tertiaries vary from segmented costae to rows of granules; striae between intervals distinctly punctate, and the punctures are often contiguous with one another to form transverse wrinkles; propleura almost smooth, though sometimes vaguely and sporadically punctate; sides of sternites weakly rugoso-punctate; metacoxa bisetose, lacking anterior seta, or trisetose; sternal sulci unclear. Genitalic features: Figs. 7-15. Penis less than half the elytral length; ostium lobe very small but narrowly elongated dorsad, with tip faintly bilobed; internal sac also narrowly elongated, hardly arcuate at base under fully everted state, nearly rectangularly protruded dorsad, with hemispherical, hairly unilobate presaccellar lobe and remarkably protruded saccellus; parasaccellar lobe small and nearly symmetrical; peripheral rim of gonopore hardly sclerotized and only weakly pigmented. Inner plate of ligular apophysis of female genitalia (Fig. 15) peach- or walnut-shaped and strongly sclerotized. In various places of Hokkaido, coloration and sculptural condition of this species are very similar in parallel with those of *Carabus (Acoptolabrus) gehinii*, showing a typical morphological convergence between these two species.

Molecular phylogeny On the phylogenetic tree of the mitochondrial ND5 gene of the Carabina of the world, this species belongs to the Eurasian lineage of so-called Procrustimorphi (= almost corresponding to the *Procrustes* subgenus-group in this book), whose principal constituents are *Megodontus*, *Procrustes*, *Lamprostus* and *Procerus*. *Carabus (Megodontus) kolbei* shows the closest affinity to *Carabus (Megodontus) vietinghoffi* from Amur and Sakhalin, *C. (M.) imperialis* from Kazakhstan and *C. (M.) schoenherri* from Siberia, though morphologically most closely allied *C. (M.) avinovi* of Sakhalin has not yet analyzed. In the Japanese Carabina, this species is considered to belong to the category of so-called invader, being either a recent immigrant from Sakhalin or from northeastern Asia. Inter-subspecific relationship of this species has not been investigated in detail, because of insufficient number of the samples and the difficulty of analysis (Osawa et al., 2002, 2004; Kim et al., 2003, etc.).

Distribution Hokkaido / Iss. Rishiri-tô, Kunashiri-tô & Etorofu-tô (Hokkaido). Outside Japan: Iss. Ushishir & Ketoj (C. Kurils).

Habitat Usually inhabiting the forest region on hills to mountains, though also visible in lowlands (especially in northern, eastern and south-central Hokkaido), grasslands, and sometimes in subalpine meadows.

Bionomical notes Autumn breeder. The larva prefers to feed on snails. For details, see Karasawa (1992) who succeeded in the captive breeding of this species.

Geographical variation Polytypical species, classified into 14 subspecies as to the population distributed in the Japanese territory. The population inhabiting Is. Ushishir of Central Kurils is described as different subspecies named *ushishirensis* (Obydov & Saldaitis, 1996).

Etymology This species is named after the German entomologist Hermann Julius Kolbe (1855~1939), a curator at the Berlin Zoological Museum from 1890 until 1921 specializing in Coleoptera, Psocoptera and Neuroptera. In the field of Coleopterology, he was specialized in Scarabaeidae and Brenthidae. He left numerous number of writings (about 350 articles) during his lifetime, and contributed mainly to the morphology and systematics of Coleoptera.

研究史　オオルリオサムシと共に北海道を代表する美麗種の双璧をなす．Rost (ロスト) (1908) はこれを *Carabus aino* という名で記載したため，我々が「アイヌキンオサムシ」と呼んでいる種の種小名には長年，この *aino* が適用されてきた．Nakane (1955) はこ

36. *Carabus (Megodontus) kolbei*

れをプロクルステス属のメゴドントゥス亜属に置き，小樽から亜種 *kosugei* を記載した．NAKANE (1960) は更に，択捉島から亜種 *chishimanus* を記載した（ただし，択捉島にアイヌキンオサムシを産することは ROST (1908) による *Carabus aino* の原記載論文中ですでに述べられており，北海道産のものよりも幅広い個体である由の記述がある）．ISHIKAWA は属・亜属の扱いについては NAKANE に倣い，1966 年に 4 亜種 (*nitidipunctatus, yubariensis, hidakamontanus, futabae*)，1969 年に 1 亜種 (*munakataorum*)，1971 年に 2 亜種 (*kuniakii, yasudai*) を記載した．MANDL (マンドル) (1975) は ROESCHKE (ロシュケ) (1900) により「中国」から記載された *Carabus kolbei* のホロタイプを調べ，その結果，我々が *aino* という種小名をあてていた種（アイヌキンオサムシ）がこれと同種であるという事実が判明した．従って，本種の種小名には記載年において先行する *kolbei* を用いなければならず，それまで長年親しまれてきた *aino* はその 1 亜種名へと降格されることになった．因みに，ROESCHKE により指定された *Carabus kolbei* のタイプ産地 "China" は恐らく誤りで，北海道（恐らく日高地方沿岸部）で得られた標本が中国産として記載されたものであろうと考えられる．こうした経緯を踏まえ，ISHIKAWA (1985) は本種の学名として *Procrustes (Pachycranion) kolbei* を用い，1993 年には日高山脈南東部の大樹町から亜種 *taikinus* を記載した．これらの他，IMURA により 1991 年には利尻島から亜種 *hanatanii* が，1994 年には増毛山地南西部から亜種 *mitsumasai* が記載されている．

タイプ産地 中国（恐らく誤りで，実際は北海道日高地方の沿岸地域から得られたものである可能性が高い）．

タイプ標本の所在 フンボルト大学自然博物館（ベルリン）［ホロタイプ，♀］．

形態 19~29 mm．背面の色彩には（黄）赤銅色から緑色，青色，青紫色に至る変異があり，一般に強い金属光沢を具える．頭部と前胸背板は鞘翅より，またそれぞれの中央部は側縁部よりやや波長の長い色彩となる．外部形態：図 1~6．細部の記載：下唇肢末端節は多毛で 4~7 本の剛毛があり，下唇基節の中央歯は前下方に強く突出する．咽頭剛毛を欠き，♂触角腹面に無毛帯はない．前胸背板の側縁剛毛は基本的に 2 対（中央と後角付近）．鞘翅は細長く，強く膨隆し，側方と後方の傾斜が急なため，箱型ないし円筒形に近い形状を呈する．鞘翅彫刻は三元異規的で，第一次間室は連続または断続する強い隆条となり，第二次および第三次間室は不規則に断続する隆条から小顆粒列まで変化する．間室間には点刻列があり，しばしば互いに連続して細かい横皺を呈する．前胸側板にはごく弱い点刻を疎に装い，腹部腹板の側面には弱い皺と点刻を装う．後基節の剛毛は通常 2 本（前方剛毛を欠く），時に 3 本．腹部腹板の横溝は不明瞭．交尾器形態：図 7~15．陰茎は鞘翅長の 1/2 弱で，葉片は小さいが細長く伸長し，先端は僅かに分葉傾向を示す．内嚢は細長く，完全膨隆下では基部で殆ど屈曲せず，陰茎先端部の主軸と 90 度近い角度をなして背側後方よりに突出する．半球形の微毛に覆われた単葉の盤前葉と内嚢盤の背側への突出が目立ち，傍盤葉は小さいながらも左右ほぼ対称に認められる．射精孔縁膜の硬化と色素沈着はごく軽度．♀交尾器膣底部節片内板は桃の実形ないしクルミ形でよく硬化する．

※オオルリオサムシとの平行変異 同所的に生息する本種とオオルリオサムシとの間に，背面の色彩や鞘翅彫刻の収斂現象がしばしば見られることは従来からよく知られている．分布域の最南端に位置する大千軒岳，積丹半島から札幌西部の山地，日高山脈の高所，大雪山，道北，道東などでは，一般に両種の色彩，体形，鞘翅彫刻が非常によく似ており，一瞥しただけでは同定に迷う場合もあるほどである．いっぽう，渡島半島基部の狩場山，ニセコ山地，夕張山地の一部，日高山脈の低地などでは，両種の外見はかなり異なっている．また，色彩の変異幅は本種よりもオオルリオサムシの方がはるかに広いため，後者に色彩多型の見られる地域では，自ずと一部の個体同士にしか収斂現象を見いだすことはできない．このように，両種の収斂現象は必ずしも道内のあらゆる地域で，しかも全ての個体において生じているわけではなく，その意味やメカニズムに関しては今後の研究に待つところが大きい．

分子系統 ミトコンドリア ND5 遺伝子を用いた世界のオサムシ亜族の分子系統樹において，本種はヨロイオサムシ群（本書で言うプロクルステス亜属群にほぼ相当する，カラブス属の中の大群）のユーラシア系亜群に属する．同亜群はユーラシア大陸に広く分布し，本種の属するメゴドントゥス亜属を始め，プロクルステス（ヨロイオサムシ），ランプロストゥス（トックリオサムシ），プロケルス（イボハダオサムシ）などの亜属からなる大群である．系統樹上で，本種アイヌキンオサムシはサハリンやアムールの *Carabus (Megodontus) vietinghoffi*，カザフスタンの *C. (M.) imperialis*，シベリアの *C. (M.) schoenherri* などに類縁が近い（分析できた種の範囲においての話であり，メゴドントゥス亜属にはこれら以外にも多数の種が知られている）．形態学的に最も近縁と思われるサハリンの *C. (M.) avinovi* が未分析であるため，確定的なことは言えないが，邦産オサムシの中で本種は恐らく侵入種のカテゴリーに属するものであり，氷期の陸橋経由でサハリンから，或いはより古い時代にユーラシア大陸北東部から渡来したものであろうと推定されている．道内における各亜種間の関係については，分子系統解析用に得られたサンプル数が少なく，また分析自体が困難であったため，まだよく解明されていない（大澤ら，2002; OSAWA *et al*., 2004; KIM *et al*., 2003 等）．

分布 北海道／利尻島，国後島，択捉島．国外：ウシシル島，ケトイ島（千島列島中部）．国内では北海道と利尻島，さらに北方領土の国後・択捉両島に分布し，石狩低地帯より東~北部の地域では標高を問わず比較的普遍的に分布しているが，同低地帯よりも南西方では山地性となり，最南端の産地である大千軒岳では山頂付近の稜線に広がる草原のみに，また渡島半島基部の狩場山では標高およそ 1,000 m 以上の地域に生息する．道北の宗谷総合振興局等では一般に平地や低標高の丘陵地などに産するのに対し，利尻島では標高 1,400~1,500 m から山頂にかけてのハイマツ・ダケカンバ帯やお花畑といった環境に生息しており，顕著な対比を示す点は興味深い．国外では千島列島中部のケトイ島（LAFER (ラーフェル)，1989）とウシシル島からも記録されており，ウシシル島のものは *ushishirensis* という亜種名で記載されている（OBYDOV (オビドフ) & SALDAITIS (サルダイティス)，1996）．

生息環境 道央部の山地では主として成熟した亜寒帯針葉樹林や針広混交林に見られ，所々に岩盤が露出し，林床には苔

むした古い朽木の点在するような西向きの斜面に多産することが多い．林床が一面笹で覆われたような単調な植生よりも，下草に各種のイネ科植物やツツジ，シャクナゲ類などの混入する環境を好む傾向にある．しかし，本種の好む生息環境は地域による差が大きく，道北・道東地方では平地の草原に見られることも多いし，日高の新冠近郊では海岸沿いのカシワ林に生息しており，道南の大千軒岳では亜高山性の草原に多産する．また，日高山系など，山麓の河川が深いV字谷を形成する地域においては，標高200~300 m程度の低地においてもかなりの高密度で採集される地点があるが，これは渓谷特有の鉄砲水により高所から流されてきた個体が漂着してコロニーを形成したものではないかと考察されている（西島，1989）．

　生態　自然状態における本種の生態に関する知見は極めて乏しいが，唐沢（1992）の飼育観察記録によれば，本種は秋繁殖型で，成虫は8~9月に活動期に入り産卵する．孵化した幼虫はその年の10~11月に終令となり，そのまま越冬して翌春の4~5月頃羽化するという．幼虫令数は3令までで，1令は食期3日半，眠期5~6日，2令は食期10~16日，眠期8~9日，3令は食期8~23日，眠期168~191日（越冬のため）であったという．3令（終令）幼虫で越冬し，翌春は摂食活動を行うことなく越冬窩の中でそのまま蛹化し，10日ほどの蛹期を経たのち羽化に至るという．野外では通常9月下旬頃，年によっては10月前半頃までトラップに入る成虫が観察され，11月初旬にはすでに越冬態勢に入っている．飼育下でも10月上旬頃に土中に潜り込んで越冬態勢に入るという．成虫はカタツムリを主食とするほか，果物なども食するが，幼虫はカタツムリを好んで食することから，基本的食性は軟体動物食と考えられる．越冬成虫は朽木からの採集例が多いようだが，少ないながら土中からの採集例もある．

　地理的変異　顕著な地理的変異を示す多型種で，本書では我が国に産する本種を以下の14亜種に整理した．道北・道東地域では分布が連続するため，各亜種間の境界が不明瞭な場合も多いが，南西部では各山塊の高所に孤立分布するため，集団間の識別も比較的容易である．♂交尾器先端の形態には地理的変異がそれほど明瞭に反映されない場合が多く，各亜種は主として外部形態，とりわけ鞘翅の彫刻の違いによって識別される．

　種小名の由来　種小名は，ドイツの昆虫学者，Hermann Julius Kolbe（ヘルマン・ユリウス・コルベ）（1855~1939）に因む．ベルリン大学動物学博物館の昆虫学部門で1890年から1921年までキュレーターを務めた甲虫類，チャタテムシ類，脈翅類の専門家で，甲虫の中では特にコガネムシ科とミツギリゾウムシ科を専門とした．生涯におよそ350編の論文を著し，形態学・系統分類学の分野に大きな足跡を残したとされる．

36-1. *C. (M.) k. hanatanii* Imura, 1991 アイヌキンオサムシ利尻島亜種

Original description　*Carabus (Megodontus) kolbei hanatanii* Imura, 1991, Illustrations of selected insects in the world, B (2), p. 32, fig. 50 (p. 27, pl. 6).

History of research　This least known subspecies was first recorded by Hanatani et al. (1968) based on a single specimen collected from the southern slope (the Oniwaki climbing route) of Mt. Rishiri-zan (at an altitude of 1,400 m) on August 13, 1965. The second specimen (female) was obtained in 1977 from the western slope (the Kutsugata climbing route) of the same mountain by the research team of Obihiro University of Agriculture and Veterinary Medicine. Finally, a male was collected also from the western slope, and it was designated as the holotype of subsp. *hanatanii* by Imura (1991). This is all that is known so far, as the formal records of this subspecies.

Type locality　Mt. Rishiri-daké, 1,400m alt., Rishiri Is., Hokkaido, northern Japan.

Type depository　National Museum of Nature and Science, Tokyo [Holotype, ♂].

Diagnosis　Approximately 24 mm. Body above yellowish coppery red with lateral margins greenish, and strongly polished. Identified by the following characters: 1) pronotum strongly rugoso-striate; 2) elytra long, moderately convex above, widest at or a little behind middle, with slight preapical emargination in female; 3) primary costae contiguous, wide and strongly raised; secondaries indicated by rows of granules; tertiaries weaker, at most forming irregularly set rows of small granules; rows of punctures between intervals fused with one another to form rough transverse wrinkles; 4) apical part of penis slender, rather strongly bent ventrad.

Distribution　Upper part of Mt. Rishiri-zan on Is. Rishiri-tô, northernmost Japan.

Conservation of the subspecies　This subspecies is endemic to the Island of Rishiri-tô, off the western coast of the northern tip of Hokkaido, and is confined to the alpine zone of Mt. Rishiri-zan. Detailed distributional range, habitat and ecology of this ground beetle are not known, since only three specimens have been formally recorded up to the present. This subspecies is designated as the threatened species (EN; endangered species) in the 4th edition of Red List (Threatened Wildlife of Japan) of Ministry of the Environment (2012).

Etymology　Named after Tatsuo Hanatani (Is. Ishigaki-jima, Okinawa Pref.) who first recorded this species from Is. Rishiri-tô.

　研究史　利尻島における本種の記録は，花谷ら（1968）により1965年8月13日に「山頂付近」から記録された1頭（性別不詳）が初のものである．採集者の花谷によれば，この標本は，実際には鬼脇コースの標高1,400 m付近で「平たい石の下」から採集されたものであるという．次いで，帯広畜産大学の調査（1977）により得られた1♀（沓形コース登山道沿い），そして新亜種として記載された際にホロタイプに指定された，沓形コース登山道沿いで得られた1♂の記録があり，これら計3頭がこれまで公式に明らかにされている本亜種の記録の全てである．未発表ながらその後，数頭の標本が得られているようだが，いずれにせよ分布は極めて局地的で，個体数は非常に少ない．対岸の北海道本島に分布するアイヌキンオサムシ道北亜種が平地にも広く分布しているのに

対し，本亜種は海岸沿いの草原や中腹の森林帯には決して見られず，利尻山高所のハイマツ・ダケカンバ帯より上の地域のみから記録されている点も特異である．

　タイプ産地　北日本，北海道，利尻島，利尻岳，標高1,400m．
　タイプ標本の所在　国立科学博物館［ホロタイプ，♂］．
　亜種の特徴　24 mm前後．背面は黄色味の強い赤銅色で，辺縁は緑色を帯び，光沢が極めて強い．以下の特徴を有する：1）前胸背板には比較的強い皺と線条を装う；2）鞘翅は細長く，本種としては中等度に膨隆し，最広部は中央ないしやや後方にあり，♀の鞘翅端は僅かにえぐれる；3）第一次間室は連続する隆条で，太く強壮．第二次間室は明瞭な顆粒列として認められるが，第三次間室の発達は弱く，不規則で弱い顆粒列を形成するにすぎない．間室間の点刻列は側方で強く融合し，顕著で粗な横皺となる；4）陰茎先端は細長く，腹側に向けてやや強く湾曲する．道北亜種に最も近いが，背面の金属光沢がより強く，前胸背板の皺と線条がより顕著で，鞘翅第一次間室の隆条はより太く強壮，陰茎先端はより強く腹側に湾曲する，といった点により識別される．
　分布　利尻島（利尻山高所）．
　保全　研究史の項で述べたように，本亜種は正式に記録されたものとしてはこれまでに3頭の記録しかなく，「改訂・日本の絶滅のおそれのある野生生物－レッドデータブック－，昆虫類（2006）」の中では，国産のオサムシとしては唯一，カテゴリーCR＋EN（絶滅危惧I類；絶滅の危機に瀕している種）に指定され，その後，環境省第4次レッドリスト（2012）では絶滅危惧IB類（カテゴリーEN）に指定されている．その生息地は利尻礼文サロベツ国立公園の特別保護地区内にあり，無許可での捕獲・殺傷が固く禁じられているとはいえ，美麗でしかも分布が限局されているため，マックレイセアカオサムシ利尻島亜種と同様，マニアによる密猟が本亜種の存続を脅かす最大の要因と思われる．また，利尻山の高所は各所で自然崩落が著しく，これにより本亜種の生息地の一部が自然消滅してしまう可能性も否定できない．
　亜種小名の由来と通称　亜種小名は利尻島から初めて本種を記録した花谷達郎に因む．通称はリシリキンオサムシ．

36-2. *C. (M.) k. futabae* (Ishikawa, 1966) アイヌキンオサムシ道北亜種

　Original description　*Procrustes* (*Megodontus*) *aino futabae* Ishikawa, 1966, Bull. natn. Sci. Mus., Tokyo, 9 (4), p. 462, figs. 19, 26 (p. 461) & 1 (pl. 3); 1971, Bull. natn. Sci. Mus., Tokyo, 14 (4), fig. 15 (pl. 5).
　Type locality　Nakatonbetsu, Sôya Co. [sic] (Nakatonbetsu does not belong to Sôya County, but belongs to Esashi County).
　Type depository　National Museum of Nature and Science, Tokyo [Holotype, ♂].
　Diagnosis　23~27 mm. Body above reddish coppery with brownish tinge, with strong golden greenish metallic luster on margins of pronotum and elytra. 1) Size larger on an average for this species; 2) pronotum widest before middle, with margins weakly sinuate behind; 3) elytra less strongly convex above, elongated and nearly parallel-sided in male, widest before middle in female; 4) primary costae almost contiguous, secondaries and tertiaries indicated by rows of granules, though the latter two intervals are often vaguely recognized; 5) transverse wrinkles between intervals not so strongly carved.
　Distribution　Teshio & Sôya Regions, N. Hokkaido. Distributional borders between this subspecies and subsp. *aino* is not yet defined properly.
　Etymology　Named after Futaba Ishikawa, the wife of the author of this subspecies.

　タイプ産地　"宗谷郡"中頓別（原記載では宗谷郡と記されているが，中頓別町は枝幸郡（えさしぐん）に属する）．
　タイプ標本の所在　国立科学博物館［ホロタイプ，♂］．
　亜種の特徴　23~27 mm．背面は褐色味を帯びた赤銅色で，辺縁には強い金緑色の金属光沢を帯びる．1）本種としてはやや大型の個体を多く含む；2）前胸背板の最広部は中央より前方にあり，側縁後方の波曲は弱い；3）鞘翅の膨らみは比較的弱く，♂では細長く，側縁が平行に近くなり，♀では中央より後方に最広部がある；4）第一次間室は連続する隆条，第二次，第三次間室は顆粒列となるが，後者二者はやや発達が弱い；5）間室間の横皺はあまり強くないものが多い．
　分布　北海道北部（天塩，宗谷地方）．道北の天塩，宗谷地方に分布し，北見山地の中部から南部あたりで道央道東亜種へと移行しているものと思われるが，両者の境界付近における調査はいまだに不十分である．タイプ産地の中頓別の他に，原記載では「エベコロベツ」（天塩郡豊富町（とよとみちょう）～原記載では「宗谷郡豊富村」），「更岸（さらぎし）」（天塩町）産の標本がパラタイプに指定されている．
　亜種小名の由来と通称　亜種小名は本亜種を記載した石川良輔の妻，双葉に因む．テシオキンオサムシ，ホソキンオサムシなどの通称がある．

36-3. *C. (M.) k. chishimanus* (Nakane, 1961) アイヌキンオサムシ択捉島亜種

　Original description　*Procrustes aino chishimanus* Nakane, 1961, Fragm. coleopterol., Kyoto, (1), p. 3
　Type locality　Etorofu (= Iturup), Kurils.
　Type depository　Hokkaido University Museum (Sapporo) [Holotype, ♂].
　Diagnosis　Approximately 26 mm. Body above reddish coppery with weak greenish tinge on elytra, with margins golden green.

Allied to subsp. *aino*, but differs from it in the following points: 1) body a little larger and broader; 2) punctures on head extending backward; 3) pronotum more transverse, with surface more coarsely punctate and more weakly and shallowly rugulose; 4) primary costae of elytra continuous.

Distribution   Is. Etorofu-tô.

Etymology   Named after Chishima which means the Kuril Islands in Japanese.

タイプ産地　千島列島，択捉．

タイプ標本の所在　北海道大学総合博物館（札幌）［ホロタイプ，♂］．

亜種の特徴　26 mm前後．背面は赤銅色で，鞘翅はやや緑色を帯び，辺縁は金緑色．道央道東亜種に似るが，以下の点で異なる：1）より大型で体形は幅広い；2）頭部の点刻はより後方にまで及ぶ；3）前胸背板は横長で，表面はより密に点刻され，横皺は弱く浅い；4）鞘翅第一次隆条は連続する．

分布　択捉島．

亜種小名の由来と通称　亜種小名は千島に因む．通称はチシマキンオサムシ．

36-4. *C. (M.) k. aino* Rost, 1908　アイヌキンオサムシ道央道東亜種

Original description   *Carabus aino* Rost, 1908, Dt. ent. Zs., Berlin, p.32.

Type locality   Hokkaido. This subspecies was described by Rost (1908) as an independent species from "Hokkaido" without details on the type locality. Judging from the morphological characters of the holotype (small-sized male with the body length ca. 20 mm; primary intervals of the elytra are indicated by contiguous costae), and the historical background, it should probably be concluded that the type specimen was collected from somewhere in the Konsen area of easternmost Hokkaido.

Type depository   Zoölogisch Museum Amsterdam [Holotype, ♂].

Diagnosis   20~25 mm. Body above reddish coppery, often with golden greenish tinge along lateral margins. Identified by the following characters: 1) size small on an average; 2) lateral margins of pronotum weakly sinuate in posterior halves; 3) primaries of elytra rather strongly carinate, forming contiguous ridges; secondaries and tertiaries much more attenuated, indicated by weakly set rows of granules; areas between intervals remarkably and transversely rugulose.

Distribution   C.~E. Hokkaido / Is. Kunashiri-tô.

Etymology   The subspecific name comes from the native inhabitants of Hokkaido, the Ainu people.

タイプ産地　北海道．本亜種は当初，Rost（1908）により"Hokkaido"から記載されたもので，場所の詳細は不明だが，タイプ標本が極めて小型（♂，20 mm）で，鞘翅第一次原線が連続するという形態学的特徴を具えていること，また記載当時（タイプとなった標本が採集されたのは1902年＝明治35年），北海道の山奥深くで得られた標本が外国人研究者の手に渡ったとは考えにくいことなどから，根釧地方の沿岸部等，道東地域のどこかで得られたものが*aino*のタイプ標本になったと考えるのが自然であろう．

タイプ標本の所在　アムステルダム動物学博物館［ホロタイプ，♂］．

亜種の特徴　20~25 mm．赤銅色で体側縁が金緑色を帯びるものが多い．以下の特徴により識別される：1）一般に小型；2）前胸背板側縁後半の波曲は比較的弱い；3）鞘翅第一次原線は連続する隆条で，強く隆起し，第二次，第三次間室は弱い顆粒列へと減弱し，間室間の横皺は顕著．

分布　北海道中～東部．大雪山塊から根釧地方一帯にかけて比較的広く分布し，釧路湿原内や根室半島からも記録されている．国後島のものも本亜種に含めてよいと思われる．

亜種小名の由来と通称　亜種小名は北海道の先住民族アイヌに因む．通称はアイヌキンオサムシ（狭義）だが，種名を指すのか亜種名を指すのか紛らわしいうえ，後述するように本種の基亜種にはコルベキンオサムシという，種和名とは異なる通称まであるため，注意が必要である．

36-5. *C. (M.) k. mitsumasai* Imura, 1994　アイヌキンオサムシ増毛山地南西部亜種

Original description   *Carabus* (*Megodontus*) *kolbei mitsumasai* Imura, 1994, Gekkan-Mushi, Tokyo, (276), p. 14, figs. 6, 11-14.

Type locality   Hattari (in Atsuta-mura, hills in the southwestern part of the Mashiké Mountain Range in Hokkaido, Northeast Japan).

Type depository   National Museum of Nature and Science, Tokyo [Holotype, ♂].

Diagnosis   22~26mm. Allied to subsp. *futabae*, but differs from that race in the following points: 1) elytra usually more strongly yellow-greenish and less strongly polished; 2) head and pronotum more remarkably rugoso-punctate; 3) pronotum widest a little behind middle, with lateral margins more strongly narrowed toward apex; 4) areas between intervals more strongly punctate, rows of granules forming secondaries and tertiaries more prominent and regularly set; 5) apical part of penis a little shorter and robuster, and a little more strongly bent ventrad.

Distribution   Hills on SW. slope of Mashiké Mts., CW. Hokkaido.

Etymology Named after Mitsumasa KAWATA (Sapporo) who first recorded this species from the southwestern foot of the Mashiké Mountains.

　タイプ産地　発足(北東日本，増毛山地南東部の丘陵，厚田村)(現在は石狩市厚田区に属する).
　タイプ標本の所在　国立科学博物館［ホロタイプ，♂］.
　亜種の特徴　22~26mm. 道北亜種に近いが，以下の点で異なる：1)鞘翅は一般により強く黄緑色味を帯び，光沢がやや鈍い；2)頭部と前胸背板には顕著な皺と点刻を有する；3)前胸背板の最広部はより後方にあることが多く，側縁は前方に向けてより強く狭まる；4)鞘翅間室間の点刻はより強く刻まれ，第二次，第三次間室の顆粒列はより顕著で，規則的に配列する；5)陰茎先端はやや太短く，腹側への湾曲がやや強いものが多い.
　分布　増毛山地南西麓の丘陵地帯(標高およそ200m以下). 増毛山地南西麓の，標高約200m以下の丘陵地帯に分布し，石狩市厚田区(あったく)の発足(はったり)，牧佐内(ぼくさない)，安瀬(やすけ)や，同市浜益区(はまますく)の毘沙別(びしゃべつ)などから記録されている. 増毛山地の北麓(増毛町(ましけちょう)，留萌市など)に分布する集団も，ほぼ同様の形態的特徴を有するため，本亜種に含めてもよいかもしれないが，同所よりさらに内陸にかけての調査は不十分である. 苫前町(とままえちょう)あたりまで北上すると道北亜種にかなり近い個体が混入してくるようで，力昼(りきびる)，初山別(しょさんべつ)などの個体は光沢が強く，前胸背板の点刻もやや減弱する傾向がある. なお，増毛山地の中央にある暑寒別岳(しょかんべつだけ)等の高所にも本種を産するが，これは形態的に見ると増毛山地南西部亜種とはやや異なっており，小型で，むしろ道央道東亜種の系列に近い集団のように思われる. 低地に産する増毛山地南西部亜種との間には明らかな分布の空白域が見られ，両者はそれぞれ異なる経路を経てこの山塊に侵入してきたものかもしれない.
　亜種小名の由来と通称　亜種小名は旧厚田村から初めて本種を記録した札幌市の歯科医，川田光政に因む. 原記載論文中で提唱された通称はマシケキンオサムシ.

36-6. *C. (M.) k. taikinus* (ISHIKAWA, 1993) アイヌキンオサムシ大樹町亜種

　Original description　*Procrustes* (*Pachycranion*) *kolbei taikinus* ISHIKAWA, 1993, Jpn. J. Ent., 61 (3), p. 585, figs. 1, 2 (p. 586).
　Type locality　Oshibanosawa, ca. 100m alt., Taiki-cho, Tokachi, Hokkaido.
　Type depository　Systematic Zoology Laboratory, Department of Biological Sciences, Tokyo Metropolitan University (Hachiôji) [Holotype, ♂].
　Diagnosis　22~26 mm. Body above golden-reddish coppery, with margins often greenish. Allied to subsp. *aino*, but identified by the following characters: 1) pronotum more strongly convex above, more remarkably cordate in dorsal view, with hind angles more strongly protruding postero-laterad; 2) elytral sculpture weaker, though rows of granules in secondary intervals more remarkably recognizable.
　Distribution　Known so far only from a part of Taiki-chô in the southernmost part of the Tokachi Region, SC. Hokkaido.
　Etymology　Named after Taiki-chô.

　タイプ産地　北海道，十勝，大樹町，オシバの沢，標高約100 m.
　タイプ標本の所在　首都大学東京動物系統分類学研究室(八王子)［ホロタイプ，♂］.
　亜種の特徴　22~26 mm. 背面は金赤銅色で，辺縁はしばしば緑色光沢を帯びる. 道央道東亜種に近いが，前胸背板の膨隆がより強く，背面から見てより強い心臓型を呈し，後角はより強く外側後方へと突出する. 鞘翅の彫刻は弱いが，第二次間室の顆粒列はより顕著.
　分布　大樹町の一部から得られているにすぎず，本亜種の分布範囲や近隣に分布する他亜種との関係などについては殆ど何もわかっていない.
　亜種小名の由来と通称　亜種小名はタイプ産地がある大樹町に因む. タイキキンオサムシという通称で呼ばれることもある.

36-7. *C. (M.) k. nitidipunctatus* (ISHIKAWA, 1966) アイヌキンオサムシトマム亜種

　Original description　*Procrustes* (*Megodontus*) *aino nitidipunctatus* ISHIKAWA, 1966, Bull. natn. Sci. Mus., Tokyo, 9 (4), pp. 459, figs. 21, 28 (p. 461) & 2, 3 (pl. 3).
　Type locality　Tomamu, Shimukappu, Kamikawa Co. (it belongs to Shimukappu-mura of Yûfutsu Co.).
　Type depository　National Museum of Nature and Science, Tokyo [Holotype, ♀].
　Diagnosis　22~28 mm. Body above golden red or golden green. Profile rather robust; primary costae of elytra mostly contiguous, and punctures between intervals fused with one another to form irregular wrinkles.
　Distribution　C. ~ N. Hidaka Mts., C. Hokkaido.
　Etymology　The subspecific name means "with shiny punctures".

　タイプ産地　上川郡，シムカップ，トマム(トマムのある占冠村(しむかっぷむら)は上川地方の最南部に位置するが，郡としては勇払郡(ゆうふつぐん)

between intervals dense and regular.
   Distribution   Niseko Mts. & Mt. Yôtei-zan, SW. Hokkaido.
   Habitat   Niseko Mts. & Mt. Yôtei-zan.
   Etymology   Named after Kuniaki SUGA (Tokyo), a member of the research group on carabid beetles in Keihin Insect Lovers' Club, Tokyo.

   タイプ産地　ニセコ, イワオヌプリ, 900 m.
   タイプ標本の所在　国立科学博物館［ホロタイプ, ♀］.
   亜種の特徴　21~27 mm. 赤褐色で体側縁が金緑色のものから, 強く緑色を帯びたものまで, 色彩変異の幅が比較的広い. 一般に金属光沢が鈍く, 褐色がかって見える個体が多い. 鞘翅は比較的幅広く, 最広部は中央より後方にある. 第一次間室の隆条は, ほぼ連続するものから頻繁に分断されるものまで変異の幅が大きい. 第二次, 第三次間室の隆起部は狩場山亜種の場合より発達が弱い.
   分布　ニセコ山地, 羊蹄山. ニセコ山地と羊蹄山の標高およそ500 m以上の地域に分布する. 生息地においては一般に個体数が多く, 殆どの場所で同所的に生息するヒメクロオサムシに次ぐ優占種となっている場合が多い. オオルリオサムシと混生していることも多いが, この地域のオオルリオサムシは青紫色型が大半を占めるため, 両者の間に収斂現象は認められない.
   亜種小名の由来と通称　亜種小名は, 京浜昆虫同好会オサムシグループのメンバーで, 本亜種を最初に採集した, 東京の須賀邦耀に因む. 通称はニセコキンオサムシ.

36-13. *C. (M.) k. yasudai* (ISHIKAWA, 1971)　アイヌキンオサムシ狩場山亜種

   Original description   *Procrustes* (*Megodontus*) *aino yasudai* ISHIKAWA, 1971, Bull. natn. Sci. Mus., Tokyo, 14 (4), p. 574, figs. 2, 3 (pl. 1) & 4, 5 (pl. 2).
   Type locality   Mt. Karibayama, Hiyama-Shiribeshi, Hokkaido.
   Type depository   National Museum of Nature and Science, Tokyo [Holotype, ♀].
   Diagnosis   19~24 mm. Relatively small-sized subspecies. Pronotum yellow-reddish coppery often bearing greenish tinge, elytra yellowish green or somewhat bluish green; entirely reddish coppery individual is rarely visible. Allied to subsp. *munakataorum*, but differs from that race in the following points: 1) neck more sparsely punctate; 2) elevated parts of secondary and tertiary intervals of elytra much narrower, so that elytral coloration much lighter.
   Distribution   Mt. Kariba-yama, SW. Hokkaido.
   Etymology   Named after Yukio YASUDA of Hakodaté.

   タイプ産地　北海道, 檜山 - 後志, 狩場山.
   タイプ標本の所在　国立科学博物館［ホロタイプ, ♀］.
   亜種の特徴　19~24 mm. 本種としては比較的小型. 前胸背板は黄赤銅色で, しばしば緑色味を帯び, 鞘翅は黄緑色ないしやや青味を帯びた緑色. 稀に全体に赤銅色を強く帯びた個体も出現する. 次の大千軒岳亜種に比し, 後頭部の点刻はより疎, 鞘翅は第二次および第三次間室の発達が弱いので, 隆起部の面積が相対的に小さくなり, 鞘翅の色調ははるかに明るい.
   分布　狩場山高所. 狩場山の高所に産するが, 稜線部の草原やお花畑のみならず, 中腹の樹林帯にも生息している. オオルリオサムシ渡島半島亜種と混生しているが, 同亜種との色彩面における明らかな収斂現象は認められない.
   亜種小名の由来と通称　亜種小名は本亜種の発見者, 安田幸夫(函館市の産婦人科医, オサムシ愛好家)に因む. 通称はカリバキンオサムシ.

36-14. *C. (M.) k. munakataorum* (ISHIKAWA, 1969)　アイヌキンオサムシ大千軒岳亜種

   Original description   *Procrustes* (*Megodontus*) *aino munakataorum* ISHIKAWA, 1969, Bull. natn. Sci. Mus., Tokyo, 12 (1), p. 33, fig. 1 (p. 34).
   Type locality   Mt. Daisengendake, Matsumae, Hokkaido.
   Type depository   National Museum of Nature and Science, Tokyo [Holotype, ♀].
   Diagnosis   21~25 mm. Pronotum reddish coppery bearing crimson tinge, elytra dark yellowish green with marginal area golden green, and coloration is rather stable. Identified by the following characters: 1) median tooth of mentum thick and robust; 2) hind angles of pronotum triangularly pointed and strongly protruding postero-laterad; 3) elytra less strongly convex above, all three intervals are indicated by frequently segmented costae, fused with one another to form reticular pattern. A typical morphological convergence is visible in coloration and sculptural condition of the elytra between this subspecies and *C.* (*Acoptolabrus*) *gehinii munakatai* that occurs sympatrically in the subalpine meadow of Mt. Daisengen-daké.
   Distribution   Mt. Daisengen-daké, SW. Hokkaido. Endemic to the subalpine meadow in the high altitudinal ridge of Mt. Daisengen-daké.

36. *Carabus (Megodontus) kolbei*

Etymology Name after Meiyô MUNAKATA and his brother, Keimei MUNAKATA (Hakodaté) who collected the holotype.

タイプ産地　北海道, 松前, 大千軒岳.

タイプ標本の所在　国立科学博物館［ホロタイプ, ♀］.

亜種の特徴　21~25 mm. 前胸背板は紅色を帯びた赤銅色, 鞘翅は辺縁が金緑色を帯びた暗黄緑色で, 色彩は安定している. 鞘翅間室隆起部の面積が大きいため, 他の亜種に比し, 鞘翅の色調が暗く見えることが特徴である. 下唇基節の中央歯は太短い. 前胸背板の後角は三角形に尖り, 外側後方へ強く突出する. 鞘翅の膨らみは比較的弱く, 3つの間室はいずれも断続する隆条で, 側方において互いに融合し, 不規則な網眼状構造を呈する. 同所に産するオオルリオサムシ渡島半島亜種との間に色彩や鞘翅彫刻の顕著な収斂現象が認められる.

分布　大千軒岳高所. 大千軒岳高所の稜線に広がる草原のみに産する.

亜種小名の由来と通称　亜種小名は函館の棟方兄弟に因む. オシマキンオサムシという通称が一般的だが, ムナカタキンオサムシと呼ばれることもある.

37. *Carabus* (*Acoptolabrus*) *gehinii* Fairmaire, 1876
オオルリオサムシ

Figs. 1-15. *Carabus* (*Acoptolabrus*) *gehinii gehinii* Fairmaire, 1876 (Chûô, Chitosé-shi, Hokkaido). — 1, ♂; 2, ♀; 3, male head & pronotum; 4, male right protarsus in ventral view; 5, male right protibia in subdorsal view; 6, male left elytron (median part); 7, penis & fully everted internal sac in right lateral view; 8, apical part of penis in right lateral view; 9, ditto in right subdorsal view; 10, ditto in dorsal view; 11, internal sac in caudal view; 12, ditto in frontal view; 13, ditto in dorsal view; 14, ditto in left sublateral view; 15, inner plate of ligular apophysis.

図 1-15. **オオルリオサムシ基亜種**(北海道千歳市中央). — 1, ♂; 2, ♀; 3, ♂の頭部と前胸背板; 4, ♂右前跗節(腹面観); 5, ♂右前脛節(斜背面観); 6, ♂左鞘翅中央部; 7, 陰茎と完全反転下の内嚢(右側面観); 8, 陰茎先端(右側面観); 9, 同(右斜背面観); 10, 同(背面観); 11, 内嚢(尾側面観); 12, 同(頭側面観); 13, 同(背面観); 14, 同(左斜側面観); 15, 膣底部節片内板.

37. *Carabus* (*Acoptolabrus*) *gehinii*

Figs. 16-32. *Carabus* (*Acoptolabrus*) *gehinii* subspp. — 16-23, *C*. (*A*.) *g*. *aereicollis* (16, 18, 20-22, ♂, 17, 19, 23, ♀; 16-17, Is. Rishiri-tô; 18, Lake Hebi-numa, Teshio-chô; 19, Left bank of Riv. Teshio-gawa, Teshio-chô; 20, Chiyoshibetsu, Hamamasu-ku, Ishikari-shi; 21, Benten, Is. Teuri-tô; 22-23, Mt. Asahi-yama, Asahikawa-shi); 24-26, *C*. (*A*.) *g*. *konsenensis* (24, 26, ♀, 25, ♂; 24, Nakashibetsu-chô; 25, Futatsu-yama, Shibecha-chô; 26, Higashi-mokoto-chikusa, Ôzora-chô); 27-29, *C*. (*A*.) *g*. *manoianus* (27-28, ♂, PT, 29, ♀, PT, Ochiai, Minami-furano-chô); 30-32, *C*. (*A*.) *g*. *radiato-costatus* (30-31, ♂, 32, ♀; 30, upper courses of Riv. Chiroro-gawa, Hidaka-chô; 31-32, Oshirabetsu, Hiro'o-chô). All from Hokkaido.

図 16-32. オオルリオサムシ各亜種. — 16-23, 道北亜種(16, 18, 20-22, ♂, 17, 19, 23, ♀; 16-17, 利尻島; 18, 天塩町ヘビ沼; 19, 天塩町天塩川左岸; 20, 石狩市浜益区千代志別; 21, 天売島弁天; 22-23, 旭川市旭山); 24-26, 根釧台地亜種(24, 26, ♀, 25, ♂; 24, 中標津町; 25, 標茶町二ツ山; 26, 大空町東藻琴千草); 27-29, 南富良野亜種(27-28, ♂, PT, 29, ♀, PT, 南富良野町落合); 30-32, 日高山脈南東部亜種(30-31, ♂, 32, ♀; 30, 日高町千呂露川上流; 31-32, 広尾町音調津). すべて北海道産.

37. *Carabus* (*Acoptolabrus*) *gehinii*

Figs. 33-49. ***Carabus*** (***Acoptolabrus***) ***gehinii*** subspp. — 33-34, *C.* (*A.*) *g. sapporensis* (33, ♂, 34, ♀; Samani-chô); 35-47, ***C.*** (***A.***) ***g. gehinii*** (35, 37, 38, 40, 41, 43, 44, 46, ♂; 36, 39, 42, 45, 47, ♀; 35-36, Rokugô, Furano-shi; 37, Iwamizawa-shi; 38, Nokanan-chô, Ashibetsu-shi; 39, Mt. Yûbari-daké, Yûbari-shi; 40-42, Bibi, Chitosé-shi, 43, Is. Ô-shima of Naka-jima Iss., Lake Tôya-ko, Sôbetsu-chô; 44-45, Niseko Mts., Kyôwa-chô; 46-47, Mikawa, Tsukikoshi-gen'ya Hills, Shimamaki-mura); 48-49, ***C.*** (***A.***) ***g. shimizui*** (48, ♂, HT, 49, ♀, PT, Pass Tômaru-tôgé, Furubira-chô). All from Hokkaido.

図 33-49. オオルリオサムシ各亜種. — 33-34, 日高山脈南西部亜種(33, ♂, 34, ♀; 様似町); 35-47, **基亜種**(35, 37, 38, 40, 41, 43, 44, 46, ♂, 36, 39, , 42, 45, 47, ♀; 35-36, 富良野市麓郷; 37, 岩見沢市; 38, 芦別市野花南町; 39, 夕張市夕張岳; 40-42, 千歳市美々; 43, 壮瞥町洞爺湖中島大島; 44-45, 共和町ニセコ連山; 46-47, 島牧村月越原野美川); 48-49, **積丹半島亜種**(48, ♂, HT, 49, ♀, PT, 古平町当丸(トーマル)峠). すべて北海道産.

37. *Carabus* (*Acoptolabrus*) *gehinii*

Figs. 50-63. *Carabus* (*Acoptolabrus*) *gehinii* subspp. — 50-51, "**Shimamaki lowland population**" (50, ♂, 51, ♀, Nagatoyo-machi, Shimamaki-mura); 52-53, *C.* (*A.*) *g. nishijimai* (52, ♂, PT, 53, ♀, PT, Mt. Obira-yama, Shimamaki-mura); 54-63, *C.* (*A.*) *g. munakatai* (54, ♂, 55, ♀, Mt. Kariba-yama, Shimamaki-mura; 56, ♀, Mt. Oshamanbé-daké, Oshamanbé-chô; 57-58, ♂, Pass Unseki-tôgé, Yakumo-chô; 59, ♂, Mt. Otobé-daké, Otobé-chô; 60, ♀, Pass Umezuké-tôgé, Assabu-chô / Hokuto-shi; 61, ♂, 62, ♀, Mt. Daisengen-daké, Kaminokuni-chô / Matsumaé-chô; 63, Mt. Shirakami-daké, Matsumaé-chô). — 64-65, Natural hybrid between *C.* (*A.*) *g. gehinii* & *C.* (*Damaster*) *blaptoides rugipennis* (64, ♂, Shima-no-shita, Furano-shi; 65, ♀, Niseko Mts., Kyôwa-chô). — 66, Natural hybrid between *C.* (*A.*) *g. munakatai* & *C.* (*D.*) *b. rugipennis* (Pass Unseki-tôgé, Yakumo-chô). All from Hokkaido.

図 50-63. **オオルリオサムシ**各亜種. — 50-51, **島牧低地個体群**(50, ♂, 51, ♀, 島牧村永豊町); 52-53, **大平山亜種**(52, ♂, PT, 53, ♀, PT, 島牧村大平山); 54-63, **渡島半島亜種**(54, ♂, 55, ♀, 島牧村狩場山; 56, ♀, 長万部町長万部岳; 57-58, ♂, 八雲町雲石峠; 59, ♂, 乙部町乙部岳; 60, ♀, 厚沢部町 / 北斗市梅漬峠; 61, ♂, 62, ♀, 上ノ国町 / 松前町大千軒岳; 63, 松前町白神岳). — 64-65, オオルリオサムシ基亜種とマイマイカブリ北海道亜種の自然雑種(64, ♂, 富良野市島ノ下; 65, ♀, 共和町ニセコ連山). — 66, オオルリオサムシ渡島半島亜種とマイマイカブリ北海道亜種の自然雑種(♀, 八雲町雲石峠). すべて北海道産.

37. *Carabus* (*Acoptolabrus*) *gehinii* Fairmaire, 1876 オオルリオサムシ

Original description   *Carabus Gehinii* Fairmaire, 1876, Pet. Nouv. ent., 2, p. 37.

History of research   This beautiful species was described by the French entomologist Leon M. H. Fairmaire (1876) on the basis of a single male specimen from "Japon" without designation of detailed type locality. It is endemic to Hokkaido and is variable in the coloration and elytral sculpture. Hauser (1921) described four varieties of this species, that is, var. *viridis*, var. *cyaneo-violaceus* [sic], var. *aereicollis* and var. *katomelas*, and recognized var. *grandis* described by Bates (1883). Of these, the third one (*aereicollis*) is currently adopted as a subspecific name for the population widely occupying the northeastern part of the distributional range of this species. Uchida & Tamanuki (1927) described a new species of the genus *Acoptolabrus* from "Sapporo" under the name of *A. sapporensis*, though it is currently regarded as a mere local race of *C.* (*A.*) *gehinii*. Its locality is not Sapporo but most probably somewhere in the southern part of the Hidaka region. Nakane (1960, 1961, 1962) used *Damaster* (*Acoptolabrus*) *gehinii* as the scientific name of this species, and described *gustavhauseri* as its variety (idem, 1961), and downgraded *A. sapporensis* to a mere variety of *D.* (*A.*) *gehinii* (idem, 1962). Ishikawa (1968a) reviewed this species, and recognized four color varieties, namely, standard type, gustavhauseri type, cyaneoviolaceus type and aereicollis type. He regarded *A. sapporensis* as one of the subspecies of *D.* (*A.*) *gehinii* and described two new subspecies under the names of *konsenensis* and *radiatocostatus*. He also described a new species *D.* (*A.*) *munakatai* from Mt. Daisengendake (= Daisengen-daké) near the southwestern tip of Hokkaido, which is regarded as a subspecies of *C.* (*A.*) *gehinii* in this book. *Carabus* (*A.*) *munakatai furumii* Mandl, 1981 is a mere color variety of *C.* (*A.*) *gehinii munakatai*, and is synonymized with the latter. Many new taxa described by Nakajima (1993, 1994) under the specific or subspecific names of *hashinaukamui*, *suganoe*, *okikulumi*, *tagawai*, *kunikanei*, *icalari*, *shiuni*, *kimunkamui*, *teunni*, *pilicasius*, *ponlui*, *pulepilica*, *ninetopa*, *origin*, *lamtui*, *kamuiterrestris* and *australofinis* are also the synonyms of *C.* (*A.*) *g. gehinii* or *C.* (*A.*) *g. munakatai*, or the name given to the natural hybrid between *C.* (*A.*) *gehinii* and *C.* (*D.*) *blaptoides rugipennis* (Imura, 1994b, c). Three more subspecies were described by Imura, namely, *manoianus* (as a subspecies of *D.* (*A.*) *gehinii* (Imura, 1989b)), *nishijimai* (as a subspecies of *C.* (*A.*) *munakatai* (idem, 1991)) and *shimizui* (as a subspecies of *C.* (*A.*) *gehinii* (idem, 1994a)). In his new higher classificatory system based on the molecular phylogeny, Imura (2002a) proposed *Yezacoptolabrus* as a subgenus of the genus *Acoptolabrus* for *A. gehinii* and *A. lopatini* to distinguish these two species from several continental species belonging to *Acoptolabrus* in a narrow sense. Some strange individuals most probably produced by crossbreeding between *C.* (*A.*) *gehinii* and *C.* (*D.*) *blaptoides rugipennis* have been reported from various localities in Hokkaido by Suzuki (1977), Imura (1989a), Yamazaki (1984), Nakajima (1993, 1994), etc. (Figs. 64~66). Such hybrids have been successfully obtained also by the captive crossbreeding of the two species (Arai, 1987a; Karasawa, 1988b; Kunikane & Kunikane, 1994, etc.).

Type locality   Japon.

Type depository   Muséum Nationale d'Histoire Naturelle, Paris [Holotype, ♂]. For the findings of the holotype specimen, see Imura (1994a).

Morphology   21~38 mm. Body above variable in color, golden red, green, blue, purple, or intermediate between them, usually with strong metallic luster, though rarely black and mat. The following color forms are known: 1) standard type (head and pronotum reddish, elytra greenish); 2) gustavhauseri type (golden red with greenish tinge); 3) cyaneoviolaceus type (blue or greenish blue); 4) aereicollis type (dark color). In addition, entirely pure green individual is called var. *viridis*, but typical *viridis* is quite rare. External features: Figs. 1~6. Description on details: retinaculum of right mandible bidentate, with anterior tooth much larger than posterior; apical segment of galea elongated and distinctly concave above like a spoon; penultimate segment of labial palpus bisetose; median tooth of mentum shorter than lateral lobes, with tip not so sharply pointed and not protruding ventrad; submentum asetose; male antennae without hairless ventral areas; pronotum basically with two pairs of marginal setae, basal foveae deep, median longitudinal line narrow but clearly curved, raised posteriorly to form a short flat-topped ridge. Elytral apices not forming mucros; elytral sculpture heterodyname; primaries indicated by contiguous costae or rows of broad, subquadrate callosities fused with tertiaries, and interrupted by primary foveae; secondaries always weaker than primaries and never costate; tertiaries much reduced but fused with primaries and recognized as their lateral protrusions, which is a peculiar characteristic of this subgenus; propleura smooth, at most irregularly scattered with shallow wrinkles; sides of sternites irregularly rugoso-punctate; metacoxa bisetose, lacking inner seta; sternal sulci not carved. Genitalic features: Figs. 7~15. Penis a little shorter than half the elytral length; ostium lobe unilobed, though showing tendency of bifurcation at tip; internal sac without basal lateral lobes, median lobe and paraligula; basal portion tube-like and strongly arcuate; saccellus strongly protruding dorsad in fully inflated condition; parasaccellar lobe not so large and almost symmetrical; peripheral rim of gonopore slightly sclerotized, weakly pigmented, forming a short terminal plate at middle. Female genitalia with outer plate of ligular apophysis oval, inner plate (Fig. 15) narrow spindle-shaped though variable according to individuals.

Molecular phylogeny   On the phylogenetic tree of the mitochondrial ND5 gene of the Carabina of the world, this species belongs to the Chinese lineage of so-called Procrustimorphi (= almost corresponding to the *Procrustes* subgenus-group in this book) (Osawa et al., 2002, 2004; Kim et al., 2003, etc.), and is most closely allied to *C.* (*A.*) *lopatini* of Sakhalin. Despite considerable morphological diversity, the difference in the ND5 gene sequence of many individuals of *C.* (*A.*) *gehinii* from various localities in Hokkaido is very small (Okamoto, 1999). This suggests that diversification of this species started very recently in Hokkaido with rapid morphological diversification. It cannot be judged whether the taxon represented by *munakatai* is a subspecies of *C.* (*A.*) *gehinii* or a distinct species only from the ND5 tree, but it should be mentioned that the diversification of *C. (A.) g. munakatai* and other subspecies of *C. (A.) gehinii* took place very recently (OSAWA *et al.*, 2002, 2004).

Distribution　Hokkaido / Iss. Rebun-tô, Rishiri-tô, Teuri-tô & Naka-jima [Lake Tôya-ko] (Hokkaido).

Habitat　This species is widely distributed from flatlands to rather high mountains, and usually prefers to inhabit the edge of rather thin forest with the floor covered by rich undergrowth. It is also visible in bamboo fields or subalpine meadows.

Bionomical notes　Spring breeder, snail feeder. For details of the life cycle of this species in captivity, see KARASAWA (1988, in Japanese).

Geographical variation　Polytypic species, morphologically classified into nine subspecies.

Etymology　Named after the French entomologist, Joseph Jean Baptiste GÉHIN (1816~1889).

　研究史　フランスの昆虫学者（専門は甲虫類と半翅類）Leon M. H. FAIRMAIRE（レオン M. H. フェーァメー；この名を片仮名で正確に表記するのは困難で，敢えて試みるなら"フェーR メーR"（R はフランス語独特の，喉の奥を震わせるような音）とでも書くしかない．よく見る「フェルメール」や「フェアメール」という仮名表記は実際の発音との乖離が大きすぎる）によって広義のカラブス属の新種として記載された，邦産オサムシ屈指の美麗種．タイプ産地の詳細は不明だが，ホロタイプはその形態から推測して，札幌ないしその周辺地域において得られたものである可能性が高い．HAUSER（ハッザー）（1921）は本種の名の下に var.（変種）という階級で四つのタクソン（*viridis*, *cyaneo-violaceus* [sic], *aereicollis*, *katomelas*）を"Teschio"（＝天塩）から記載し，BATES（ベイツ）により札幌から記載された *grandis* を同じく var. として認めた．HAUSER が記載したこれらの変種は，国際動物命名規約第 4 版条 45.6.4 により原公表の時点から亜種の階級とみなされるが，*viridis* と *cyaneoviolaceus* はカラブス属の中に同じ名のタクソンがすでにあって新参一次同名となるため，残った *aereicollis* と *katomelas* のうち，前者がオオルリオサムシ道北亜種の亜種名として用いられている．UCHIDA & TAMANUKI（1927）はアコプトラブルス属の新種 *Acoptolabrus sapporensis* を記載したが，タイプ産地に指定された"札幌"は明らかに誤りで，この名は現在，日高地方南西部に分布するオオルリオサムシの亜種に適用されている．NAKANE（1960, 1961）及び中根（1962）は本種に *Damaster (Acoptolabrus) gehinii* という学名をあて，その一色彩型として var. *gustavhauseri* を記載し（NAKANE, 1961），UCHIDA & TAMANUKI の *A. sapporensis* を本種の一変種とみなした（中根，1962）．ISHIKAWA（1968a）は本種に再検討を加え，標準型（standard type），赤銅色型（gustavhauseri type），青紫色型（cyaneoviolaceus type），暗色型（aereicollis type）の 4 色彩型に分類したうえで，*A. sapporensis* を本種の一亜種とみなし，二つの亜種 *konsenensis* と *radiatocostatus* を記載した．また，道南の大千軒岳から新種 *D. (A.) munakatai*（本書ではオオルリオサムシの一亜種 *Carabus (Acoptolabrus) gehinii munakatai* とみなす）を記載した．MANDL（マンドル）（1981）により記載された *C. (A.) munakatai furumii* はその一色彩型（青紫色型）に対して与えられた名称とみなされ，一般には異名として扱われている．中嶋（1993, 1994）は渡島半島から極めて不適切な方法によって多数のタクソン（*hashinaukumui*, *suganoe*, *okikulumi*, *tagawai*, *kunikanei*, *icalari*, *shiuni*, *kimunkamui*, *teunni*, *pilicasius*, *ponlui*, *pulepilica*, *ninetopa*, *origin*, *lamtui*, *kamuiterrestris*, *australofinis*）を新種又は新亜種として「記載」したが，いずれもオオルリオサムシ基亜種ないし同渡島半島亜種の異名，或いはオオルリオサムシとマイマイカブリ北海道亜種の雑種に対して与えられた名とみなされている（IMURA, 1994b；井村，1994c）．本種の亜種としては他に，南富良野の *manoianus*（井村，1989b），渡島半島北部大平山の *nishijimai*（同，1991）及び積丹半島の *shimizui*（同，1994a）が記載されている．IMURA（2002a）は分子系統に基づくオサムシ亜族の新しい上位分類体系を発表し，本種とカラフトクビナガオサムシの 2 種に対し，大陸産の狭義のアコプトラブルス類と区別するため，イェザコプトラブルス亜属を設立した．本種とマイマイカブリ北海道亜種の自然雑種と思われる個体はこれまでに数例報告されている．図 64 は鈴木（1977），図 65~66 は IMURA（1989a）によって発表されたもので，この他に夕張岳東側標高 800 m から得られた 1♂（西島 浩氏所蔵），札幌市定山渓標高約 200 m から得られた 1♀（山崎，1984），島牧村千走川（ちはせがわ）温泉から得られた 1♀（中嶋，1993, 1994；新種として記載されたが，井村（1994b），IMURA（1994c）によりオオルリオサムシとマイマイカブリ北海道亜種の雑種として処理されている）等がある．また，飼育下において人為的に作り出された本種とマイマイカブリ北海道亜種の種間雑種も報告されている（荒井，1987a；唐沢，1988b；国兼・国兼，1994 等）．

　タイプ産地　日本．

　タイプ標本の所在　パリ自然史博物館［ホロタイプ，♂］．ホロタイプ標本に関しては，井村（1994a）に詳しい解説と再記載，図示がある．

　形態　21~38 mm. 背面の色彩は多彩で，銅赤色から緑，青，紫，更に黒色に至る多様な変化を示し，一般に強い金属光沢を具えるが，黒色に近い色彩型では殆ど光沢を欠く場合がある．本種の示す色彩型は ISHIKAWA（1968a）によって以下のように分類されている：1) 標準型 standard type（頭胸部が赤銅色，鞘翅が金緑色のもの）；2) 赤銅色型 gustavhauseri type（赤銅色~銅褐色で緑色光沢を帯びるもの）；3) 青紫色型 cyaneoviolaceus type（青色で時に暗化ないし緑化するもの）；4) 暗色型 aereicollis type（頭胸部が銅緑色系，鞘翅が暗色で光沢の鈍いもの）．これらの他に，HAUSER（1921）によって記載された緑色型 f. *viridis* がある．これは全身が金緑色に輝く個体に対して与えられた名であるが，典型的なものはかなり珍しく，ISHIKAWA（1968a）は標準型に含めている．実際には，これら 4~5 通りの名称のみで本種に出現する全ての色彩を表現できるわけではなく，さまざまな中間的色合いやいずれにも属さない微妙な色合いを持つ個体も現れる．一般に，頭部と前胸背板は鞘翅よりも長波長となり（前二者が赤なら後者は緑，前二者が緑なら後者は青），原則としてその逆の例はない．また，斜めから観察すると全体に一段階短い波長の色彩を帯びて見える場合が多い．各色彩型の出現比率は地域により異なるが，一般に道北や道東地域では赤銅色型の比率が高い．外部形

態：図1~6. 細部の記載：右大顎の抱歯は先端が二股に分かれ，前方突起は後方突起よりはるかに大きい．小顎外葉の先端節は長く，内側が匙状に窪む．下唇肢亜端節の剛毛は2本．下唇基節の中央歯は側葉より短く，先端はあまり鋭く尖らない．下唇亜基節は無毛．♂触角腹面に無毛部を欠く．前胸背板の形には地理的変異が見られ，一般に分布域の北東部に産するものでは全体の幅が広く，側縁前方の丸みが強く，頚部の外側縁と一定の段差をもって接続するのに対し，南西部のものでは全体的に細長く見え，側縁前方の丸みが弱く，頚部の外側縁と比較的滑らかに接続することが多いが，例外もある．最も細長く見える場合でも前胸背板の長幅比は1を越えない．すなわち，幅よりも前後の長さが長くなることは通常ない．前胸背板の側縁剛毛は基本的に2対．基部凹陥は深く，正中線は細いが明瞭で，後方で低く隆起し，平坦な稜を形成する．鞘翅彫刻は異規的で，第一次間室は亜種，集団，或いは個体によって「連続する隆条」→「孔点を含む隆条」→「第三次間室の顆粒によって側方を輪状に縁どられた第一次回陥により分断された鎖線」→「分断面が竹の節状を呈する破線」の順に変化し，隆起部の分断は鞘翅基部よりも翅端部から，また内側よりも外側から起こりやすい．第二次間室は隆条とならず，さまざまな程度の顆粒列或いは小瘤列となり，時として痕跡的．第三次間室の顆粒は第一次間室の隆起部と融合し，後者の側突起として認められる．第一次隆条の分断度が大きいほど第二次，第三次間室の顆粒は発達する傾向にある．走査型電子顕微鏡で観察すると，鞘翅の金属光沢を有する部分は蜂窩様の六角形の微細構造が鱗状に折り重なって構築されていることがわかる．鞘翅端は尖らず，尾状突起は形成しない．前胸側板は滑らかで，せいぜい不規則な浅い皺を有する程度．腹板側面には不規則な皺があり，時に小点刻を伴う．腹板の横溝を欠く．後基節の剛毛は2本で，内側剛毛を欠く．交尾器形態：図7~15．葉片は単葉だが，先端はごく僅かに分葉傾向を示す．内嚢には基部側葉，側舌，中央葉などを欠き，基部は筒状で大きく湾曲し，内嚢盤は背側へ強く突出し，傍盤葉はそれほど大きくなく，左右ほぼ対称．射精孔縁膜は僅かに硬化，色素沈着し，中央にごく短い頂板を形成する．♂交尾器には若干の変異が認められるものの，しばしば個体変異の範疇にとどまり，外部形態に見られるほど明らかな地理的変異をみいだすことは困難である．♀交尾器膣底部節片の外板は卵形，内板（図15）は中央部が僅かに広がった細長い紡錘形で，個体変異がかなり見られる．

分子系統 ミトコンドリアND5遺伝子を用いた世界のオサムシ亜族の分子系統樹において，本種はヨロイオサムシ群（本書で言うプロクルステス亜属群にほぼ相当する，カラブス属の中の大群）の中の中国系亜群に属し，（大澤ら，2002; OSAWA et al., 2004; KIM et al., 2003等），サハリンのカラフトクビナガオサムシに最も類縁が近い．本種は北海道内においてかなり顕著な地理的変異を示すものの，各集団（亜種）間における塩基配列の差は僅少であり，別種オシマルリオサムシとして記載された渡島半島亜種との間にもそれほど大きい差は見られない（岡本，1999）．このことから，北海道における本種の亜種分化は，急速な外部形態上の変化を伴いつつも，時期的には比較的最近起こったものであろうと推定されている．本書ではオオルリオサムシ渡島半島亜種として扱った*munakatai*という名に代表されるタクソンが，オオルリオサムシの一亜種なのか，或いは独立種とみなすべきものなのかをND5系統樹の結果のみから判定することは困難であるが，いずれにせよ両者は系統的に非常に近く，その分化はかなり最近になって起きたものと考えられている．少なくとも，一部の研究者によって想定されていたような，ユーラシア大陸東部に分布するアコプトラブルス亜属のいずれかの種（シュレンククビナガオサムシ等）との間に直接の類縁関係があるといった推論には無理がありそうである．

分布 北海道／礼文島，利尻島，天売島，中島［洞爺湖］（以上，北海道）．北海道のほぼ全域に産するが，知床半島と亀田半島からは確実な記録がない．離島では礼文，利尻，天売の3島から記録があり，洞爺湖に浮かぶ中島にも生息している．

生息環境 低地から山地まで幅広く分布し，一般に明るく下草の豊富な環境を好む．本種の生息地でよく見られる植物として，マイヅルソウ，シダ類，アザミ，フキ，ウド，各種ユリ科植物，イチリンソウ，オクエゾサイシン，オオカメノキ等がある．一般に落葉広葉樹の疎林や針広混交林の林床ないし林縁部が本種の多産地となっている場合が多く，深い森林の内部や笹の密生した単調な植生の場所には少ない．しかし，本種はカタツムリを主な食餌としているためか，その生息環境の幅は非常に広く，波打ち際に近い海岸草原（礼文島，利尻島，雄武町など），樹木のない一面のチシマザサの笹原にアザミ，ウド，ツタウルシ，イワガラミ，小型の単子葉植物などが混在するような環境（島牧村の月越原野等），さらに亜高山性の草原（大千軒岳）等，一見すると意外と思われるような場所に多産している場合もある．

生態 春繁殖型で，成虫・幼虫双方の越冬態を持つ，軟体動物（カタツムリ）食のオサムシ．融雪直後（低地で4月下旬から5月上旬，高地・寒冷地で6月以降）より活動を開始し，同所的に生息するマイマイカブリ北海道亜種より一般に啓蟄は早い．活動開始から1~2ヶ月ほどの間は個体数が多く，新鮮な個体の占める比率が高い傾向にあるが，盛夏に向かうにつれて古く擦れた個体の比率が高まり，採集される個体数も次第に減少する．分布域の南部や平野部の産地では，たとえ生息していても夏季には採集されなくなることも多い．交尾は啓蟄直後より始まり，以後夏まで比較的長期間にわたって産卵が続く．基本的には春繁殖型のオサムシと言えるが，幼虫による越冬も確認されており，環境条件によってかなり融通性に富む生活環を有する種のように思われる（井村，1985）．唐沢（1988a, b）による飼育下での観察によれば，1頭の♀は平均約10回産卵し，一度に産む卵数は約6個，従って総産卵数は60個近くに達するという．卵期間は10~20日間．幼虫令数は3令までで，眠期（平均50日）に比べ，摂食日数（平均10日）の短いことが特徴である．3令（終令）幼虫の入眠から羽化までの平均所要日数は40日弱．土中に体長の2倍以上に及ぶ蛹室を形成したのち，10日から2週間ほどで前蛹になり，2~3日の前蛹期間の後に蛹化し，更に1週間から10日ほどで羽化に至る．これらの期間には個体による差が大きく，早いものでは1ヶ月，遅いと2ヶ月に及ぶこともある．羽化直後の未熟成虫は，野外では8~9月頃に見られることが多いが，テネラル個体の採集記録は散発的で，極端なピークを形成することは少ない．これは，母虫の産卵が長期に亘ることや幼虫の発育速度にばらつきが大きいことなどに起因する現象であろう．6月頃にも稀に未熟個体が

採集されることがあるが，これはおそらく幼虫で越冬し，春先に蛹化した個体であろうと思われる．羽化した成虫は初め真白で，5~6時間経つと前胸背板，鞘翅に虹色の光沢が出現し，全体が一旦黒色になった後，一昼夜ほどして固有の色彩が現れてくる．越冬に入る時期は9月下旬から10月頃で，マイマイカブリ北海道亜種の場合よりも一般に遅い．越冬中の成虫の観察記録は多くなく，朽木や土中からの散発的な記録があるにすぎない．北海道では，一般にオサムシ類が越冬場所として好むような朽木は通常，マイマイカブリ北海道亜種によって占拠されており，本種の主たる越冬場所は土中，それも崖の肩部などではなく，地面に直接もぐって到達する植物の根際等ではないかと推測される．成虫は比較的雑食性で，各種の動・植物質を食するが，やはりカタツムリを最も好む傾向にある．井村は1981年7月，ニセコ山群ニトヌプリ中腹において本種がカタツムリを捕食している様子を観察したことがあるが，直径3 cmほどのヒメマイマイの殻に頭胸部を差し込んださまはマイマイカブリさながらであった．また，島牧村一帯では，同地に多産する特徴的な形態をもったカドバリヒメマイマイを主な食餌としているようである．前出の唐沢によれば，飼育下では1令幼虫の餌として直径1.5 cmほどのオナジマイマイが最適で，幼虫は殻の中にもぐり込み，2~3日かけて食い尽くすという．令数が増すにつれて次第により大型のカタツムリを摂食するようになり，餌が豊富な場合には発育が極めて早い．

地理的変異　顕著な地理的変異と個体変異を示すが，外部形態の変化の幅の大きさに比し，各集団間における交尾器形態やミトコンドリアDNAの塩基配列の差はごく僅かである．本書では計9亜種に分類したが，各集団は移行帯をもって互いに連続する場合が多く，明確な境界線を引いて区別することが困難な場合も多い．

種小名の由来　フランスの昆虫研究家（本業は薬剤関係の仕事であったという），Joseph Jean Baptiste Géhin（ジョセフ・ジャン・バプティスト・ジェアン）(1816~1889)に因む．我が国では本種の種名を「ゲヒニィ」と読む人が多いが，Géhinの正しい発音は「ジェアン」に近く，「ゲヒン」や「ゲーヒン」ではない．オオルリオサムシという和名が最も一般的だが，オオルリクビナガオサムシという名が使われたこともある．

37-1. *C. (A.) g. aereicollis* (Hauser, 1921) オオルリオサムシ道北亜種

Original description　*Acoptolabrus gehinii* var. *aereicollis* Hauser, 1921, Zool. Jahrb., Abt. Syst., 45, p. 152.
Type locality　Teschio.
Type depository　Museum für Naturkunde der Humboldt-Universität zu Berlin ?
Morphology　22~32 mm. Medium to small sized subspecies. Body above predominantly coppery reddish (gustavhauseri type), sometimes entirely greenish (var. *viridis*), dark bluish (cyaneoviolaceus type) or dark colored (aereicollis type). Identified by the following characters: 1) pronotum rather broad, with lateral margins usually strongly rounded in anterior halves; 2) primary costae of elytra narrow but strongly raised, almost continuous excepting apical and lateral portions; secondaries often vestigial, tertiaries also vestigial, fused with primaries.
Distribution　N. & E. Hokkaido / Iss. Rebun-tô, Rishiri-tô, Teuri-tô.
Geographical variation　The populations inhabiting Iss. Rebun-tô and Rishiri-tô are predominantly reddish coppery (gustavhauseri type), and no bluish color form (cyaneoviolaceus type) has been known. The Teuri-tô population is also small, usually coppery reddish or somewhat greenish. The population inhabiting Hamamasu-ku of Ishikari-shi along the coastal area of the Sea of Japan is unique in having very small and narrow body, characteristic coloration (reddish coppery bearing golden greenish tinge along elytral margins, sometimes entirely greenish or yellowish, rarely dark purplish blue) and elytral sculpture (secondary intervals are usually recognized as clearly set rows of granules).
Etymology　The subspecific name means "having a collar made of copper".

タイプ産地　テシオ（北海道天塩）．
タイプ標本の所在　フンボルト大学自然博物館（ベルリン）？Hauserコレクションは現在，同博物館に所蔵されているようだが，本タクソンのタイプ標本の所在は確認できなかった．
亜種の特徴　22~32 mm. 本種としては中型からやや小型の個体が多い．赤銅色型が優勢となる地域が多いが，時に全身が金緑色に輝く典型的な緑色型や，青紫色型，紫褐色を帯びた暗色型も見られる．本亜種において出現する青紫色型は基亜種に見られるそれとは若干異なり，前胸背板が緑色味を帯びず，全身がいわゆるナス紺色を呈するものが多い．青紫色型については，一定比率で出現する地域と全く現れない地域がある．1) 前胸背板側縁前半部は丸みが強く，一定の角度をもって頭部側縁と連続する；2) 鞘翅第一次原線は細いが強く隆起し，翅端付近まで連続するが，側方や後方では分断されやすい傾向にある．第二次原線はしばしば痕跡的で，第三次間室の隆起部は第一次原線の辺縁に融合して目立たない．
分布　北海道北~東部／礼文島，利尻島，天売島．
地理的変異　礼文・利尻両島のものは基本的に赤銅色型で中型からやや小型，利尻島の高所などでは稀に緑色光沢を強く帯びた個体も見られるが，青紫色型は知られていない．天売島のものはやや小型で赤銅色型か標準型のものが多い．石狩市浜益区（旧浜益村）の日本海沿岸部一帯に産するものも小型で，特に♂では小型化が著しく，細身の個体が多い．色彩としては辺縁の金緑色光沢の強い赤銅色型が大部分を占め，全体に緑色や黄色味を帯びたものも見られる．青紫色型も稀に出現するようである．また，鞘翅第二次原線を明瞭な点刻列として認めうる個体の出現比率が比較的高い．旭川から北見，知床半島あたりから南では他亜種へと移行していくようだが，境界線は明確にされていない．

亜種小名の由来と通称　亜種小名は「銅(製)の襟の」を意味する．通称はキタオオルリオサムシ．

37-2. *C. (A.) g. konsenensis* (Ishikawa, 1968)　オオルリオサムシ根釧台地亜種

Original description　*Damaster* (*Acoptolabrus*) *gehinii konsenensis* Ishikawa, 1968, Bull. natn., Sci. Mus., Tokyo, 11 (1), p. 145, fig. 2 (p. 146).
Type locality　Yoroushi, Nemuro Prov.
Type depository　National Museum of Nature and Science, Tokyo [Holotype, ♀].
Morphology　24~35 mm. Body above predominantly coppery reddish, sometimes dark coppery brown, and rarely dark bluish. Closely allied to subsp. *aereicollis*, but identified by the following characters: 1) size larger on an average, with robuster body; 2) pronotum broad and strongly cordate; 3) elytra broad and strongly convex above, with primary costae almost uninterrupted.
Distribution　Konsen Plateau, E. Hokkaido.
Geographical variation　The populations inhabiting the north of Konsen Plateau (Abashiri area etc.) and southern part of the same plateau near the Pacific coast are smaller and the cyaneoviolaceus type sometimes appears.
Etymology　Named after the Konsen Plateau of eastern Hokkaido.

タイプ産地　根室支庁, 養老牛(現在の所属は中標津町(なかしべつちょう)).
タイプ標本の所在　国立科学博物館［ホロタイプ, ♀］.
亜種の特徴　24~35 mm. 背面の色彩は赤銅色型主体で安定しているが, 汚銅褐色を呈する個体も見られ, ごく稀に青紫色型も出現する. 原記載に示された本亜種の特徴は以下の通り: 1) 大型で体格は強壮; 2) 前胸背板は幅広く, 顕著な心臓形を呈する; 3) 鞘翅は強く膨らみ, 幅が広く, 第一次隆条は殆ど分断されない; 4) 色彩は赤銅色で道北亜種のそれと大差ないが, やや大型で鞘翅のの膨らみの強い個体が多い.
分布　根釧台地. タイプ産地は台地北部の中標津町養老牛で, 根釧台地に産するものが典型的とされているが, 北は網走付近のものまで本亜種に含めてよいと思われる.
地理的変異　阿寒湖から摩周湖付近の山地に産するものは小型で鞘翅の膨らみが弱く, 青紫色型も一定比率で混入するようになり, 道北亜種に近くなる. また, 厚岸から釧路にかけての太平洋沿岸部に産するものも小型で青紫色型が一定比率で混入する. いずれにせよ, 道北亜種との形態的な差異は軽微で, 分布も連続しており, 両者の間にはっきりとした分布境界線を設けることは難しい. また, 十勝平野から日高山脈東麓にかけての地域も調査が不十分である.
亜種小名の由来と通称　亜種小名は根釧台地(または根釧地方)に由来する. コンセンオオルリオサムシ, コンセンルリオサムシ等の通称がある.

37-3. *C. (A.) g. manoianus* (Imura, 1989)　オオルリオサムシ南富良野亜種

Original description　*Damaster* (*Acoptolabrus*) *gehinii manoianus* Imura, 1989, Illustrations of selected insects in the world, B, (1), Mushi-sha, Tokyo, p. 13, figs. 1~48 (pl. 7).
Type locality　Ecchû-dantai, 400 m alt., on the left bank of Riv. Sorachi-gawa, Ochiai, Minami-furano-chô, Sorachi-gun.
Type depository　National Museum of Nature and Science, Tokyo [Holotype, ♂].
Diagnosis　21~31 mm. Body above variable in color. Predominant color form is so-called standard type, but the aereicollis type and intermediate form between them are also visible; typical gustavhauseri and cyaneoviolaceus types are relatively few. Allied to subsp. *radiatocostatus*, but distinguished from that race by the following points: 1) size apparently smaller on an average; 2) apical halves of pronotal margins usually more strongly rounded; 3) primary costae of elytra narrower, less frequently interrupted, usually continuous near elytral base, and more weakly bordered by transverse ridges that indicate tertiaries; 4) secondaries usually less developed; 5) areas between intervals more weakly uneven.
Distribution　S. Furano Basin~N. Hidaka Mts., C. Hokkaido.
Etymology　Named after Susumu Mano (Hokkaido), an amateur entomologist who once lived in Furano.

タイプ産地　空知郡, 南富良野町, 落合, 空知川左岸, 越中団体, 標高400 m.
タイプ標本の所在　国立科学博物館［ホロタイプ, ♂］.
亜種の特徴　21~31 mm. 背面の色彩は非常に多彩で, 標準型の他に汚緑褐色, 赤紫色を帯びたものなど, 暗色型との中間的色彩を呈する個体の出現頻度が高いが, 典型的な赤銅色型や青紫色型は比較的少ない. 日高山脈南東部亜種に近いが, 以下の点で異なる: 1) より小型(本種中, 平均して最も小型の集団); 2) 前胸背板側縁前方の丸みがより強い; 3) 鞘翅第一次隆条はより細く, 分断される頻度はより少ない(特に鞘翅基部近くでは連続する場合が多い); 4) 第一次隆条の側縁を放射状に縁取る第三次間室隆起部の発達がより弱い; 5) 第二次原線の発達も一般により弱い; 6) 間室間の鞘翅基面は日高山脈南東部亜種のそれほど強い粗面とならず, しばしば基亜種と同程度の状態を示す.
分布　富良野盆地南部～日高山脈北端部一帯. 日高山脈北端部の狭い範囲に分布し, 典型的なものは南富良野町落合から

占冠村(しむかっぷむら)上(かみ)トマム付近で見られる．西限はほぼ金山湖(かなやまこ)〜下(しも)トマム〜日高町日高付近を結ぶ線で，これより西方には基亜種(富良野型)が分布する．南限は沙流川(さるがわ)上流にほぼ一致し，これより南方には日高山脈南東部亜種に相当する集団が分布する．東限は日勝峠(にっしょうとうげ)から狩勝峠(かりかちとうげ)一帯と思われる．北限はやや不明瞭ながら空知川(そらちがわ)を境に形態形質の変化が見られるようで，基亜種の富良野型に近いものやさまざまな移行型の介在を経て道北亜種の分布圏へと連続する．

亜種小名の由来と通称　亜種小名は，かつて富良野に住み，本亜種の発見に寄与した北海道のオサムシ愛好家，真野 進に因む．原記載論文中で提唱された通称はヒメオオルリオサムシ．

37-4. *C. (A.) g. radiatocostatus* (Ishikawa, 1968) オオルリオサムシ日高山脈南東部亜種

Original description　*Damaster* (*Acoptolabrus*) *gehinii radiatocostatus* Ishikawa, 1968, Bull. natn. Sci. Mus., Tokyo, 11 (1), p. 144, fig. 1 (p. 144).
Type locality　Oshirabetsu, Tokachi.
Type depository　National Museum of Nature and Science, Tokyo [Holotype, ♀].
Diagnosis　24~34 mm. Body above usually reddish (gustavhauseri type) or greenish (standard type), sometimes bluish (cyaneoviolaceus type) or dark colored (aereicollis type). Characterized by the following points: 1) pronotum a little more elongated than in other subspecies (0.9 to 1.0 times as long as broad), with front angles not stepped to neck, sides less strongly rounded in front, occasionally feebly subangulate at middle and strongly sinuate behind; 2) primary costae broad, frequently segmented by deep primary foveoles, and bordered by distinct transverse ridges indicating tertiaries; secondaries also distinct, forming rows of small tubercles which are often contiguous to form short costae; areas between elevated parts of intervals strongly transversely rugulose and coarsely sculptured.
Distribution　SE. Hidaka Mts., SC. Hokkaido.
Etymology　The subspecific name means "of radiating (elytral) costae".

タイプ産地　十勝，音調津(現在の所属は広尾町)．
タイプ標本の所在　国立科学博物館［ホロタイプ，♀］．
亜種の特徴　24~34 mm．赤銅色型ないし標準型の比率が高いが，青紫色型や暗色型も出現する．鞘翅隆起部の面積が大きいため，全体に暗褐色を帯びて見える個体が多い．以下の特徴により他亜種から識別される：1)前胸背板の幅に対する長さの比率がやや大きく，細長く見える個体が多い．前角は段差なく比較的スムースに頸部へと連続し，側縁は前方で弱く弧を描き，最広部では時にやや角張り，後方の波曲は顕著；2)鞘翅の第一次間室は幅広く，深い第一次凹陥によって頻繁に分断され，顕著な側突起(第三次間室)によって縁取られる．第二次間室もよく発達し，小結節列を形成し，しばしば部分的に連続して短い隆条を形成する．各間室隆起部間の基面は粗く，顕著な横皺を有する．
分布　日高山脈南東部．タイプ産地は広尾町の音調津で，典型的なものは日高山脈南東部一帯に分布しており，北限は同山脈北部の沙流川上流一帯と思われる．日高山脈東麓から十勝平野南西部にかけての地域で道北亜種ないし根釧台地亜種へ移行するものと思われるが，この地域の調査は不十分である．
亜種小名の由来と通称　亜種小名は「放射状の(鞘翅)隆条の」を意味する．アラメオオルリオサムシ，キレスジルリオサムシ等の通称がある．

37-5. *C. (A.) g. sapporensis* (Uchida et Tamanuki, 1927) オオルリオサムシ日高山脈南西部亜種

Original description　*Acoptolabrus sapporensis* Uchida et Tamanuki, 1927, Ins. Matsumurana, Sapporo, 2 (2), p. 103, fig. 1.
Type locality　"Hokkaido (Sapporo)" (This locality is most probably erroneous. The type specimen is assumed to be collected from somewhere near the southwestern part of the Hidaka Mountains).
Type depository　Hokkaido University Museum (Sapporo) [Holotype, ♀].
Diagnosis　26~35 mm. Predominant color form is aereicollis type, though there are two types in pronotal coloration, golden green and reddish coppery, or intermediate state between them; elytra purplish black, bearing red-purplish tinge along margins; rarely entirely dark greenish or bearing indigo. Pronotum not very long, with sides strongly rounded in apical halves. elytral intervals not so strongly raised, but secondaries rather constantly recognized as segmented costae.
Distribution　SW. Hidaka Mts. (Mt. Apoi-daké and its nearby regions).
Etymology　Named after Sapporo, the prefectural capital of Hokkaido, though the present subspecies is not distributed in that area.

タイプ産地　「北海道(札幌)」．原記載に記されたこのタイプ産地名は恐らく誤りで，当該標本が得られたのは日高山脈南西部ないしその近隣地域であろうと推察される．
タイプ標本の所在　北海道大学総合博物館(札幌)［ホロタイプ，♀］．

い．原型と富良野型が東西から出合って形成された集団と思われ，勇払原野から千歳市一帯の丘陵地帯に産するものがこれに相当する．一般に産地における個体数が極めて多いこともこの地域の特徴である．

6) 石狩型　24~35 mm．ほぼ全ての色彩が様々な比率で出現するが，赤銅色型と暗色型の比率が比較的高いようである．本群に見られる暗色型には，Hauserが記載した典型的なaereicollisに相当するような，紫褐色を帯びた光沢の鈍いものの出現比率が高い．勇払型に見られるような紫色の個体は少なくなる．一般に第一次間室の隆条は原型よりも連続する傾向にあるが，隣接する各型からの影響を受けて変異が大きい．原型，富良野型及び道北亜種の三者によって形成された一種の移行群と考えられ，旭川盆地南西部から石狩平野北東部にかけての石狩川中流左岸一帯の広い範囲に産するものがこれに相当する．

37-7. *C. (A.) g. shimizui* Imura, 1994　オオルリオサムシ積丹（しゃこたん）半島亜種

Original description　*Carabus (Acoptolabrus) gehinii shimizui* Imura, 1994, Gekkan-Mushi, Tokyo, (276), p. 13, fig. 2.
Type locality　Pass Tômaru-tôgé, ca. 600 m alt.
Type depository　National Museum of Nature and Science, Tokyo [Holotype, ♂].
Diagnosis　23~32 mm. So-called standard type is dominant; cyaneoviolaceus type or gustavhauseri type is hardly visible. Allied to the nominotypical subspecies, but distinguished from it in the following points: 1) sides of pronotum less strongly divergent before hind angles which are less strongly protruded posteriad with apices more sharply pointed; 2) elytra a little less strongly convex above, above all in male; 3) primary costae wider and more frequently and regularly interrupted; 4) areas between intervals more conspicuously uneven. Also allied to subsp. *radiatocostatus*, but distinguished from that race by barely subangulated widest part of pronotum and much less developed elevated parts of secondary and tertiary intervals of elytra.
Distribution　Shakotan Peninsula, SW. Hokkaido.
Etymology　Named after Shôhei Shimizu who first recorded this species from Mt. Shakotan-daké.

タイプ産地　当丸（トーマル）峠，標高約600 m（所属は古平町（ふるびらちょう））．
タイプ標本の所在　国立科学博物館［ホロタイプ，♂］．
亜種の特徴　23~32 mm．色彩は標準型のものが多く，時に緑色或いは赤銅色を強く帯びるが，青紫色型や暗色型は見られないようである．新鮮な個体では一般に背面の光沢が強い．基亜種に比し，1) 前胸背板の側縁は後角の前方であまり強く側方へ広がらず，後角は短く，先端はより鋭く尖るものが多い．2) 鞘翅の膨らみはやや弱く，この傾向はとりわけ♂においてより顕著である．3) 鞘翅第一次間室の隆条は幅広く，頻繁に，かつ比較的均等に分断される．3) 基面の凹凸はより顕著である．鞘翅彫刻の特徴から，日高山脈南東部亜種にも近いと言えるが，本亜種の場合，前胸背板の側縁が最広部において角張ることは殆どなく，鞘翅第二次，第三次間室隆起部の発達がはるかに弱いため，識別は容易である．
分布　積丹半島．半島基部の仁木町（にきちょう）然別（しかりべつ）などの個体では鞘翅隆条の連続性が高くなり，色彩的にも青紫色型が混入するようになるので，余市と岩内を結ぶ線が基亜種との分布境界に相当する可能性が高い．
亜種小名の由来と通称　亜種小名は積丹岳から初めてオオルリオサムシを記録した清水昭平に因む．原記載論文中で提唱された通称はシャコタンオオルリオサムシ．

37-8. *C. (A.) g. nishijimai* Imura, 1991　オオルリオサムシ大平山亜種

Original description　*Carabus (Acoptolabrus) munakatai nishijimai* Imura, 1991, *Illustrations of selected insects in the world*, B, (2), p. 21, figs. 11–30.
Type locality　Mt. Ôbira-yama (correct pronunciation is Obira-yama or Ôhira-yama), 800-900 m alt. on SW slope, Shimamaki-gun, Southwest Hokkaido, northern Japan.
Type depository　National Museum of Nature and Science, Tokyo [Holotype, ♂].
Diagnosis　23~33 mm. Body above predominantly coppery reddish, sometimes greenish, but no bluish form has been known. Allied to *C. g. munakatai*, but differs from that race in the following points: 1) size smaller on an average; 2) elytra with primary costae narrower and less frequently interrupted, secondaries and tertiaries less strongly developed.
Distribution　Mt. Obira-yama & its nearby regions, SW. Hokkaido.
Geographical variation
Etymology　Named after Yutaka Nishijima, a former professor at the Obihiro University of Agriculture and Veterinary Medicine, who discovered this subspecies in the early 1980s.

タイプ産地　北日本，南西北海道，島牧郡，大平山（おびらやま/おおひらやま），南西斜面，標高800-900 m．
タイプ標本の所在　国立科学博物館［ホロタイプ，♂］．
亜種の特徴　23~33 mm．色彩は赤銅色味を帯びるものが多く，緑色型や中間的色彩を有するものも一定の比率で出現するが，青紫色型は知られていない．渡島半島亜種に近いが，1) 一般により小型で，2) 第一次原線はより細く，より連続する傾向にあり，第二次，第三次間室の発達も弱い．すなわち，鞘翅彫刻がより単純化し，基亜種のそれに近くなる．

分布　大平山（北海道島牧村）.

生息環境　タイプ産地の大平山は石灰岩からなる山塊で，特産のウスユキソウが自生するなど，植物の分野でもよく知られている．中腹から山頂にかけてセリ科，キク科を主体とする草原が広がっており，独特の形状をもつカドバリヒメマイマイ（カタツムリ）を豊産するため，本亜種の個体数は多いが，発見当初の密度に比し，近年はかなり減少しているようである．

亜種小名の由来と通称　亜種小名は本亜種を発見した西島 浩（元帯広畜産大学教授）に因む．原記載論文中で提唱された通称はオオビラルリオサムシだが，タイプ産地の大平山は「おびらやま」もしくは「おおひらやま」と読むのが正しいようである．

※ *C. (A.) gehinii* "Shimamaki lowland population" オオルリオサムシ島牧低地個体群

In the low altitudinal area of Shimamaki-mura at the northwestern foot of Mt. Obira-yama, a unique population of *C. (A.) gehinii*, called "Shimamaki lowland population" (IMURA, 1991, p. 21) is distributed. They are large-sized (27~37 mm in body length), body above predominantly reddish or greenish coppery (gustavhauseri type or standard type), rarely purplish blue (cyaneoviolaceus type). Elytral sculpture is variable distinctly according to individuals, changes drastically from the *gehinii*-type to the *munakatai*-type, suggesting that it is an intergrading or hybridized population between the nominotypical *gehinii* and subsp. *munakatai*. This population is narrowly restricted to a part of Shimamaki-mura facing the Sea of Japan.

大平山北西麓の島牧村低地には，大型（体長27~37 mm）で変化に富む特異な本種の一群が生息している．背面の色彩は赤銅色型と標準型ないし緑色型が主体であるが，大平山亜種には見られないような独特の色彩を有する個体が現れ，また稀ながら青紫色型も出現する．体長や体形，鞘翅彫刻も変化に富み，渡島半島亜種と変わらないものから基亜種に準じたものまで様々な程度の個体が見られる．恐らくは基亜種から渡島半島亜種への移行型ないし両者の交雑集団であろうと推測されるが，その由来に関しては不明な点が多い．分布は非常に局地的で，宮内（ぐうない）温泉，泊，永豊，大平，床丹（とこたん）等，日本海に面した島牧村のごく一角のみから知られる．

37-9. *C. (A.) g. munakatai* (ISHIKAWA, 1968) オオルリオサムシ渡島半島亜種

Original description　*Damaster (Acoptolabrus) munakatai* ISHIKAWA, 1968, Bull. natn. Sci. Mus., Tokyo, 11 (1), p. 146, fig. 3 (p. 147).
Type locality　Mt. Daisengendake.
Type depository　National Museum of Nature and Science, Tokyo [Holotype, ♂].
Diagnosis　24~38 mm. Body above coppery reddish (standard type) or purplish blue with head and pronotum greenish (cyaneoviolaceus type). Pronotum almost as long as wide, with sides less strongly arcuate in anterior halves, not stepped to neck, strongly constricted before hind angles which are triangular, rather prominently protruding laterad but hardly so posteriad, propleura are partly visible from above between widest part and narrowest constriction. Elytra distinctly uneven, with primaries very broad, irregularly and frequently interrupted by large and deep primary foveoles to form rows of callosities; secondaries indicated by rows of smaller callosities; tertiaries indicated by arborescent processes bordering primary callosities, further fused with secondaries to form reticular pattern;; elevated part of each interval rather flat-topped.
Distribution　Oshima Peninsula, SW. Hokkaido.
Etymology　Named after Meiyô MUNAKATA of Hakodate.

タイプ産地　北海道（松前），大千軒岳．
タイプ標本の所在　国立科学博物館［ホロタイプ，♂］．
亜種の特徴　24~38 mm. 背面の色彩は標準型或いは青紫色型．タイプ産地の大千軒岳では，頭部と前胸背板が赤銅色，鞘翅が金緑色の標準型が大半を占め，全体にかなり青みを強く帯びたものや暗化した個体も出現するが，出現頻度は低い．大千軒岳以外の渡島半島各地では，頭部と前胸背板が緑色，鞘翅が青紫色となる，いわゆる青紫色型の個体が多いが，さまざまな程度の中間的色彩を有する個体も出現する．前胸背板は長幅ほぼ等しく，側縁の前半部は丸みに乏しく，前方で段差なく比較的滑らかに頚部と接続し，後方では後角前方で顕著にくびれ，最広部と最狭部の間では背面から前胸側板の一部が見える．後角は鋭角でその後縁は押圧され，側方に向けて突出するが後方への突出は弱い．鞘翅は凹凸が著しい．第一次間室は幅広く，大きく深い第一次凹陥によって不規則かつ頻繁に分断された隆起列，第二次間室はそれよりやや小さく，第三次間室は第一次間室の隆起部周囲を放射状に取り囲む小突起となり，第二次間室の隆起部とも連結して複雑な網眼状構造を呈する．各間室隆起部の頂部は比較的平坦．

分布　渡島半島．松前半島の大千軒岳から記載されたもので，当初は同山の特産と思われていたが，その後の調査で渡島半島各地に広く分布していることが判明している．ただし，太平洋側の内浦湾に沿った低地や亀田半島からは記録されていない．渡島半島基部において基亜種と側所的に棲み分けており，両者の分布境界（G-M line）は，太平洋側の長万部川から黒松内町中ノ川～旭野の間を通り，日本海側の島牧村折川～大平川間を結ぶ線であることが判明している（井村，1984）．タイプ産地の大千軒岳では標高900 m以上の亜高山性草原が主な生息地となっており，生息密度は高い．狩場山や遊楽部岳などは1,000 m以上の

37. *Carabus* (*Acoptolabrus*) *gehinii*

標高があるにもかかわらず，大千軒岳のような草原が発達していないため，本種の生息密度ははるかに低い．世代別岳や黒松内町各地では純然たるササ原から得られることもあり，島牧村の低地では海岸に近いイタドリの茂みなどにも生息している．タイプ産地の大千軒岳を除き，一般に局地的で個体数もそれほど多くない．

亜種小名の由来と通称 亜種小名は函館の棟方明陽に因む．最も一般的な通称はオシマルリオサムシだが，ムナカタルリオサムシ，ダイセンゲンオサムシ等と呼ばれたこともある．

37. *Carabus* (*Acoptolabrus*) *gehinii*

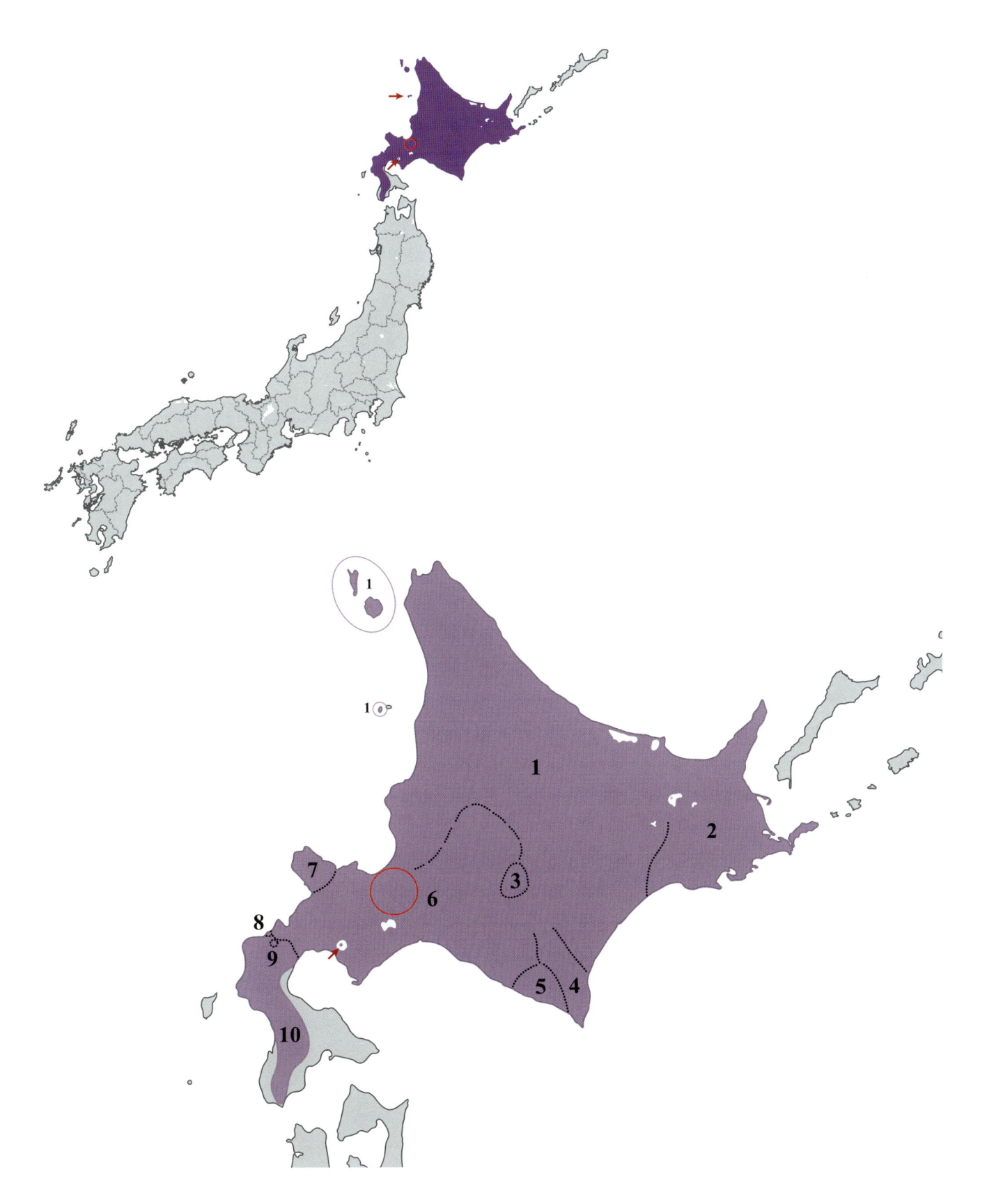

Fig. 67. Map showing the distributional range of ***Carabus*** (***Acoptolabrus***) ***gehinii*** Fairmaire, 1876. — 1, Subsp. ***aereicollis***; 2, subsp. ***konsenensis***; 3, subsp. ***manoianus***; 4, subsp. ***radiatocostatus***; 5, subsp. ***sapporensis***; 6, subsp. ***gehinii***; 7, subsp. ***shimizui***; 8, Shimamaki lowland population; 9, subsp. ***nishijimai***; 10, subsp. ***munakatai***. ○: Most likely type area of the species (Sapporo or its nearby regions).

図 67. **オオルリオサムシ**分布図. — 1, 道北亜種; 2, 根釧台地亜種; 3, 南富良野亜種; 4, 日高山脈南東部亜種; 5, 日高山脈南西部亜種; 6, 基亜種; 7, 積丹半島亜種; 8, 島牧低地個体群, 9, 大平山亜種; 10, 渡島半島亜種. ○: 最も可能性の高い種のタイプ地域（札幌ないしその周辺地域）.

329

38. *Carabus* (*Damaster*) *blaptoides* (Kollar, 1836)
マイマイカブリ

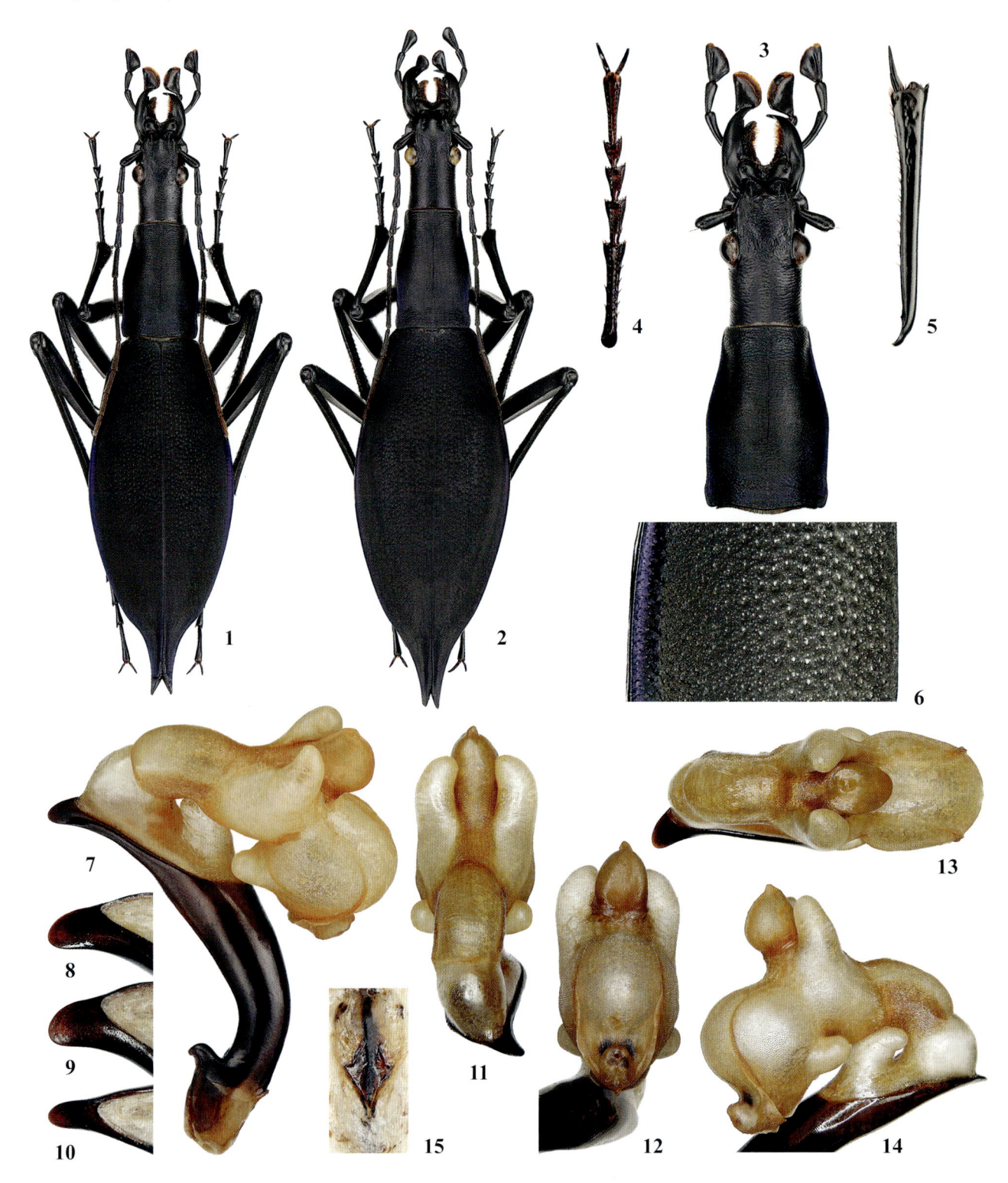

Figs. 1-15. *Carabus* (*Damaster*) *blaptoides blaptoides* (Kollar, 1836) (Unzen-onsen, Unzen-shi, Nagasaki Pref.). — 1, ♂; 2, ♀; 3, male head & pronotum; 4, male right protarsus in ventral view; 5, male right protibia in subdorsal view; 6, male left elytron (median part); 7, penis & fully everted internal sac in right lateral view; 8, apical part of penis in right lateral view; 9, ditto in right subdorsal view; 10, ditto in dorsal view; 11, internal sac in caudal view; 12, ditto in frontal view; 13, ditto in dorsal view; 14, ditto in left sublateral view; 15, inner plate of ligular apophysis.

図 1-15. **マイマイカブリ基亜種**(長崎県雲仙市雲仙温泉). — 1, ♂; 2, ♀; 3, ♂の頭部と前胸背板; 4, ♂右前跗節(腹面観); 5, ♂右前脛節(斜背面観); 6, ♂左鞘翅中央部; 7, 陰茎と完全反転下の内嚢(右側面観); 8, 陰茎先端(右側面観); 9, 同(右斜背面観); 10, 同(背面観); 11, 内嚢(尾側面観); 12, 同(頭側面観); 13, 同(背面観); 14, 同(左斜側面観); 15, 膣底部節片内板.

38. *Carabus* (*Damaster*) *blaptoides*

Figs. 16-27. *Carabus* (*Damaster*) *blaptoides* subspp. — 16-21, *C.* (*D.*) *b. rugipennis* (16-18, 21, ♂, 19, 20, ♀; 16, Cape Sôya-misaki, Wakkanai-shi; 17, Mt. Asahi-yama, Asahikawa-shi; 18, Is. Etorofu-tô (= Iturup); 19, Mt. Moiwa-yama, Sapporo-shi; 20, Is. Okushiri-tô; 21, Dôzan-chô, Hakodaté-shi; all from Hokkaido); 22-24, *C.* (*D.*) *b. viridipennis* (22, 24, ♂, 23, ♀; 22, Aomori-shi, Aomori Pref.; 23, Iwasaki, Fukaura-machi, Aomori Pref.; 24, Futatsui-machi, Noshiro-shi, Akita Pref.); 25-27, *C.* (*D.*) *b. babaianus* (25, 27, ♂, 26, ♀; 25-26, Yonezawa-shi, Yamagata Pref.; 27, Urasa, Minami-uonuma-shi, Niigata Pref.). — a, Male right protarsus in ventral view; b, penis & fully everted internal sac in right lateral view.

図 16-27. マイマイカブリ各亜種. — 16-21, **北海道亜種**(16-18, 21 ♂, 19, 20, ♀; 16, 稚内市宗谷岬; 17, 旭川市旭山; 18, 択捉島; 19, 札幌市藻岩山; 20, 奥尻島; 21, 函館市銅山町; 以上すべて北海道産); 22-24, **東北地方北部亜種**(22, 24, ♂, 23, ♀; 22, 青森県青森市; 23, 青森県深浦町岩崎; 24, 秋田県能代市二ツ井町); 25-27, **東北地方南部亜種**(25, 27, ♂, 26, ♀; 25-26, 山形県米沢市; 27, 新潟県南魚沼市浦佐). — a, ♂右前跗節(腹面観); b, 陰茎と完全反転下の内嚢(右側面観).

38. *Carabus* (*Damaster*) *blaptoides*

Figs. 28-39. ***Carabus*** (***Damaster***) ***blaptoides*** subspp. — 28-29, ***C.*** (***D.***) ***b. fortunei*** (28, ♂, 29, ♀; Is. Awa-shima, Niigata Pref.); 30-31, ***C.*** (***D.***) ***b. capito*** (30, ♂, Mt. Myôken-san; 31, ♀, Umezu; both in Is. Sado-ga-shima, Niigata Pref.); 32-33, intergrading population between *C.* (*D.*) *b. babaianus* & *C.* (*D.*) *b. oxuroides* (so-called *cyanostolus*) (32, ♂, Nakadogawa, Takahagi-shi, Ibaraki Pref.; 33, ♂, Ryûzu-no-taki, Nikkô-shi, Tochigi Pref.); 34-39, ***C.*** (***D.***) ***b. oxuroides*** (34, 36-39, ♂, 35, ♀; 34, Mukaikoga, Kazo-shi, Saitama Pref.; 35, Omonma, Toridé-shi, Ibaraki Pref.; 36, Shinohara-chô, Kôhoku-ku, Yokohama-shi, Kanagawa Pref.; 37, Yamanakako-mura, Yamanashi Pref.; 38, Mt. Yatsu-ga-také, Koumi-machi, Nagano Pref.; 39, Kuma, Tenryû-ku, Hamamatsu-shi, Shizuoka Pref.). — a, Male right protarsus in ventral view; b, penis & fully everted internal sac in right lateral view.

図 28-39. **マイマイカブリ各亜種**. — 28-29, **粟島亜種**(28, ♂, 29, ♀; 新潟県粟島); 30-31, **佐渡島亜種**(30, ♂, 新潟県佐渡島妙見山, 31, ♀, 同県同島梅津); 32-33, 東北地方南部亜種と関東・中部地方亜種の交雑集団(いわゆるミヤママイマイカブリ)(32, ♂, 茨城県高萩市中戸川; 33, ♂, 栃木県日光市竜頭の滝); 34-39, **関東・中部地方亜種**(34, 36-39, ♂, 35, ♀; 34, 埼玉県加須市向古家; 35, 茨城県取手市小文間; 36, 神奈川県横浜市港北区篠原町; 37, 山梨県山中湖村; 38, 長野県小海町八ヶ岳; 39, 静岡県浜松市天竜区熊). — a, ♂右前跗節(腹面観); b, 陰茎と完全反転下の内嚢(右側面観).

38. *Carabus* (*Damaster*) *blaptoides*

Figs. 40-51. ***Carabus*** (***Damaster***) ***blaptoides*** subspp. — 40-41, ***C***. (***D***.) ***b***. ***brevicaudus*** (40, ♂, 41, ♀, HT; 40, Oki-kokubunji Temple; 41, Fukuura; both from Is. Dôgo-jima, Iss. Oki, Shimane Pref.); 42-51, ***C***. (***D***.) ***b***. ***blaptoides*** (42, 44, 46, 48, 49, ♂, 43, 45, 47, 50, 51, ♀; 42, Kutsuki-murai, Takashima-shi, Shiga Pref.; 43, Kameyama-jô Castle Ruins, Gobô-shi, Wakayama Pref.; 44, Jôgéchô-yano, Fuchû-shi, Hiroshima Pref.; 45, Yugé, Kumenan-chô, Okayama Pref.; 46, Chisei, Uchiko-chô, Ehime Pref.; 47, Mt. Yokokura-yama, Ochi-chô, Kochi Pref.; 48, Mt. Konpira-san, Kita-kyûshû-shi, Fukuoka Pref.; 49, Satochôsato, Is. Kami-koshiki-jima, Kagoshima Pref.; 50, Tomié-machi, Is. Fukué-jima, Nagasaki Pref.; 51, Shiratani, Is. Yaku-shima, Kagoshima Pref.). — a, Male right protarsus in ventral view; b, penis & fully everted internal sac in right lateral view.

図 40-51. **マイマイカブリ各亜種**. — 40-41, **隠岐亜種**(40, ♂, 41, ♀, HT; 40, 隠岐国分寺; 41, 福浦; ともに島根県隠岐島後島産); 42-51, **基亜種** (42, 44, 46, 48, 49, ♂, 43, 45, 47, 50, 51, ♀; 42, 滋賀県高島市朽木村井; 43, 和歌山県御坊市亀山城跡; 44, 広島県府中市上下町矢野; 45, 岡山県久米南町弓削; 46, 愛媛県内子町知清; 47, 高知県越知町横倉山; 48, 福岡県北九州市金比羅山; 49, 鹿児島県上甑島里町里; 50, 長崎県福江島富江町; 51, 鹿児島県屋久島白谷). — a, ♂右前跗節(腹面観); b, 陰茎と完全反転下の内嚢(右側面観).

38. *Carabus* (*Damaster*) *blaptoides* (KOLLAR, 1836) マイマイカブリ

Original description   *Damaster blaptoides* KOLLAR, 1836, Ann. Wien Mus., 2, p. 334, tab. 31, fig. 1.

History of research   This taxon was described by KOLLAR (1836) as a new genus and species of the subtribe Cychrodea from "Japonia ?". Since then till 1880s, not a few taxa of the genus *Damaster* were described as independent species. Of these, the representative examples are as follows: *D. rugipennis* MOTSCHULSKY, 1861 from "Khokodady" (= Hakodaté of SW. Hokkaido), *D. Fortunei* [sic] ADAMS, 1861 from "Awa-Sima" (= Is. Awa-shima of Niigata Pref.), *D. oxuroides* SCHAUM, 1862 from "Japan", *D. auricollis* WATERHOUSE, 1867 (= *D. rugipennis*) from "Japan (Hakodady)", *D. Lewisii* [sic] RYE, 1872 (= *D. blaptoides*) from "Hiogo in insula Nipon et Simabara in insula Kiushiu Japanorum", *D. pandurus* BATES, 1873 (= *D. oxuroides*) from "Yokohama", *D. viridipennis* LEWIS, 1880 from "Awomori, Japan", *D. capito* LEWIS, 1881 from "ins. Sado", and *D. pandurus* var. *cyanostola* LEWIS, 1882 (= intergrading population between *Carabus* (*D.*) *b. babaianus* and *C.* (*D.*) *b. oxuroides* in the present sense) from "the mountains of Chiuzenji". However, KOLLAR's race is currently regarded as a single, polytypic species belonging to the subtribe Carabina, and is generally classified into the following eight subspecies; *rugipennis*, *viridipennis*, *fortunei*, *babaianus*, *capito*, *oxuroides*, *brevicaudus* (described by IMURA & MIZUSAWA from the Oki Islands in 1995) and the nominotypical *blaptoides*.

Type locality   "Japonia ?" (most probably Nagasaki or its nearby regions).

Type depository   Naturhistorisches Museum Wien [Holotype, ♂].

Morphology   26~65 mm. Head and pronotum variable in color, metallic green, golden coppery, reddish purple, bluish purple, purple, dark indigo or black; elytra black and mat, though usually bearing purplish, yellow greenish or greenish tinge in individuals occurring in northern Japan. External features: Figs. 1~6. Description on details: penultimate segment of labial palpus with two to six setae; median tooth of mentum much shorter than lateral lobes, with tip weakly pointed and hardly protruded ventrad; submentum asetose; male antennae without hairless areas on ventral side; pronotum asetose; elytral sculpture triploid and very weak; primaries and secondaries are indicated by very low segmented costae or rows of granules; tertiaries vestigial, not fused with primaries; striae between intervals indicated by rows of punctures, areas between intervals coarsely scattered with small granules; propleura smooth or vaguely rugoso-punctate; sides of sternites irregularly wrinkled and uneven; metacoxa bisetose, lacking anterior seta; sternal sulci clearly carved. Genitalic features: Figs. 7~15. Penis 0.3~0.4 times as long as elytra; ostium lobe large and robust, with tip faintly bilobed; internal sac simple in basal portion, median portion with strongly protruded parasaccellar lobes and saccellus, apical portion strongly inflated with well-recognized podian lobes; peripheral rim of gonopore short and barely projecting. Female genitalia with inner plate of ligular apophysis (Fig. 15) rhomboidal, strongly sclerotized and pigmented.

Molecular phylogeny   On the molecular phylogenetic tree of the mitochondrial ND5 gene of the Carabina of the world, this species belongs to the Chinese lineage of so-called Procrustimorphi (= almost corresponding to the *Procrustes* subgenus-group in this book), and forms an independent cluster. On the tree, this species is first divided into two major lineages whose distributional areas are the northeast Japan and southwest Japan. The first lineage is further divided into three sublineages, each distributed in Hokkaido including the northernmost part of Honshu, northern Tohoku and southern Tohoku including Iss. Awa-shima and Sado-ga-shima. The second lineage is divided into five sublineages distributed in central Honshu (Kanto + NE. Chubu), Chubu, Kii Peninsula, SW. Honshu (W. Kinki + Chugoku) + Shikoku and Kyushu, respectively. This result does not necessarily run parallel with the morphological classification of this species. For details, refer to SU *et al.* (1996), SU *et al.* (1998), KIM, SAITO *et al.* (1999), Osawa & Su (2001b), OSAWA *et al.* (2002, 2004), etc.

Distribution   Hokkaido, Honshu, Shikoku & Kyushu / Iss. Naka-jima [Lake Tôya-ko], Daikoku-jima, Okushiri-tô, Shikotan-tô, Kunashiri-tô, Etorofu-tô (Hokkaido), Ô-shima of Kesennuma-shi, Izu-shima, Kinkasan, Tashiro-jima, Aji-shima, Miyato-jima, Sabusawa-jima, Nono-shima, Katsura-shima (Miyagi Pref.), Awa-shima, Sado-ga-shima (Niigata Pref.), Noto-jima (Ishikawa Pref.), Izu-ô-shima, Nii-jima (Tokyo Met.), Saru-shima, Jô-ga-shima (Kanagawa Pref.), Chikubu-jima & Oki-shima (or Oki-no-shima) [Lake Biwa-ko] (Shiga Pref.), Tôshi-jima (Mie Pref.), Oki-no-shima (Wakayama Pref.), Awaji-shima (Hyogo Pref.), Dôgo, Naka-no-shima, Nishi-no-shima, Chiburi-jima (Shimane Pref.), In-no-shima, Mukai-shima, Ikuchi-jima, Ôsaki-kami-jima, Ôsaki-shimo-jima, Kami-kamagari-jima, Shimo-kamagari-jima, Itsuku-shima, Ujina-jima, Eta-jima+Nômi-shima, Kurahashi-jima (Hiroshima Pref.), Yashiro-jima, Heigun-tô, Naga-shima, Iwai-shima, Kasado-jima, Futaoi-jima (Yamaguchi Pref.), Shôdo-shima, (Maé-jima (= Ômi Penins.)), Awa-shima, (Ya-shima), (Kagawa Pref.), Shimada-jima, Ôgé-jima (Tokushima Pref.), Ikina-jima, Hakata-jima, Ômi-shima, Ô-shima of Imabari-shi, Tsuwaji-jima, Nuwa-jima, Naka-jima, Futagami-jima, Muzuki-jima, Nogutsuna-jima, Gogo-shima (Ehime Pref.), Noko-no-shima, Ô-shima of Munakata-shi (= Chikuzen-ô-shima) (Fukuoka Pref.), Iki, Ikitsuki-shima, Hirado-shima, Nakadôri-jima, Wakamatsu-jima, Naru-shima, Hisaka-jima, Fukué-jima, Taka-shima, Fuku-shima, Maki-shima (Nagasaki Pref.), Ô'nyû-jima (Oita Pref.), Amakusa-kami-shima, Amakusa-shimo-jima, Ôyano-jima, Tobasé-jima, Gosho-no-ura-jima (or Gosho-ura-jima) (Kumamoto Pref.), Shimaura-tô (Miyazaki Pref.), Shishi-jima, Ikara-jima, Naga-shima, Kami-koshiki-jima, Naka-koshiki-jima, Shimo-koshiki-jima, Tané-ga-shima & Yaku-shima (Kagoshima Pref.). Outside Japan: S.~C. Kurils.

Habitat   This is the species with a high adaptability to various environment, and widely distributed in Japan including many attached islands.

Bionomical notes   Spring breeder, snail feeder. This species overwinters in both adult and larva in the soil or dead decayed trunk.

Geographical variation   Morphologically classified into eight subspecies.

Etymology   Specific name means "similar to *Blaps* (a genus belonging to the family Tenebrionidae)".

38. *Carabus (Damaster) blaptoides*

研究史 本タクソンはオーストリアのKOLLAR（コラー）(1836)によりセダカオサムシ亜族に属する新属新種*Damaster blaptoides*として「日本？」から記載された．以来，1880年代前半に至るまでのおよそ半世紀の間に，いくつもの*Damaster*属の「新種」が海外の研究者により記載された．それらのうち代表的なものを記載年順に挙げると以下のようになる：*D. rugipennis* MOTSCHULSKY, 1861（函館；以下，学名末尾の地名はその種のタイプ産地を表す），*Damaster Fortunei* ADAMS, 1861（粟島），*D. oxuroides* SCHAUM, 1862（日本，恐らく横浜付近？），*D. auricollis* WATERHOUSE, 1867（函館，*D. rugipennis*の異名），*D. Lewisii* RYE, 1872（兵庫と島原，*D. blaptoides*の異名），*D. pandurus* BATES, 1873（横浜，*D. oxuroides*の異名），*D. viridipennis* LEWIS, 1880（青森），*D. capito* LEWIS, 1881（佐渡），*D. pandurus* var. *cyanostola* LEWIS, 1882（中禅寺，本書で言うマイマイカブリ東北地方南部亜種と関東・中部地方亜種の移行型）．現在では，マイマイカブリを多型を示す1種とみなす扱いが一般的で，これらの「種」はいずれも亜種に降格されるか，異名として扱われている．本書でもマイマイカブリを多型を示す1種*Carabus (Damaster) blaptoides*として扱い，*rugipennis*, *viridipennis*, *fortunei*, *babaianus*, *capito*, *oxuroides*, *brevicaudus*, *blaptoides*の8亜種に分類しておく（他にも，ダマステル属（あるいはカラブス属のダマステル亜属）の下にいくつもの種や亜種が記載されているが，産地の間違いや人為的変更に基づく怪しげなものも含まれているうえ，一般にそのすべてが上記8亜種のいずれかの異名とされているため，詳細については省略する）．

タイプ産地 「日本？」．タイプ産地の詳細は不明だが，タイプ標本の形態と本種が記載された当時の時代背景（1836年＝天保7年）から考えて，長崎ないしその周辺地域から得られたものである可能性が高い．ホロタイプ（♂）は大顎を含めた体長が50 mm（上唇先端から鞘翅端までは47mm），鞘翅最広部の幅が11 mmと，比較的細く，小型の個体である．長崎市そのものから得られる本種には比較的大型で太い個体が多いので，タイプ標本は少し離れた山岳地帯，例えば古くから湯治場として開けていた雲仙温泉などから得られたものかもしれない．筆者（井村）はかつてウィーン自然史博物館を訪れた際に本種のホロタイプを借り受け，同博物館の許可を得て胸部の筋肉を摘出し，生命誌研究館の共同研究者に依頼してDNAの分析を試みたが，残念ながら塩基配列を読み取ることはできなかった．

タイプ標本の所在 ウィーン自然史博物館［ホロタイプ，♂］．

形態 26~65 mm．頭部と前胸背板の色彩は亜種又は個体によって金緑，金銅，紫紅，青紫，紫，暗藍，黒などに変化し，鞘翅は光沢を欠く黒色だが，北方に産するものでは黄緑色，緑色，紫色等を帯びる．体側面と腹面にも前胸背板と同様ないし紫色の金属光沢を帯びる．外部形態：図1~6．細部の記載：下唇肢亜端節の剛毛は2~6本．下唇基節の中央歯は側葉よりはるかに短く，先端の前方への突出は弱く，腹側に向けて殆ど突出しない．下唇亜基節は無毛．♂触角腹面に無毛部はない．前胸背板の側縁に剛毛を欠く．鞘翅端は多少なりとも腹端を越えて後方へ突出し，尾状突起を形成するが，南のものほど長く顕著になる傾向がある．鞘翅彫刻は三元的で，第一次，第二次間室は破線状の極めて低い隆条列ないし顆粒列，第三次間室は退化的でしばしば不明瞭だが，第一次間室の隆起列と融合することはない．間室間の線条は点刻列となり，鞘翅基面には微小な顆粒を比較的密に装う．前胸側板は滑らかだが，ごく弱い皺ないし点刻を装う場合がある．腹部腹板の側面は凹凸が著しく，表面には不規則な皺を装う．後基節の剛毛は2本で，前方剛毛を欠く．腹部腹板の先端3節には顕著な横溝がある．交尾器形態：図7~15．陰茎長は鞘翅長の0.3~0.4倍．葉片は大きく，太く，先端は僅かに分葉傾向を示す．内嚢基部は単純で，突起や隆起を欠き，同中部には背側に強く突出した傍盤葉と内嚢盤を具え，内嚢先端部は強く膨らみ，脚葉が目立つ．射精孔縁膜の硬化と色素沈着は弱く，殆ど突出しない．♀交尾器膣底部節片内板（図15）は菱形で，硬化と色素沈着が見られる．

分子系統 ミトコンドリアDNAによる世界のオサムシ亜族の分子系統樹上で，本種はオオルリオサムシやツシマカブリモドキと共にヨロイオサムシ群（本書で言うプロクルステス亜属群にほぼ相当する，カラブス属の中の大群）の中の中国系亜群に属し，種としても単系統でまとまりのよいクラスターを形成する．種内ではまず東西の2系統に大きく分かれ，前者はさらに北海道（本州北端を含む）・東北地方北部・東北地方南部（粟島と佐渡島を含む）に分布する3亜系統に，後者は本州中央部（関東地方と中部地方の北東部）・中部地方・紀伊半島・本州西部（近畿地方西部と中国地方）と四国・九州に分布する5亜系統に分かれる．こうした分子系統解析の結果は，外部形態の相違に基づく現行の亜種分類との間に部分的にかなりのギャップが見られ，「マイマイカブリ」の系統分類に関しては再検討を要するが，ここでは便宜上，従来の形態学から見た一般的な亜種分類を踏襲しておくことにする．本種の分子系統に関しては，SU, OHAMA *et al*.(1996)，SU *et al*.(1998)，KIM, SAITO *et al*.(1999)，大澤・蘇(2001b)，大澤ら(2002)，OSAWA *et al*.(2004)等に詳しいので参照されたい．

分布 北海道，本州，四国，九州／中島［洞爺湖］，大黒島，奥尻島，色丹島，国後島，択捉島（以上，北海道），大島［気仙沼市］，出島，金華山，田代島，網地島，宮戸島，寒風沢島，野々島，桂島（以上，宮城県），粟島，佐渡島（以上，新潟県），能登島（石川県），伊豆大島，新島（以上，東京都），猿島，城ヶ島（以上，神奈川県），竹生島，沖島［琵琶湖］（以上，滋賀県），答志島（三重県），沖ノ島（和歌山県），淡路島（兵庫県），島後，中ノ島，西ノ島，知夫里島（以上，島根県），因島，向島，生口島，大崎上島，大崎下島，上蒲刈島，下蒲刈島，厳島，宇品島，江田島＋能美島，倉橋島（以上，広島県），屋代島［＝周防大島］，平郡島，長島，祝島，笠戸島，藍井島（以上，山口県），小豆島，（前島（＝大深半島）），粟島，（屋島），（以上，香川県），島田島，大毛島（以上，徳島県），生名島，伯方島，大三島，大島［今治市］，津和地島，怒和島，中島，二神島，睦月島，野忽那島，興居島（以上，愛媛県），能古島，大島（＝筑前大島）［宗像市］（以上，福岡県），壱岐，生月島，平戸島，中通島，若松島，奈留島，久賀島，福江島，鷹島，福島，牧島（以上，長崎県），大入島（大分県），天草上島，天草下島，大矢野島，戸馳島，御所浦島（以上，熊本県），島浦島（宮崎県），獅子島，伊唐島，長島，上甑島，中甑島，下甑島，種子島，屋久島（以上，鹿児島県）．国外：千島列島中～南部（列島中部のブラト・チルポィエフ島からは確実な記録がある（Su *et al*., 2000）が，その他の島々における正確な分布

状況はよくわかっていない).

生息環境 様々な環境に広く適応している種で，日本列島のほぼ全域に分布しており，周辺島嶼からの記録も多い．しかし，近年とみにその個体数を減じており，例えば冬季のオサ掘りによる採集では，河川敷の朽木等からはいまだにまとまった数が採集できる場所も多いが，一般の山道の崖崩しで得られる個体は以前より明らかに少なくなっている印象を受ける．

生態 春繁殖型で，成虫・幼虫双方の越冬態を持つ，軟体動物(カタツムリ)食のオサムシ．冬季には土中・朽木中双方で越冬し，同一の越冬窩内に複数の個体が集団で越冬しているケースもよく観察される．

地理的変異 外部形態のうえからは顕著な地理的変異を示し，以下の8亜種に分類されるが，これらの各集団間において，陰茎および内嚢の基本的形態には殆ど差が見られない．

種小名の由来 種小名は「*Blaps*(ゴミムシダマシ科の中の一属)の形をした，*Blaps*のような」を意味する．黒くて琵琶のような形をした*Blaps*属の種を連想させることから，こう名付けられたものであろう．因みに，亜属名の*Damaster*はギリシア語のdamastēsの語尾を書き換えたもので，「征服者」を意味する．

38-1. *C. (D.) b. rugipennis* (MOTSCHULSKY, 1861) マイマイカブリ北海道亜種

Original description *Damaster rugipennis* MOTSCHULSKY, 1861, Etudes Ent., 10, p. 6.
Type locality Des environs de Khokodady.
Type depository Zoological Museum of Moscow University [Syntype ?].
Diagnosis 26~44 mm. Head above and pronotum green, greenish red or reddish coppery bearing metallic luster; elytra black and mat, often with purplish tinge, rarely greenish. Neck relatively short for this species, with the length less than twice of longitudinal diameter of eye; pronotum relatively robust for this species, about 1.1 to 1.2 times as long as wide, widest nearly at middle, narrowed toward front and base, with margins almost parallel-sided behind front angles, slightly divergent toward hind angles; apices of elytra short, slightly extending apical end of sternites, with tips not sharply pointed; primary intervals of elytra clearly recognized, secondaries weaker than primaries, tertiaries sometimes unclear; striae between intervals indicated by rows of large and deep punctures fused with foveoles of intervals to form reticular pattern or transverse wrinkles; sternites smooth, associated with punctures and setae; basal three segments of male protarsus dilated with hair pads on ventral surfaces. Apical part of penis often triangularly pointed; internal sac with basal portion robust and strongly inflated dorsad, parasaccellar lobes usually rather small and short.
Distribution Hokkaido / Iss. Naka-jima [Lake Tôya-ko], Daikoku-jima, Okushiri-tô, Shikotan-tô, Kunashiri-tô & Etorofu-tô (Hokkaido).
Etymology Subspecific name means "having wrinkled elytra".

タイプ産地 ココダディ(＝函館)．

タイプ標本の所在 モスクワ大学動物学博物館[シンタイプ？]．同館のN. NIKITSKIJ(ニキツキィ)によれば，MOTSCHULSKY(モッチュルスキィ)が記載に使用したと思われる標本の他に，同時期にコレクションされたと思しき複数の本タクソンの個体が保管されているという．シンタイプとみなすべきものであれば，いずれレクトタイプの指定が必要となろう．

亜種の特徴 26~44 mm. 頭部と前胸背板の背面は緑色，緑赤色，赤銅色などのものがあり，金属光沢を帯びる．鞘翅は黒色で，しばしば紫色，稀に緑色を帯び，光沢は鈍い．頸部は本種としては比較的短く，長さは複眼の長径の2倍以下．前胸背板は長さが幅の1.1~1.2倍前後で，最広部はほぼ中央にあり，前後に向けて狭まり，側縁は前角の後方で左右ほぼ平行になり，後角に向けてやや広がる．鞘翅端は腹節の先端を僅かに越えるが，先端は鈍く尖るか又は全く角をなさず，顕著な尾状突起は形成しない．鞘翅第一次間室は常に明瞭だが，第二次間室はそれより弱く，第三次間室は時に不明瞭．間室間の線条は比較的顕著な点刻列となり，各間室の孔点と融合して網眼状ないし横皺状となる．腹板は滑らかで孔点があり，剛毛を装う．♂前跗節は基部3節が広がり，腹面に絨毛盤を有する．陰茎先端はやや尖ったものが多く，内嚢は基部が太短く，同部の背面は膨らみが強く，傍盤葉は小さく短めだが，地域や個体による変異も大きい．

分布 北海道／中島[洞爺湖]，大黒島，奥尻島，色丹島，国後島，択捉島(以上，北海道)．

亜種小名の由来と通称 亜種小名は「皺のある羽(鞘翅)の」を意味する．通称はエゾマイマイカブリ．

38-2. *C. (D.) b. viridipennis* (LEWIS, 1880) マイマイカブリ東北地方北部亜種

Original description *Damaster viridipennis* LEWIS, 1880, Ent. month. Mag., 17, p. 161.
Type locality Awomori, Japan.
Type depository Natural History Museum, London [Holotype, ♂].
Diagnosis 28~44 mm. Head above, pronotum and scutellum reddish purple with metallic luster; elytra black, more or less bearing greenish tinge, or sometimes yellowish or bluish, and rather mat; lateral and ventral sides of body also bearing red-purplish metallic tinge; pronotum a little slenderer than in subsp. *rugipennis*, with lateral margins apparently divergent before hind angles which are sharper than in subsp. *rugipennis*; elytra also slenderer and less strongly convex above than in subsp. *rugipennis*, with shoulders less effaced to form spindle-shaped profile in dorsal view; apical tips of elytra sharply pointed to form mucros; elytral sculpture almost as

in subsp. *rugipennis*, though less strongly uneven in population distributed in southern part of distributional range; abdominal sternites smooth and shiny, scattered with small pores; basal three segments of male protarsus dilated, with hair pads on ventral surfaces. Male genitalia almost as in subsp. *rugipennis*.

Distribution　N. Tohoku Region (Aomori Pref., Akita Pref., Iwate Pref. & E. Miyagi Pref.).

Etymology　Subspecific name means "having green elytra".

タイプ産地　日本，青森．

タイプ標本の所在　ロンドン自然史博物館［ホロタイプ，♂］．現存するホロタイプは，大きな台紙に触角と脚を一杯に伸ばして貼り付けられており，頭胸部の紫紅色と鞘翅の緑色が非常に鮮烈な，美しい個体である．

亜種の特徴　28~44 mm．頭部の背面と前胸背板及び小盾板は金属光沢を帯びた紫紅色，鞘翅は多少なりとも緑色がかり，時に黄色ないし青色味を帯び，光沢は弱い．体側面，腹面にも赤紫～紫色の金属光沢を具える．前胸背板は北海道亜種よりもやや細く，側縁は後角の手前で明らかに広がり，後角はより鋭角になる．鞘翅も北海道亜種に比しより細く，肩部の張り出しが弱く，紡錘形となり，膨隆はより弱い．鞘翅端は鋭く尖り，尾状突起を形成する．鞘翅彫刻は北海道亜種のそれに似るが，分布域南部のものではより滑らかになる．腹板は光沢が強く滑らかで，孔点がある．♂前附節は基部3節が広がり，腹面に絨毛盤を有する．交尾器形態は北海道亜種のそれに比較的近い．

分布　東北地方北部（青森，秋田，岩手各県と宮城県東部）．

亜種小名の由来と通称　亜種小名は「緑色の羽（鞘翅）の」を意味する．通称はキタカブリまたはキタマイマイカブリ．

38-3. *C. (D.) b. babaianus* (Ishikawa, 1985)　マイマイカブリ東北地方南部亜種

Carabus (*Damaster*) *rugipennis* subsp. *fortunei* var. *montanus* Baba, 1939, Kontyû, 13, p. 205.
Damaster (s. str.) *blaptoides montanus*: Nakane, 1962, Ins. Jap., 2 (3), p. 69.
Carabus (*Damaster*) *ruipennis babaianus*: Ishikawa, 1984, Proc. Jap. Soc. syst. Zool., 28, p. 57.
Damaster (*Damaster*) *blaptoides babaianus*: Ishikawa, 1984, Proc. Jap. Soc. syst. Zool., 28, p. 57.
Damaster (*Damaster*) *blaptoides babaianus* Ishikawa, 1985, Coleopt. Japan Col., Osaka, 2, p. 48, figs. 3f~h (pl. 9).

Type locality　"Nagoya" according to the original description, though this locality is obviously wrong. The type locality of this taxon should be regarded as Yonezawa of Yamagata Prefecture inferred from the context of the original description. This mistake is thought to be due to careless proofreading.

Type depository　Unknown.

Diagnosis　27~44 mm. Closely allied to subsp. *viridipennis*, but differs from that subspecies in coloration and sculptural condition of elytra. Head and pronotum reddish crimson, purplish crimson, purple, bluish purple, indigo blue or blue; elytra black, though sometimes bearing faint bluish or greenish tinge. Elytral intervals usually not costate but reduced to rows of granules, punctures in and between intervals small and shallow. Apical tips of elytra (mucros) almost as in subsp. *viridipennis*. Apical three segments of male protarsus also as in subsp. *viridipennis*, though they become narrower with hair pads only in basal one segment or two segments in individuals from hybrid zone between subspp. *babaianus* and *oxuroides*.

Distribution　S. Tohoku Region (Miyagi Pref., Yamagata Pref. & N. Fukushima Pref.) & N. Chubu Region (Niigata Pref. & N. Nagano Pref.) / Iss. Ô-shima of Kesennuma-shi, Izu-shima, Kinkasan, Tashiro-jima, Aji-shima, Miyato-jima, Sabusawa-jima, Nono-shima & Katsura-shima (Miyagi Pref.).

Etymology　Named after Kintaro Baba (Niigata), a physician and entomologist who first described this subspecies under the name of "*Carabus* (*Damaster*) *rugipennis* subsp. *fortunei* var. *montanus*".

研究史　本タクソンはBaba（1939）によりCarabus (Damaster) rugipennis subsp. fortunei var. montanusという名で記載された．しかし，この学名の中のmontanusは三語名に付加した第四名であり，亜種よりも低位の実体に対して提唱された名であることが明らかなので適格ではない（国際動物命名規約（以下「規約」と略す）第4版条45.5）．中根（1962）はこれを亜種の有効名としてDamaster (s. str.) blaptoides montanus (Baba, 1939)という形で使用し，その形態学的な特徴についても言及した．これにより，montanusは最初にDamaster属に置かれた適格な亜種小名となり（規約第4版条10.2），その著者権は中根（1962）に帰することになる（規約第4版条45.5.1）．一方，Ishikawa（1984a）は，Baba（1939）の第四名montanusがNakane（1960と1962）によって適格な亜種小名になったものと想定し（"Baba published in 1939, *Carabus* (*Damaster*) *rugipennis* subsp. *fortunei* var. *montanus*, which is now regarded as a subspecies of *blaptoides* as above (Nakane, 1960 and 1962)" ～ Ishikawa（1984a）より抜粋．ただし，Nakane（1960）ではmontanusという名が亜種小名として用いられてはいるものの，このタクソンを識別するための形質を言葉で示した記載もしくは定義を伴っておらず，すでに公表されている言明への文献参照も伴っていないので，適格ではない（規約第4版条13.1）），その時点での有効名はNakaneの示したDamaster属に属することを支持しながらも，当時，無効名と考えていたと思われるCarabus blaptoides montanusをCarabus punctatoauratus montanus Géhin, 1882の同名とみなし（従って，Ishikawaにとって同名関係は成立していなかったことになるのだが），置換名としてCarabus (Damaster) rugipennis babaianus Ishikawaを提唱した．Ishikawaは当時，この新学名を有効名と考えておらず，その証拠に，直後にその時点での学名は"*Damaster* (*Damaster*) *blaptoides babaianus* (Ishikawa)" だと述べている．学名が適格であるためには，提唱された時点で有効名でなければならない（規約第4版条11.5）ので，Ishikawa（1984a）の *babaianus* は不適格である．本書では，国際的に広く認められている上位分類体系に従って，マイマイカブリを広義の *Carabus* 属の中の *Damaster* 亜属の一員として扱っている．しかし，この東北地方南部亜種に関しては，最初に *Damaster* 属に置かれた適格な亜種小名 *montanus* を用いて *Carabus* (*Damaster*) *blaptoides montanus* (Nakane, 1962) とすると，*Carabus* (*Chrysocarabus*) *punctatoauratus montanus* Géhin, 1882 の新参二次同名になってしまうので，*montanus* に代わる亜種小名が必要となる．Ishikawa（1984a）により与えられた *babaianus* は前述の如く不適格なので使えないが，石川により翌1985年に原色日本甲虫図鑑IIの中で与えられた *babaianus* (*Damaster blaptoides babaianus* という形で使用され，このタクソンを識別するための形質を言葉で示した記載文も伴っている) は適格名とみなしうる．従って，これを用いて *Carabus* (*Damaster*) *blaptoides babaianus* (Ishikawa, 1985) という学名で表記されることになる．

　説明が長く複雑になったが，要するに本タクソン（マイマイカブリ東北地方南部亜種）の学名は，

　1）マイマイカブリを *Damaster* 属とみなす場合には *Damaster blaptoides montanus* Nakane, 1962

　2）マイマイカブリを *Carabus* 属とみなす場合には *Carabus blaptoides babaianus* (Ishikawa, 1985)

となる．我が国では中根，石川らの影響によりマイマイカブリを *Damaster* 属とみなす著者が多い（世界的に見れば極めて少数派）が，その場合に *Damaster blaptoides babaianus* Ishikawa, 1984 を用いるのは誤りである．

　タイプ産地とタイプ標本の所在　Baba（1939）の提示した学名 *Carabus* (*Damaster*) *rugipennis* subsp. *fortunei* var. *montanus* が不適格であることにより，彼によって "Holotypus"，"Allotypus" と指定された標本は命名規約で規定されるタイプシリーズの資格がない．この *montanus* という名が客観異名や同名であれば自動的かつ強制的に古参のタクソンの標本がホロタイプになるが，本ケースではそのような自動化の規定もない．しかし，その後の Nakane（1962），Ishikawa（1984a），石川（1985）を初め，本タクソンを扱った著者の大半が，その学名上の取り扱いがどうであれ，実質的には Baba（1939）の原著のタクソン（元は var. すなわち変種としてのタクソンであるが）を支持していることは明らかなので，特別な理由がない限り，Baba（1939）が "Holotypus" に指定した標本をもってマイマイカブリ東北地方南部亜種のホロタイプとするのが理念的かつ合理的である．要するにホロタイプを後指定する必要がある．ところが，ここに一つ大きな問題がある．それは，Baba（1939）の論文中で述べられた *montanus* の形態学的特徴は東北地方南部のマイマイカブリのそれに一致しており，本文中でもドイツ語で「私の知る限り，この型は米沢周辺の山地のみから発見されている」という意味の記述がなされているにもかかわらず，Holotypus, Allotypus の産地は "Nagoya" と明記されている点である．つまり，Baba の *montanus* は米沢付近の集団に対して命名されたことが文脈からは明らかなのに，タイプシリーズの産地は「名古屋」となっており，記載文の内容に著しい矛盾が見られるのである．恐らく，同じ論文で記載された *Carabus* (*Damaster*) *blaptoides* var. *paraoxuroides*（タイプ産地は名古屋）との混同による産地の誤記あるいは誤植である可能性が高く，このことは胎内昆虫の家（新潟県胎内市）に保管されている馬場 金太郎の著作集の中に，問題の記載文のタイプ標本の産地と採集者の項（"Nagoya (Kol. K. Doi)"）が横一本線で消され，馬場の肉筆と思われる文字によって "Yonezawa (Y. Kurosawa)" と修正されているものがあることからも強く支持される．しかし，最終的な結論は，Baba により指定されたタイプシリーズ（もしも現存するものであるならば）を検し，その形態学的特徴と，付されているラベルを調べなければ出すことができない．この問題を解決することは本書の主旨とするところではないため，ここでは問題点を指摘するに留めておくが，マイマイカブリ東北地方南部亜種，いわゆるコアオマイマイカブリのタイプ産地とタイプ標本に関しては，このように大きな問題が残されているという点を一言記しておきたい．

　亜種の特徴　27～44 mm．東北地方北部亜種に近いが，頭部と前胸背板背面の色彩は紅赤色，紅紫色，紫色，青紫色，青藍色，青色などに変化し，一般に分布域北部に産するものは長波長の，分布域南部のものは短波長の色彩を帯びることが多い．鞘翅は基本的に黒色だが，僅かに青味や緑色味を帯びる場合もある．間室は一般に隆条とはならず，顆粒列へと退化し，間室内および間室間の点刻も小さく浅くなる．鞘翅端は東北地方北部亜種のそれに近い．♂前跗節は東北地方北部亜種と同様，基部3節が広がり，腹面に絨毛を有するが，分布域南部の関東・中部地方亜種との交雑帯のものでは退化傾向を示し，各節がより細くなり，基部1～2節に絨毛盤を有する個体が混じる．内嚢は基部の背側方向への膨らみがやや弱く，傍盤葉がやや大きくなる個体が混じる．

　分布　東北地方南部（宮城県，山形県，福島県北部）と中部地方北部（新潟県，長野県北部）／大島［気仙沼市］，出島，金華山，田代島，網地島，宮戸島，寒風沢島，野々島，桂島（以上，宮城県）．

　亜種小名の由来と通称　亜種小名は医師であり著名なアマチュア昆虫研究家でもあった新潟の馬場 金太郎に因む．コアオマイマイカブリという通称が一般的だが，ヤママイマイカブリという名が提唱されたこともある．

38-4. *C. (D.) b. fortunei* (Adams, 1861) マイマイカブリ粟島亜種

Original description　*Damaster Fortunei* Adams, 1861, Annls. Mag. Nat. Hist., London, (3), Ser. 8, p. 59.
Type locality　Awa-Sima, Japan.
Type depository　Natural History Museum, London ?
Diagnosis　33～46 mm, Head above and pronotum usually purplish red, sometimes bluish purple, and less strongly polished than in subsp. *babaianus*; elytra black, sometimes with faint bluish or purplish tinge; pronotum with lateral margins narrowly rimmed and not strongly curved above, above all near hind angles which are not remarkably lobate, basal foveae shallow; mucros longer than those of

subsp. *capito* but shorter than those of subsp. *babaianus*; elytral sculpture almost as in subsp. *capito*, though punctures forming striae between intervals a little smaller and shallower; abdominal sternites smooth and strongly shiny, bearing shallow transverse wrinkles and micropores with setae; basal three segments of male protarsus dilated with hair pads on ventral surface; male genitalia similar to those of subsp. *viridipennis* or of subsp. *babaianus*.

Distribution   Is. Awa-shima (Niigata Pref.).

Etymology   This subspecies is named after the Scottish botanist, Robert FORTUNE (1812-1880), the famous plant hunter and traveler, best known for introducing tea plants from China to India. He stayed in China from 1848 to 1851, and subsequently visited Taiwan and Japan.

タイプ産地　日本，粟島．
タイプ標本の所在　ロンドン自然史博物館？
亜種の特徴　33~46 mm．頭部背面と前胸背板は紫紅色のものが多いが，時に青紫色で．金属光沢は東北地方南部亜種よりもやや弱いものが多い．鞘翅は黒色で，時に僅かに青～紫色を帯びる．前胸背板は側縁部の縁取りが細く，背側へ向かって殆ど反らず，特に後角付近において顕著．後角は目立たず，基部凹陥は浅い．鞘翅の尾状突起は佐渡島亜種よりやや長いが，東北地方南部亜種より短い．鞘翅彫刻は佐渡島亜種のそれに近いが，間室間線条の点刻列はやや小さく浅いものが多い．腹板は光沢が強く滑らかだが，浅い横皺が認められ，剛毛を伴う孔点がある．♂前跗節は基部3節が広がり，腹面に絨毛盤を有する．♂交尾器は東北地方北部亜種，東北地方南部亜種のそれに近い．

分布　粟島．
亜種小名の由来と通称　亜種小名は，スコットランド人のRobert FORTUNE（ロバート・フォーチュン）（1812~1880）に因む．1800年代に活躍した，いわゆるプラントハンターの一人で，1848年から1851にかけてロンドン園芸協会から中国へ派遣され，台湾と日本にも足跡を残している．彼の最大の功績は中国から英領インドへ茶の栽培技術を紹介したこととされるが，その他の植物学の分野全般にも大きく貢献した．亜種の通称はアオマイマイカブリで，これは「青いマイマイカブリ」の意味ではなく，粟島が以前，アヲシマと呼ばれていたことに因んでつけられた「アヲマイマイカブリ」から転じたものだという．

38-5. *C. (D.) b. capito* (LEWIS, 1881)　マイマイカブリ佐渡島亜種

Original description   *Damaster capito* LEWIS, 1881, Ent. month. Mag., 17, p. 197.
Type locality   ins. Sado.
Type depository   Natural History Museum, London [Holotype (?), ♀].
Diagnosis   32~44 mm. Head above and pronotum purple and not so shiny; elytra black and mat, bearing weak purplish tinge along shoulders. Head and pronotum hypertrophic, with eyes relatively small and weakly protruded; neck broad, about twice as long as longitudinal diameter of eye; pronotum almost as broad as long, widest a little before middle, rather acutely narrowed toward apex and base, divergent posteriad before hind angles which are slightly protruded postero-laterad and pointed at tips; lateral margins strongly reflexed, above all behind middle; elytra with apical ends slightly extending that of terminal sternites and not pointed at tips; intervals weakly raised, indicated by rows of longitudinally elongated granules, punctures forming striae between intervals a little larger than those of subsp. *fortunei* though rather vaguely carved; abdominal sternites smooth and polished, bearing shallow transverse wrinkles and micropores with setae; male protarsus widest of all subspecies of *C. (D.) blaptoides*, with hair pads on ventral surface of basal three segments; apical lobe of penis triangular, robust and rounded at tip; basal portion of internal sac strongly inflated dorsad; both parasaccellar and podian lobes developed.

Distribution   Is. Sado-ga-shima (Niigata Pref.).
Etymology   Subspecific name means "large-headed".

タイプ産地　佐渡．
タイプ標本の所在　ロンドン自然史博物館［ホロタイプ（？），♀］．
亜種の特徴　32~44 mm．頭部背面と前胸背板は紫色で，中部地方以北に産する各亜種より光沢は鈍い．鞘翅は光沢を欠く黒色だが，肩部側縁には紫色の光沢を有する．頭部と前胸背板が肥大し，相対的に複眼の突出が弱い特異な体型が特徴で，頚部は太く，長さは複眼の長径の約2倍．前胸背板は長幅ほぼ等しく，最広部は中央よりやや前方にあり，前後に強く狭まり，後方は後角に向けて広がる．後角は外側後方へ僅かに突出し，先端は尖る．側縁部は本種の亜種の中で最も強く背側へ反り，中央から後方で特に顕著．鞘翅端は僅かに腹節末端を越える程度で尖らない．間室の隆起は弱く，前後にやや伸長した顆粒の列からなり，間室間線条内の点刻は粟島亜種のそれよりやや大きいが輪郭は不鮮明．腹部腹板は滑らかで光沢があり，浅い横皺と，剛毛を伴う小孔点がある．♂前跗節は本種の亜種の中で最も幅広く，基部3節に絨毛盤がある．陰茎先端部は太短い三角形で先端は丸まり，内囊基部背面の膨隆は強く，傍盤葉，脚葉が発達する．

分布　佐渡．
亜種小名の由来と通称　亜種小名は「頭の大きい人」を意味し，本亜種の頭部が肥大することに因む．通称はサドマイマイカブリ．

38. *Carabus* (*Damaster*) *blaptoides*

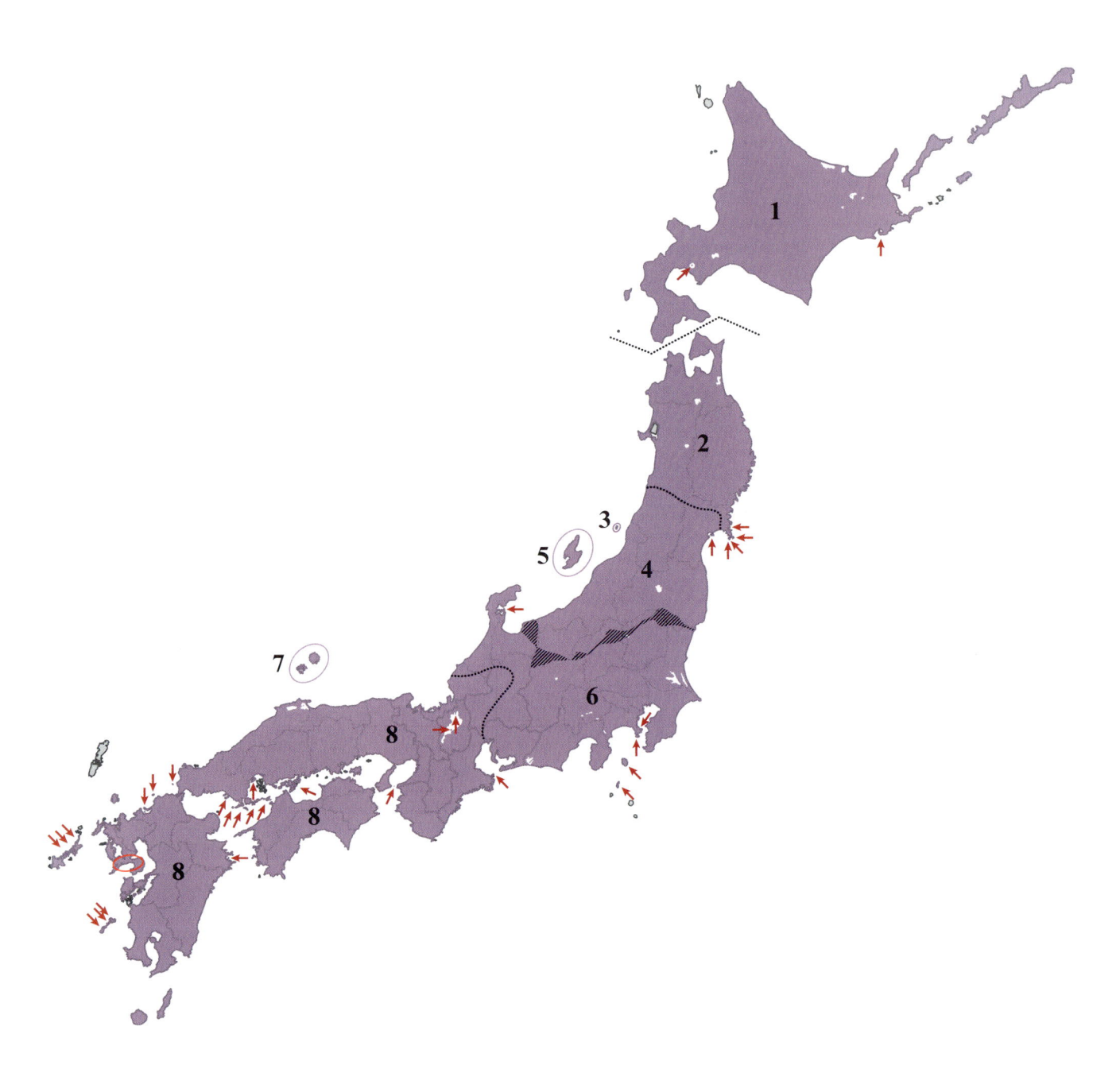

Fig. 52. Map showing the distributional range of **Carabus** (**Damaster**) **blaptoides** (Kollar, 1836). — 1, Subsp. ***rugipennis***; 2, subsp. ***viridipennis***; 3, subsp. ***fortunei***; 4, subsp. ***babaianus***; 5, subsp. ***capito***; 6, subsp. ***oxuroides***; 7, subsp. ***brevicaudus***; 8, subsp. ***blaptoides***. Shaded area indicates the range of intergrading population between *babaianus* and *oxuroides*. ◯: Most likely type area of the species (Nagasaki or its nearby regions).

図 52. **マイマイカブリ**分布図. — 1, 北海道亜種; 2, 東北地方北部亜種; 3, 粟島亜種; 4, 東北地方南部亜種; 5, 佐渡島亜種; 6, 関東・中部地方亜種; 7, 隠岐亜種; 8, 基亜種. 斜線部は東北地方南部亜種と関東・中部地方亜種の移行地帯. ◯: 最も可能性の高い種のタイプ地域（長崎ないしその周辺）.

38. *Carabus* (*Damaster*) *blaptoides*

Fig. 53. *Carabus* (*Damaster*) *blaptoides rugipennis* passing the winter in rotten wood (Niikappu-chô, Hokkaido).

図53. 朽木中で越冬しているマイマイカブリ北海道亜種(北海道新冠町).

Fig. 54. *Carabus* (*Damaster*) *blaptoides babaianus* (Urasa, Minami-uonuma-shi, Niigata Pref.).

図54. マイマイカブリ東北地方南部亜種(新潟県南魚沼市浦佐).

Fig. 55. *Carabus* (*Damaster*) *blaptoides oxuroides* passing the winter in rotten wood (Watarasé retarding basin, Fujioka-machi-uchino, Tochigi-shi, Tochigi Pref.).

図55. 朽木中で越冬しているマイマイカブリ関東・中部地方亜種(栃木県栃木市藤岡町内野渡良瀬遊水地).

39. *Carabus* (*Coptolabrus*) *fruhstorferi* (Roeschke, 1900)
ツシマカブリモドキ

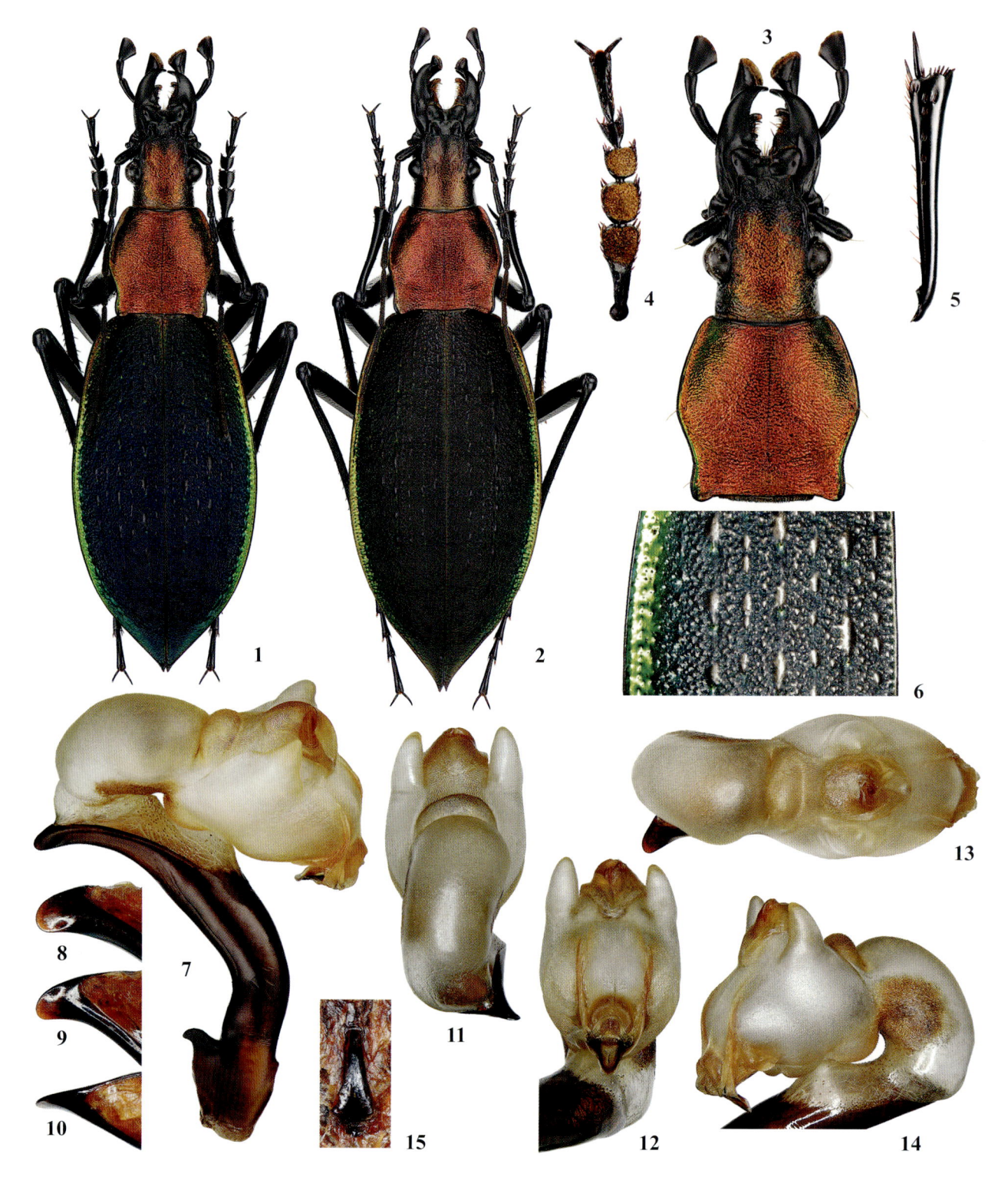

Figs. 1-15. *Carabus* (*Coptolabrus*) *fruhstorferi* (Roeschke, 1900) (Toyotama-machi, Is. Tsushima, Nagasaki Pref.). — 1, ♂; 2, ♀; 3, male head & pronotum; 4, male right protarsus in ventral view; 5, male right protibia in subdorsal view; 6, male left elytron (median part); 7, penis & fully everted internal sac in right lateral view; 8, apical part of penis in right lateral view; 9, ditto in right subdorsal view; 10, ditto in dorsal view; 11, internal sac in caudal view; 12, ditto in frontal view; 13, ditto in dorsal view; 14, ditto in left sublateral view; 15, inner plate of ligular apophysis.

図 1-15. ツシマカブリモドキ（長崎県対馬豊玉町）. — 1, ♂; 2, ♀; 3, ♂の頭部と前胸背板; 4, ♂右前跗節（腹面観）; 5, ♂右前脛節（斜背面観）; 6, ♂左鞘翅中央部; 7, 陰茎と完全反転下の内嚢（右側面観）; 8, 陰茎先端（右側面観）; 9, 同（右斜背面観）; 10, 同（背面観）; 11, 内嚢（尾側面観）; 12, 同（頭側面観）; 13, 同（背面観）; 14, 同（左斜側面観）; 15, 膣底部節片内板.

39. *Carabus* (*Coptolabrus*) *fruhstorferi* (Roeschke, 1900) ツシマカブリモドキ

Original description   *Coptolabrus Fruhstorferi* Roeschke, 1900, Ent. Nachr., 26 (11), p. 162.
History of research   This taxon was originally described by Roeschke (1900) from Is. Tsushima, off the northwestern coast of Kyushu, and has been consistently regarded as an independent, monotypical species.
Type locality   Japonica insula Tsushima.
Type depository   Zoölogisch Museum Amsterdam [Syntype]. Only one female is now preserved in this museum.
Morphology   29~45 mm. Upper surface of head and pronotum reddish coppery or golden coppery with greenish tinge along margins; elytra black and mat excepting lateral margins which are golden green or golden coppery with strong metallic luster, their depressed parts often dark green and bottom of primary foveoles as in lateral margins; venter black, though bearing purplish tinge in most part. External features: Figs. 1~6. Description on details: penultimate segment of labial palpus bisetose; median tooth of mentum much shorter than lateral lobes, with tip sharply pointed and not protruding ventrad; submentum asetose; retinaculum of right mandible with anterior tooth a little smaller than posterior; male antennae without hairless ventral areas; elytral apices protruded posteriad to form mucros; elytral sculpture heterodyname; primaries and secondaries indicated by rows of spindle-shaped low tubercles, the former always larger and broader than the latter, primary foveoles very shallow and vaguely outlined, tertiaries reduced to rows of granules, areas between intervals coarsely scattered with small granules; propleura strongly and coarsely punctate and partly wrinkled; sides of sternites distinctly rugoso-punctate; metacoxa bisetose, lacking anterior seta. Genitalic features: Figs. 7~14. Penis a little shorter than half the elytral length, with apical part similar to that of *C.* (*D.*) *blaptoides* though much more sharply pointed in dorsal view; internal sac also as in *C.* (*D.*) *blaptoides*, but a little shorter and robuster as a whole, with a little more distinct presaccellar lobe, thinner and more sharply pointed parasaccellar lobes and less strongly developed podian lobes; peripheral rim of gonopore strongly sclerotized and pigmented, forming a triangular terminal plate. Inner plate of ligular apophysis of female genitalia widest near base, gradually narrowed toward apex, strongly sclerotized and pigmented.
Molecular phylogeny   On the phylogenetic tree of the mitochondrial ND5 gene of the Carabina of the world, this species belongs to the Chinese lineage of so-called Procrustimorphi (= almost corresponding to the *Procrustes* subgenus-group in this book), together with other *Coptolabrus* species of eastern Eurasia. Its closest relative is not superficially similar *Carabus* (*Coptolabrus*) *jankowskii*, but is *C.* (*C.*) *smaragdinus* of the Korean Peninsula (Osawa *et al.*, 2002, 2004).
Distribution   Is. Tsushima (Nagasaki Pref.), SW. Japan.
Habitat   Usually inhabiting dry, thin forest or around there. Also visible in the open place such as Japanese pampas grass fields.
Bionomical notes   Spring breeder, snail feeder.
Etymology   Named after Hans Fruhstorfer (1866~1922), the famous lepidopterologist in Germany.

研究史　Roeschke（ロシュケ）（1900）により対馬から記載された，我が国唯一のカブリモドキ．対馬の固有種で，記載以来単型の独立種として扱われている．

タイプ産地　日本，対馬．

タイプ標本の所在　アムステルダム動物学博物館［シンタイプ］．原記載で使われた6頭の中の1♀が保管されている．

形態　29~45 mm．頭部背面と前胸背板は紅～金銅色で，主として辺縁部に緑色の金属光沢を帯び，鞘翅は光沢を欠く黒色だが，側縁部は金緑～金銅色で，基面の凹部は暗緑色を，また第一次凹陥の底部は側縁部と同様の金属色を帯びる．腹面には弱い紫色の金属光沢を帯びることが多い．外部形態：図1~6．細部の記載：下唇肢亜端節の剛毛は2本．下唇基節の中央歯は側葉よりはるかに短く，先端は鋭く尖るが，腹側へ突出しない．下唇亜基節は無毛．右大顎抱歯の前方突起は後方突起よりやや小さい．♂触角腹面に無毛部を欠く．前胸背板の側縁剛毛は2対．鞘翅端は腹端を越えて鋭く尖り，尾状突起を形成する．鞘翅彫刻は異規的で，第一次，第二次間室は共に紡錘形の低い瘤状小隆起列となり，前者は後者より常に大きく幅広く，第一次凹陥は極めて浅く，輪郭は不完全．第三次間室は顆粒列へと減退し，間室間の基面には小顆粒を密に装う．前胸側板は強く密に点刻され，部分的に皺状．腹部腹板の側面は不規則な皺と点刻が著しい．後基節の剛毛は2本で前方剛毛を欠く．腹部腹板の横溝は深く明瞭に刻まれ，各節に剛毛を伴う孔点がある．交尾器形態：図7~14．陰茎は鞘翅長の1/2弱．先端部はマイマイカブリのそれに近いが，背面から見ると先端に向けてはるかに鋭く尖る．内嚢もマイマイカブリのものに近いが，全体により太短く，盤前葉がより目立ち，傍盤葉はより細く，先端はより鋭く尖り，脚葉の発達は弱く，射精孔縁膜は強く硬化して中央に顕著な1個の三角形の頂板を形成する．♀交尾器膣底部節片内板（図15）は後方で最も幅広く，前方に向けて狭まり，硬化と色素沈着が顕著．

分子系統　世界のオサムシ亜族の分子系統樹上で，本種はマイマイカブリと共にいわゆるヨロイオサムシ群（本書におけるプロクルステス亜属群にほぼ相当する，カラブス属の中の大群）の中国系亜群に属し，カブリモドキ類のなかでは外部形態が似ているヤンコウスキィカブリモドキよりもアオカブリモドキに類縁が近いと想定されている（大澤ら，2002；Osawa *et al.*, 2004）．

分布　対馬（地峡の人工的開削により主島が分断された後の，現在で言う上島（かみじま），下島（しもじま）双方に産する）．

生息環境　乾燥した疎林や林縁に多く，ススキなどの生えた草原的環境にも見られる．

生態　春繁殖・成虫越冬型で，軟体動物（カタツムリ）食のオサムシ．

種小名の由来　ドイツの昆虫研究家 Hans Fruhstorfer（ハンス・フルーストォファー）（1866~1922）に因む．専門は鱗翅類で，多くの業績を残したほか，ベルリンで標本商も営んでいた．晩年は癌を患い，50歳代半ばの若さで世を去ったという．

39. *Carabus* (*Coptolabrus*) *fruhstorferi*

Fig. 16. Map showing the distributional range of ***Carabus*** (***Coptolabrus***) ***fruhstorferi*** (Roeschke, 1900).
図 16. **ツシマカブリモドキ**分布図.

39. *Carabus* (*Coptolabrus*) *fruhstorferi*

Fig. 17. Habitat of *Carabus* (*Coptolabrus*) *fruhstorferi* (Mt. Ariaké-yama, Is. Tsushima, Nagasaki Pref.). The adult usually pass the winter in such an ocherous clay soil.

図 17. ツシマカブリモドキの生息地（長崎県対馬有明山）．成虫はこのような黄土色をした粘土質の崖の肩〜垂直面で越冬することが多い．

Fig. 18. *Carabus* (*Coptolabrus*) *fruhstorferi* (Mt. Ariake-yama, Is. Tsushima, Nagasaki Pref.).

図 18. ツシマカブリモドキ♂（長崎県対馬有明山）．

Fig. 19. *Carabus* (*Coptolabrus*) *fruhstorferi* passing the winter in the soil (Mt. Ariaké-yama, Is. Tsushima, Nagasaki Pref.).

図 19. 土中で越冬しているツシマカブリモドキの成虫（長崎県対馬有明山）．

マイマイカブリのホロタイプを調査中の主著者，井村（ウィーン自然史博物館にて，2000年10月2日撮影）.

Yûki IMURA, the main author of this book, is examining the holotype of *Carabus* (*Damaster*) *blaptoides* (KOLLAR, 1886) (Naturhistorisches Museum Wien, October 2, 2000).

マイマイカブリのホロタイプを調査中の主著者，井村（ウィーン自然史博物館にて，2000年10月2日撮影）.

文献REFERENCES/索引INDEX

文献　REFERENCES

[A]

ADAMS, A., 1861. Notice of a new species of *Damaster* from Japan. *Ann. Mag. nat. Hist.*, *London*, (3), Ser. 8: 59.

荒井充朗, 1987a. オオルリオサムシ♂とオシマルリオサムシ♀の交配実験. *jezoensis*（北海道昆虫同好会会誌）, (14): 100-105. [ARAI, M., 1987a]

荒井充朗, 1987b. 翔べコブスジアカガネ！オサムシマップ, (30): 688. [ARAI, M., 1987b]

[B]

BABA, K., 1939. Beschreibung von zwei neuen Formen des *Damaster* (Coleoptera, Carabidae, *Carabus*). Untersuchung von *Damaster*. *Kontyû, Tokyo*, 13: 203-207.

BATES, H. W., 1873. On the geodephagous Coleoptera of Japan. *Trans. ent. Soc. Lond.*, 1873: 219-322.

BATES, H. W., 1883. Supplement to the geodephagous Coleoptera of Japan, chiefly from the collection of Mr. George LEWIS, made during his second visit, from February, 1880, to September, 1881. *Trans. ent. Soc. Lond.*, 1883: 205-290, 1 pl.

BORN, W., 1922. Beitrag zur Kenntnis der Carabenfauna von Ostasien (Col.). *Ent. Mitt., Berlin*, 11: 166-174.

BREUNING, S., 1932-1937. Monographie der Gattung *Carabus* L. *Best. -Tab. eur. Coleopt.*, (104-110): 1-1610, 41 pls. Reitter, Troppau.

BREUNING, S., 1934. Ueber Carabini. *Folia zool. hydrobiol., Riga*, 1: 29-40.

BREUNING, S., 1957. Weiterer Beitrag zur Kenntnis der Gattung *Carabus* L. *Ent. Arb. Mus. Frey, Tutzing*, 8: 18.

[C]

CHAUDOIR, M., DE, 1848. Mémoires sur famile des Carabiques. *Bull. Soc. Imp. Nat. Moscou*, 21 (1): 452.

CHAUDOIR, M., DE, 1862. Descriptions, sommaires d'espèces nouvelles de Cicindélètes et de Carabiques de la collection de M. le baron DE CHAUDOIR. *Rev. Mag. Zool.*, 14: 484-490.

CHAUDOIR, H. DE, 1869. Descriptions des Cicindélètes et de Carabiques nouveaux. *Rev. Mag. Zool.*, 21: 22-28.

CSIKI, E., 1927. *Coleopterorum Catalogus Carabidae: Carabinae I*. (ed. S. SCHENKLING), (Col. Cat., 91) pp. 1-648. Junk, Berlin.

[D]

DEJEAN, P., 1826. *Species général des Coléoptères de la collection de M. le comte DEJEAN*, 2, VIII＋501 pp. Paris-Crenot.

DEJEAN, M. le Comte & J.-A. BOISDUVAL, 1829. *Iconographie et histoire naturelle des Coléoptères d'Europe*, Tome I. xiv＋400 pp., 60 pls. Paris: Mequignon-Marvis Père et Fils.

DEUVE, Th., 1994. Une classification du genre *Carabus*. *Bibliothèque ent.*, 5: 1-296, 115 figs.

DEUVE, Th., 2004. *Illustrated catalogue of the genus Carabus of the world (Coleoptera, Carabidae)*. 461 pp. Pensoft, Sofia-Moscow.

DEUVE, Th., 2012. *Une nouvelle classification du genre Carabus L., 1758. Liste Blumenthal 2011-2012*. 54 pp. Association Magellanes.

DEUVE, Th., A. CRUAUD, G. GENSON, J.-Y. RASPLUS, 2012. Molecular systematics and evolutionary history of the genus *Carabus* (Col. Carabidae). *Mol. Phylogenet. Evol.*, 66: 259-275.

[E]

EMDEN, F. VAN, 1932. Einige neue Carabinae des Staatlichen Museums für Tierkunde zu Dresden. *Neue Beitr. syst. Insektenkunde*, 5: 62-69, 1 pl.

[F]

FAIRMAIRE, L., 1876. Description d'une nouvelle espèce du genre *Carabus*. *Pet. Nouv. Ent.*, 2 (148): 37.

FISCHER, G., 1820-1822. *Ent. imp. Russ.*, 1, VIII+208 pp.

FISCHER VON WALDHEIM, G., 1825-1828. Cicindeletas et Carabicorum partem continens, cum XVIII tabulis aeneis. *Ent. imp. Ross.*, 3, VIII+314 pp.

船越崇嗣・高見泰興・氏家昌行・曽田貞滋, 1998. ホソアカガネオサムシの生活史. 昆虫と自然（東京）, 33 (9): 38-43.［FUNAKOSHI *et al.*, 1998］

[G]

GÉHIN, J. B., 1885. *Catalogue synonymique et systématique des Coléoptères de la tribu Carabides, avec des planches dessinées par Ch. HAURY*. xxxviii+104 pp., 10 pls. Remiremont & Prague.

[**H**]

花谷達郎・小沼 篤・酒井 香, 1968. 利尻島の昆虫 (II), 鱗翅目を除くその他の昆虫. 中村武久 (編), 利尻島動植物調査報告 (利尻島動植物調査の記録, 東京農業大学第一高等学校): 79-91. [HANATANI et al., 1968]

春沢 圭太郎, 1978. 奈良県大峰山脈のコクロナガオサムシ. 大阪市立自然科学博物館研究報告, (31): 39-46. [HARUSAWA, 1978]

長谷川 道明・小鹿 亨・四方 圭一郎, 2001. ミカワオサムシの色彩変異. 穂積俊文博士記念論文集発行事業会 (編), 東海甲虫誌 (穂積俊文博士記念論文集): 281-291. [HASEGAWA et al., 2001]

HAUSER, G., 1921. Die *Damaster-Coptolabrus*-Gruppe der Gattung *Carabus*. *Zool. Jahrb., Abt. Syst.*, 45: 1-394, 11 pls.

HERBST, J. F. W., 1784. Archiv der Insectengeschichte. Fünftes Heft. Zwote Abtheilung. Zürich, 1784". *Fueßly, Arch. Insekteng.*, 5: 74-151.

東日本オサムシ研究会, 1982-1989. オサムシマップ, (1)～(41) (1~3号は「東日本オサムシ研究会連絡紙」, 4, 5号は「治虫地図」, 6号以降は「オサムシマップ」). [East Japan Research Group of Carabid Beetles, 1982-1989]

東日本オサムシ研究会 (編), 1989. 東日本のオサムシ－地域の特徴をつなぎ合わせる－. 224 pp. ぶなのき出版 (米沢). [East Japan Research Group of Carabid Beetles (ed.), 1989]

平山洋人, 1981. スルガオサムシの西限について. 月刊むし (東京), (123): 32. [HIRAYAMA, 1981]

平山洋人, 1985. マークオサムシの後翅発達個体の記録. 月刊むし (東京), (177): 41-42. [HIRAYAMA, 1985]

日浦 勇, 1965. 金剛生駒山脈のオオオサムシ属. 大阪市立自然科学博物館研究報告, (18): 49-68. [HIURA, 1965]

日浦 勇, 1966. 金剛生駒山地における歩行虫数種の分布. 大阪市立自然科学博物館研究報告, (19): 47-52. [HIURA, 1966]

日浦 勇・桂 孝次郎, 1971. 信楽山地のマヤサンオサムシ. 大阪市立自然科学博物館研究報告, (24): 15-27. [HIURA & KATSURA, 1971]

日浦 勇・桂 孝次郎・谷 幸三・春沢 圭太郎・冨永 修, 1971. 近畿地方におけるオサムシの地理的分布 (予報). 大阪市立自然科学博物館研究報告, (25): 27-42. [HIURA, et al., 1971]

[**I**]

井村有希, 1971. 湯ノ丸山合宿記. ノープリウス (私立駒場東邦中高等学校生物部誌), (9): 78-82. [IMURA, 1971]

井村有希, 1984. G-M Lineを探る！オサムシマップ, (15): 245-252. [IMURA, 1984]

井村有希, 1985. 北海道地方のオサムシに関する知見. 1) オオルリオサムシ, 2) アイヌキンオサムシーその3. 周年経過と生活史に関する報告－. オサムシマップ, (19): 324-333. [IMURA, 1985]

IMURA, Y., 1989a. Natural hybrids of the *Damaster* species (Coleoptera, Carabidae) in Hokkaido, northern Japan. *Jpn. J. Ent.,* 57: 67-71.

井村有希, 1989b. オサムシ亜族の地理的変異と個体変異 (1), オオルリオサムシ. 猪又敏男 (編), 図説・世界の重要昆虫, B (1): 1-16. むし社 (東京). [IMURA, 1989b]

井村有希, 1991. オサムシ亜族の地理的変異と個体変異 (2), オシマルリオサムシ・アイヌキンオサムシ. 猪又敏男 (編), 図説・世界の重要昆虫, B (2): 17-32. むし社, 東京. [IMURA, 1991]

井村有希, 1994a. 北海道産オサムシの2新亜種－付・オオルリオサムシ正基準標本の再記載－. 月刊むし (東京), (276): 10-14. [IMURA, 1994a]

IMURA, Y., 1994b. Taxonomic notes on *Acoptolabrus* (Coleoptera, Carabidae) recently described from the Oshima Peninsula in southwestern Hokkaido, Northeast Japan. *Elytra, Tokyo*, 22: 15-20.

井村有希, 1994c. 北海道渡島半島から最近記載された*Damaster*属の2新分類単位に対する分類学的評価. 月刊むし (東京), (285): 24-25. [IMURA, 1994c]

IMURA, Y., 2002a. Proposal of eighteen new genera and subgenera of the subtribe Carabina (Coleoptera, Carabidae). *Spec. Bull. Jpn. Soc. Coleopterol., Tokyo*, (5): 129-147.

IMURA, Y., 2002b. Classification of the subtribe Carabina (Coleoptera, Carabidae) based on molecular phylogeny. *Elytra, Tokyo*, 30: 1-28.

IMURA, Y., 2003a. An isolated population of *Homoeocarabus maeander* (Coleoptera, Carabidae) discovered from a palsa bog on the Daisetsu-zan Mountains in Hokkaido, Northeast Japan. *Elytra, Tokyo*, 31: 439-445.

IMURA, Y., 2003b. Occurrence of *Ohomopterus chugokuensis* (Coleoptera, Carabidae) in the eastern part of the Sanuki Hills in northeastern Shikoku, Southwest Japan. *Elytra, Tokyo*, 31: 447-460.

IMURA, Y., 2004a. Lectotype designation of *Carabus vanvolxemi* (Coleoptera, Carabidae). *Elytra, Tokyo*, 32: 1-3.

IMURA, Y., 2004b. An isolated new subspecies of *Ohomopterus yamato* (Coleoptera, Carabidae) discovered from the southeastern part of the Kii Peninsula. *Elytra, Tokyo*, 32: 3-4.

IMURA, Y., 2004c. Discovery of *Hemicarabus macleayi* (Coleoptera, Carabidae) from the alpine zone of the Island of Rishiri-tô, Northeast Japan. *Elytra, Tokyo*, 32: 235-240.

IMURA, Y., 2005a. Lectotype designation of *Ohomopterus kiiensis* (Coleoptera, Carabidae). *Elytra, Tokyo*, 33: 311-312.

IMURA, Y., 2005b. Lectotype designation of *Ohomopterus daisen* (Coleoptera, Carabidae). *Elytra, Tokyo*, 33: 485-486.

IMURA, Y., 2005c. Additional notes on *Hemicarabus macleayi amanoi* (Coleoptera, Carabidae) recently discovered from the Island of Rishiri-tô northern Japan. *Elytra, Tokyo*, 33: 679-688.

IMURA, Y., 2008. Records of *Ohomopterus arrowianus* (Coleoptera, Carabidae) in the southwestern part of Shizuoka Prefecture, Central Japan, with description of a new subspecies. *Elytra, Tokyo*, 36: 387-395.

IMURA, Y., 2011. Two new subspecies of the subtribe Carabina from Japan (Coleoptera, Carabidae). *Coléoptères, Paris*, 17 (3): 21-28.

IMURA, Y., 2012. A new species of *Ohomopterus* from northern Shikoku in southwestern Japan (Coleoptera, Carabidae). *Coléoptères, Paris*, 18 (7): 57-62.

IMURA Y., K. AKITA, M. OKAMOTO, O. TOMINAGA, N. KASHIWAI, Z.-H. SU, T. OJIKA & S. OSAWA, 2005. A probable origin of *Ohomopterus arrowianus kirimurai* (Coleoptera, Carabinae:Carabidae) as deduced from phylogenetic trees of mitochondrial ND5 gene and nuclear ITS I DNA as well as morphology of genital organs. *Ent. Rev. Japan*, 60 (1): 35-38.

井村有希・出嶋利明・水沢清行, 1993. ヒメオサムシの6新亜種. 月刊むし（東京），(264)：10-16＋1 pl. [IMURA *et al.*, 1993]

井村有希・平井克男, 2003. 赤石山脈南端部のホソヒメクロオサムシ. 月刊むし（東京），(388)：45-46. [IMURA & HIRAI, 2003]

IMURA, Y., C.-G. KIM, Z.-H. SU & S. OSAWA, 1998. An attempt at the higher classification of the Carabina（Coleoptera, Carabidae）based on morphology and molecular phylogeny, with special reference to *Apotomopterus, Limnocarabus and Euleptocarabus*. *Elytra, Tokyo*, 26: 17-35.

IMURA, Y. & M. MATSUNAGA, 2010. Taxonomic status of *Ohomopterus komiyai* and descriptions of two new subspecies of *Ohomopterus* from Shizuoka Prefecture, Central Japan (Coleoptera, Carabidae). *Coléoptères, Paris*, 16 (14): 141-147.

IMURA, Y. & M. MATSUNAGA, 2011. Discovery of an isolated population of *Limnocarabus* (*Euleptocarabus*) *porrecticollis* (Coleoptera, Carabidae) from the southern foot of Mt. Fuji in Shizuoka Prefecture, Central Japan. *Elytra, Tokyo*, N.S., 1 (1): 1-14.

井村有希・水沢清行, 1994. ヒメオサムシの正基準標本に関する知見と3新亜種の記載. 月刊むし（東京），(282)：14-21. [IMURA & MIZUSAWA, 1994]

井村有希・水沢清行, 1995. 西日本に産するオサムシ3種の地理的変異と6新亜種の記載. 月刊むし（東京），(293)：9-14. [IMURA & MIZUSAWA, 1995]

井村有希・水沢清行, 1996. 世界のオサムシ大図鑑. 261 pp., 84 pls. むし社, 東京. [IMURA & MIZUSAWA, 1996]

井村有希・水沢清行, 1999. 西南日本におけるヒメオサムシの2新亜種. 月刊むし（東京），(338)：2-5. [IMURA & MIZUSAWA, 1999]

井村有希・水沢清行, 2002a. ホソヒメクロオサムシの3新亜種. 月刊むし（東京），(380)：2-6. [IMURA & MIZUSAWA, 2002]

IMURA, Y. & K. MIZUSAWA, 2002b. Lectotype designation of *Ohomopterus yamato* (Coleoptera, Carabidae), with description of four new subspecies. *Elytra, Tokyo*, 30: 363-383.

井村有希・永幡嘉之, 2006. リシリノマックレイセアカオサムシ, さいはての島の小さな奇跡. 120 pp. 昆虫文献六本脚. [IMURA & NAGAHATA, 2006]

IMURA, Y., O. TOMINAGA, N. KASHIWAI, M. OKAMOTO, Z.-H. SU, K. AKITA, T. OJIKA & S. OSAWA, 2005. Phylogenetic properties of *Ohomopterus iwawakianus* (Coleoptera, Carabidae) as evidenced by the sequence comparisons of mitochondrial ND5 Gene and nuclear internal transcribed spacer I: Extensive participation of *O. iwawakianus* in the faunal establishment of the Genus *Ohomopterus* in the Kinki District. *Elytra, Tokyo*, 33: 13-24.

井上 壽, 1953a. セアカオサムシの生活史. 新昆蟲, 6(4)：43-44. [INOUE, 1953a]

井上 壽, 1953b. エゾアカガネオサムシの生活史. 新昆蟲, 6(12)：32. [INOUE, 1953b]

井上 壽, 1953c. ヒメクロオサムシの生活史. 新昆蟲, 6(12)：32-33. [INOUE, 1953c]

ISHIKAWA, R., 1960. A new species of *Apotomopterus* from Japan (Coleoptera, Carabidae). *Kontyû, Tokyo*, 24(2): 168-169.

ISHIKAWA, R., 1962. Studies on the subgenus *Leptocarabus* (1). *Kontyû, Tokyo*, 30 (2): 110-115.

ISHIKAWA, R., 1966a. Studies on some species of Japanese Carabina (Coleoptera, Carabidae). *Bull. natn. Sci. Mus., Tokyo*, 9: 9-26.

ISHIKAWA, R., 1966b. Descriptions of new subspecies in the Japanese Carabina (Coleoptera, Carabidae). *Bull. natn. Sci. Mus., Tokyo*, 9: 451-464, 4 pls.

ISHIKAWA, R., 1968a. A study on *Damaster* (*Acoptolabrus*) *gehinii* (FAIRMAIRE) with a description of a new species (Coleoptera, Carabidae). *Bull. natn. Sci. Mus., Tokyo*, 11: 141-148.

ISHIKAWA, R., 1968b. Three new subspecies of Japanese Carabina (Coleoptera, Carabidae). *Bull. natn. Sci. Mus., Tokyo*, 11: 263-267.

ISHIKAWA, R., 1969a. Descriptions of two new subspecies of Japanese Carabina with notes on the synonymy of *Carabus fureoiensis* KANO (Coleoptera, Carabidae). *Bull. natn. Sci. Mus., Tokyo*, 12: 33-37.

ISHIKAWA, R., 1969b. A taxonomic study on *Apotomopterus japonicus* (MOTSCHULSKY) and its allied species (Coleoptera, Carabidae). *Bull. natn. Sci. Mus., Tokyo*, 12: 517-530.

ISHIKAWA, R., 1971. Notes on *Procrustes* (*Megodontus*) *aino* ROST with descriptions of new subspecies (Coleoptera, Carabidae). *Bull. natn. Sci. Mus., Tokyo*, 14: 571-577, 5 pls.

ISHIKAWA, R., 1972. Studies on *Leptocarabus* and its related subgenera of the genus *Carabus* L. (Coleoptera, Carabidae). *Bull. natn. Sci. Mus., Tokyo*, 15: 19-27.

ISHIKAWA, R., 1973. Notes on some basic problems in the taxonomy and the phylogeny of the subtribe Carabina (Coleoptera, Carabidae). *Bull. natn. Sci. Mus., Tokyo*, 16: 191-215.

ISHIKAWA, R., 1981. On the subspecies of *Carabus* (*Ohomopterus*) *lewisianus* BREUNING (Coleoptera, Carabidae). *Kontyû, Tokyo*, 49: 498-501.

ISHIKAWA, R., 1984a. Short report. *Proc. Jap. Soc. syst. Zool.*, (28): 57.

ISHIKAWA, R., 1984b. Two new subspecies of Japanese Carabina (Coleoptera, Carabidae). *Akitu, Kyoto*, (N.S.), (68): 1-5.

石川良輔, 1985. オサムシ科（オサムシ亜科）. 原色日本甲虫図鑑, 2: 54-88. 保育社（大阪）. [ISHIKAWA, 1985]

ISHIKAWA, R., 1986b. Taxonomic studies on *Leptocarabus harmandi* (LAPOUGE) (Coleoptera: Carabidae). *Trans. Shikoku Ent. Soc.*, (17): 221-238.

石川良輔, 1991. オサムシを分ける錠と鍵. 8 pls. XXXII+295 pp., 八坂書房（東京）. [ISHIKAWA, 1991]

ISHIKAWA, R., 1992. Taxonomic studies on *Leptocarabus* (*Adelocarabus*) *arboreus* (LEWIS) (Coleoptera, Carabidae). *TMU Bull. nat. Hist., Tokyo*, (1): 1-40, 39 pls.

ISHIKAWA, R., 1993. A new subspecies of *Procrustes* (*Pachycranion*) *kolbei* (ROESCHKE) (Coleoptera, Carabidae) from Tokachi, Hokkaido. *Jpn. J. Ent.*, 61: 585-587.

ISHIKAWA, R. & K. KUBOTA, 1984. *Carabus arrowianus* (BREUNING) and its new subspecies from Mie Prefecture, Japan (Coleoptera, Carabidae). *Akitu, Kyoto*, (N. S.), 65: 1-6.

ISHIKAWA, R. & K. KUBOTA, 1994a. Geographical races of *Carabus yaconinus* BATES (Coleoptera, Carabidae): A tentative revision. *Proc. Jap. Soc. syst. Zool.*, (50): 28-40.

ISHIKAWA, R. & K. KUBOTA, 1994b. Geographical races of *Carabus maiyasanus* BATES and *C. arrowianus* (BREUNING) in Honshu, Japan: A tentative revision (Coleoptera, Carabidae). *Bull. Biogeogr. Soc. Japan*, 49: 105-128.

ISHIKAWA, R. & K. KUBOTA, 1995. Geographical races of *Carabus iwawakianus* (NAKANE) in Honshu, Japan: A tentative revision (Coleoptera, Carabidae). *Bull. Biogeogr. Soc. Jap., Tokyo*, 50: 39-50, 55-58.

ISHIKAWA, R. & K. MIYASHITA, 2000. A revision of *Leptocarabus* (*Aulonocarabus*) *kurilensis* (LAPOUGE) in Hokkaido, Japan (Coleoptera, Carabidae). *Jpn. J. syst. Ent.*, 6: 63-77.

ISHIKAWA, R. & Y. TAKAMI, 1996. New subspecies of *Carabus albrechti* MORAWITZ (Coleoptera, Carabidae) from Japan. *Spec. Div.*, 1: 39-48

ISHIKAWA, R. & M. UJIIE, 2000. A revision of *Carabus*（*Ohomopterus*）*insulicola* CHAUDOIR, 1869 (Coleoptera, Carabidae) in Honshu, Japan. *Jpn. J. syst. Ent.*, 6: 253-297.

[**K**]

神吉正雄, 1963. *Apotomopterus* の雌交尾器について. 昆虫科学（大阪）, (13): 1-7. [KAMIYOSHI, 1963]

神吉弘視・溝口 修, 1960. 四国産オサムシの二新種について. 昆虫科学（大阪）, 1960: 5-7. [KAMIYOSHI & MIZOGUCHI, 1960]

菅 晃, 1981. 愛媛県のオサムシ. 愛媛県立博物館研究報告, (11), 33 pp. [KAN, 1981]

唐沢安美, 1988a. オオルリオサムシの飼育観察から[Ⅰ]. インセクタリウム, 25(2): 4-9. [KARASAWA, 1988a]

唐沢安美, 1988b. オオルリオサムシの飼育観察から[Ⅱ]. インセクタリウム, 25(3): 18-22. [KARASAWA, 1988b]

唐沢安美, 1992. アイヌキンオサムシの飼育から. インセクタリウム, 29(12): 4-9. [KARASAWA, 1992]

桂 孝次郎・冨永 修・日浦 勇・土井 仲治郎・春沢 圭太郎・谷 幸三, 1978. 伊勢・志摩地方のオサムシ分布. 大阪市立自然科学博物館研究報告, (31): 47-60. [KATSURA et al., 1978]

京浜昆虫同好会（編）, 1971. *Ins. Mag.*, (76), オサムシ特集号. 214 pp. [Keihin Insect Lovers' Club, Tokyo (ed.), 1971]

KIM, C.-G., S. SAITO, O. TOMINAGA, Z.-H. SU & S. OSAWA, 1999. Distributional boundaries of the geographic races of *Damaster blaptoides* in northeastern Japan as deduced from mitochondrial ND5 gene sequences. *Elytra, Tokyo*, 27: 635-641.

KIM, C.-G., Z.-H. SU, Y. IMURA, M. OKAMOTO & S. OSAWA, 2003. Phylogeny and evolution of the division Procrustimorphi (Coleoptera, Carabidae) of the world as deduced from the mitochondrial ND5 gene sequences. *Elytra, Tokyo*, 31: 263-284.

KIM, C.-G., O. TOMINAGA, Z.-H. SU & S. OSAWA, 1999. Origin and diversification of *Euleptocarabus porrecticollis* (Coleoptera, Carabidae) in the Japanese Islands inferred from mitochondrial ND5 gene sequences. *Mol. Phylogen. Evol.*, 13: 440-444.

KIM, C.-G., O. TOMINAGA, Z.-H. SU & S. OSAWA, 2000. Differentiation within the genus *Leptocarabus* (excl. *L. kurilensis*) in the Japanese Islands as deduced from mitochondrial ND5 gene sequences (Coleoptera, Carabidae). *Genes. Genet. System.*, 75: 335-342.

KIM, C.-G., H.-Z. ZHOU, Y. IMURA, O. TOMINAGA, Z.-H. SU & S. OSAWA, 2000. Pattern of morphological diversification in the *Leptocarabus* ground beetles (Coleoptera, Carabidae) as deduced from mitochondrial ND5 gene and nuclear 28S rDNA sequences. *Mol. Biol. Evol.*, 17: 137-145.

木村欣二, 1971. クロナガオサムシ亜属(*Leptocarabus*)を見直す. *Ins. Mag.*,(76): 65-86. [KIMURA, 1971]

KIMURA, K. & J. KOMIYA, 1974. Description of a new species allied to *Leptocarabus procerulus* (CHAUDOIR) (Coleoptera, Carabidae), with notes on the taxonomy of the related species. *Kontyû, Tokyo*, 42: 395-400.

近畿オサムシ研究グループ(冨永 修・桂 孝次郎・春沢 圭太郎・日浦 勇・谷 幸三・土井 仲治郎), 1974-1978. オサムシ層群, (1)-(100). [Kinki Research Group of Carabid Beetles (TOMINAGA, O., K. KATSURA, K. HARUSAWA, I. HIURA, K. TANI & N. DOI), 1974-1978]

近畿オサムシ研究グループ(冨永 修・桂 孝次郎・春沢 圭太郎・日浦 勇・谷 幸三・土井 仲治郎), 1978-1984. HANENASHI 段丘(=はねなし段丘, 無翅段丘(号により表記が変化)),(1)-(96). [Kinki Research Group of Carabid Beetles (TOMINAGA, O., K. KATSURA, K. HARUSAWA, I. HIURA, K. TANI & N. DOI), 1978-1984]

近畿オサムシ研究グループ(冨永 修・桂 孝次郎・春沢 圭太郎・日浦 勇・谷 幸三・土井 仲治郎), 1979. 近畿地方のオサムシ. 大阪市立自然科学博物館収蔵資料目録第11集, 83 pp. [Kinki Research Group of Carabid Beetles (TOMINAGA, O., K. KATSURA, K. HARUSAWA, I. HIURA, K. TANI & N. DOI), 1979]

KOLLAR, V., 1836. Species insectorum coleopterorum novae. *Ann. Wien Mus.*, 2: 327-336, 1 pl.

小宮次郎, 1971. オオオサムシ属(*Apotomopterus*)の分類. *Ins. Mag.*, (76): 22-64. [KOMIYA, 1971]

河野廣道, 1936. 大雪山の甲蟲類. *Biogeographica*, 1(1): 75-104. [KÔNO, 1936]

久保田 正秀, 1993. IV. 動物. 5. 昆虫類. 自然公園内環境調査, 八溝山地域(東京電力株式会社・東電設計株式会社): 165-219. [KUBOTA, 1993]

KUBOTA, K., 2010. Discovery of *Carabus* (*Ohomopterus*) *dehaanii* CHAUDOIR, 1848 (Coleoptera, Carabidae) from the Ôsumi Peninsula, Kyushu, Japan: description of a new subspecies. *Biogeography*, 12: 49-52.

KUBOTA, K. & R. ISHIKAWA, 2004. Geographical differentiation of *Leptocarabus* (*Leptocarabus*) *kumagaii* KIMURA & KOMIYA, 1974 (Coleoptera, Carabidae) in Honshu, Japan. *Biogeography*, 6: 27-38.

KUBOTA, K. & K. YAHIRO, 2003. Description of an isolated and specialized population of *Carabus arrowianus* (BREUNING, 1934) (Coleoptera, Carabidae) discovered in the southernmost part of Mie Prefecture, Japan, as a new subspecies, with analyses of its morphological features. *Biogeography*, 5: 9-15.

国兼信之・国兼正明, 1994. マイマイカブリ属の異種交配－エゾマイマイカブリ×オオルリオサムシ. *Celastrina*(津軽昆虫同好会会誌),(29): 9-16. [KUNIKANE & KUNIKANE, 1994]

黒澤良彦, 1988. アキタクロナガオサムシをめぐって. 甲虫ニュース(東京),(81): 1-3. [KUROSAWA, 1988]

草刈広一, 1984. 鳥海山にヒメクロオサの越冬を見る！オサムシマップ,(16): 268. [KUSAKARI, 1984]

[L]

LAFER, G. Sh., 1989. Fam. Carabidae. *In* LEHR, P. A. (eds.). *Key to the Insects of Far East U.S.S.R.*, 3 (1): 71-222. Nauka, Sankt-Petersburg. (In Russian.)

LAPOUGE, G. V. DE, 1909. Tableaux de détermination des formes du genre 《*Carabus*》. *Échange*, 25: 189-190.

LAPOUGE, G. V. DE, 1913-1928. Carabes nouveaux ou mal connus. *Misc. Ent.*, 21, 26, 28, 30: 241 pp.

LAPOUGE, G. V. DE, 1929-1953. Coleoptera Adephaga, Fam. Carabidae, Subfam. Carabinae. *In* WYTSMAN, P. (ed.), *Gen. ins.*, (192): ME, A-C, E+1-747, 7 maps, 10 pls. Wytsman, Bruxelles.

LEWIS, G., 1880. On the distribution of *Damaster*, with description of a new species. *Ent. monthly Mag.*, 17: 159-161.

LEWIS, G., 1881. Description of another new species of *Damaster*. *Ent. month. Mag.*, 17: 197.

LEWIS, G., 1882. A supplementary note on the specific modifications of Japan Carabi, and some observations on the mechanical action of solar rays in relation to colour during the evolution of species. *Trans. ent. Soc. Lond.*, 1882: 503-530.

LINNAEUS, C., 1758. *Systema naturae per regna tria naturae, secundum classes, ordines, genera, species, cum characteribus, differentiis, synonymis, locis*. Tomus I. ed. 10, reformata. IV+824. Holmiae (Laurentii Salvii).

[M]

MANDL, K., 1975. Beitere Beiträge zur Kenntnis der Carabini (Carabidae, Col.). Koleopt. Rdsch., 52: 61-85

MANDL, K., 1981. Eine neue *Acoptolabrus*-Subspezies aus Japan: *munakatai furumii* (Col., Carabidae). *Z. ArbGem. öst. Ent.*, 33: 89-91.

松本堅一, 2011. 釧路川流域のセスジアカガネオサムシの分布について. *SYLVICOLA*(釧路昆虫同好会会誌),(29): 19-24. [MATSUMOTO, 2011]

三村義友, 1988. 採集記録表. オサムシマップ, (38): 862-864. [MIMURA, 1988]
Ministry of the Environment (ed.), 2012. 4th edition of Red List (Threatened Wildlife of Japan). Published on the Web.
MORAWITZ, A., 1862a. Vorläufige Diagnosen neuer Coleopteren aus Südost-Sibirien. *Mélang. Biol., Bull. Acad. imp. Sci. St.-Pétersb.*, 4: 180-228.
MORAWITZ, A., 1862b. Vorläufige Diagnosen neuer Carabiciden aus Hakodate. *Bull. Acad. imp. Sci. St.-Pétersb.*, 4: 321-328.
森岡 茂・東 浩司, 2013. ドウキョウオサムシの新分布. 大昆 *Crude*（大阪昆虫同好会会誌）, (57): 1-5. [MORIOKA & AZUMA, 2013]
MOTSCHULSKY V. DE, 1858. Insectes du Japon. VI. Nouveautés. *Étud. ent., Helsingfors*, 6 [1857]: 108-112.
MOTSCHULSKY V. DE, 1861. Insectes du Japon. Coléoptères. Entomophages. *Etud. ent.*, 10: 1-7.
MOTSCHULSKY V. DE, 1865. Enumeration de nouvelles espèces de Coléoptères rapportés de ses voyages. *Bull. Soc. imp. Moscou*, 2: 227-300.

[N]

永幡嘉之, 1995. 鳥取平野周辺のオサムシの分布資料. すかしば（山陰むしの会会誌）(41・42): 1-9. [NAGAHATA, 1995]
NAGATA N., K. KUBOTA & T. SOTA, 2007. Phylogeography and introgressive hybridization of ground beetles *Carabus yamato* in Japan based on mitochondrial gene sequences. *Zool. Sci.*, 24: 465-474.
NAGATA N., K. KUBOTA, Y. TAKAMI & T. SOTA, 2009. Historical divergence of mechanical isolation agents in the ground beetle *Carabus arrowianus* as revealed by phylogeographical analyses. *Mol. Ecol.*, 18: 1408-1421.
中嶋康二（編）, 1993. 渡島半島の *Acoptolabrus* 亜属. 105 pp. 函館昆虫同好会（函館）. [NAKAJIMA, 1993]
中嶋康二, 1994. 渡島半島から発見された *Damaster* 属の1新種1新亜種. 月刊むし（東京）, (283): 9-11. [NAKAJIMA, 1994]
中根猛彦, 1952a. 邦産オサムシの分布と変異，特にアオオサムシ近縁種について. 新昆蟲, 5(6): 12-13. [NAKANE, 1952a]
中根猛彦, 1952b. 日本の甲虫 (4). 新昆蟲, 5(11): 46-51, 1 pl. [NAKANE, 1952b]
NAKANE, T., 1953. New or little known Coleoptera from Japan and its adjacent regions. IX. Caraboidea II. *Sci. Rep. Saikyo Univ.*, (*Nat. Sci. & Liv. Sci.*), 1 (2): 93-102.
NAKANE, T., 1955. New or little known Coleoptera from Japan and its adjacent regions. XII. *Sci. Rep. Saikyo Univ.* (*Nat. Sci. & Liv. Sci.*), 1 (1): 24-42.
NAKANE, T., 1957. New or little-known Coleoptera from Japan and its adjacent regions, XIV. *Sci. Rep. Saikyo Univ.* (*Nat. Sci. & Liv. Sci.*), 1 (4): 235-239.
NAKANE, T., 1960. Studies in the Carabidae (Insecta, Coleoptera). *Sci. Rep. Kyoto Pref. Univ.* (*Nat. Sci. & Liv. Sci.*), 3 (2), A: 17-44.
NAKANE, T., 1961. New or little-known Coleoptera from Japan and its adjacent regions. XV. *Fragm. coleopterol., Kyoto*, (1): 1-6.
中根猛彦, 1962. 鞘翅目・オサムシ科 [I]. 日本昆虫分類圖説, 2(3), 2+98 pp., 6 pls., 北隆館（東京）. [NAKANE, 1962]
NAKANE, T., 1968. New or little-known Coleoptera from Japan and its adjacent regions. XXVIII. *Fragmenta Coleopterologica, Kyoto*, (21): 85-86.
中根猛彦, 1977. 日本の甲虫 (44)，おさむし科 6. 昆虫と自然（東京）, 12(10): 6-12. [NAKANE, 1977]
西島 浩, 1989. 北海道のオサムシ分布概要. 東日本オサムシ研究会（編），東日本のオサムシ: 11-14. [NISHIJIMA, 1989]
西川協一・奥村 尚, 1971. *Damaster* (s. str.) マイマイカブリについて. *Ins. Mag.*, (76): 87-96. [NISHIKAWA & OKUMURA, 1971]

[O]

OBYDOV D. & A. SALDAITIS, 1996. A new subspecies of Carabus (Ainocarabus) kolbei Roeschke, 1897 from the Ushishir Island. *Coleoptera (Schwanfeld. coleopterol. Mitteil.), Schwanfeld*, 22: 1-5.
尾形洋一, 1980. 絶滅に瀕するマークオサムシ－生かされない市の自然環境調査－. みやこわが町, 4(31): 43-45. [OGATA, 1980]
尾形洋一, 1982. マークオサムシの観察. インセクタリウム, 19: 58-61. [OGATA, 1982]
岡本宗裕, 1999. ミトコンドリア DNA からみたクビナガオサムシ類の系統関係. 昆虫と自然（東京）, 34 (2): 11-14. [OKAMOTO, M., 1999]
岡本宗裕・冨永 修・井村有希・蘇 智慧・大澤省三, 2007. DNA からみたヒメオサムシ群の分化. 甲虫ニュース, (159): 1-5. [OKAMOTO, et al. 2007]
OKAMOTO, M., O. TOMINAGA, Z.-H. SU, Y. IMURA, N. KASHIWAI, T. OJIKA, K. AKITA & S. OSAWA, 2005. Differentiation of *Ohomopterus yaconinus* and its "subspecies" (Coleoptera, Carabidae) inferred from the DNA sequences of mitochondrial ND5 Gene and internal transcribed spacer I (ITS I). *Elytra, Tokyo*, 33: 363-370.
奥村 尚, 1971. 北海道地方オサムシ相の特徴. *Ins. Mag.*, (76): 119-126. [OKUMURA, 1971]
奥村 尚, 1972. オサ堀り苦戦記 I 奥村尚の巻－その第2話 日高に雪を掘る－. 月刊むし（東京）, (11): 10-12. [OKUMURA, 1972]
大倉正文・後藤光男, 1960. セアカオサムシの1新変種. 昆虫学評論, 11(1): 18. [OHKURA & GOTÔ, 1960]

オオオサムシ属の分子系統研究グループ（冨永修・岡本宗裕・井村有希・蘇智慧・小鹿亨・秋田勝己・柏井伸夫・永幡嘉之．大澤省三），2005．オオオサムシ属（*Ohomopterus*）における雑種集団の安定化・分布域拡大・種分化．ねじればね，(115)：1-14．［Research group of molecular phylogeny on the *Ohomopterus* ground beetles (Tominaga, O, M. Okamoto, Y. Imura, Z.-H. Su, T. Ojika, K. Akita, N. Kashiwai, Y. Nagahata & S. Osawa), 2005］

オオオサムシ属の分子系統研究グループ（冨永修・岡本宗裕・井村有希・蘇智慧・小鹿亨・秋田勝己・柏井伸夫・永幡嘉之．大澤省三），2006a．オオオサムシ属（*Ohomopterus*）をめぐるさらなる問題．ねじればね，(116)：6-12．［Research group of molecular phylogeny on the *Ohomopterus* ground beetles (Tominaga, O, M. Okamoto, Y. Imura, Z.-H. Su, T. Ojika, K. Akita, N. Kashiwai, Y. Nagahata & S. Osawa), 2006a］

オオオサムシ属の分子系統研究グループ（冨永修・岡本宗裕・井村有希・蘇智慧・小鹿亨・秋田勝己・柏井伸夫・永幡嘉之．大澤省三），2006b．DNAからみたマヤサンオサムシの"亜種"について．ねじればね，(118)：18-24．［Research group of molecular phylogeny on the *Ohomopterus* ground beetles (Tominaga, O, M. Okamoto, Y. Imura, Z.-H. Su, T. Ojika, K. Akita, N. Kashiwai, Y. Nagahata & S. Osawa), 2006b．］

長内 淳, 1986. マークオサムシの飼育. 多摩虫（東京），9(19)：1-10. ［Osanai, 1986］

大澤省三・蘇 智慧, 2000. 日本のオサムシ相の形成---分子系統樹からの推定---(1). ねじればね，(90)：1-5. ［Osawa & Su, 2000］

大澤省三・蘇 智慧, 2001a. 日本のオサムシ相の形成---分子系統樹からの推定---(2). ねじればね，(91)：1-5. ［Osawa & Su, 2001a］

大澤省三・蘇 智慧, 2001b. 日本のオサムシ相の形成---分子系統樹からの推定---(5). ねじればね，(94)：1-9. ［Osawa & Su, 2001b］

大澤省三・蘇 智慧・井村有希, 2002. DNAでたどるオサムシの系統と進化. 264 pp., 哲学書房（東京）. [Osawa *et al*., 2002]

Osawa, S., Z.-H. Su & Y. Imura, 2004. Molecular phylogeny and evolution of carabid ground beetles. 191 pp. 119 figs. Springer-Verlag, Tokyo.

Osawa, S., Z.-H. Su, C.-G. Kim, M. Okamoto, O. Tominaga & Y. Imura, 1999. Evolution of the carabid ground beetles. *Adv. Biophys.*, 36: 65-106.

Oyama, K., 蘇 智慧, 冨永 修, 斎藤秀生, 柏井伸夫, 大澤省三, 2000. ミトコンドリアDNAによるクロオサムシ（*Ohomopterus albrechti*）とその関連種の分子系統. 甲虫ニュース, (131)：13-14.[Oyama *et al*., 2000]

[P]

Putzeys, J., 1875. Notice sur carabiques recueillis par M. Jean van Volxem a Ceylan, a Manille, en Chine et au Japon (1873-1874). *Ann. Soc. Ent. Belg., Bruxelles*, 18: 45-55.

[R]

Reitter, E., 1896. Bestimmungs-Tabelle der europäischen Coleopteren, Carabidae, 1, Carabini, gleichzeitig mit einer systematischen Darstellung sammtlicher Subgenera der Gattung *Carabus* L. *Verh. naturf. Ver. Brünn*, 34: 36-198.

Roeschke, H., 1897. Ein neuer *Carabus* aus China. *Ent. Nachr., Berlin*, (8): 116-117.

Roeschke, H., 1900. Carabologische Notizen VII. *Ent. Nachr., Berlin*, (26): 162-163.

Rost, C., 1908. Ein neuer *Carabus* aus Japan (Col.). *Dt. ent. Zs., Berlin*, 1908: 32-33.

Rye, E, C,, 1872. Descriptions of a new species of *Damaster* from Japan. *Ent. month. Mag., Nov.*: 131-132.

[S]

桜井俊一・鳥羽明彦・山谷文仁, 1989. セアカオサムシ *Carabus* (*Hemicarabus*) *tuberculosus* (Dejean et Boisduval). 東日本オサムシ研究会（編），東日本のオサムシ：54-58. ［Sakurai *et al*., 1989］

佐野信雄, 1993. 四国におけるオオオサムシの分布南限. へりぐろ（瀬戸内むしの会会誌），(14)：24-25. [Sano, 1993]

Schaum, H. R., 1862. Espèce nouvelle du genre *Damaster*. *Ann. Soc. ent. Fr.*, (4), Ser. 2: 68, pl. 2, fig. 1.

清水勝幸・山谷文仁・荒井充朗, 1989. ヒメオサムシ種群 *Carabus* (*Ohomopterus*) *japonicus* group. 東日本オサムシ研究会（編），東日本のオサムシ：84-88. [Shimizu *et al*., 2000]

曽田貞滋, 2000a. オサムシの分子系統－オオオサムシ亜属において平行進化は本当に起こったのか？インセクタリウム, 37：176-185. ［Sota, 2000a］

曽田貞滋, 2000b. オサムシの春夏秋冬－生活史の進化と種多様性. 4 pls. VII＋248 pp. 京都大学学術出版会（京都）. ［Sota, 2000b］

曽田貞滋, 2000c. 核・ミトコンドリア遺伝子の分子系統解析からみたオオオサムシ亜属の進化. *Shinka*, 10: 14-20. ［Sota, 2000c］

曽田貞滋, 2002a. DNAが語る日本列島のオサムシの分化. 遺伝, 56(2)：67-73. ［Sota, 2002a］

Sota T., 2002b. Radiation and reticulation: extensive introgressive hybridization in the carabid beetles *Ohomopterus* inferred from mitochondrial gene genealogy. *Popul. Ecol.*, 44: 145-156.

曽田貞滋, 2003. ミトコンドリアDNA分析に基づく西日本のオオオサムシ亜属の系統地理学的研究. ホシザキグリーン財団研究報告, (6): 153-166. [Sota, 2003]

曽田貞滋, 2011. オサムシ学の新展開－生態から進化まで: 概説. 昆虫と自然 (東京), 46(3): 2-7. [Sota, 2011]

曽田貞滋 (編), 2013. 新オサムシ学. 8 pls.＋209 pp. 北隆館 (東京). [Sota (ed.), 2013]

Sota T. & R. Ishikawa, 2004. Phylogeny and life-history evolution in *Carabus* (subtribe Carabina: Coleoptera, Carabidae) based on sequences of two nuclear genes. *Biol. J. Linn. Soc.*, 81: 135-149.

Sota T., R. Ishikawa, M. Ujiie, F. Kusumoto & A. P. Vogler, 2001. Extensive trans-species mitochondrial polymorphisms in the carabid beetles *Carabus* subgenus *Ohomopterus* caused by repeated introgressive hybridization. *Mol. Ecol.*, 10: 2833-2847.

Sota, T., M. Kusumoto & K. Kubota, 2000. Consequences of hybridization between *Ohomopterus insulicola* and *O. arrowianus* (Coleoptera, Carabidae) in a segmented river basin: parallel formation of hybrid swarms. *Biolog. J. Linn. Soc.*, 71: 297-313.

Sota, T. & N. Nagata (2008). Diversification in a fluctuating island setting: rapid radiation of *Ohomopterus* ground beetles in the Japanese Islands. *Phil. Trans. Roy. Soc. B.*, 363: 3377-3390.

Sota, T. & A. P. Vogler 2001. Incongruence between mitochondrial and nuclear gene trees in the carabid beetles *Ohomopterus*. *Syst. Biol.*, 50: 39-50.

Sota, T. & A. P. Vogler 2003. Reconstructing species phylogeny of the carabid beetles *Ohomopterus* using multiple nuclear DNA sequences: heterogeneous information content and the performance of simultaneous analysis. *Mol. Phylogenet. Evol.* 26: 139-154.

Su, Z.-H., Y. Imura, M. Okamoto, C.-M. Kim, H.-Z. Zhou, J.-C. Paik & S. Osawa, 2004. Phylogeny and evolution of Digitulati ground beetles (Coleoptera, Carabidae) inferred from mitochondrial ND5 gene sequences. *Mol. Phylogenet. Evol.*, 30: 152-166.

Su, Z.-H., Y. Imura, O. Tominaga & S. Osawa, 2000. Phylogeny in the Crenolimbi ground-beetles (Coleoptera, Carabidae) as deduced from mitochondrial ND5 gene sequences. *Elytra, Tokyo*, 28: 229-233.

Su, Z.-H., T. Ohama, T. S. Okada, K. Nakamura, R. Ishikawa & S. Osawa, 1996. Geography-linked phylogeny of the *Damaster* ground beetles inferred from mitochondrial ND5 gene sequences. *J. mol. Evol.*, 42: 130-134.

Su, Z.-H., M. Ôhara, Y. Imura & S. Osawa, 2000. *Damaster blaptoides* (Coleoptera, Carabidae) from Brat Chirpoyev Island of the Kurils, Russia. *Elytra, Tokyo*, 28: 233-234.

Su, Z-H., M. Okamoto, O. Tominaga, K. Akita, N. Kashiwai, Y. Imura, T. Ojika, Y. Nagahata & S. Osawa, 2006. Establishment of hybrid-derived offspring populations in the *Ohomopterus* ground beetles through unidirectional hybridization. *Proc. Jpn. Acad., Ser. B*, 82: 232-250.

Su, Z.-H., O. Tominaga, T. Ohama, E. Kajiwara, R. Ishikawa, T. S. Okada, K. Nakamura & S. Osawa, 1996. Parallel evolution in radiation of *Ohomopterus* ground beetles inferred from mitochondrial ND5 gene sequences. *J. mol. Evol.*, 43: 662-671.

Su, Z.-H., O. Tominaga, M. Okamoto & S. Osawa, 1998. Origin and diversification of hindwingless *Damaster* ground beetles within the Japanese Islands as deduced from mitochondrial ND5 gene sequence (Coleoptera, Carabidae). *Mol. Biol. Evol.*, 15: 1026-1039.

杉江 昇・藤原敏幸, 1981. 北海道に生息する数種オサムシ類の生態と幼虫形態について. 51 pp. (帯広畜産大学畜産環境学科畜産環境学研究室の卒業論文として提出されたもので, 正式に出版はされていない) [Sugie & Fujiwara, 1981]

鈴木竣策, 1977. オサムシの異常型について. *Jezoensis*, 4: 40-41. [Suzuki, 1977]

[**T**]

多賀敏正, 2002. マヤサンオサムシの新分布地について. 月刊むし (東京), (276): 10-14. [Taga, 2002]

Takami Y. & R. Ishikawa, 1997. Subspeciation and distribution pattern of *Carabus albrechti* Morawitz in Japan (Coleoptera, Carabidae). *TMU Bull. nat. Hist, Tokyo*, (3): 55-99.

Takami Y. & H. Suzuki, 2005. Morphological, genetic and behavioural analyses of a hybrid zone between the ground beetles *Carabus lewisianus* and *C. albrechti* (Coleoptera, Carabidae): Asymmetrical introgression caused by movement of the zone？*Biol. J. Linn. Soc.*, 86: 79-94.

玉貫光一, 1972. ホソコバネカミキリとクビナガオサムシ. 昆虫と自然 (東京), 7(10): 2-6. [Tamanuki, 1972]

冨永 修, 1984. 四国のオサムシ (13). 無翅 (はねなし) 段丘, (93): 1461-71. [Tominaga, 1984]

Tominaga, O., Y. Imura, M. Okamoto, Z.-H. Su, T. Ojika, N. Kashiwai & S. Osawa, 2005. Origin of *Ohomopterus uenoi* (Coleoptera, Carabinae: Carabidae) as deduced from comparisons of DNA sequences of mitochondrial ND5 gene and nuclear internal transcribed spacer I (ITS I) with morphological characters. *Ent. Rev. Japan*, 60 (1): 23-33.

冨永 修・岡本宗裕・井村有希・蘇 智慧・大澤省三・小鹿 亨・柏井伸夫・秋田勝己, 2005. 日本のオオオサムシ属 (*Ohomopterus*) 相の形成, 特に近畿, 中部日本系について～分子系統からの推定. 昆虫DNA研究会ニュースレター, (2):

7-24. [TOMINAGA, OKAMOTO *et al.*, 2005]

TOULGOËT, H. DE, 1975a. Les types du genre *Carabus* (s. l.) du Muséum d'Histoire Naturelle de Paris (Coléoptères Carabidae Carabinae). *Nouv. Rev. Ent.*, 5 (1): 13-30.

TOULGOËT, H. DE, 1975b. Les types du genre *Carabus* (s. l.) du Muséum d'Histoire Naturelle de Paris (Coléoptères Carabidae Carabinae). *Nouv. Rev. Ent.*, 5 (3): 221-237.

土屋裕志, 1982. 新潟県佐渡ヶ島でマークオサムシの上翅を拾う. 月刊むし（東京）,(137)：30. [TSUCHIYA, 1982]

TURIN, H., L. PENEV & A. CASALE (eds.), 2003. The genus *Carabus* in Europe, a synthesis. *Fauna Europaea Evertebrata* (2). 511 pp. Pensoft, Sofia-Moscow.

[U]

UCHIDA T. & K. TAMANUKI, 1927. Zwei neue Arten von *Acoptolabrus* (Coleopt.). *Ins. Mats.*, 2 (2): 102-104.

UJIIE M., R. ISHIKAWA & K. KUBOTA, 2010. Geographical variation of *Carabus* (*Ohomopterus*) *esakii* CIKI, 1927 (Coleoptera, Carabidae) in Japan. *Biogeography*, (12): 53-64.

UJIIE M., K. KUBOTA, T. SOTA & R. ISHIKAWA, 2005. Parallel formation of hybrid swarms of ground beetles in the genus *Carabus* (Coleoptera: Carabidae) in adjacent river basins. *Entomol. Sci.*, 8: 429-437.

[V]

VAN EMDEN, F., 1932. Einige neue Carabinae des Staatlichen Museums für Tierkunde zu Dresden. *Neue Beitr. syst. Insektenkunde*, 5: 62-69, 1 pl.

[Y]

山地 治・脇本 浩, 1979. 岡山県のオサムシ採集記録. すずむし（倉敷昆虫同好会機関紙）,(116)：39-42. [YAMAJI & WAKIMOTO, 1979]

山谷文仁, 1989. アキタクロナガオサムシ *Carabus* (*Euleptocarabus*) *porrecticollis* BATES. 東日本オサムシ研究会（編）, 東日本のオサムシ：127-130. [YAMAYA, 1989]

山谷文仁・草刈広一, 1989. ヒメクロオサムシ *Carabus* (*Aulonocarabus*) *opaculus* PUTZEYS. 東日本オサムシ研究会（編）, 東日本のオサムシ：111-117. [YAMAYA & KUSAKARI, 1989]

山崎亮一, 1984. エゾマイマイカブリとオオルリオサムシとの中間的性格を持った個体の検討. オサムシマップ,(13)：207-208. [YAMAZAKI, 1984]

[Z]

ZHANG, A. B. & T. SOTA, 2007. Nuclear gene sequences resolve species phylogeny and mitochondrial introgression in *Leptocarabus* beetles showing trans-species polymorphisms. *Mol. Phylogenet. Evol.*, 45: 534-546.

索引　INDEX

Acoptolabrus 314~329
adatarasanus 59, 63, 69
aereicollis 315, 318, 319, 321, 329
aino 299, 306, 312
akaishiensis 60, 67, 69
akitanus 104, 111, 119
albrechti (sp.) 164~177
albrechti (subsp.) 164, 165, 169, 176
amanoi 38, 39, 41~43
apoi 121, 125, 127
aquatilis 24~29
arboreus (sp.) 102~119
arboreus (subsp.) 102, 109, 119
arcensis 130
arrowianus (sp.) 278~287
arrowianus (subsp.) 278, 283, 286
arvensis 128~133
Asthenocarabus 120~127
auricollis 334, 335
Aulonocarabus 70~77
awajiensis 199, 203, 209
awakazusanus 159, 161~163
awashimae 166, 172, 176
awashimaensis 263, 267, 272

babai 104, 114, 119
babaianus 331, 337, 342, 343
blairi 241, 244, 249
blaptoides (sp.) 330~343
blaptoides (subsp.) 330, 333, 341, 342
botchan 211, 213~215
brevicaudus 333, 340, 342

capito 332, 339, 342
Carabus (subgenus) 128~145
canaliculatus 72
cerberus 79, 81, 83
chishimanus 299, 305, 312
chotaroi 200, 206, 209
chugokuensis (sp.) 184~191
chugokuensis (subsp.) 184, 185, 187, 190
Coptolabrus 344~347
conciliator 130, 131
corvinus 202
cupidicornis 241, 245, 249
cyaneo-violaceus 318, 319
cyanostola 334, 335

daisen (sp.) 178~183
daisen (subsp.) 178, 179, 181~183
daisetsuzanus 71, 75, 76
Damaster 330~343
dehaanii (sp.) 222~231
dehaanii (subsp.) 227

echigo 166, 172, 176
esakianus 166, 173, 176, 177
esakii (sp.) 167, 173, 256~261
esakii (subsp.) 256, 257, 259, 260
Euleptocarabus 30~37
exilis 113, 118, 119

fortunei 332, 338, 342

freyi 166, 173, 176
fruhstorferi 344~347
fujisan 60, 66, 69
fujisanus 105, 115, 119
furumii 318, 319
futabae 299, 305, 312

gehinii (sp.) 314~329
gehinii (subsp.) 314, 316, 324, 329
gracillimus 105, 116, 119
grandis 318, 319
granulatus 134~139
gustavhauseri 318, 319

hagai 165, 171, 176
hakusanus 105, 117, 119
hanatanii 299, 304, 312
harmandi (sp.) 58~69
harmandi (subsp.) 58, 64, 69
Hemicarabus 38~49
hida 91, 93, 95
hidakamontanus 300, 308, 312
hidakanus 165, 169, 176, 177
hidamontanus 116
hidaosa 279, 283, 286
hiradonis 200, 205, 209
hiraii 212
hiurai 84~89
hokkaidensis 128, 129, 131, 133
hokurikuensis 289, 291, 296
Homoeocarabus 50~57
horioi 105, 116, 119

ikiensis 200, 205, 209
imafukui 225, 226
incompletus 53
insulicola (sp.) 262~273
insulicola (subsp.) 262, 268, 272
ishidai 223, 229, 231
ishikarinus 103, 108, 119
ishizuchianus 217, 220, 221
itoi 165, 168, 176
iwawakianus (sp.) 232~239
iwawakianus (subsp.) 232, 236, 238, 239

japonicus (sp.) 198~209
japonicus (subsp.) 198, 199, 204, 209

kansaiensis 31, 35, 36
kantoensis 263, 267, 272, 273
karasawai 60, 65, 69
karatsuanus 204
karibanus 103, 108, 119
katomelas 318, 319
katsumai 223, 228, 231
kawaharai 79, 82, 83
kawanoi (sp.) 210~215
kawanoi (subsp.) 210, 211, 213, 214
kiiensis 233, 237, 238
kimurai 154~157
kinkimontanus 147, 152, 153
kirimurai 279, 284, 286
kiso 264, 270, 272, 273

kita 263, 266, 272
kitai 147, 152, 153
kitakamisanus 104, 111, 119
kojii 59, 64, 69
kolbei (sp.) 298~313
kolbei (subsp.) 298, 308, 312, 313
komaganensis 264, 270, 272, 273
komiyai (sp.) 250~255
komiyai (subsp.) 250, 251, 253, 255
konsenensis 315, 322, 329
koshikicola 223, 229, 231
kosugei 301, 309, 312
kumagaii (sp.) 90~95
kumagaii (subsp.) 90, 93, 95
kumaso 223, 229, 231
kuniakii 301, 309, 312
kurilensis (sp.) 70~77
kurilensis (subsp.) 71, 74, 76
kurosawai 121, 124, 127
kyushuensis (sp.) 78~83
kyushuensis (subsp.) 78, 79, 82, 83

Leptocarabus 78~119
lewisianus (sp.) 158~163
lewisianus (subsp.) 158, 159, 161~163
Lewisi 258
Lewisii 334, 335
Limnocarabus 24~29

maacki 24~29
macleayi 38~43
maeander 50~57
maetai 241, 246, 249
maiyasanus (sp.) 288~297
maiyasanus (subsp.) 288, 291, 296, 297
manoianus 315, 322, 329
matsunagai 251, 253, 255
Megodontus 298~313
mikianus 185, 187, 190, 191
minoensis 279, 284, 286
mitsumasai 300, 306, 312
miyakei 46, 47, 97, 99, 101
mizunumai 59, 65, 69
mochizukii 185, 188, 190
montanus 337, 338
moritai 60, 67, 69
munakatai 317, 327, 329
munakataorum 301, 310, 312, 313
murakii 279, 284, 286, 287
muro 233, 236, 238

nagoyaensis 265, 280, 281
nakagomei 224, 230, 231
nakamurai 62, 279, 281, 286
nakatomii 79, 81, 83
narukawai 233, 235, 238
nepta 103, 109, 119
nishi 91, 93, 95
nishijimai 317, 326, 329
nishikawai 263, 268, 272
nitidipunctatus 300, 307, 312
nobukii 51, 54, 56, 57
noesskei 141, 143, 144
nozakicola 200, 206, 209

ogurai 105, 115, 119
ohkawai 289, 294, 296
ohminensis 105, 117, 119
Ohomopterus 146~297
ojikai 147, 150, 153
oki 241, 245, 249
okianus 179, 181, 182
okinoshimanus 199, 204, 209
okumurai 166, 175, 176
okutamaensis 60, 66, 69
okutonensis 263, 268, 272
onodai 200, 207, 209
opaculus (sp.) 120~127
opaculus (subsp.) 120, 121, 125, 127
oxuroides 332, 340, 342, 343

pacificus 31, 34, 36, 37
paludis 50, 51, 53, 56, 57
palustris 53
pandurus 334, 335
pararboreus 103, 107, 119
parexilis 104, 111, 119
Pentacarabus 58~69
porrecticollis (sp.) 30~37
porrecticollis (subsp.) 30, 31, 33, 36, 37
procerulus (sp.) 96~101
procerulus (subsp.) 96, 97, 99, 101
pronepta 103, 110, 119
pseudinsulicola 264~266
punctatostriatus 223, 227, 231

quinquecatellatus 59, 64, 69

radiatocostatus 315, 323, 329
rausuanus 71, 74, 76
rishiriensis 70, 73, 76, 77
rugipennis 331, 336, 342, 343

sado 264, 269, 272
sapporensis 316, 323, 329
seizaburoi 185, 187, 190, 191
seto 241, 247, 249
shichinoi 279, 282, 286, 287
shigaraki 289, 293, 296
shikatai 147, 150, 153
shikokuensis 86
shima 233, 235, 238
Shimamaki lowland population 317, 327, 329
shimizui 316, 326, 329
shimoheiensis 103, 110, 119
shinanensis 104, 114, 119
shinano 264, 269, 272
shirahatai 122, 126, 127
shizuokaensis 258, 265
sotai 241, 244, 249
strenuus 223, 228, 231
sue 192~197
sugai 71, 73, 76
suruganus 257, 259~261
suzukanus 289, 292, 296
syui 59, 63, 69

taikinus 300, 307, 312
takanonis 147, 151, 153
takiharensis 289, 293, 296

tanzawaensis 60, 66, 69
telluris 134, 135, 138, 139
Tenryû-gawa population 91, 94, 95
tenuiformis 104, 112, 119
tohokuensis 165, 170, 176
tosanus (sp.) 216~221
tosanus (subsp.) 216, 217, 219, 221
tsukubanus 166, 174, 176
tsushimae 199, 205, 209
tuberculatus 46
tuberculosus 44~49

uenoi 274~277
umekii 185, 189, 190
ushishirensis 303

vanvolxemi (sp.) 140~145
vanvolxemi (subsp.) 140, 141, 143~145
viridipennis 331, 336, 342
viridis 318, 319

wakamatsuensis 200, 207, 209

yaconinus (sp.) 240~249
yaconinus (subsp.) 240, 245, 249
yamaokai 241, 247, 249
yamato (sp.) 146~153
yamato (subsp.) 146, 151, 153
yamauchii 165, 170, 176
yamazumiensis 251, 253, 255
yasudai 301, 310, 312
Yezacoptolabrus 318
yezoensis 135, 137, 139
yoroensis 289, 292, 296
yoshiyukii 199, 203, 209
yoteizanus 71, 75, 76
yubariensis 300, 308, 312
yudanus 59, 62, 69

アイヅクロナガオサムシ 112
アイヌキンオサムシ 298~313
アイヌキンオサムシ択捉島亜種 299, 305, 312
アイヌキンオサムシ狩場山亜種 301, 310, 312
アイヌキンオサムシ基亜種 298, 308, 312, 313
アイヌキンオサムシ積丹半島亜種 301, 309, 312
アイヌキンオサムシ大樹町亜種 300, 307, 312
アイヌキンオサムシ大千軒岳亜種 301, 310, 312, 313
アイヌキンオサムシ道央道東亜種 299, 306, 312
アイヌキンオサムシ道北亜種 299, 305, 312
アイヌキンオサムシトマム亜種 300, 307, 312
アイヌキンオサムシニセコ亜種 301, 309, 312
アイヌキンオサムシ日高山脈亜種 300, 308, 312
アイヌキンオサムシ増毛山地南西部亜種 300, 306, 312
アイヌキンオサムシ夕張山地亜種 300, 308, 312
アイヌキンオサムシ利尻島亜種 299, 304, 312
アウロノカラブス亜属 70~77
アオオサムシ 262~273
アオオサムシ粟島亜種 263, 267, 272
アオオサムシ奥利根亜種 263, 268, 272
アオオサムシ関東平野多摩川以北亜種 263, 267, 272, 273
アオオサムシ基亜種 262, 268, 272
アオオサムシ木曽亜種 264, 270, 272, 273
アオオサムシ駒ヶ根亜種 264, 270, 272, 273
アオオサムシ佐渡島亜種 264, 269, 272
アオオサムシ信濃亜種 264, 269, 272
アオオサムシ東北地方亜種 263, 266, 272
アオオサムシ房総半島南部亜種 263, 268, 272
アオマイマイカブリ 339
アカイシアルマンオサムシ 67
アカイシクロナガオサムシ 116
アカイシホソヒメクロオサムシ 67
アカオサムシ 268
アカガネオサムシ 134~139
アカガネオサムシ北海道亜種 135, 137, 139
アカガネオサムシ本州亜種 134, 135, 138, 139
アキオサムシ 184~191
アキオサムシ基亜種 184, 185, 187, 190
アキオサムシ芸予諸島大島亜種 185, 188, 190
アキオサムシ讃岐山脈東部亜種 185, 187, 190, 191
アキオサムシ小豆島亜種 185, 187, 190, 191
アキオサムシ周防亜種 185, 189, 190
アキタクロナガオサムシ 30~37
アキタクロナガオサムシ岩湧亜種 31, 35, 36
アキタクロナガオサムシ基亜種 30, 31, 33, 36, 37
アキタクロナガオサムシ富士亜種 31, 34, 36, 37
アコプトラブルス亜属 314~329
アステノカラブス亜属 120~127
アダタラアルマンオサムシ 63
アダタラホソヒメクロオサムシ 63
アブクマホソヒメクロオサムシ 63
アラメオオルリオサムシ 323
アルマンオサムシ 62
アワオサムシ 210~215
アワオサムシ基亜種 210, 211, 213, 214
アワオサムシ高縄半島亜種 211, 213~215
アワカズサオサムシ 161
アワジヒメオサムシ 203
アワシマアオオサムシ 267

アワシマクロオサムシ 173
アヲマイマイカブリ 339

イェザコプトラブルス亜属 319
イキオサムシ 205
イキツキヒメオサムシ 206
イシカリクロナガオサムシ 108
イシヅチオサムシ 220
イセオサムシ 284
イブシキンオサムシ 309
イワタオサムシ 253
イワワキオサムシ 232~239
イワワキオサムシ基亜種 232, 236, 238, 239
イワワキオサムシ紀伊半島亜種 233, 237, 238
イワワキオサムシ志摩半島亜種 233, 235, 238
イワワキオサムシ布引山地亜種 233, 235, 238
イワワキオサムシ室生山地亜種 233, 236, 238

ウガタオサムシ 295

エウレプトカラブス亜属 30~37
エサキオサムシ 174
エゾアカガネオサムシ 138
エゾオサムシ 138
エゾクロナガオサムシ 107
エゾマイマイカブリ 336
エチゴクロオサムシ 172

オオオサムシ 222~231
オオオサムシ大隅半島亜種 224, 230, 231
オオオサムシ上甑島亜種 223, 229, 231
オオオサムシ基亜種 227
オオオサムシ九州山地南部亜種 223, 229, 231
オオオサムシ熊本県南西部亜種 223, 228, 231
オオオサムシ御所浦島亜種 223, 229, 231
オオオサムシ四国東部亜種 223, 228, 231
オオオサムシ本州中部亜種 223, 227, 231
オオクロナガオサムシ 90~95
オオクロナガオサムシ基亜種 90, 93, 95
オオクロナガオサムシ近畿・中部地方亜種 91, 93, 95
オオクロナガオサムシ天竜川個体群 91, 94, 95
オオクロナガオサムシ飛騨高山亜種 91, 93, 95
オオスミオオオサムシ 230
オオビラルリオサムシ 327
オオミネクロナガオサムシ 118
オオルリオサムシ 314~329
オオルリオサムシ大平山亜種 317, 326, 329
オオルリオサムシ渡島半島亜種 317, 327, 329
オオルリオサムシ基亜種 314, 316, 324, 329
オオルリオサムシ根釧台地亜種 315, 322, 329
オオルリオサムシ島牧低地個体群 317, 327, 329
オオルリオサムシ積丹半島亜種 316, 326, 329
オオルリオサムシ道北亜種 315, 321, 329
オオルリオサムシ日高山脈南西部亜種 316, 323, 329
オオルリオサムシ日高山脈南東部亜種 315, 323, 329
オオルリオサムシ南富良野亜種 315, 322, 329
オオルリクビナガオサムシ 321
オキオサムシ 181
オキノシマヒメオサムシ 204

オキマイマイカブリ 341
オキヤコンオサムシ 246
オクエゾクロオサムシ 169
オクエゾクロナガオサムシ 108
オクタマアルマンオサムシ 66
オクタマホソヒメクロオサムシ 66
オクトネアオオサムシ 269
オクムラクロオサムシ 175
オシマキンオサムシ 311
オシマルリオサムシ 328
オニクロナガオサムシ 82
オオホモプテルス亜属 146~297
オンタケクロナガオサムシ 117

カケガワオサムシ 250~255
カケガワオサムシ磐田原亜種 251, 253, 255
カケガワオサムシ基亜種 250, 251, 253, 255
カケガワオサムシ山住峠亜種 251, 253, 255
カラツオサムシ 204
カラブス亜属 128~145
カリバキンオサムシ 310
カリバクロナガオサムシ 109
カントウアオオサムシ 268

キイオサムシ 237
キソアオオサムシ 270
キタアオオサムシ 267
キタオオルリオサムシ 322
キタカブリ 337
キタカミクロナガオサムシ 111
キタクロオサムシ 170
キタクロナガオサムシ 107
キタマイマイカブリ 337
キタヤマトオサムシ 151
キムラクロオサムシ 155
キュウシュウオサムシ 228
キュウシュウクロナガオサムシ 78~83
キュウシュウクロナガオサムシ基亜種 78, 79, 82, 83
キュウシュウクロナガオサムシ中国地方亜種 79, 81, 83
キュウシュウクロナガオサムシ広島県東部亜種 79, 81, 83
キュウシュウクロナガオサムシ山口県亜種 79, 82, 83
キューピーヤコンオサムシ 245
キレスジルリオサムシ 323

クマソオオオサムシ 229
クマノヤマトオサムシ 152
クロオサムシ 164~177
クロオサムシ粟島亜種 166, 172, 176
クロオサムシ越後亜種 166, 172, 176
クロオサムシ関東地方北西部亜種 166, 173, 176, 177
クロオサムシ関東地方北東部亜種 166, 174, 176
クロオサムシ基亜種 164, 165, 169, 176
クロオサムシ佐渡島亜種 166, 173, 176
クロオサムシ白神山地亜種 165, 170, 176
クロオサムシ東北地方中部亜種 165, 171, 176
クロオサムシ東北地方東部亜種 165, 170, 176
クロオサムシ道東亜種 165, 168, 176
クロオサムシ日高亜種 165, 169, 176, 177
クロオサムシ山梨長野亜種 166, 175, 176

クロナガオサムシ 96~101
クロナガオサムシ基亜種 96, 97, 99, 101
クロナガオサムシ九州亜種 97, 99, 101

ケンザンクロナガオサムシ 87

コアオマイマイカブリ 338
コクロナガオサムシ 102~119
コクロナガオサムシ青森県亜種 103, 109, 119
コクロナガオサムシ赤石山脈亜種 105, 116, 119
コクロナガオサムシ秋田県亜種 104, 111, 119
コクロナガオサムシ石狩亜種 103, 108, 119
コクロナガオサムシ岩手県北部亜種 103, 110, 119
コクロナガオサムシ大峰山脈亜種 105, 117, 119
コクロナガオサムシ奥秩父亜種 105, 115, 119
コクロナガオサムシ渡島半島北部亜種 103, 108, 119
コクロナガオサムシ基亜種 102, 109, 119
コクロナガオサムシ北上山地亜種 104, 111, 119
コクロナガオサムシ北関東上越亜種 104, 112, 119
コクロナガオサムシ佐渡島亜種 113, 118, 119
コクロナガオサムシ下閉伊亜種 103, 110, 119
コクロナガオサムシ東北地方南部亜種 104, 111, 119
コクロナガオサムシ道央道東道北亜種 103, 107, 119
コクロナガオサムシ白山亜種 105, 117, 119
コクロナガオサムシ飛騨御嶽木曽亜種 105, 116, 119
コクロナガオサムシ富士箱根丹沢亜種 105, 115, 119
コクロナガオサムシ妙高連峰亜種 104, 114, 119
コクロナガオサムシ八ヶ岳亜種 104, 114, 119
コシキオオサムシ 230
コシキヒメオサムシ 208
ゴショノウラオオオサムシ 229
コスゲアイヌキンオサムシ 309
ゴスジアルマンオサムシ 65
ゴスジホソヒメクロオサムシ 65
コブスジアカガネオサムシ 128~133
コブスジアカガネオサムシ北海道亜種 128, 129, 131, 133
コプトラブルス亜属 344~347
コマガネアオオサムシ 271
コルベキンオサムシ 306, 309
コンゴウオサムシ 276
コンセンオオルリオサムシ 322
コンセンルリオサムシ 322

サッポロクビナガオサムシ 324
サドアオオサムシ 269
サドオサムシ 143, 173
サドクロオサムシ 173
サドクロナガオサムシ 114
サドホソアカガネオサムシ 143
サドマイマイカブリ 339
サヌキアキオサムシ 188
サンインヤコンオサムシ 246
サンプククロナガオサムシ 116

シガラキオサムシ 293
シコクオオサムシ 228
シコクオサムシ 213, 219
シコククロナガオサムシ 84~89
シコクヤコンオサムシ 247

シズオカオサムシ 167, 173, 256~261
シズオカオサムシ基亜種 256, 257, 259, 260
シズオカオサムシ富士川流域以西亜種 257, 259~261
シナノアオオサムシ 270
シナノクロナガオサムシ 115
シマオサムシ 236
シモイナヤマトオサムシ 150
シモヘイクロナガオサムシ 110
シャコタンオオルリオサムシ 326
ショウドヒメオサムシ 187
シラカミクロオサムシ 171

スオウヒメオサムシ 189
スズカオサムシ 293
スルガオサムシ 154~157

セアカオサムシ 44~49
セスジアカガネオサムシ 50~57
セスジアカガネオサムシカムチャツカ北海道亜種 50, 51, 53, 56, 57
セスジアカガネオサムシ大雪山パルサ湿原亜種 51, 54, 56, 57
セスジクロナガオサムシ 72
セトヒメオサムシ 188
セトヤコンオサムシ 248

ソタヤコンオサムシ 245

タイキキンオサムシ 307
ダイセツオサムシ 73, 75
ダイセツヒメクロオサムシ 125
ダイセンオサムシ 178~183
ダイセンオサムシ隠岐亜種 179, 181, 182
ダイセンオサムシ基亜種 178, 179, 181~183
ダイセンゲンオサムシ 328
ダイボサツクロナガオサムシ 115
タカサゴカタビロオサムシ 28
タカネセスジアカガネオサムシ 55
タキハラオサムシ 294
ダマステル亜属 330~343
タンザワアルマンオサムシ 66
タンザワホソヒメクロオサムシ 66

チシマオサムシ 70~77
チシマオサムシ基亜種 71, 74, 76
チシマオサムシ大雪山亜種 71, 75, 76
チシマオサムシ道央道東亜種 71, 74, 76
チシマオサムシ羊蹄山亜種 71, 75, 76
チシマオサムシ利尻島亜種 70, 73, 76, 77
チシマオサムシ礼文島亜種 71, 73, 76
チシマキンオサムシ 306
チチブクロナガオサムシ 115
チチブホソクロナガオサムシ 115
チュウゴクオサムシ 187
チュウゴククロナガオサムシ 81
チュウブオオオサムシ 227
チョウカイヒメクロオサムシ 126

ツクシオサムシ 228
ツクバクロオサムシ 175

ツシマオサムシ　205
ツシマカブリモドキ　344~347
ツルギオサムシ　213

テシオキンオサムシ　305
デワクロナガオサムシ　111
テンリュウオサムシ　282

トウキョウオサムシ　174
ドウキョウオサムシ　274~277
トウホクアルマンオサムシ　63
トウホククロオサムシ　170
トウホククロナガオサムシ　112
トウホクヒメクロオサムシ　126
トウホクホソヒメクロオサムシ　63
トサオサムシ　216~221
トサオサムシ石鎚山脈亜種　217, 220, 221
トサオサムシ基亜種　216, 217, 219, 221
トマムキンオサムシ　308
トヤマオサムシ　244

ナゴヤオサムシ　281

ニシアルマンオサムシ　65
ニシオオクロナガオサムシ　94
ニシシズオカオサムシ　259
ニシホソヒメクロオサムシ　65
ニセコキンオサムシ　310

ヌノビキオサムシ　235

ノザキヒメオサムシ　207

ハガクロオサムシ　172
ハクサンクロナガオサムシ　117
ハクサンホソヒメクロオサムシ　65
ハコネオサムシ　161

ヒダオオクロナガオサムシ　93
ヒダオサムシ　283
ヒダカキンオサムシ　308
ヒダカクロオサムシ　169
ヒダクロナガオサムシ　117
ヒメオオルリオサムシ　323
ヒメオサムシ　198~209
ヒメオサムシ淡路島四国亜種　199, 203, 209
ヒメオサムシ壱岐亜種　200, 205, 209
ヒメオサムシ生月島亜種　200, 206, 209
ヒメオサムシ沖の島亜種　199, 204, 209
ヒメオサムシ基亜種　198, 199, 204, 209
ヒメオサムシ下甑島亜種　200, 207, 209
ヒメオサムシ対馬亜種　199, 205, 209
ヒメオサムシ野崎島亜種　200, 206, 209
ヒメオサムシ平戸島亜種　200, 205, 209
ヒメオサムシ若松島亜種　200, 207, 209
ヒメオサムシ鷲尾山亜種　199, 203, 209
ヒメクロオサムシ　120~127
ヒメクロオサムシ基亜種　120, 121, 125, 127
ヒメクロオサムシ道央道東道北亜種　121, 124, 127

ヒメクロオサムシ東北地方亜種　122, 126, 127
ヒメクロオサムシ日高山脈南端部亜種　121, 125, 127
ヒメマイマイカブリ　340
ヒメミカワオサムシ　283
ヒラドヒメオサムシ　206

フクミオサムシ　213
フジアキタクロナガオサムシ　35
フジアルマンオサムシ　67
フジクロナガオサムシ　115
フジホソヒメクロオサムシ　67

ヘミカラブス亜属　38~49
ペンタカラブス亜属　58~69

ホウオウオサムシ　192~197
ホクリクオサムシ　244, 291
ホソアオクロナガオサムシ　35
ホソアカガネオサムシ　140~145
ホソアカガネオサムシ基亜種　140, 141, 143~145
ホソアカガネオサムシ佐渡島亜種　141, 143, 144
ホソアキタクロナガオサムシ　35
ホソキンオサムシ　305
ホソクロナガオサムシ　113
ホソヒメクロオサムシ　58~69
ホソヒメクロオサムシ赤石山脈亜種　60, 67, 69
ホソヒメクロオサムシ阿武隈高地亜種　59, 63, 69
ホソヒメクロオサムシ安倍峠亜種　60, 67, 69
ホソヒメクロオサムシ奥羽山脈亜種　59, 62, 69
ホソヒメクロオサムシ奥秩父亜種　60, 66, 69
ホソヒメクロオサムシ基亜種　58, 64, 69
ホソヒメクロオサムシ丹沢亜種　60, 66, 69
ホソヒメクロオサムシ東北地方南西部亜種　59, 63, 69
ホソヒメクロオサムシ白山飛騨御嶽木曽亜種　59, 65, 69
ホソヒメクロオサムシ飛騨山脈北部亜種　59, 64, 69
ホソヒメクロオサムシ富士山亜種　60, 66, 69
ホソヒメクロオサムシ八ヶ岳亜種　60, 65, 69
ホソヒメクロオサムシ八溝山亜種　59, 64, 69
ホソムネクロナガオサムシ　100
ボッチャンオサムシ　213
ホモエオカラブス亜属　50~57

マークオサムシ　24~29
マークオサムシ本州亜種　24~29
マイマイカブリ　330~343
マイマイカブリ粟島亜種　332, 338, 342
マイマイカブリ隠岐亜種　333, 340, 342
マイマイカブリ関東・中部地方亜種　332, 340, 342, 343
マイマイカブリ基亜種　330, 333, 341, 342
マイマイカブリ佐渡島亜種　332, 339, 342
マイマイカブリ東北地方南部亜種　331, 337, 342, 343
マイマイカブリ東北地方北部亜種　331, 336, 342
マイマイカブリ北海道亜種　331, 336, 342, 343
マシケキンオサムシ　307
マックレイセアカオサムシ　38~43
マックレイセアカオサムシ利尻島亜種　38, 39, 41~43
マヤサンオサムシ　288~297
マヤサンオサムシ鵜方亜種　289, 294, 296
マヤサンオサムシ基亜種　288, 291, 296, 297

マヤサンオサムシ信楽亜種 289, 293, 296
マヤサンオサムシ鈴鹿山脈南東部亜種 289, 292, 296
マヤサンオサムシ滝原亜種 289, 293, 296
マヤサンオサムシ北陸地方亜種 289, 291, 296
マヤサンオサムシ養老山地亜種 289, 292, 296
マルバネオサムシ 175
マルバネクロオサムシ 175
マルバネヒメオサムシ 175

ミカワオサムシ 278~287
ミカワオサムシ基亜種 278, 283, 286
ミカワオサムシ岐阜県中北部亜種 279, 283, 286
ミカワオサムシ岐阜県南部亜種 279, 284, 286
ミカワオサムシ佐久間亜種 279, 282, 286, 287
ミカワオサムシ志摩半島北部亜種 279, 284, 286, 287
ミカワオサムシ天竜川中流域亜種 279, 281, 286
ミカワオサムシ御浜町亜種 279, 284, 286
ミカワヤマトオサムシ 151
ミチノクアルマンオサムシ 63
ミチノクホソヒメクロオサムシ 63
ミナミホソヒメクロオサムシ 68
ミナミヤマトオサムシ 152
ミノオサムシ 284
ミハマオサムシ 285
ミヤケクロナガオサムシ 100
ミョウコウクロナガオサムシ 114
ミョウコウホソクロナガオサムシ 114

ムツクロナガオサムシ 110
ムナカタキンオサムシ 311
ムナカタルリオサムシ 328
ムロウオサムシ 237

メゴドントゥス亜属 298~313

モンローマイマイ 340

ヤコンオサムシ 240~249
ヤコンオサムシ隠岐亜種 241, 245, 249
ヤコンオサムシ基亜種 240, 245, 249
ヤコンオサムシ近畿地方中部亜種 241, 245, 249
ヤコンオサムシ近畿地方北部亜種 241, 244, 249
ヤコンオサムシ忽那諸島西部亜種 241, 247, 249
ヤコンオサムシ山陰地方亜種 241, 246, 249
ヤコンオサムシ四国和歌山亜種 241, 247, 249
ヤコンオサムシ北陸地方亜種 241, 244, 249
ヤツアルマンオサムシ 66
ヤツクロナガオサムシ 115
ヤツホソクロナガオサムシ 115
ヤツホソヒメクロオサムシ 66
ヤマウチクロオサムシ 171
ヤマズミオサムシ 254
ヤマトオサムシ 146~153
ヤマトオサムシ基亜種 146, 151, 153
ヤマトオサムシ近畿地方中東部亜種 147, 152, 153
ヤマトオサムシ熊野亜種 147, 152, 153
ヤマトオサムシ下伊那亜種 147, 150, 153
ヤマトオサムシ北陸地方亜種 147, 151, 153
ヤマトオサムシ三河亜種 147, 150, 153

ヤママイマイカブリ 338
ヤミゾホソヒメクロオサムシ 64

ユウバリキンオサムシ 308

ヨウテイオサムシ 75
ヨウロウオサムシ 292

ラウスオサムシ 75

リクチュウクロナガオサムシ 110
リシリオサムシ 74
リシリキンオサムシ 305
リシリノマックレイセアカオサムシ 42
リムノカラブス亜属 24~29

ルイスオサムシ 158~163
ルイスオサムシ基亜種 158, 159, 161~163
ルイスオサムシ房総半島南部亜種 159, 161~163

レプトカラブス亜属 78~119
レブンオサムシ 73

ワカマツヒメオサムシ 207
ワシオヒメオサムシ 203

分担　ROLE SHARING

■標本
収集（水沢清行・井村有希）
展脚・交尾器標本作成（井村有希）

■デザイン・構成
レイアウト全般及び表紙（井村有希）
（表紙基本デザイン：川井信矢）
中表紙「朽木から掘り出されたマイマイカブリ北海道亜種」
（井村有希）

■画像
全ての標本写真撮影（井村有希）
用語解説・図解検索・日本の地方区分と県・日本の島・
及び分布図作成（井村有希）
生態写真撮影（井村有希）
（セスジアカガネオサムシの図 31 を除く）

■解説
総論・各論本文執筆（井村有希）
情報・文献収集（水沢清行・井村有希）

■撮影機材
カメラ
Canon EOS 5D Mark III
Canon EOS Kiss X4
レンズ
Canon EF 100mm f/2.8L Macro IS USM
Canon EF 70-200mm f/4L USM
　　+ Nikon Nikkor 50mm f/1.4 (reverse)
　　or Nikon Nikkor 85mm f/1.8 (reverse)
顕微鏡システム
Leica Z 6 APO + PLANAPO 1.0x, 2.0x
RICOH Caplio R6, RICOH CX4

■深度合成ソフト
Helicon Focus 5.3 for Windows
Zerene Stacker

■編集
全般（井村有希）
画像処理全般（Robert Lizler）

著者　AUTHORS

AUTHOR

井 村　有 希
Yûki IMURA

1954年11月20日生
神奈川県横浜市在住

■現職
医師

　徳島県徳島市に生まれる．幼少時より昆虫に対して強い興味を持ち，小学生の頃から甲虫と蝶を中心に採集と標本収集を始める．本格的な学術論文の執筆開始は1989年．以来，四半世紀に亘って世界のオサムシと中国のルリクワガタを主な対象とした調査研究活動を続けている．中国奥地への調査行を最も得意とし，同国における滞在期間はこれまでに延べ400日近くに及ぶ．虫以外の目下の趣味は中国語の学習．
　著書に「世界のオサムシ大図鑑」（むし社），「DNAでたどるオサムシの系統と進化」（哲学書房），同英語版（Springer-Verlag, Tokyo），「リシリノマックレイセアカオサムシ，さいはての島の小さな奇跡」（昆虫文献六本脚）（以上，共著），「東アジアのルリクワガタ属」（昆虫文献六本脚）などがある．他に論文多数．東京慈恵会医科大学医学部卒業．医学博士．

AUTHOR

水 沢　清 行
Kiyoyuki MIZUSAWA

1938年8月16日生
神奈川県横須賀市在住

■現職
㈱サガミ　会長

　新潟県柏崎市に生まれる．東京農業大学農学部卒業．在日米陸軍医学総合研究所員，㈱サガミ代表取締役社長を経て現在は同社会長．本社の2階に設けた「蝶のひろば」には，多数の昆虫標本を常時展示し，一般の人々への啓蒙活動も行っている．主な収集・研究対象は日本ならびに世界のオサムシ．とりわけ離島に産するオサムシの調査と収集には長年，力を注いでいる．所有するオサムシコレクションはドイツ型標本箱（大型）で1000箱以上．
　著書に「日本と朝鮮の蚊類」（英文．共著，Contr. Amer. Ent. 研究所），「世界のオサムシ大図鑑」（共著，むし社）等がある．
　「蝶のひろば」：〒238-0025　神奈川県横須賀市衣笠町45-19　㈱サガミ2階　Tel. 046-837-6060

The *Carabus* of Japan
ISBN 978-4-902649-14-7

Date of publication : September 20th, 2013   1st print
Authors : Yûki IMURA and Kiyoyuki MIZUSAWA
Printed by TAITA Publishers (Czech Republic)
Published by Roppon-Ashi Entomological Books (Tokyo, Japan)
  Sanbanchō MY building, Sanbanchō 24-3, Chiyoda-ku, Tokyo, 102-0075 JAPAN
  Phone: +81-3-6825-1164   Fax: +81-3-5213-1600
  URL: http://kawamo.co.jp/roppon-ashi/
  E-MAIL: roppon-ashi@kawamo.co.jp
Retail price: JPY 27,000

Copyright©2013 Roppon-Ashi Entomological Books
All rights reserved. No part or whole of this publication may be reproduced without written permission of the publisher.

日本産オサムシ図説
ISBN 978-4-902649-14-7

発行日： 2013年9月20日 第1刷
著 者： 井村 有希・水沢 清行 共著
印 刷： TAITA Publishers (Czech Republic)
発行者： 川井 信矢
    昆虫文献 六本脚
    〒102-0075　東京都千代田区三番町24-3　三番町MYビル
    TEL: 03-6825-1164   FAX: 03-5213-1600
    URL: http://kawamo.co.jp/roppon-ashi/
    E-MAIL: roppon-ashi@kawamo.co.jp
定 価： 27,000円（消費税込）

　本書の一部あるいは全部を無断で複写複製することは，法律で認められた場合を除き，著作権者および出版社の権利侵害となります．あらかじめ小社あて許諾をお求め下さい．